Stochastic Mechanics
Random Media
Signal Processing and Image Synthesis
Mathematical Economics and Finance
Stochastic Optimization
Stochastic Control
Stochastic Models in Life Sciences

Applications of Mathematics

Stochastic Modelling and Applied Probability

52

Springer
Berlin
Heidelberg
New York
Hong Kong
London
Milan
Paris
Tokyo

Applications of Mathematics

1 Fleming/Rishel, **Deterministic and Stochastic Optimal Control** (1975)
2 Marchuk, **Methods of Numerical Mathematics** 1975, 2nd. ed. 1982)
3 Balakrishnan, **Applied Functional Analysis** (1976, 2nd. ed. 1981)
4 Borovkov, **Stochastic Processes in Queueing Theory** (1976)
5 Liptser/Shiryaev, **Statistics of Random Processes I: General Theory** (1977, 2nd. ed. 2001)
6 Liptser/Shiryaev, **Statistics of Random Processes II: Applications** (1978, 2nd. ed. 2001)
7 Vorob'ev, **Game Theory: Lectures for Economists and Systems Scientists** (1977)
8 Shiryaev, **Optimal Stopping Rules** (1978)
9 Ibragimov/Rozanov, **Gaussian Random Processes** (1978)
10 Wonham, **Linear Multivariable Control: A Geometric Approach** (1979, 2nd. ed. 1985)
11 Hida, **Brownian Motion** (1980)
12 Hestenes, **Conjugate Direction Methods in Optimization** (1980)
13 Kallianpur, **Stochastic Filtering Theory** (1980)
14 Krylov, **Controlled Diffusion Processes** (1980)
15 Prabhu, **Stochastic Storage Processes: Queues, Insurance Risk, and Dams** (1980)
16 Ibragimov/Has'minskii, **Statistical Estimation: Asymptotic Theory** (1981)
17 Cesari, **Optimization: Theory and Applications** (1982)
18 Elliott, **Stochastic Calculus and Applications** (1982)
19 Marchuk/Shaidourov, **Difference Methods and Their Extrapolations** (1983)
20 Hijab, **Stabilization of Control Systems** (1986)
21 Protter, **Stochastic Integration and Differential Equations** (1990)
22 Benveniste/Métivier/Priouret, **Adaptive Algorithms and Stochastic Approximations** (1990)
23 Kloeden/Platen, **Numerical Solution of Stochastic Differential Equations** (1992, corr. 3rd printing 1999)
24 Kushner/Dupuis, **Numerical Methods for Stochastic Control Problems in Continuous Time** (1992)
25 Fleming/Soner, **Controlled Markov Processes and Viscosity Solutions** (1993)
26 Baccelli/Brémaud, **Elements of Queueing Theory** (1994, 2nd ed. 2003)
27 Winkler, **Image Analysis, Random Fields and Dynamic Monte Carlo Methods** (1995, 2nd. ed. 2003)
28 Kalpazidou, **Cycle Representations of Markov Processes** (1995)
29 Elliott/Aggoun/Moore, **Hidden Markov Models: Estimation and Control** (1995)
30 Hernández-Lerma/Lasserre, **Discrete-Time Markov Control Processes** (1995)
31 Devroye/Györfi/Lugosi, **A Probabilistic Theory of Pattern Recognition** (1996)
32 Maitra/Sudderth, **Discrete Gambling and Stochastic Games** (1996)
33 Embrechts/Klüppelberg/Mikosch, **Modelling Extremal Events for Insurance and Finance** (1997, corr. 4th printing 2003)
34 Duflo, **Random Iterative Models** (1997)
35 Kushner/Yin, **Stochastic Approximation Algorithms and Applications** (1997)
36 Musiela/Rutkowski, **Martingale Methods in Financial Modelling** (1997)
37 Yin, **Continuous-Time Markov Chains and Applications** (1998)
38 Dembo/Zeitouni, **Large Deviations Techniques and Applications** (1998)
39 Karatzas, **Methods of Mathematical Finance** (1998)
40 Fayolle/Iasnogorodski/Malyshev, **Random Walks in the Quarter-Plane** (1999)
41 Aven/Jensen, **Stochastic Models in Reliability** (1999)
42 Hernandez-Lerma/Lasserre, **Further Topics on Discrete-Time Markov Control Processes** (1999)
43 Yong/Zhou, **Stochastic Controls. Hamiltonian Systems and HJB Equations** (1999)
44 Serfozo, **Introduction to Stochastic Networks** (1999)
45 Steele, **Stochastic Calculus and Financial Applications** (2001)
46 Chen/Yao, **Fundamentals of Queuing Networks: Performance, Asymptotics, and Optimization** (2001)
47 Kushner, **Heavy Traffic Analysis of Controlled Queueing and Communications Networks** (2001)
48 Fernholz, **Stochastic Portfolio Theory** (2002)
49 Kabanov/Pergamenshchikov, **Two-Scale Stochastic Systems** (2003)
50 Han, **Information-Spectrum Methods in Information Theory** (2003)

(continued after index)

Philippe Robert

Stochastic Networks and Queues

Springer

Author

Philippe Robert

INRIA
Domaine de Voluceau
Rocquencourt BP 105
78153 Le Chesnay, France

e-mail: Philippe.Robert@inria.fr

This book was first published in French in the *Mathématiques et Applications* series (3-540-67872-7). This edition is an extensively revised translation of the French version.

Mathematics Subject Classification (2000): 60G55, 60K25, 60H10, 60K30, 90B15, 60F17

Cover pattern by courtesy of Rick Durrett (Cornell University, Ithaca)

Cover art by Natalia Robert

Cataloging-in-Publication Data applied for

A catalog record for this book is available from the Library of Congress.

Bibliographic information published by Die Deutsche Bibliothek
Die Deutsche Bibliothek lists this publication in the Deutsche Nationalbibliografie;
detailed bibliographic data is available in the Internet at http://dnb.ddb.de

ISSN 0172-4568

ISBN 3-540-00657-5 Springer-Verlag Berlin Heidelberg New York

Springer-Verlag Berlin Heidelberg New York
a member of BertelsmannSpringer Science + Business Media GmbH

http://www.springer.de

Cover design: Erich Kirchner, Heidelberg
Typesetting by the author using a Springer TEX macro package
Printed on acid-free paper 41/3142DB-543210

Соне, Нате и Ирине

Preface to the English Edition

This book is translated from the French edition published in 2000 by Springer-Verlag. One of its goals is to show that mathematical models of stochastic networks can be analyzed by using general results on stochastic processes. It could be argued that, since stochastic networks are "applied", general results concerning Markov processes or martingales (for example) are not of a real use in practice and therefore that ad-hock results are more helpful. Recent developments in the study of stochastic networks have shown that this point of view is flawed. Technical tools such as scaling methods, large deviations and rare events, requiring a good understanding of some fundamental results concerning stochastic processes, are now used in the analysis of these stochastic models.

Background: Some of the general results on Markov processes and martingales are summarized in the appendices. A good knowledge of basic aspects of probability theory is nevertheless assumed, Grimmett and Stirzaker [**25**], Williams [**55**] and Durrett [**17**] are good references in this domain. Asmussen's book [**3**] covers most of aspects concerning queues. At a more advanced level, the books by Rogers and Williams [**43, 44**] are (highly) recommended.

Chapter 7 on queues with Poisson input and Chapter 9 on fluid limits have been entirely rewritten for this edition. The chapter on Poisson approximation has not been included in this edition for the main reason that the so-called "Chen-Stein" method is quite well known in the English literature (the book by Barbour *et al.* [**6**] gives an idea of the potential of this set of methods.)

It is a pleasure to thank Konstantin Avrachenko, Laurent Decreusefond, Christine Fricker, Isi Mitrani, Mike Veatch and Ruth Williams for their remarks on a preliminary version of this edition. In a quite different domain, several visits in Beaune, Fuissé, Givry and Nuits Saint-Georges have also been helpful during the writing of this book.

Pontoise, January 2003 Philippe Robert

Preface to the French Edition

This book is a set of lectures on a class of stochastic models gathered under the name of queues or stochastic networks which can be thought, in first approximation, as sets of processing units submitted to flows of requests. These models are used to describe communication networks (since Erlang, 1917 !), computer networks or production lines. From a theoretical point of view, they are related to a large class of problems concerning random walks, diffusions, asymptotic properties of Markov processes and point processes. The various behaviors of these models are investigated : Existence of equilibrium, characterization of the stationary regime, analysis of the transient behavior (rare events, hitting times) and critical regimes (saturation), etc... In this book, queues and stochastic networks are seen as a rich field of applications for the classical tools of probability theory in the context of Markov processes, martingales or ergodic theory.

The book is organized as follows. Chapter 1 introduces the definitions concerning point processes which are used to describe the probabilistic structure of arrival processes of requests. The *very important* Poisson point processes are presented in a quite general framework. Basic results concerning renewal point processes (including renewal Theorem) are recalled.

Chapter 2 and 3 consider the $GI/GI/1$ queue where the arrival process is a renewal point process. One-dimensional random walks play the main role in the study of this queue. Wiener-Hopf factorization is used to determine the distribution of the maximum of the associated random walk. The distribution of various quantities like waiting time, number of customers,... at equilibrium are then derived. Chapter 3 studies limit theorems for the $GI/GI/1$ queue: the tail distribution of waiting time at equilibrium and the behavior of the queue just below the saturation regime. The change of probability distribution is the main probabilistic tool in the proof of these results.

Classical queueing networks are analyzed in Chapter 4 : Jackson networks, loss networks, Gordon-Newel networks and Kelly's networks. The reversibility properties of the associated Markov processes give a satisfactory description of their stationary behavior. The invariant probabilities of these Markov processes have a product form.

Chapters 5 and 6 present a detailed analysis of two important queues: The $M/M/1$ queue and the $M/M/\infty$ queue. The $M/M/1$ queue is the basic element of a Jackson network and the $M/M/\infty$ queue the basic element of a loss network. To these two queues correspond two classical stochastic processes: The reflected random walk on $\mathbb{N}$ for the first one and a discrete Ornstein-Uhlenbeck process for the second queue. Results concerning conditioning, hitting times, rare events, rate of convergence, large deviations (for the $M/M/1$ queue) and functional central limit theorems are proved in these chapters. The fluid limit scaling, analyzed in full generality in Chapter 9, is introduced for these simple examples. If the results presented in Chapters 5 and 6 are interesting in their own right, the purpose of these chapters is also of showing the efficiency of elementary stochastic calculus methods.

Chapter 7 studies several queueing systems with a Poisson arrival point process: the $M/G/1$ queue with the FIFO, LIFO, Processor-Sharing disciplines and the $M/G/\infty$ queue. This chapter is an occasion to work with Markov processes with values in non-finite dimensional state space: the set of finite sequences for LIFO queue and the set of Radon measures for Processor-Sharing queue. Some generic properties of these queueing systems are also presented: the PASTA (Poisson Arrivals See Time Averages) property and the phenomenon of insensitivity.

Chapter 8 is a quick presentation of the main ergodicity and transience criteria for Markov processes on countable state spaces. These criteria are based on Lyapunov functions.

Chapter 9 introduces the fluid limit scaling of multi-dimensional Markov processes. If $\|x\|$ is the size of the initial state, this scaling consists in speeding up time with a factor $\|x\|$ and renormalizing the Markov process (space variable) by $1/\|x\|$. A fluid limit is one the limiting points of the rescaled process when $\|x\|$ tends to infinity. The rescaled process may converge locally to the solution of an ordinary deterministic differential equation, in which case the fluid limit scaling has the effect of erasing the stochastic perturbation around this deterministic trajectory. This scaling is used to get a first order approximation of queueing networks more general than those seen in Chapter 4. Fluid limit scaling is also a natural framework to understand the connection, noticed in Chapter 8, between the ergodicity of Markov processes and the stability of some deterministic differential equations.

Chapters 11 and 12 study the single server queue in the context of ergodic theory: Only a weak condition is assumed for the distribution of interarrival intervals and services of customers. Chapter 10 gives a quick presentation of ergodic theory: Measure preserving transformations, ergodic theorems and special flows. This is an important chapter since most of the results concerning stationary point processes are easier to understand and more natural within that framework. Stationary point processes on $\mathbb{R}$ are obtained from the sequence of the interarrival intervals via a simple construction. They are presented in detail in Chapter 11. The single server queue whose arrival process is a stationary point process is analyzed in Chapter 12.

Some results concerning Markov processes, martingales and the convergence of stochastic processes are recalled in Appendix A, B and C, and for Skorohod problems, Appendix D.

Many of the results presented are quite classical: Wiener-Hopf factorization of random walks in $\mathbb{R}$, product form networks, stationary point processes, etc ... Several parts are more "modern", i.e. still significantly evolving: The different scalings of stochastic processes, the estimation of tail distributions, rare events, large deviations results, ... Most of the proofs rely on classical probabilistic results: Markovian methods, martingale results or ergodic theory are used throughout this book. The results presented in this book are not, in some cases, stated under the most general assumptions. This gives a better picture of the results and, moreover, some marginal technical complications can then be avoided.

This book owes much to Jacques Neveu through his lectures and his work in the domain. Most of the chapters have been taught at the Laboratoire de Probabilités of the University of Paris VI, I am grateful to Jean Jacod who invited me to give these lectures. Numerous readers and students influenced these lectures. During them, more than once, Philippe Bougerol asked embarrassing questions. Franck Delcoigne, Vincent Dumas and Fabrice Guillemin read and made comments on several chapters. I am especially grateful to Christine Fricker and Danielle Tibi who made many remarks, corrected many errors in most of chapters. I thank Thierry Jeulin who convinced me of the interest of writing such a book. An anonymous referee made several remarks which clearly improved several parts of the original version. This work has been done while I was in the dynamic and friendly research team ALGORITHMES at INRIA-Rocquencourt, I am particularly grateful to Philippe Flajolet for such a stimulating environment.

Pontoise, August 2000 Philippe Robert

Some Definitions and Notations

A queue is a *service unit* where jobs or customers arrive and require the use of the service unit for some duration of time called *service time*. Customers can be accommodated in one or several queues while being served or waiting for service. A queue can accept only a finite number of customers (the *capacity of the queue*). In this case, when the maximal number is reached, all arriving customers are rejected during that time. The *residual service time* of a given customer is the remaining amount of service required by the customer to the server. When the residual service time of a customer is 0, it leaves the queue. The server may serve in parallel several customers at different service rates. Mathematically, a queue is given by

- An arrival process, i.e. a non-decreasing sequence (t_n), where t_n is the arrival instant of the nth customer.
- A non-negative sequence (σ_n), where σ_n is the service time required by the nth customer.
- A sequence (m_n) of marks attached to customers. For $n \in \mathbb{N}$ m_n can be a priority level, a deadline or a sequence of subsequent services.
- A service discipline.

  Examples.
  1. FIFO (First In First Out), customers are served in their order of arrival.
  2. LIFO (Last In First Out), the last arrived in the queue is served.
  3. Processor-Sharing. When n customers are in the queue, the residual service time of each of them decreases at rate $1/n$.
  4. Priority queue : If two classes of customers arrive at the queue, the red and the green say, the server gives priority to red customers. A green customer is served only if there are no red customers in the queue.

Kendall's Terminology. Queues are usually represented under the form $A/S/N_s/N_c$, where

- A gives an indication on the distribution of the arrival process.
  1. G (General) : General distribution for interarrival intervals
  2. GI (General Independent): The sequence $(t_{n+1} - t_n)$ is i.i.d.
  3. M (Markov) : The variables $(t_{n+1} - t_n)$ are i.i.d. and exponentially distributed. The arrival process is Poisson in this case.

4. D (Deterministic) : $t_{n+1} - t_n = C$ for any $n \in \mathbb{N}$, this corresponds to periodic arrivals.

– S gives an indication on the distribution of service times. Its values are the same as for the arrivals, the sequence $(t_{n+1} - t_n)$ is replaced by (σ_n).
– N_s is the number of servers and N_c is the capacity of the queue.

Stochastic processes are generally represented under the form $(X(t))$ or (X_n) instead of $(X(t);\ t \geq 0)$ or $(X_n;\ n \in \mathbb{N})$.

Bibliography. Books are referenced at the end of the book. Relevant research papers are mentioned within the text in footnotes, see also page 389 for a complete list of them.

Table of Contents

Dependence relations between chapters

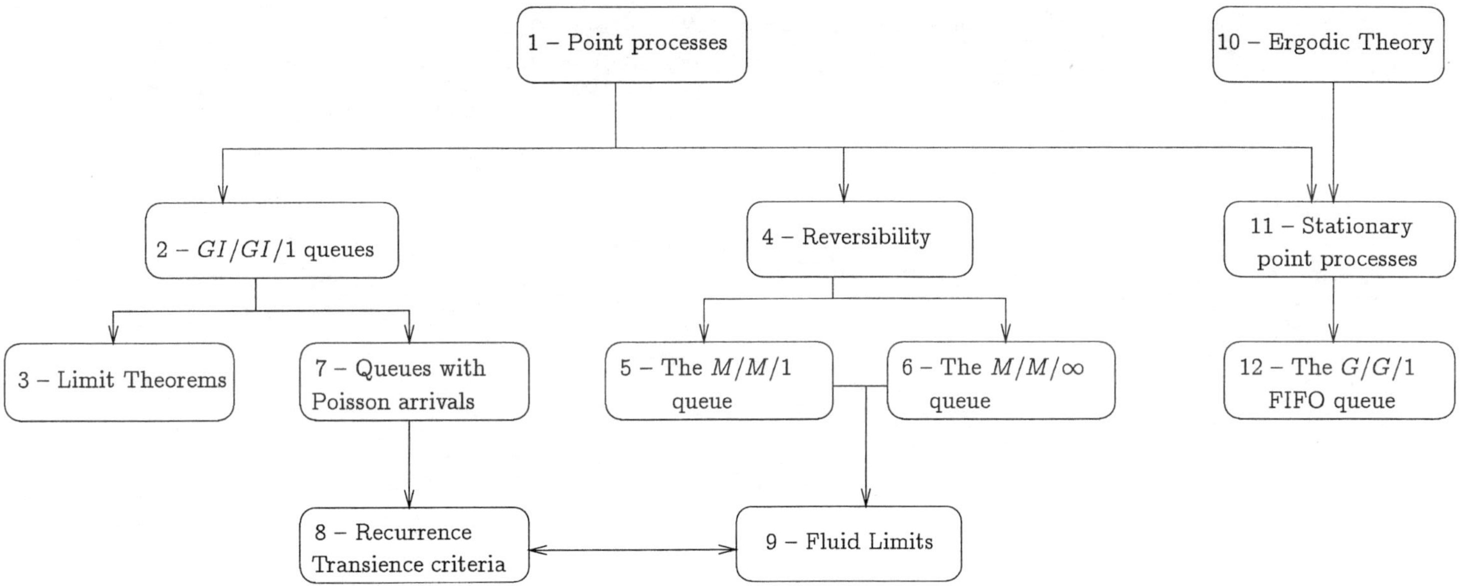

1. Point Processes

Contents

This chapter introduces the basic definitions and results concerning point processes. A point process is the mathematical object used to describe the flow of customers in a queueing system: the arrival instants and the value of services they require. It can be an arrival process as well as a departure process. General definitions and properties of point processes are briefly presented. An important subclass is analyzed in detail: Poisson point processes. The Poisson point process is, with Brownian motion, an ubiquitous object in probability theory. It shows up in numerous limit theorems and the explicit form of the distribution of many of its functionals can be easily derived. Poisson point processes are presented in a quite general framework (a locally compact space or a complete metric space) because their main properties are simple and independent of the particular structure of $\mathbb{R}^d$ where they are generally considered. Moreover, this level of generality is required to study the important *marked Poisson point processes.* The last part of this chapter is a short summary of the main properties of renewal point processes, including the renewal theorem.

1.1 General Definitions

The state space H is a locally compact metric space which is σ-compact (i.e. H can be covered by a countable family of compact subsets). Most of the results of this chapter are also valid in the case of a complete metric space. The set H is endowed with the Borelian σ-field. A Radon measure μ on H is a positive Borelian measure on H such that $\mu(K) < +\infty$ for any compact subset of H (see Rudin [**45**]). The set of all non-negative Radon measures on H is denoted by $\mathcal{M}(H)$ and $\mathcal{M}_p(H)$ is the set of *point measures* on H, i.e. the subset of elements m of $\mathcal{M}(H)$ which can be represented as

$$m = \sum_n \delta_{u_n}, \tag{1.1}$$

where $u_n, n \in \mathbb{Z}$ is a sequence of elements of H and δ_a is the Dirac measure at the point a, i.e.

$$\int_H f(x)\, \delta_a(dx) = f(a),$$

for any measurable function f. If m does not have multiple points, that is, if $u_n \neq u_p$ when $n \neq p$, this measure can be represented as a set $m = \{u_n\}$. A measure of the form (1.1) has the Radon property if and only if every compact contains a finite number of u_n, $n \in \mathbb{Z}$ or, equivalently, if the sequence has no limiting point in H.

If f is a function on H, integrable with respect to the point measure $m = \{u_n\}$, the integral of f with respect to m is denoted by

$$m(f) = \int_H f(x)\, m(dx) = \sum_n f(u_n),$$

and, if A is a subset of H, $m(A)$ is the number of points of m in A,

$$m(A) = m(1_A) = \sum_n 1_{\{u_n \in A\}}.$$

The space $\mathcal{M}_p(H)$ is endowed with the topology of weak convergence of measures and the associated σ-field. Thus, a sequence (m_n) of point measures converges to $m \in \mathcal{M}_p(H)$ if and only if

$$\lim_{n \to +\infty} m_n(f) = m(f),$$

for any continuous function $f : H \to \mathbb{R}_+$ with compact support. For any Borelian subset A of H, the function $m \longrightarrow m(A)$ is measurable for this σ-field. The underlying probability space is assumed to be $(\Omega, \mathcal{F}, \mathbb{P})$ in the following.

Definition 1. *A* point process *is a random variable with values in the set of Radon point measures* $\mathcal{M}_p(H)$.

The distribution of a point process

$$\begin{aligned} N : (\Omega, \mathcal{F}, \mathbb{P}) &\to \mathcal{M}_p(H) \\ \omega &\to N(\omega, dx) \end{aligned}$$

is completely determined by the distribution of the $N(\omega, f)$ where f is a continuous function on H with compact support. Another class of functions also determines the distribution of N, *simple functions*, i.e. functions of the form $f(x) = \alpha_1 1_{A_1}(x) + \alpha_2 1_{A_2}(x) + \cdots + \alpha_n 1_{A_n}(x)$, $x \in H$, where $n \in \mathbb{N}$ and, for any index $1 \leq i \leq n$, $\alpha_i \in \mathbb{R}$ and A_i is a Borelian subset of H. Using the fact that a point measure N is integer valued on Borelian subsets, it is easily seen that the distribution of N is completely determined by the quantities

$$\mathbb{P}(N(\omega, A_1) = k_1, \ldots, N(\omega, A_n) = k_n),$$

where $n \geq 1$, (A_i) are Borelian subsets of H and (k_i) are integers.

If N is a point process on H and f a measurable, non-negative function on H, the integral of f with respect to N is denoted by

$$\int f(x)\, N(\omega, dx) \quad \text{or} \quad \int f(x)\, N(dx),$$

in the same way, if A is a Borelian subset of H, $N(A)$ denotes the variable $N(\omega, A)$.

Definition 2. *A point process N is* simple *if, $\mathbb{P}$-almost surely, the equation $N(\omega, \{a\}) = 0$ or 1 holds for all $a \in H$.*

The intensity *of a point process is the non-negative Borel measure on H defined by $\mu(A) = \mathbb{E}(N(A))$, for a Borelian subset A of H.*

Notice that the measure μ is not necessarily Radon. If T is a non integrable integer valued random variable, the variable $N = \delta_1 + \delta_{1/2} + \delta_{1/3} + \cdots + \delta_{1/T}$ is a point process on $\mathbb{R}$ and satisfies the relation $\mathbb{E}(N([0, a])) = +\infty$ for any positive number a.

The Laplace Transform of a point process

Definition 3. *If $N : \Omega \to \mathcal{M}_p(H)$ is a point process and f is a non-negative function on H, the Laplace transform of N at f is defined by*

$$\psi_f(N) = \mathbb{E}\left[\exp\left(-\int f(x)\, N(\omega, dx)\right)\right].$$

If N_1 and N_2 are two independent point processes, the Laplace transform of the superposition $N_1 + N_2$ of these point processes at f is given by

$$\begin{aligned} \psi_f(N_1 + N_2) &= \mathbb{E}\left[\exp\left(-\int f(x)\, N_1(\omega, dx) - \int f(x)\, N_2(\omega, dx)\right)\right] \\ &= \mathbb{E}\left[\exp\left(-\int f(x)\, N_1(\omega, dx)\right)\right] \mathbb{E}\left[\exp\left(-\int f(x)\, N_2(\omega, dx)\right)\right] \\ &= \psi_f(N_1)\psi_f(N_2). \end{aligned}$$

Proposition 1.1. *The Laplace transform $f \to \psi_f$ characterizes the distribution of N.*

Proof. If $A_1, \dots, A_k$ are Borelian subsets of H and

$$\phi(\alpha) = \mathbb{E}\left[\exp\left(-\sum \alpha_i N(A_i)\right)\right] = \psi_f(N),$$

with $f(x) = \alpha_1 1_{A_1}(x) + \dots + \alpha_n 1_{A_n}(x)$, then ϕ (and therefore $f \to \psi_f$) determines completely the distribution of the random vector $(N(A_1), \dots, N(A_k))$ and consequently the distribution of N. □

As for the case of real valued random variables, the Laplace transform is a convenient tool to prove the convergence in distribution of a sequence of point processes. The next proposition gives a useful criterion for the convergence in distribution of point processes. (See Neveu [**37**] for a proof.)

Proposition 1.2. *If (N_n) is a sequence of point processes such that the associated sequence of Laplace transforms $\psi_f(N_n)$ converges to some $L(f)$ and $L(\alpha f)$converges to 1 when $\alpha \searrow 0$ for any non-negative function f with compact support on H, then (N_n) converges in distribution to a point process whose Laplace transform is given by $f \to L(f)$.*

The books by Dawson [**15**] and Neveu [**37**] are recommended for more details on general point processes and, more generally, for results on random measures on a metric space.

A technical result concludes this section. It concerns *diffuse* Radon measures i.e. Radon measures without mass on points. The ball of center x and radius r is denoted by $B(x, r)$.

Lemma 1.3. *If μ is a diffuse Radon measure on a locally compact metric space H and K is a compact of H, then for any $\varepsilon > 0$, there exist $n \geq 1$, $x_1, \dots, x_n \in H$ and $r_1, \dots, r_n < \varepsilon$ such that, for $i = 1, \dots, n$, the inequality $\mu(\bar{B}(x_i, r_i)) < \varepsilon$ holds and $K \subset \cup_1^n B(x_i, r_i)$.*

Proof. Indeed, for $x \in K$, the local compactness implies that there exists $r_x > 0$ such that $\bar{B}(x, r)$ is compact for all $r \leq r_x$ ($\bar{B}(x, r)$ is the closure of $B(x, r)$). In particular, $\mu(\bar{B}(x, r)) < +\infty$, Lebesgue's Theorem applied to the indicator function of the ball $\bar{B}(x, r)$ gives the convergence

$$\lim_{r \to 0} \mu\left(\bar{B}(x, r)\right) = \mu(\{x\}).$$

This last quantity is 0 since the measure is diffuse. By taking $r_x < \varepsilon$ such that $\mu(\bar{B}(x, r_x)) < \varepsilon$, since $\cup_{x \in K} B(x, r_x)$ is a covering of K, the compactness property implies that a finite number of these sets covers K. The lemma is proved. □

1.2 Poisson Processes

An Elementary Result

Poisson random variables occur in many convergence theorems. The result below is typical in this domain.

PROPOSITION 1.4 (Law of Small Numbers). *If I_i^n, $i \geq 1$ are independent random variables with values in $\{0,1\}$ such that*

a) $\lim_{n\to+\infty} \sum_i \mathbb{E}(I_i^n) = \lim_{n\to+\infty} \sum_i \mathbb{P}(I_i^n = 1) = \lambda < +\infty$;

b) $\lim_{n\to+\infty} \sup_i \mathbb{E}(I_i^n) = 0$.

The variable $S_n = \sum_i I_i^n$ converges in distribution to a Poisson random variable with parameter λ.

PROOF. For $|u| < 1$ and $p_i^n = \mathbb{E}(I_i^n)$, the generating function of S_n taken at $u \in [0,1[$ is given by

$$\mathbb{E}\left(u^{S_n}\right) = \prod_i (1 - p_i^n + p_i^n u) = \exp\left(\sum_i \log\left[1 - p_i^n(1-u)\right]\right).$$

The elementary inequality $-x - x^2 \leq \log(1-x) \leq -x$, valid for $0 \leq x \leq 1/2$, and condition b) give, for n sufficiently large,

$$\left|\sum_i \log\left(1 - p_i^n(1-u)\right) + \sum_i p_i^n(1-u)\right| \leq \sum_i (p_i^n)^2 \leq \sup_i \mathbb{E}(I_i^n) \sum_i \mathbb{E}(I_i^n).$$

When n tends to infinity, the right hand side of this inequality is equivalent to $\lambda \sup_i \mathbb{E}(I_i^n)$, hence

$$\lim_{n\to+\infty} \sum_i \log\left(1 - p_i^n(1-u)\right) = \lim_{n\to+\infty} -\sum_i p_i^n(1-u) = -\lambda(1-u),$$

$$\lim_{n\to+\infty} \mathbb{E}\left(u^{S_n}\right) = \exp(-\lambda(1-u)) = \sum_{k\geq 0} \frac{\lambda^k}{k!} e^{-\lambda} u^k.$$

Therefore, the sequence (S_n) converges in distribution to a Poisson random variable with parameter λ. □

1.2.1 Definition of a Poisson Point Process

As before the state space H is supposed to be a locally compact and σ-compact metric space. Throughout this section μ is some positive Radon measure on H. The notion of Poisson point process can be now introduced. To be consistent with the usual denomination when $H = \mathbb{R}$, the terminology "Poisson process" will be sometimes used.

DEFINITION 4. *A Poisson point process N with intensity μ is a point process on H such that:*

a) If A is a Borelian subset of H the distribution of the random variable $N(\omega, A)$ is Poisson with parameter $\mu(A)$, i.e. for $k \geq 0$,

$$\mathbb{P}(N(\omega, A) = k) = \frac{\mu(A)^k}{k!} e^{-\mu(A)};$$

b) If $A_1, \ldots, A_n$ are disjoint Borelian subsets of H, the random variables $N(A_1), \ldots, N(A_n)$ are independent.

For any compact subset K of H, $N(K)$ is a Poisson random variable with parameter $\mu(K) < +\infty$, in particular $\mathbb{P}(N(K) \in \mathbb{N}) = 1$. A Poisson point process is necessarily a point process. The intensity of such a point process is given by the measure μ since $\mathbb{E}(N(A)) = \mu(A)$, the average value of a Poisson distribution with parameter $\mu(A)$ is $\mu(A)$. The following proposition shows that a) and b) determine completely the distribution of a Poisson point process.

PROPOSITION 1.5. *If N is a Poisson point process with intensity μ on H, its Laplace transform is given by*

$$F_f(N) = \exp\left[-\int \left(1 - e^{-f(x)}\right) \mu(dx)\right],$$

for any non-negative Borelian function f. Moreover, N is a simple point process if and only if μ is diffuse.

PROOF. It is sufficient to consider simple functions of the form

$$f = \sum_1^n \lambda_i 1_{A_i},$$

where λ_i are non-negative and the A_i's are disjoint Borelian subsets of H. The Laplace transform of N at f is given by

$$\mathbb{E}\left[\exp\left(-N(f)\right)\right] = \mathbb{E}\left[\exp\left(-\sum_1^n \lambda_i N(A_i)\right)\right],$$

by independence Assumption b) it can be written as

$$\mathbb{E}\left[\exp\left(-N(f)\right)\right] = \prod_1^n \mathbb{E}\left[\exp\left(-\lambda_i N(A_i)\right)\right] = \prod_1^n \exp\left[-\mu(A_i)\left(1 - e^{-\lambda_i}\right)\right]$$

because the distribution of $N(A_i)$ is Poisson according to a), finally

$$\mathbb{E}\left[\exp\left(-N(f)\right)\right] = \exp\left[-\sum_1^n \mu(A_i)\left(1 - e^{-\lambda_i}\right)\right]$$
$$= \exp\left[-\int \left(1 - e^{-f(x)}\right) \mu(dx)\right].$$

If the measure μ has a mass $\mu(\{x\}) > 0$ on $x \in H$, the variable $N(\{x\})$ is Poisson with parameter $\mu(\{x\})$, in particular $\mathbb{P}(N(\{x\}) = 2) > 0$, hence N cannot be simple.

Conversely, if μ is diffuse and K is a compact subset of H, Lemma 1.3 shows the existence of a finite partition of K with Borelian subsets (A_n) such that $\mu(A_n) \leq \varepsilon$. If X is a Poisson variable with parameter λ,

$$\mathbb{P}(X \geq 2) = \sum_{k\geq 2} \frac{\lambda^k}{k!} e^{-\lambda} = \lambda^2 \sum_{k\geq 0} \frac{\lambda^k}{(k+2)!} e^{-\lambda} \leq \lambda^2, \tag{1.2}$$

thus, the inequality

$$\begin{aligned}\mathbb{P}(N \text{ is not simple on } K) &\leq \sum_{n\geq 1} \mathbb{P}(N(A_n) \geq 2) \\ &\leq \sum_{n\geq 1} \mu(A_n)^2 \leq \varepsilon \sum_{n\geq 1} \mu(A_n) \leq \varepsilon\mu(K),\end{aligned}$$

holds. By letting ε go to 0 one gets

$$\mathbb{P}(N \text{ is not simple on } K) = 0.$$

The point process N is therefore $\mathbb{P}$-almost surely simple on K. If (K_n) is a non-decreasing sequence of compact subsets converging to H, i.e. $\cup_n K_n = H$, since $\mathbb{P}(N \text{ is not simple on } K_n) = 0$, for any $n \geq 1$, by using Lebesgue's Theorem, one gets that N is $\mathbb{P}$-almost surely simple. □

The next proposition shows that Property b) of Definition 4 characterizes, modulo an assumption, Poisson point processes.

Proposition 1.6. *If a simple point process N on H with a diffuse Radon intensity measure μ satisfies the independence property*

- *for any sequence $A_1, \dots, A_n$ of disjoint Borelian subsets of H, the random variables $N(A_1), \dots, N(A_n)$ are independent,*

it is necessarily a Poisson point process with intensity μ.

Proof. If K is a compact subset of H, because the measure μ is diffuse, for $n \geq 1$, from Lemma 1.3 there exists a finite partition (A_i^n) of K of subsets whose closure is compact such that, for $i \in \mathbb{N}$,

$$A_i^n \subset \bar{B}\left(x_i^n, r_i^n\right), \; \mu\left(A_i^n\right) < \frac{1}{n} \text{ and } r_i^n < \frac{1}{n}.$$

By induction, it can be assumed that for all n and $k \geq 1$, there exists a subset I_k^n of $\mathbb{N}$ such that $(A_i^{n+1}; i \in I_k^n)$ is a partition of A_k^n. The variable $N(K)$ can be decomposed in the following way

$$N(K) = \sum_i N\left(A_i^n\right) = \sum_i 1_{\{N(A_i^n)=1\}} + \sum_i N\left(A_i^n\right) 1_{\{N(A_i^n)\geq 2\}}. \tag{1.3}$$

The inequality

$$N\left(A_i^n\right)1_{\{N(A_i^n)\geq 2\}} = 1_{\{N(A_i^n)\geq 2\}}\sum_{j\in I_i^n} N\left(A_j^{n+1}\right)$$
$$\geq \sum_{j\in I_i^n} N\left(A_j^{n+1}\right)1_{\{N(A_j^{n+1})\geq 2\}}$$

shows that the sequence

$$\left(\sum_i N\left(A_i^n\right)1_{\{N(A_i^n)\geq 2\}}\right)$$

is non-increasing. Its limit is almost surely 0 at infinity. If not, with positive probability there should exist a sequence (i_n) such that $A_{i_{n+1}}^{n+1}\subset A_{i_n}^n$ and $N(A_{i_n}^n)\geq 2$ for all n. The $A_{i_n}^n$'s being within balls whose radius tends to 0, the limit of $A_{i_n}^n$ is either empty or reduced to a point. The mass of such a possible point should be at least 2 for the point process, which is in contradiction with the fact that the process N is simple. The almost sure convergence to 0 is therefore true. Since the sequence is bounded by the integrable variable $N(K)$, it converges also in L_1 to 0, hence

$$\lim_{n\to+\infty}\mathbb{E}\left(\sum_i 1_{\{N(A_i^n)=1\}}\right) = \mu(K).$$

Since $\mathbb{E}(1_{\{N(A_i^n)=1\}})\leq \mathbb{E}(N(A_i^n)) = \mu(A_i^n) < 1/n$, the variables $1_{\{N(A_i^n)=1\}}$ satisfy the assumptions of Proposition 1.4, one concludes that the variable

$$\sum_i 1_{\{N(A_i^n)=1\}}$$

converges in distribution to a Poisson distribution with parameter $\mu(K)$. Equation (1.3) shows that the distribution of the variable $N(K)$ is Poisson with parameter $\mu(K)$.

If A is a Borelian subset with finite measure, there exists a sequence of compact subsets $K_n\subset A$ such that $\mu(A)$ is the limit if the sequence $(\mu(K_n))$. For $n\geq 1$, the distribution of the variable $N(K_n)$ is Poisson with parameter $\mu(K_n)$. Therefore, it converges in distribution to a Poisson distribution with parameter $\mu(A)$. The quantity $\mathbb{E}(|N(A)-N(K_n)|) = \mu(A-K_n)$ converges to 0 when n goes to infinity, the variable $N(A)-N(K_n)$ converges to 0 in L_1. The variable $N(A)$ has the same distribution as the limit of the sequence $(N(K_n))$, i.e. a Poisson distribution with parameter $\mu(A)$. □

If μ is not diffuse, the above proposition is not true anymore. The trivial point process $N=\delta_0$ on $\mathbb{R}$ satisfies clearly the independence condition without being Poisson.

If μ is diffuse but N is not simple, the proposition also fails : If $\mathbb{P}=\{u_n\}$ is a Poisson process with intensity $\mu/2$ (the existence of such a process is proved in the next section), then the point process $N = 2\delta_{u_0}+2\delta_{u_1}+\cdots+2\delta_{u_n}+\cdots$

has intensity μ, satisfies the independence condition (because it is true for $\mathbb{P}$), but this is not a Poisson process (all Borelian subsets have an even number of points).

1.2.2 Construction of a Poisson Point Process

LEMMA 1.7 (Superposition of Poisson Point Processes). *If (μ_n) is a sequence of Radon measures on H such that their sum μ is also Radon and if (N_n) is a sequence of independent Poisson point processes whose intensities are given by (μ_n), then*

$$N = \sum_n N_n$$

is a Poisson point process with intensity μ.

PROOF. For $n \geq 1$, the Laplace of the point process $N_1 + \cdots + N_n$ at f is given by

$$\exp\left[-\int_H \left(1 - e^{-f(x)}\right)(\mu_1 + \cdots + \mu_n)(dx)\right],$$

therefore, this point process is Poisson with intensity $\mu_1 + \cdots + \mu_n$. Proposition 1.2 concludes the proof. □

The following proposition shows that Poisson point processes indeed exist.

PROPOSITION 1.8. *If μ is a positive Radon measure on H, there exists a Poisson point process with intensity μ.*

PROOF. If A is a Borelian subset of H such that $0<\mu(A)<+\infty$ and $\mu_A(dx)$ denotes the probability distribution $1_A(x)\mu(dx)/\mu(A)$ and (X_i) an i.i.d. sequence of random variables with distribution μ_A. If ν is a Poisson random variable with parameter $\mu(A)$ independent of (X_i), N_A is a point process on A defined by $N_A = \delta_{X_1} + \cdots + \delta_{X_\nu}$. This point process N_A is Poisson with intensity $1_A(x)\mu(dx)$. Indeed, if f is a non-negative measurable function

$$\begin{aligned}
\mathbb{E}\left(e^{-N_A(f)}\right) &= \sum_{n\geq 0} \frac{\mu(A)^n}{n!} e^{-\mu(A)} \mathbb{E}\left[\exp\left(-\sum_{i=1}^n f(X_i)\right)\right] \\
&= \sum_{n\geq 0} \frac{\mu(A)^n}{n!} e^{-\mu(A)} \left(\int_A e^{-f(x)}\,\mu_A(dx)\right)^n \\
&= \exp\left[-\int_A \left(1 - e^{-f(x)}\right)\mu(dx)\right].
\end{aligned}$$

The point process N_A is therefore Poisson with intensity $1_A(x)\mu(dx)$.

The hypothesis that H is σ-compact gives a partition (A_n) of Borelian subsets of H such that $0 < \mu(A_n) < +\infty$ for all $n \geq 1$. It is therefore possible to build a sequence of independent Poisson point process (N_{A_i}) whose intensities are respectively $1_{A_i}(x)\mu(dx)$. The above lemma shows that the sequence $(N_{A_1} + \cdots + N_{A_n})$ converges in distribution to a Poisson process with intensity

$$\sum_{1}^{+\infty} 1_{A_i}(x)\mu(dx) = \mu(dx).$$

The proposition is proved. □

The above construction gives also the classical property of conditioning for Poisson point processes.

Corollary 1.9 (Conditioning on the Number of Points in a Set). *If N is a Poisson point process with intensity μ and A a Borelian subset of H such that $0 < \mu(A) < +\infty$,*

a) the point process N restricted to A has the Poisson property with intensity $\mu_A = 1_A(x)\mu(dx)$;
b) conditionally on the event $\{N(A) = n\}$, $n \geq 1$, the points of the point process N in A has the same distribution as n independent random variables with distribution $1_A(x)\mu(dx)/\mu(A)$.

Proof. The properties a) and b) are direct consequences of the previous proof : The point process N can be represented as the sum $N_A + N_{H-A}$ of two independent Poisson point process on the disjoint subsets A and $H - A$, N_A is N restricted to A. The point process N_A is Poisson with intensity $1_A(x)\mu(dx)$, a) is proved. The variable N_A has been expressed as $\delta_{X_1} + \cdots + \delta_{X_\nu}$, where the variables (X_i) are i.i.d. with distribution $1_A(x)\mu(dx)/\mu(A)$ and independent of ν, a Poisson random variable with parameter $\mu(A)$. Therefore, the distribution of N conditioned on the event $\{N(A) = n\}$ is the distribution of $\delta_{X_1} + \cdots + \delta_{X_n}$, hence b) holds. □

A Poisson point process on $\mathbb{R}^p$, $p \geq 1$ with intensity λdx can be simply constructed as follows : If (X_i) is an i.i.d. sequence of uniform random variables on $[-n/2, n/2]^p$, the first $\lfloor \lambda n^p \rfloor$ elements of this sequence are taken to define the point process N_n (where $\lfloor a \rfloor$, is the integer part of a). When n goes to infinity, N_n is asymptotically Poisson with intensity $\lambda\, dx$ on $\mathbb{R}^p$. Indeed, if f is a non-negative function with compact support on $\mathbb{R}^p$, the Laplace transform of N_n at f is given by

$$\psi_{N_n}(f) = \mathbb{E}\left[\exp\left(-\sum_{i=1}^{\lfloor \lambda n^p \rfloor} f(X_i)\right)\right] = \left(\frac{1}{n^p}\int_{[-n/2,n/2]^p} e^{-f(x)}\, dx\right)^{\lfloor \lambda n^p \rfloor},$$

if n is sufficiently large so that the support of f is contained in $[-n/2, n/2]^p$,

$$\psi_{N_n}(f) = \left(1 - \frac{1}{n^p}\int_{[-n/2,n/2]^p}\left(1 - e^{-f(x)}\right) dx\right)^{\lfloor \lambda n^p \rfloor},$$

when n tends to infinity this expression converges to

$$\exp\left(-\lambda \int_{\mathbb{R}^p} \left(1 - e^{-f(x)}\right) dx\right),$$

the Laplace transform of a Poisson point process with intensity $\lambda\, dx$ at f. The point process N_n converges in distribution to a Poisson point process with intensity $\lambda\, dx$. A (non-rigorous) formulation of this result is the following: The points of a Poisson point process are randomly distributed on $\mathbb{R}^p$.

1.2.3 General Properties of Poisson Point Processes

The classical general properties of Poisson point processes are presented in this section.

PROPOSITION 1.10. *If N is a Poisson point process with intensity μ on H and A is a Borelian subset of H such that $\mu(A)$ is infinite, $\mathbb{P}$-almost surely, the number $N(A)$ of points of N in A is also infinite.*

PROOF. There exists a non-decreasing sequence of compact subsets $K_n \subset A$ such that $\mu(K_n)$ tends to infinity as n goes to infinity. For $n \in \mathbb{N}$, the distribution of $N(K_n)$ is Poisson with parameter $\mu(K_n)$, if $p \in \mathbb{N}$,

$$\mathbb{P}(N(K_n) = p) = \frac{\mu(K_n)^p}{p!} \exp\left[-\mu(K_n)\right],$$

and therefore, the sequence $[\mathbb{P}(N(K_n) \leq p)]$ converges to 0 for all $p \geq 0$. For $n \in \mathbb{N}$, K_n is a subset of A, the inequality $\mathbb{P}(N(A) \leq p) \leq \mathbb{P}(N(K_n) \leq p)$ gives the identity $\mathbb{P}(N(A) \leq p) = 0$ for all $p \in \mathbb{N}$. The variable $N(A)$ is $\mathbb{P}$-a.s. infinite. □

EXAMPLES.

1. If $H = \mathbb{R}$ and $\mu(dx) = \lambda\, dx$, the points of N can be represented by a doubly infinite sequence (t_n) such that
$$\lim_{n \to -\infty} t_n = -\infty \text{ and } \lim_{n \to +\infty} t_n = +\infty.$$
2. If $\mu(dx) = dx/x^2$ on $H =]0, +\infty[$, the associated Poisson point process has almost surely a finite number of points on the interval $]a, +\infty]$ for $a > 0$ and any neighborhood of 0 has an infinite number of points.

PROPOSITION 1.11 (Marked Poisson point processes). *If $N = \{u_n\}$ is a Poisson point process with intensity μ on H and (X_n) an i.i.d sequence of random variables with distribution ν in a locally compact, σ-compact space G, then*

$$N_X = \sum \delta_{(u_n, X_n)}$$

is a Poisson point process on $H \times G$ with intensity $\mu \otimes \nu$.

Marked Poisson point processes are important in practice. For $n \in \mathbb{N}$, u_n is the arrival time of the nth customer and the corresponding mark X_n may be the sequence of queues that this customer visits in the queueing system and/or services that it requires at each of them.

PROOF. If $f : H \times G \to \mathbb{R}_+$ is continuous with compact support, the Laplace transform of the point process N_X at f is given by

$$\mathbb{E}\left[\exp\left(-\sum_n f(u_n, X_n)\right)\right] = \mathbb{E}\left[\mathbb{E}\left(\exp\left(-\sum_n f(u_n, X_n)\right)\middle| N\right)\right],$$

where " $| N$" is the conditioning with respect to the σ-field generated by the point process. The independence assumptions of the sequences (u_n) and (X_n) give the identities

$$\mathbb{E}\left[\exp\left(-\sum_n f(u_n, X_n)\right)\right] = \mathbb{E}\left[\prod_{n\in\mathbb{N}} E\left(e^{-f(u_n,X_n)}\middle| N\right)\right]$$

$$(1.4) \qquad = \mathbb{E}\left[\prod_n \int e^{-f(u_n,y)}\,\nu(dy)\right].$$

If, for $x \in H$, $h(x)$ is defined by

$$e^{-h(x)} = \int e^{-f(x,y)}\,\nu(dy),$$

$\{u_n\}$ being a Poisson point process with intensity μ, Relationship (1.4) can be rewritten as follows

$$\mathbb{E}\left[\exp\left(-\sum_n h(u_n)\right)\right] = \exp\left(-\int\left(1 - e^{-h(x)}\right)\mu(dx)\right)$$

$$= \exp\left(-\int\left(1 - e^{-f(x,y)}\right)\mu(dx)\,\nu(dy)\right).$$

The proposition is proved. □

PROPOSITION 1.12. *If N is a Poisson point process on H with intensity μ, ϕ is a measurable mapping from H into a locally compact metric space G, if the measure μ_ϕ, induced on G by μ and ϕ is Radon, the point process N_ϕ on G induced by N and ϕ is Poisson with intensity μ_ϕ.*

PROOF. If f is continuous with compact support on G and $N = \{u_n\}$, then $N \circ \phi = \{\phi(u_n)\}$ consequently,

$$\mathbb{E}\left[\exp\left(-N_\phi(f)\right)\right] = \mathbb{E}\left[\exp\left(-N(f\circ\phi)\right)\right]$$

$$= \exp\left(-\int\left(1 - e^{-f\circ\phi(x)}\right)\mu(dx)\right) = \exp\left(-\int\left(1 - e^{-f(x)}\right)\mu_\phi(dx)\right),$$

The proposition is proved. □

PROPOSITION 1.13. *If $N = \{u_n\}$ is a Poisson point process with intensity μ on H and (X_n) an i.i.d. sequence of random variables with values in a locally compact, σ-compact metric space G, independent of N and with distribution ν then,*

a) Thinning : If A is a Borelian subset of G,

$$N_A = \sum_n 1_{\{X_n \in A\}} \delta_{u_n}$$

is a Poisson point process on H with intensity $\nu(A)\mu$;

b) Translation (Dobrushin) : If $H = G = \mathbb{R}^p$, $p \geq 1$ and

$$T^X(N) = \sum_n \delta_{u_n + X_n},$$

and the convolution $\mu * \nu$ is Radon on $\mathbb{R}^p$ then, the point process $T^X(N)$ is Poisson with intensity $\mu * \nu$. In particular, if the intensity of N is Lebesgue's Measure $\lambda\, dx$, the point process $T^X(N)$ is also Poisson with intensity $\lambda\, dx$.

PROOF. Proposition 1.11 shows that the point process $M = \{u_n, X_n\}$ is Poisson with intensity $\mu \otimes \nu$. By using the definition of the Poisson property it is easily seen that the point process N_A has the independence property for the number of points in Borelian disjoint subsets. The number of points of N_A in a Borelian subset B of H is the number of points of M in $B \times A$ therefore, the distribution of this variable is Poisson with parameter $\mu(B)\nu(A)$, a) is proved.

Proposition 1.12 and the fact that the function $(u, x) \to u + x$ is measurable prove the first part of b). Finally, it is sufficient to remark that Lebesgue's Measure Leb_p on $\mathbb{R}^p$ is invariant by convolution with a probability distribution. Indeed, if f is continuous with compact support on $\mathbb{R}^p$,

$$\int_{\mathbb{R}^p} f(z)\, \text{Leb}_p * \nu(dz) = \int_{\mathbb{R}^p \times \mathbb{R}^p} f(x+y)\, dx\, \nu(dy)$$
$$= \int_{\mathbb{R}^p \times \mathbb{R}^p} f(x)\, dx\, \nu(dy) = \int_{\mathbb{R}^p} f(x)\, dx\, \nu(\mathbb{R}^p) = \int_{\mathbb{R}^p} f(x)\, dx,$$

hence $\lambda\, \text{Leb}_p * \nu = \lambda\, \text{Leb}_p$. The proposition is proved. □

COROLLARY 1.14. *The distribution of a Poisson point process $\{u_n\}$ on $\mathbb{R}^p$ with intensity $\lambda\, dx$ is invariant by translation: For $t \in \mathbb{R}^p$, the point process $\{u_n + t\}$ is also Poisson with intensity $\lambda\, dx$.*

1.2.4 Conditioning in the Neighborhood of 0

In this part $M = \{u_n\}$ is a Poisson point process with intensity $\lambda\, dx$ on $\mathbb{R}^p$ and (X_n) is a sequence of i.i.d. random variables, independent of M, with distribution ν on G, a locally compact, σ-compact metric space. Proposition 1.11 shows that $N = \{(u_n, X_n)\}$ is a Poisson point process with intensity $\lambda\, dx \otimes \nu$. If $B(0, \varepsilon)$ is the ball centered in 0 with radius ε,

$$B(0, \varepsilon) = \{x = (x_i) \in \mathbb{R}^p : |x_i| < \varepsilon, i = 1, \ldots, p\},$$

when A is a Borelian subset of $\mathbb{R}^p$ and f an integrable function on $\mathbb{R}^p$, $N(A)$ denotes, with a slight abuse of notation, $N(A \times G)$ and

$$\int_A f(x)\, N(dx) = \int_{A\times G} f(x)\, N(dx, dy).$$

The following proposition studies the distribution of N conditioned to have a point near 0. It shows in particular that conditioning of a Poisson point process to have a point at 0 does not change the distribution of the point process outside 0.

PROPOSITION 1.15. *If N is a Poisson point process on $\mathbb{R}^p \times G$ with intensity $\lambda dx \otimes \nu(dy)$, the distribution of N conditioned on the event $\{N(B(0,\varepsilon)) \neq 0\}$ converges to the distribution of $\widetilde{N} + \delta_{0,X}$ when ε tends to 0. The process $\widetilde{N}$ is a Poisson point process with the same distribution as N and X is a random variable with distribution $\nu(dx)$ independent of the point process $\widetilde{N}$.*

PROOF. For the proof, the criterion of Proposition 1.2 with Laplace transform is used. If f is a non-negative continuous function with compact support on $\mathbb{R} \times G$,

$$\mathbb{E}\left[\exp\left(-\int f(x)\, N(dx)\right) \,\middle|\, N(B(0,\varepsilon)) \neq 0\right] =$$
$$\mathbb{E}\left[\exp\left(-\int_{\mathbb{R}^p - B(0,\varepsilon)} f(x)\, N(dx) - \int_{B(0,\varepsilon)} f(x)\, N(dx)\right) \,\middle|\, N(B(0,\varepsilon)) \neq 0\right],$$

since the Poisson point process N restricted to $\mathbb{R}^p - B(0,\varepsilon)$ is independent of the restriction of N on $B(0,\varepsilon)$, one has

$$\mathbb{E}\left[\exp\left(-\int f(x)\, N(dx)\right) \,\middle|\, N(B(0,\varepsilon)) \neq 0\right] = \mathbb{E}\left[\exp\left(-\int_{\mathbb{R}^p - B(0,\varepsilon)} f(x)\, N(dx)\right)\right] \times \mathbb{E}\left[\exp\left(-\int_{B(0,\varepsilon)} f(x)\, N(dx)\right) \,\middle|\, N(B(0,\varepsilon)) \neq 0\right]. \tag{1.5}$$

If $t_1 = \inf\{x > 0 : N(B(0,x)) \neq 0\}$ and X is the variable such that (t_1, X) is a point of N, then X is a random variable independent of t_1 and its distribution is ν. Inequality (1.2) is used again so that

$$\left| \mathbb{E}\left[\exp\left(-\int_{B(0,\varepsilon)} f(x)\, N(dx) \right) 1_{\{N(B(0,\varepsilon))\neq 0\}} \right.\right.$$
$$\left.\left. - \exp\left(-f(t_1, X)\right) 1_{\{N(B(0,\varepsilon))=1\}} \right] \right| \leq \mathbb{P}\left(N(B(0,\varepsilon)) \geq 2\right) \leq \lambda^2 \varepsilon^{2p}.$$

Since $\mathbb{P}(N(B(0,\varepsilon)) \neq 0) = 1 - \exp(-\lambda\varepsilon^p) \sim \lambda\varepsilon^p$, the last inequality shows that it is sufficient to study the asymptotic behavior of

$$\text{(1.6)}\qquad \frac{\mathbb{E}\left(\exp\left(-f(t_1, X)\right) 1_{\{N(B(0,\varepsilon))=1\}}\right)}{\mathbb{P}(N(B(0,\varepsilon)) \neq 0)} = \frac{\mathbb{E}\left(\exp\left(-f(t_1, X)\right) 1_{\{t_1 < \varepsilon\}}\right)}{\mathbb{P}(t_1 < \varepsilon)}.$$

For $t > 0$,

$$\mathbb{P}(t_1 > t) = \mathbb{P}(N(B(0,t) = 0) = \exp(-\lambda t^p),$$

the density of t_1 is therefore the function $\lambda p u^{p-1} e^{-\lambda u^p}$ on $\mathbb{R}_+$. The right hand side of Equation (1.6) is thus given by

$$\int_0^{\varepsilon} \lambda p u^{p-1} e^{-\lambda u^p} \mathbb{E}\left[\exp\left(-f(u, X)\right)\right] du \,/ (1 - \exp(-\lambda\varepsilon^p))$$
$$= \int_0^{\varepsilon^p} \lambda e^{-\lambda u} \mathbb{E}\left[\exp\left(-f(u^{1/p}, X)\right)\right] du \,/ (1 - \exp(-\lambda\varepsilon^p))$$
$$= \frac{\lambda\varepsilon^p}{1 - \exp(-\lambda\varepsilon^p)} \frac{1}{\varepsilon^p} \int_0^{\varepsilon^p} e^{-\lambda u} \mathbb{E}\left[\exp\left(-f\left(u^{1/p}, X\right)\right)\right] du.$$

By using the continuity of f, when ε tends to 0, this term converges to $\exp(-f(0, X))$. Lebesgue's Theorem shows therefore the convergence

$$\lim_{\varepsilon\to 0} \mathbb{E}\left[\exp\left(-\int_{B(0,\varepsilon)} f(x)\, N(dx)\right) \middle| N(B(0,\varepsilon)) \neq 0\right] = \mathbb{E}\left[\exp\left(-f(0, X)\right)\right].$$

From Relationship (1.5), one gets

$$\lim_{\varepsilon\to 0} \mathbb{E}\left[\exp\left(-\int f(x)\, N(dx)\right) \middle| N(B(0,\varepsilon)) \neq 0\right] = \mathbb{E}\left[\exp\left(-\int f(x)\, N(dx)\right)\right] \mathbb{E}\left[\exp\left(-f(0, X)\right)\right],$$

and this last expression is

$$\mathbb{E}\left[\exp\left(-\int f(x)\, \left(\widetilde{N}(dx) + \delta_{(0,X)}\right)\right)\right].$$

The proposition is proved. □

1.3 Poisson Point Processes on the Real Line

1.3.1 Convergence to a Poisson Point Process

If (X_n) is a sequence of random variables on $\mathbb{R}$ and $a > 0$, the instants at which this sequence is above a, it can be represented as the points of a point process

$$\sum_n 1_{\{X_n > a\}} \delta_n.$$

In practice, if (X_n) is the number of customers of a queue at some discrete instants, this point process describes the moments of overflow of the level a. In general, it is not easy to get the distribution of such a point process. In some cases, it can be shown that, when a goes to infinity, this point process is asymptotically Poisson. The simple case where the variables (X_n) are i.i.d.

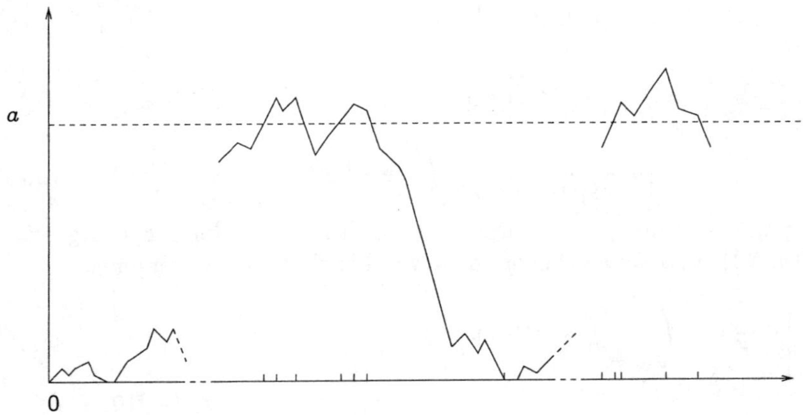

FIG. 1.1. Hitting Times of a

gives an hint of the kind of result one might expect. The following proposition corresponds to the analogue of Proposition 1.4 for point processes.

PROPOSITION 1.16. *If there exist α and $\lambda > 0$ such that*

$$\lim_{a \to +\infty} \exp(\alpha a) \mathbb{P}(X_0 > a) = \lambda,$$

when a tends to infinity, the point process

$$N_a = \sum_n 1_{\{X_n > a\}} \delta_{n \exp(-\alpha a)}$$

converges in distribution to a Poisson point process with parameter λ.

PROOF. If f is a continuous function with compact support on $\mathbb{R}$, the Laplace transform of this point process at f is given by

$$\mathbb{E}\left[\exp\left(-\sum_{n\geq 0} f(ne^{-\alpha a})1_{\{X_n\geq a\}}\right)\right]$$
$$= \prod_n \left[1-\mathbb{P}(X_0>a)\left(1-\exp\left[-f\left(ne^{-\alpha a}\right)\right]\right)\right],$$

according to the independence property and the fact that the (X_n) have the same distribution. By using the elementary inequality

$$|\log(1-x)+x|\leq x^2, \quad 0\leq x\leq 1/2,$$

if a is sufficiently large, the inequality

$$\left|\sum_n \log\left[1-\mathbb{P}(X_0>a)\left(1-\exp\left[-f\left(ne^{-\alpha a}\right)\right]\right)\right]\right.$$
$$\left.+\sum_n \mathbb{P}(X_0>a)\left(1-\exp\left[-f\left(ne^{-\alpha a}\right)\right]\right)\right|$$
$$\leq \mathbb{P}(X_0>a)^2\sum_n\left(1-\exp\left[-f\left(ne^{-\alpha a}\right)\right]\right)^2$$

is obtained. Since $\mathbb{P}(X_0>a)\sim\lambda\exp(-\alpha a)$, this last expression is equivalent to a Riemann sum, and is consequently equivalent to

$$\lambda^2 e^{-\alpha a}\int\left(1-e^{-f(x)}\right)^2\,dx,$$

and thus converges to 0 when a tends to infinity. In the same way, one gets

$$\lim_{a\to+\infty}\sum_n \mathbb{P}(X_0>a)\left(1-\exp\left[-f(ne^{-\alpha a})\right]\right)=\lambda\int\left(1-e^{-f(x)}\right)\,dx.$$

The Laplace transform at f of the point process N_n converges to

$$\exp\left(-\lambda\int\left(1-e^{-f(x)}\right)\,dx\right),$$

which is the Laplace transform of a Poisson process with intensity λ. □

In the last proposition, the independence property is a major ingredient to prove convergence. In practice this assumption is not true in general, nevertheless this kind of convergence still holds, see Aldous [**1**] on this subject. Under some mixing assumptions (asymptotic independence), it is possible to show this convergence (see Leadbetter [**32**]). Such a convergence is proved for the number of customers of the $M/M/1$ queue and $M/M/\infty$ queue in Chapter 5 and Chapter 6 respectively.

1.3.2 Poisson Marked Point Processes on $\mathbb{R}$

DEFINITION 5. *A Poisson marked point process on* $\mathbb{R} \otimes H$ *is a Poisson point process whose intensity is given by* $\lambda\, dx \otimes \nu(dy)$ *on* $\mathbb{R} \times H$ *where* ν *is a probability measure on a locally compact metric space* H *which is* σ*-compact.*

According to Proposition 1.11, such a Poisson point process can be represented as $\{(t_n, X_n)\}$ where $\{t_n\}$ is a Poisson point process on $\mathbb{R}$ with intensity $\lambda\, dx$ and (X_n) is an i.i.d. sequence of random variables with distribution ν, independent of $t_n, n \in \mathbb{Z}$. The variables (X_n) mark the points (t_n). In the sequel $N = \{(t_n, X_n)\}$ is a Poisson marked point process, with the numbering convention

$$\cdots < t_{-n-1} < t_{-n} < \cdots < t_0 \leq 0 < t_1 < \cdots < t_n < t_{n+1} < \cdots$$

With a slight abuse of notation, if A is a Borelian subset of $\mathbb{R}$ the notation $N(A)$ will be used to represent $N(A \times H)$. The first coordinate (t_n) of this point process can be interpreted as a time variable. It is therefore natural to define the associated filtration $(\mathcal{F}_t)$: For $t \in \mathbb{R}$, $\mathcal{F}_t$ is the σ-field generated by the random variables $N(A \times U)$, where A is a Borelian subset of $]-\infty, t]$ and U is a Borelian subset of H, or

$$\mathcal{F}_t = \sigma \left\langle N(]a, b] \times U) : a, b \leq t,\, U \in \mathcal{B}(H) \right\rangle.$$

LEMMA 1.17. *For* $n \geq 1$ *and* $t \geq 0$*, the random variable* $X_n 1_{\{t_n \leq t\}}$ *is* $\mathcal{F}_t$*-measurable.*

PROOF. Indeed, for $a \geq 0$, the event

$$\{X_n 1_{\{t_n \leq t\}} \geq a\} = \bigcup_{s \leq s' \in \mathbb{Q} \cap [0,t]} \Big\{ N\left([s, s'] \times \mathbb{R}_+\right) = 1,$$
$$N\left([0, s'] \times \mathbb{R}_+\right) = n, N\left([s, s'] \times [a, +\infty[\right) = 1 \Big\}$$

is clearly $\mathcal{F}_t$-measurable. □

In other words, the marks of customers arrived before time t are $\mathcal{F}_t$-measurable variables.

DEFINITION 6. *The process* $(N(]0, t]))$ *is the* counting measure *associated with* N*, the jumps of this process are of size* 1 *and occur at the points of the point process* N.

On the real line it is equivalent to consider the counting measure or the point process.

PROPOSITION 1.18. *If* N *is a marked Poisson point process on* $\mathbb{R}$ *and* T *is a finite stopping time, the translated point process by* T, $(N(]T, T+t]))$, *is independent of* $\mathcal{F}_T$ *and has the same distribution as* $(N(]0, t]))$.

PROOF. The space of measurable functions from $\mathbb{R}_+$ to $\mathbb{R}$ is denoted by $F(\mathbb{R}_+, \mathbb{R})$, it is endowed with the topology of pointwise convergence and its associated σ-field. If $N(]T, T+\cdot\,])$ denotes the process $(N(]T, T+t]))$, it is sufficient to prove that the identity

$$\mathbb{E}\left[f\left(N\left(]T, T+\cdot\,]\right)\right) \mid \mathcal{F}_T\right] = \mathbb{E}\left[f\left(N\left(]0, 0+\cdot\,]\right)\right)\right],$$

holds almost surely for any measurable function f on $F(\mathbb{R}_+, \mathbb{R})$ depending only on a finite number of coordinates. Indeed, this identity implies the independence property of $(N(]T, T+\cdot\,]))$ and $\mathcal{F}_T$, because the conditional expectation is constant. Furthermore, it also implies that the distributions of $(N(]T, T+\cdot\,]))$ and $(N(]0, \cdot\,]))$ are the same.

Notice first that $N(]a, a+x]) \to N(]b, b+x])$ when $a \searrow b$. If f is a function depending only of finite number of integer variables, for $a_k^n = k/2^n$, $k \geq 0$, and $n \geq 1$,

$$\lim_{n\to+\infty} \sum_{k\in\mathbb{Z}} f\left(N\left(]a_k^n, a_k^n+\cdot\,]\right)\right) 1_{\{a_{k-1}^n < T \leq a_k^n\}} = f(N]T, T+\cdot\,]),$$

thus,

$$\lim_{n\to+\infty} \sum_{k\in\mathbb{Z}} \mathbb{E}\left[f\left(N\left(]a_k^n, a_k^n+\cdot\,]\right)\right) 1_{\{a_{k-1}^n < T \leq a_k^n\}} \,\middle|\, \mathcal{F}_T\right] = \mathbb{E}\left[f\left(N\left(]T, T+\cdot\,]\right)\right) \mid \mathcal{F}_T\right], \tag{1.7}$$

according to Lebesgue's convergence Theorem.

The stopping time property of T implies that the set $\{T \in]a_{k-1}^n, a_k^n]\}$ is $\mathcal{F}_{a_k^n}$-measurable for k and $n \in \mathbb{N}$, hence

$$\begin{aligned}\mathbb{E}\Big[f\left(N\left(]a_k^n, a_k^n+\cdot\,]\right)\right) 1_{\{T\in]a_{k-1}^n, a_k^n]\}} \,\Big|\, \mathcal{F}_T\Big] \\ = \mathbb{E}\Big[\mathbb{E}\left(f\left(N\left(]a_k^n, a_k^n+\cdot\,]\right)\right) \mid \mathcal{F}_{a_k^n}\right) 1_{\{T\in]a_{k-1}^n, a_k^n]\}} \mid \mathcal{F}_T\Big].\end{aligned}$$

The independence property of the Poisson point process before a_k^n and after a_k^n gives the identity

$$\mathbb{E}\left[f\left(N\left(]a_k^n, a_k^n+\cdot\,]\right)\right) \mid \mathcal{F}_{a_k^n}\right] = \mathbb{E}\left[f\left(N\left(]a_k^n, a_k^n+\cdot\,]\right)\right)\right],$$

this last term is $\mathbb{E}[f(N(]0, 0+\cdot\,]))]$ due to the translation invariance seen in Corollary 1.14. The event $\{T \in]a_{k-1}^n, a_k^n]\}$ being $\mathcal{F}_T$-measurable, one gets

$$\begin{aligned}\mathbb{E}\Big[f\left(N\left(]a_k^n, a_k^n+\cdot\,]\right)\right) 1_{\{T\in]a_{k-1}^n, a_k^n]\}} \,\Big|\, \mathcal{F}_T\Big] \\ = \mathbb{E}\Big[f\left(N\left(]0, 0+\cdot\,]\right)\right) 1_{\{T\in]a_{k-1}^n, a_k^n]\}}\Big],\end{aligned}$$

by summing this last equation and using Relationship (1.7), one obtains

$$\mathbb{E}\left[f(N(]T, T+\cdot\,])) \mid \mathcal{F}_T\right] = \mathbb{E}\left[f(N(]0, 0+\cdot\,]))1_{\{T<+\infty\}}\right].$$

The theorem is proved. □

PROPOSITION 1.19. *For $n \geq 1$, t_n is a stopping time, the variables $-t_0$, t_1, $t_{n+1} - t_n$, $n \neq 0$ are i.i.d. exponentially distributed random variables with parameter λ.*

PROOF. Since $\mathbb{P}(t_1 \geq t) = \mathbb{P}(N(]0,t]) = 0) = \exp(-\lambda t)$, the variable t_1 is exponentially distributed with parameter λ.

The identity $\{t_n \leq t\} = \{N(]0,t]) \geq n\} \in \mathcal{F}_t$ shows the stopping time property of the variable t_n. The process $(N(]t_n, t_n+t]))$ is independent of the σ-field $\mathcal{F}_{t_n}$ and with the same distribution as $(N(]0,t]))$ according to Proposition 1.18. In particular $t_{n+1}-t_n$, the coordinate of the first non-negative point of the point process $(N(]t_n, t_n+t]))$ is exponentially distributed with parameter λ and is independent of $t_k - t_{k-1}, 2 \leq k \leq n$. The variables $t_{n+1} - t_n$, $n \geq 0$ and t_1 are therefore exponentially distributed with parameter λ, and independent of the variables $(t_{n+1} - t_n; n < 0)$ and $-t_0$.

Is is easily seen, via Laplace transform, that the point process $\{-t_n\}$ is also Poisson with intensity $\lambda\, dx$. Consequently, the variables $(t_{n+1}-t_n; n < 0)$ and $-t_0$ are also i.i.d. exponentially distributed with parameter λ. □

For $n \in \mathbb{N}$, the identity $\{N([0,t]) = n\} = \{t_n \leq t < t_{n+1}\}$ and the independence of t_n and $t_{n+1} - t_n$ give

$$\frac{(\lambda t)^n}{n!} e^{-\lambda t} = \int_0^{+\infty} \mathbb{P}(t_n \leq t < t_n + x)\lambda e^{-\lambda x}\, dx,$$

hence, by using Fubini's Theorem,

$$\frac{(\lambda t)^n}{n!} = \mathbb{E}\left(e^{\lambda t_n} 1_{\{t_n < t\}}\right) = \int_0^t e^{\lambda x} h_n(x)\, dx,$$

where h_n is the density function of the variable t_n on $\mathbb{R}_+$. By differentiating this equation with respect to t, one gets

$$h_n(x) = \lambda \frac{(\lambda x)^{n-1}}{(n-1)!} e^{-\lambda x}.$$

COROLLARY 1.20. *If T is a stopping time and $t_1(T)$ is the distance between T and the first point of N after T, $t_1(T) = \inf\{t_n - T : t_n > T\}$, then $t_1(T)$ is independent of $\mathcal{F}_T$ and is exponentially distributed random variable with parameter λ.*

The next proposition summarizes the main properties of a Poisson point process as a stochastic process.

PROPOSITION 1.21. *If F is a Borelian subset of H, the process*

$$(M(t)) = (N(]0,t] \times F) - \lambda\nu(F)t)$$

has stationary independent increments, in particular it is a $(\mathcal{F}_t)$-martingale with the strong Markov property. The increasing process of this martingale is given by $(\lambda\nu(F)t)$.

Proof. If $s \leq t$,

$$N(]0, s+t] \times F) = N(]0, s] \times F) + N(]s, s+t] \times F),$$

the independence of $N(]s, s+t] \times F)$ and $\mathcal{F}_s$ shows that

$$\mathbb{E}(N(]0, s+t] \times F) - \lambda\nu(F)(s+t) \mid \mathcal{F}_s) = N(]0, s] \times F) - \lambda\nu(F)s.$$

The martingale property is proved. For the increasing process the proof is similar. It is easily seen that

$$\mathbb{E}\left(M(t+s)^2 \mid \mathcal{F}_s\right) = M(s)^2 + \lambda\nu(F)t,$$

for non-negative s and t. The process $(M(t)^2 - \lambda\nu(F)t)$ is therefore a martingale and consequently $(\lambda\nu(F)t)$ is the increasing process of $(M(t))$. The strong Markov property is a direct consequence of Proposition 1.18. □

Proposition 1.22. *If N is a marked Poisson point process on $\mathbb{R}_+ \times H$ with intensity $\lambda\, dt \otimes \nu(dy)$ and f a measurable function from H into $\mathbb{R}_+$ such that f is square integrable for ν, then the process*

$$\left(\int_{]0,t] \times H} f(y) \left[N(ds, dy) - \lambda\nu(dy)\right]\right)$$

is a square integrable martingale whose increasing process is given by

$$\left(\lambda t \int_H f(y)^2 \nu(dy)\right)$$

Proof. The martingale property is proved in the same way as before.

If f is the indicator function of a measurable set, in this case the proposition is precisely the result proved in the previous proposition. If $f = \alpha\, 1_A + \beta\, 1_B$ where A and B are two disjoint measurable subsets of H, the independence property of Poisson point processes shows that the random variables

$$\int_{]0,t] \times H} 1_A(y) \left[N(ds, dy) - \lambda\nu(dy)\right], \int_{]0,t] \times H} 1_B(y) \left[N(ds, dy) - \lambda\nu(dy)\right]$$

are independent with zero mean, consequently the quantity

$$\mathbb{E}\left(\int_{]0,t] \times H} f(y) \; \left[N(ds, dy) - \lambda\nu(dy)\right]\right)^2$$

is given by

$$\alpha^2 \mathbb{E}\left(\int_{]0,t] \times H} 1_A(y) \; \left[N(ds, dy) - \lambda\nu(dy)\right]\right)^2 + \beta^2 \mathbb{E}\left(\int_{]0,t] \times H} 1_B(y) \; \left[N(ds, dy) - \lambda\nu(dy)\right]\right)^2,$$

therefore,

$$\mathbb{E}\left(\int_{]0,t]\times H} f(y)\,\big[N(ds,dy)-\lambda\nu(dy)\big]\right)^2$$
$$= \lambda t\left(\alpha^2\nu(A)+\beta^2\nu(B)\right) = \lambda t\int_H f(y)^2\,\nu(dy).$$

By induction, one gets that last identity holds for all simple measurable functions. The density of by simple functions shows that the identity is valid for an arbitrary bounded measurable function f. With the same argument as in the proof of Proposition 1.21, it is then easy to conclude the assertion on the increasing process. □

1.4 Renewal Point Processes

A Poisson point process on $\mathbb{R}_+$ with intensity λ is a non decreasing sequence of points separated by i.i.d. random variables whose distribution is exponential with parameter λ. The point processes on $\mathbb{R}_+$ considered in this section are such that the distances between the consecutive points form a general i.i.d. sequence $(\tau_i; i \in \mathbb{Z})$, τ_0 is a non-negative integrable variable. In the following the sequence (τ_i) is fixed. The next proposition shows that the convergence in distribution of distances between points implies the convergence in distribution of corresponding point processes.

PROPOSITION 1.23. *If* $(N_n) = (\{t_k^n; k \geq 0\})$ *is a sequence of point processes on* $\mathbb{R}_+$ *such that*

- $t_0^n = 0$ *for* $n \in \mathbb{N}$ *;*
- *for* $n \in \mathbb{N}$, *the sequence of non-negative random variables* $(t_{k+1}^n - t_k^n)$ *is i.i.d. ;*
- *the variable* $t_1^n - t_0^n = t_1^n$ *converges in distribution to* t_1 *when* n *tends to infinity,*

if $(t_{k+1} - t_k)$ *is an i.i.d. sequence with the same distribution as* t_1, *the point process* N_n *converges in distribution to* $N = \{t_k\}$ *as* n *tends to infinity.*

PROOF. The criterion of convergence of Proposition 1.2 is used. If f is a continuous, non negative function with compact support on $[0, K]$, the convergence

$$\lim_{n\to+\infty}\mathbb{E}\left[\exp\left(-N_n(f)\right)\right] = \lim_{n\to+\infty}\mathbb{E}\left[\exp\left(-\sum_k f(t_k^n)\right)\right]$$
$$= \mathbb{E}\left[\exp\left(-\sum_k f(t_n)\right)\right] = \mathbb{E}\left[\exp\left(-N(f)\right)\right]$$

has to proved. For $\varepsilon > 0$, since (t_k) converges in probability to infinity, there exists $k_0 > 0$ such that $\mathbb{P}\left(t_{k_0} \leq K\right) \leq \varepsilon$ and there exists A such that if $n \geq A$,

$$|\mathbb{P}(t^n_{k_0} \leq K) - \mathbb{P}(t_{k_0} \leq K)| \leq \varepsilon.$$

For $n \geq A$,

$$\left|\mathbb{E}\left[\exp\left(-\sum_k f\left(t^n_k\right)\right)\right] - \mathbb{E}\left[\exp\left(-\sum_k f(t_k)\right)\right]\right|$$
$$\leq \left|\mathbb{E}\left[\exp\left(-\sum_k f\left(t^n_k\right)\right) 1_{\{t^n_{k_0}>K\}}\right] - \mathbb{E}\left[\exp\left(-\sum_k f(t_k)\right) 1_{\{t_{k_0}>K\}}\right]\right|$$
$$+ \mathbb{P}(t_{k_0} \leq K) + \mathbb{P}(t^n_{k_0} \leq K),$$

The first term of the right hand side converges to 0 as n tends to infinity since the distribution of $(t^n_k; k \leq k_0)$ converges and, the function f being 0 outside the interval $[0, K]$, the variables under the expected value depend only of a finite number of coordinates. The two other terms are bounded by 3ε. The proposition is proved. □

Definition 7. *If $x \geq 0$, the point process $N^x = \{t^x_n\}$ with a first point at x is defined by*

– $t^x_1 = x$ *and* $t^x_{n+1} = t^x_n + \tau_n$, *for* $n \geq 0$.

The point process N^x is the renewal point process associated with the sequence (τ_n) whose first point is located at x. For $t \geq 0$, $T^t N^x$ denotes the point process N^x translated from t, i.e. $T^t N^x = \{t^x_n - t\}$ and $N^x(B\cap\cdot)$, the restriction of N^x to a Borelian subset B.

Proposition 1.24. *If F is a random variable independent of (τ_i) whose distribution has the density $\mathbb{P}(\tau_0 \geq x)/\mathbb{E}(\tau_0)$ on $\mathbb{R}_+$ i.e.*

$$\mathbb{P}(F \in A) = \int_A \frac{\mathbb{P}(\tau_0 \geq x)}{\mathbb{E}(\tau_0)}\, dx, \tag{1.8}$$

for any Borelian subset A of $\mathbb{R}_+$, the point process $\mathcal{N} = N^F$ on $\mathbb{R}_+$ associated with this random variable F is invariant with respect to the non-negative translations, i.e. the point processes $T^t\mathcal{N} = \{t^F_n - t\}$ and $\{t^F_n\} = \mathcal{N}$ have the same distribution on $\mathbb{R}_+$ for $t \geq 0$.

Proof. To simplify the notations, the exponent F in $\{t^F_n\}$ is omitted in this proof. If $\nu_t = \mathcal{N}([0,t]) + 1$, and if f and g are non-negative Borelian functions on $\mathbb{R}$ and $\mathbb{R}^{\mathbb{N}}$ respectively, by definition of the variable ν, one gets

$$\mathbb{E}\left[f(t_{\nu_t} - t)g((t_{\nu_t+k+1} - t_{\nu_t+k}))\right]$$
$$= \sum_{n\geq 1} \mathbb{E}\left[1_{\{t_{n-1}\leq t<t_n\}} f(t_n - t) g((\tau_{n+k}))\right].$$

For $n \in \mathbb{N}$, since the variable t_n depends on the variables $\{\tau_i,\ i < n\}$, the last term can be expressed as

$$\mathbb{E}\left[g((\tau_{n+k}))\right] \sum_{n\geq 1} \mathbb{E}\left[1_{\{t_{n-1}\leq t<t_n\}} f(t_n - t)\right] = \mathbb{E}\left[g((\tau_k))\right] \mathbb{E}\left[f(t_{\nu_t} - t)\right].$$

The sequence $(t_{\nu_t+k+1} - t_{\nu_t+k}) = (\tau_{\nu_t+k})$ is therefore i.i.d. with the same distribution as (τ_k) and independent of $t_{\nu_t} - t$. The point process $\mathcal{N}$ translated by t and restricted to $\mathbb{R}_+$ is given by $\{t_{\nu_t+k} - t; k \geq 0\}$. To prove that this point process has the same distribution as $\mathcal{N}$, it is sufficient to show that the distribution of the first point $t_{\nu_t} - t$ is given by Equation(1.8), or that

$$\mathbb{E}\left[\exp\left(-\xi(t_{\nu_t} - t)\right)\right] = \widetilde{F}(\xi) \stackrel{\text{def.}}{=} \mathbb{E}\left(\exp\left(-\xi F\right)\right) = \frac{1 - \widetilde{\tau}(\xi)}{\xi \mathbb{E}(\tau_0)},$$

holds for any $\xi \geq 0$, with $\widetilde{\tau}(\xi) = \mathbb{E}(\exp(-\xi\tau_0))$. The left hand side being a function of t bounded by 1, if one proves that, for any $z \geq 0$, the identity

$$(1.9) \quad \int_0^{+\infty} \mathbb{E}\left[\exp\left(-\xi(t_{\nu_t} - t)\right)\right] \exp\left(-zt\right) dt = \int_0^{+\infty} \mathbb{E}\left(\exp\left(-\xi F\right)\right) \exp\left(-zt\right) dt = \frac{1}{z}\mathbb{E}\left(\exp\left(-\xi F\right)\right),$$

holds, then the identity is obtained because of the uniqueness of the Laplace transform of a real function.

If the real line is decomposed according to the points of $\mathcal{N}$, with the convention that $t_0 = 0$, one has

$$\begin{aligned}\int_0^{+\infty} & \mathbb{E}\left[\exp\left(-\xi(t_{\nu_t} - t)\right)\right] \exp\left(-zt\right) dt \\ &= \mathbb{E}\left(\sum_{n\geq 0} \int_{t_n}^{t_{n+1}} \exp\left(-\xi(t_{n+1} - t)\right) \exp\left(-zt\right) dt\right) \\ &= \mathbb{E}\left(\sum_{n\geq 0} e^{-zt_n} \frac{\exp\left(-\xi(t_{n+1} - t_n)\right) - \exp\left(-z(t_{n+1} - t_n)\right)}{z - \xi}\right).\end{aligned}$$

By using the independence of increments, one gets

$$\int_0^{+\infty} \mathbb{E}\left[\exp\left(-\xi(t_{\nu_t} - t)\right)\right] \exp\left(-zt\right) dt = \frac{\widetilde{F}(\xi) - \widetilde{F}(z)}{z - \xi} + \sum_{n\geq 1} \widetilde{F}(z)\widetilde{\tau}(z)^{n-1} \frac{\widetilde{\tau}(\xi) - \widetilde{\tau}(z)}{z - \xi},$$

and since $\widetilde{F}(\xi) = (1 - \widetilde{\tau}(\xi))/(\xi\mathbb{E}(\tau_0))$,

$$\begin{aligned}\int_0^{+\infty} & \mathbb{E}\left[\exp\left(-\xi(t_{\nu_t} - t)\right)\right] \exp\left(-zt\right) dt \\ &= \frac{1}{z - \xi}\left(\frac{1 - \widetilde{\tau}(\xi)}{\xi\mathbb{E}(\tau_0)} - \frac{1 - \widetilde{\tau}(z)}{z\mathbb{E}(\tau_0)}\right) + \frac{\widetilde{\tau}(\xi) - \widetilde{\tau}(z)}{(z - \xi)z\mathbb{E}(\tau_0)} = \frac{1 - \widetilde{\tau}(\xi)}{z\xi\mathbb{E}(\tau_0)},\end{aligned}$$

therefore Relationship (1.9) is established and the invariance by translation of the Poisson point process $\mathcal{N}$ holds. □

Poisson Point Processes. If τ_0 is exponentially distributed with parameter λ, the variable F defined by Relationship (1.8) has also an exponential distribution with parameter λ. The point process $\mathcal{N}$ has the same distribution as N^0, a Poisson point process with intensity λ on $\mathbb{R}_+$. Proposition 1.13 also shows the invariance by translation of Poisson point processes.

Deterministic Point Processes. If τ_0 is constant equal to D, the associated random variable F has the distribution

$$\mathbb{P}(F \in dx) = \frac{1}{D} 1_{\{0 \leq x \leq D\}}\, dx,$$

i.e. F is uniformly distributed on $[0, D]$. The associated point process $\mathcal{N}$ is simply the periodic sequence $\{F + nD\}$.

Going back to the general case, the expected value of the random variable F is given by

$$\mathbb{E}(F) = \int_0^{+\infty} x \frac{\mathbb{P}(\tau_0 \geq x)}{\mathbb{E}(\tau_0)}\, dx = \frac{1}{\mathbb{E}(\tau_0)} \mathbb{E}\left(\int_0^{\tau_0} x\, dx\right) = \frac{\mathbb{E}(\tau_0^2)}{2\mathbb{E}(\tau_0)},$$

with the help of Fubini's Theorem for the first equality. The random variable F has a first moment if and only if τ_0 is square integrable. The first point of $\mathcal{N}$ is thus located very far if τ_0^2 is not integrable (see also Chapter 11 page 318 for a discussion of this phenomenon).

Definition 8. *A probability distribution μ on $\mathbb{R}$ is non-lattice if there does not exist any $\delta > 0$ such that $\mu(\delta\mathbb{Z}) = 1$.*

A probability distribution μ on $\mathbb{R}$ is spread-out *if there exists $n_0 \in \mathbb{N}$ such that the n_0th convolution μ^{*n_0} of μ has a non-trivial Lebesgue component, i.e. if there exists a non-negative, non-trivial Borelian function h such that*

$$\mu^{*n_0}(f) = \int f(x_1 + \cdots + x_{n_0}) \prod_1^{n_0} \mu(dx_i) \geq \int f(x)h(x)\, dx,$$

for any non-negative Borelian function f.

The following proposition shows that, under some assumptions, if $x \geq 0$ the distance to the first point of the point process N^x after $t > 0$ converges in distribution to F as t gets large.

Proposition 1.25. *If τ_0 has a non-lattice distribution and for $x \geq 0$, S_t^x is the first point of N^x after t,*

$$S_t^x = \inf\{u > t : N^x([t, u]) \neq 0\} = \inf\{t_n^x : t_n^x > t\},$$

when t tends to infinity, the non-negative random variable $S_t^x - t$ converges in distribution to the distribution with density $x \to \mathbb{P}(\tau_0 \geq x)/\mathbb{E}(\tau_0)$ on $\mathbb{R}_+$.

Proof. The classical proof of this theorem uses the renewal theorem. See Durrett [**17**] for example. □

The following results concern the convergence in distribution of translated renewal processes, $\mathcal{N}$ denotes the point process N^F invariant by non-negative translations (Proposition 1.24).

1.4.1 Coupling

Under the assumptions of the above proposition, the first point of $T^t N^x$ converges in distribution to the first point of $\mathcal{N}$. The theorem below gives a stronger result: it is possible to construct a probability space where the point processes $T^t N^x(\mathbb{R}_+ \cap \cdot\,)$ and $T^t \mathcal{N}(\mathbb{R}_+ \cap \cdot\,)$ are identical for t sufficiently large. (Notice that the point process $T^t \mathcal{N}(\mathbb{R}_+ \cap \cdot\,)$ has the same distribution as $\mathcal{N}$ according to Proposition 1.24.)

THEOREM 1.26 (Coupling of Renewal Processes). *If the distribution of τ_0 is spread-out, for $x \geq 0$, there exists a probability space on which the point processes $T^t N^x(\mathbb{R}_+ \cap \cdot\,)$ and $T^t \mathcal{N}(\mathbb{R}_+ \cap \cdot\,)$ have a coupling: There exists a random variable H, $\mathbb{P}$-almost surely finite such that if $t \geq H$,*

$$T^t N^x(\mathbb{R}_+ \cap \cdot\,) = T^t \mathcal{N}(\mathbb{R}_+ \cap \cdot\,).$$

PROOF. See Asmussen [3], Theorem 2.3 page 146. □

The Case of Lattice Distributions. If τ_0 is integer valued and $x \in \mathbb{N}$, the points of the associated renewal process N^x are located in $\mathbb{N}$. The corresponding point process $\mathcal{N}$ is given by $\mathcal{N} = N^F$, with $\mathbb{P}(F = n) = \mathbb{P}(\tau_0 \geq n)/\mathbb{E}(\tau_0)$, for $n \geq 1$. It is easy to check that $\mathcal{N}$ is invariant by integer translations. The next proposition is the analogue of the previous theorem.

PROPOSITION 1.27. *If τ_0 is an integer valued random variable such that $\mathbb{P}(\tau_0 \in k\mathbb{N}) < 1$ for any $k \in \mathbb{N}$, and if, for $x \in \mathbb{N}$, the point processes N^x and $\mathcal{N}$ are independent renewal processes associated with the random variable τ_0, the variable $H = \inf\{k : N^x(\{k\}) = \mathcal{N}(\{k\}) = 1\}$ is $\mathbb{P}$-almost surely finite.*

PROOF. See Lindvall [**33**]. □

1.4.2 Mixing Properties

The previous results show that, with an assumption on the support of τ_0, the point process N^x seen from the location t converges in distribution to the point process $\mathcal{N}$.

PROPOSITION 1.28. *If the distribution of the variable τ_0 is non-lattice, for $x \geq 0$ the point process $T^t N^x(\mathbb{R}_+ \cap \cdot\,)$ converges in distribution to the renewal process $\mathcal{N}$ as t tends to infinity.*

Moreover, if the distribution of τ_0 is spread-out then, for $n \in \mathbb{N}$, when t tends to infinity, the vector

$$(\tau_0, \ldots, \tau_n, T^t N^x(\mathbb{R}_+ \cap \cdot\,))$$

converges in distribution to the vector $(\tau_0',\dots,\tau_n',\mathcal{N})$ *if the variables* $(\tau_i', i=0,\dots n)$ *are i.i.d. with the same distribution as* τ_0 *and independent of the point process* $\mathcal{N}$.

PROOF. With the notations of the proof of Proposition 1.24 and Proposition 1.25, if $\nu_t = N^x([0,t])+1$ the quantity $t^x_{\nu_t}$ is given by S^x_t and the sequence $(t^x_{\nu_t+k+1}-t^x_{\nu_t+k})$ has the same distribution as the i.i.d. sequence (τ_i) and is independent of $t^x_{\nu_t}$. Consequently, Theorem 1.25 implies that the sequence $(t^x_{\nu_t+n+1}-t^x_{\nu_t+n})$ of the distances between the points of N^x converges in distribution to $(t^F_{n+1}-t^F_n)$, the corresponding sequence for the stationary renewal process. Hence, Proposition 1.23 shows the convergence in distribution of the corresponding point processes. The first part of the proposition is proved.

The second part is proved in the same way. It is sufficient to establish the convergence in distribution of $(\tau_0,\dots,\tau_n,S^x_t-t)$ to $(\tau_0',\dots,\tau_n',F)$ where F is a random variable independent of the variables $\tau_0',\dots,\tau_n'$ and its distribution is given by Relationship (1.8). It is easily verified that $Z(t)=(S^x_t-t)$ is a Markov process, that F is an invariant distribution for $(Z(t))$ and with the coupling property according to Theorem 1.26. According to Proposition 3.13 from Asmussen [**3**] this Markov process is Harris ergodic and mixing (Theorem 16.1.5 from Meyn and Tweedie [**34**]), i.e. if $K\geq 0$, g is a bounded measurable function on $\mathbb{R}_+$ and f is a measurable functional bounded on the functions from $\mathbb{R}_+$ to $\mathbb{R}$ depending only of the coordinates whose index is smaller than K,

$$\lim_{t\to+\infty}\mathbb{E}[f(Z)g(Z(t))]=\mathbb{E}[f(Z)]\mathbb{E}[g(F)].$$

This is sufficient to establish our assertion. Indeed, if f is a bounded measurable function on $\mathbb{R}^n$, the convergence implies that

$$\lim_{t\to+\infty}\mathbb{E}\left[f(\tau_0,\dots,\tau_n)1_{\{t_n\leq K\}}g(Z(t))\right] = \mathbb{E}\left[f(\tau_0,\dots,\tau_n)1_{\{t_n\leq K\}}\right]\mathbb{E}[g(F)].$$

For $\varepsilon>0$ there exists some constant K such that $\mathbb{P}(t_n\geq K)\leq\varepsilon$ and an integer n_0 such that if $n\geq n_0$,

$$\left|\mathbb{E}\left[f(\tau_0,\dots,\tau_n)1_{\{t_n\leq K\}}g(Z(t))\right]-\mathbb{E}\left[f(\tau_0,\dots,\tau_n)1_{\{t_n\leq K\}}\right]\mathbb{E}[g(F)]\right|\leq\varepsilon,$$

one gets the inequality

$$\left|\mathbb{E}\left[f(\tau_0,\dots,\tau_n)g(Z(t))\right]-\mathbb{E}[f(\tau_0,\dots,\tau_n)]\,\mathbb{E}[g(F)]\right| \leq (1+2\|f\|_\infty\|g\|_\infty)\,\varepsilon.$$

The proposition is proved. □

2. GI/GI/1 FIFO Queues and Random Walks

Contents

The arrival process of this queue is a renewal point process, services of the customers form an i.i.d. sequence and the service discipline is First In First Out (FIFO). It works as follows: When a customer requiring a service $x \geq 0$ arrives at the queue, it joins the queue and as soon as it is at the head of the queue (i.e. all the customers arrived before him have left the queue), it is served for a duration of time x by the server.

In this chapter, the equilibrium properties of this queue are analyzed: the distribution of waiting time of customers, the number of customers in the queue, or the workload of the queue. The probabilistic setting for this queue is the framework of random walks.

2.1 General results on GI/GI/1 FIFO Queues

The sequence of the instants of arrivals of customers is given by (t_n), the customer with index $n \in \mathbb{N}$ arrives at time t_n and requires the duration of service σ_{n+1}. The sequences of interarrival intervals $(\tau_n) = (t_n - t_{n-1})$ and services (σ_n) are assumed to be i.i.d. and independent ; τ_n is the interarrival time between the customers with respective index $n-1$ and n. The arrival point process is assumed to be simple, i.e. $\mathbb{P}(\tau_0 > 0) = 1$, under this assumption this is a renewal point process. The average load of the queue per time unit is defined as $\rho = \mathbb{E}(\sigma)/\mathbb{E}(\tau) = \lambda\mathbb{E}(\sigma)$, where λ is the arrival rate $1/\mathbb{E}(\tau)$. For $n \geq 0$, W_n is the waiting time of the nth customer when the waiting time of the customer with index 0 is w. At time $t_n + W_n + \sigma_{n+1}$ the nth customer completes its service and leaves the queue. Therefore the customer with index $n+1$ begins its service at time $\max(t_n + W_n + \sigma_{n+1}, t_{n+1})$. Hence the sequence (W_n) satisfies the recurrence equation

$$W_0 = w, \quad W_{n+1} = (W_n + \sigma_{n+1} - \tau_{n+1})^+, \tag{2.1}$$

for $n \geq 1$, with $a^+ = \max(0, a)$ if $a \in \mathbb{R}$.

If $X_n = \sigma_n - \tau_n$, the sequence (S_n) is the associated random walk defined by, for $n \geq 1$,

$$S_n = X_1 + \cdots + X_n,$$

with and $S_0 = 0$. The σ-field generated by the variables $\sigma_1, \ldots, \sigma_n$ and $\tau_1, \ldots, \tau_n$ is denoted by $\mathcal{F}_n$.

The asymptotic behavior of the sequence (W_n) and the distribution of its possible limit is the main object of this chapter. The condition for the existence of this limit is given by the following proposition. A similar result is true in a much more general context (see Proposition 12.3 page 334 in the chapter on the $G/G/1$ queue).

PROPOSITION 2.1.

1. *If $\rho < 1$, the sequence (W_n) is an ergodic Markov chain converging in distribution to the unique random variable W satisfying*

$$W \overset{dist.}{=} (W + X_0)^+, \tag{2.2}$$

where W and $X_0 = \sigma_0 - \tau_0$ are independent.
Moreover, $\mathbb{P}(W = 0) > 0$ and W has the same distribution as the maximum of the random walk (S_n),

$$W \overset{dist.}{=} \sup_{n \geq 0} S_n.$$

2. *If $\rho > 1$, the Markov chain is transient and $\mathbb{P}$-almost surely,*

$$\lim_{n \to +\infty} W_n/n = \mathbb{E}(\sigma) - \mathbb{E}(\tau).$$

PROOF. Relationship (2.1) gives

$$W_n = (W_{n-1} + X_n)^+,$$

by induction one gets that the variables (X_n, W_{n-1}) are independent, and W_n is $\mathcal{F}_n$-measurable. The sequence (W_n) is therefore a Markov chain. By iterating Identity (2.1) one gets, by induction, the representation

$$W_n = \sup_{2 \leq k \leq n+1} \left(\sum_{i=k}^{n} X_i \right) \vee \left(w + \sum_{i=1}^{n} X_i \right), \tag{2.3}$$

where $a \vee b = \max(a, b)$ for a, $b \in \mathbb{R}$. The law of large numbers shows that, almost surely, the sequence $((X_1 + \cdots + X_n)/n)$ converges to the expected value of X_1. Consequently, if $\mathbb{E}(X_1) = \mathbb{E}(\sigma - \tau) < 0$, $\mathbb{P}$-almost surely, the quantity $w + S_n = w + X_1 + \cdots + X_n$ tends to $-\infty$. Therefore, $w + S_n$ does not contribute to the supremum (2.3) for n sufficiently large. In particular, $\mathbb{P}$-almost surely, the variable W_n does not depend on w for n large.

The independence assumptions of the sequences (σ_i) and (τ_i) implies that

$$(X_1, X_2, \ldots, X_n) \stackrel{\text{dist.}}{=} (X_n, X_{n-1}, \ldots, X_1),$$

from Relationship (2.3), one gets

$$W_n \stackrel{\text{dist.}}{=} \sup_{0 \leq k \leq n-1} S_k \vee (w + S_n). \tag{2.4}$$

The sequence (W_n) converges in distribution to the variable $\sup\{S_n : n \geq 0\}$ which is $\mathbb{P}$-a.s. finite since the random walk (S_n) converges a.s. to $-\infty$. The Markov chain (W_n) has therefore an invariant probability distribution and all its sample paths couple independently of the initial point w. Proposition 3.13 of Asmussen [**3**] shows that (W_n) is an ergodic Markov chain. Equation (2.2) is the equation satisfied by the invariant measure of this Markov chain. If $\mathbb{P}(W = 0) = 0$, Equation (2.2) can be written as

$$W \stackrel{\text{dist.}}{=} (W + X_0)^+ \stackrel{\text{dist.}}{=} W + X_0,$$

since $\mathbb{P}(W + X_0 \leq 0) = \mathbb{P}(W = 0) = 0$. By taking Fourier transform, one gets $\mathbb{E}(\exp(\xi X)) = 1$ for $\mathrm{Re}(\xi) = 0$. Consequently, by uniqueness of Fourier transform, $X = 0$, $\mathbb{P}$-almost surely, hence $\mathbb{E}(X) = 0$ so $\rho = 1$. Contradiction. Part a) has been proved.

If $\rho > 1$ by using induction, Identity (2.3) shows that the inequality $W_n \geq w + X_1 + \cdots + X_n$ holds, again, with the law of large numbers, $\mathbb{P}$-a.s.

$$\liminf_{n \to +\infty} \frac{W_n}{n} \geq \mathbb{E}(\sigma) - \mathbb{E}(\tau) > 0.$$

In particular, $\mathbb{P}$-a.s. $W_n > 0$ after some finite (random) index n_0, by summing Relationship (2.1) between n_0 and n, one gets for $n \geq n_0$

$$W_n = W_{n_0} + X_{n_0+1} + X_{n_0+2} + \cdots + X_n.$$

The proposition is proved. □

If $\rho < 1$ and $G(x) = \mathbb{P}(W \leq x)$, the equation

$$\begin{cases} G(x) = 0, & x < 0, \\ G(x) = \int_{-\infty}^{x} G(x-y) X(dy), & x \geq 0, \end{cases} \tag{2.5}$$

where $X(dy)$ is the distribution of $X_0 = \sigma_0 - \tau_0$ is the analytic version of Equation (2.2). Complex analysis tools can be used to solve Equation (2.5), see for example Gakhov [**24**]. The problem with this kind of approach is that the probabilistic origin of the problem is somewhat lost, in particular the interpretation of Equation (2.5) in terms of the equilibrium of a Markov chain. In the next section, a random walk approach is used to solve this equation via a probabilistic approach.

The critical case $\rho = 1$ is not covered by the previous proposition. When services and interarrival intervals are i.i.d. and independent, the following proposition shows that the queue in this case is not stable if the variables σ or τ are not constant (otherwise the queue is trivial). The difference with the case $\rho > 1$ is in the growth rate of the waiting time, when $\rho > 1$, the waiting time of the nth customer is linear in n. In the critical case $\rho = 1$ the growth is only of the order of $\sqrt{n}$.

PROPOSITION 2.2 (The $GI/GI/1$ Queue : The Critical Case). *If $\rho = 1$ and if the variables σ_0 and τ_0 are square integrable and at least one of the two variables is non-degenerated, the sequence $(W_n/\sqrt{n})$ converges in distribution to the absolute value of a centered Gaussian variable with the same variance as $\sigma_0 - \tau_0$.*

PROOF. Assume $W_0 = 0$ (the general case is similar), Relationship (2.4) shows that W_{n+1} has the same distribution as

$$V_n = \sup_{0\leq k\leq n} S_k.$$

If $\eta = \sqrt{\mathbb{E}((\sigma_0 - \tau_0)^2)}$, for $n \geq 0$ the continuous function $Y_n(t)$ on $[0,1]$ which is $S_k/(\eta\sqrt{n})$ at k/n for $k = 0, \ldots, n$ and linear between these points,

$$Y_n(t) = \frac{1}{\eta\sqrt{n}}\left(S_{\lfloor nt\rfloor} + (nt - \lfloor nt\rfloor)X_{\lfloor nt\rfloor+1}\right),\ t\in[0,1],$$

where $\lfloor t\rfloor$ is the integer part of t. The maxima of the function $t\to Y_n(t)$ occur necessarily at the points k/n, $k = 0, \ldots, n$, thus one gets

$$\frac{V_n}{\eta\sqrt{n}} = \sup_{0\leq s\leq 1} Y_n(s).$$

Donsker's Theorem , see Billingsley [7] for example, shows that the sequence $(Y_n(t))$ of processes on $[0,1]$ converges in distribution to a standard Brownian motion $(B(t))$. If $C[0,1]$ is the space of continuous functions on $[0,1]$ endowed with the topology of the uniform convergence, the functional

$$g \to \sup_{0\leq s\leq 1} g(s),$$

is clearly continuous on $C[0,1]$ for this topology. Consequently, for any continuous bounded function f on $\mathbb{R}$, the functional

$$g \to f\left(\sup_{0\leq s\leq 1} g(s)\right),$$

is continuous and bounded on $C[0,1]$, therefore

$$\lim_{n\to+\infty}\mathbb{E}\left[f\left(\sup_{0\leq s\leq 1} Y_n(s)\right)\right] = \mathbb{E}\left[f\left(\sup_{0\leq s\leq 1} B(s)\right)\right].$$

Therefore, the variable $\sup\{Y_n(s) : 0 \le s \le 1\}$ converges in distribution to the variable $\sup\{B(s) : 0 \le s \le 1\}$, which has the same distribution as $|B(1)|$, see Rogers and Williams [44] for example. The proposition is proved. □

2.2 Wiener-Hopf Factorization

A classical result concerning random walks is presented in this section. It will be used to get an expression of the stationary distribution of the waiting time process. This probabilistic approach is due to Feller. Neveu's presentation[Nev84] of this topic is followed. As before, for $n \ge 0$, the variable S_n denotes the partial sum $X_1 + \cdots + X_n$ of the i.i.d. variables (X_i).

The following elementary lemma is just the strong Markov property for a random walk.

LEMMA 2.3. *If $\nu : \Omega \to \mathbb{N} \cup \{+\infty\}$ is a stopping time with respect to the filtration $(\mathcal{F}_n)$ such that $\mathbb{P}(\nu < +\infty) > 0$, conditionally on the event $\{\nu < +\infty\}$, the sequence $(S_{\nu+n} - S_\nu)$ is independent of (S_ν, ν) and with the same distribution as the initial random walk (S_n).*

PROOF. Indeed, if $p \in \mathbb{N}$ and f and g are non-negative Borelian functions on, respectively, $\mathbb{R}^{\mathbb{N}}$ and $\mathbb{R}$, then

$$\mathbb{E}\left[f((S_{\nu+n} - S_\nu))g(S_\nu)1_{\{\nu=p\}}\right] = \mathbb{E}\left[f((S_{p+n} - S_p))g(S_p)1_{\{\nu=p\}}\right].$$

The variable $g(S_p)1_{\{\nu=p\}}$ is thus $\mathcal{F}_p$-measurable and $f((S_{p+n} - S_p))$ depends only of the variables X_k with index $k \ge p+1$; the independence of the variables (X_n) yields

$$\mathbb{E}\left[f((S_{\nu+n} - S_\nu))g(S_\nu)1_{\{\nu=p\}}\right] = \mathbb{E}\left[f((S_n))\right]\mathbb{E}\left[g(S_\nu)1_{\{\nu=p\}}\right],$$

the lemma is therefore proved. □

THEOREM 2.4 (Wiener-Hopf Factorization). *For any complex number u such that $|u| < 1$, there exists a unique couple $(\phi_+(u,\cdot), \phi_-(u,\cdot))$ of functions satisfying the conditions,*

a) for $\xi \in \mathbb{C}$ such that $\mathrm{Re}(\xi) = 0$,

$$\frac{1}{1 - u\mathbb{E}(e^{-\xi X_1})} = \phi_+(u,\xi)\phi_-(u,\xi); \tag{2.6}$$

b) the functions $\phi_+(u,\cdot)$ and $\phi_-(u,\cdot)$ are holomorphic on, respectively, the right half-plane $\{\mathrm{Re}(\xi) > 0\}$ and in the left half-plane $\{\mathrm{Re}(\xi) < 0\}$. They are also continuous and bounded as well as their inverses on the closure of these domains, moreover

$$\lim_{\mathrm{Re}(\xi)\to+\infty} \phi_+(u,\xi) = 1.$$

[Nev84] J. Neveu, *Files d'attente*, 1983-1984, Cours de troisième cycle, Laboratoire de Probabilités de l'Université de Paris VI.

PROOF. The variable ν_- denotes the hitting time of $\mathbb{R}_-$,

$$\nu_- = \inf\{k > 0 : S_k \le 0\},$$

with the usual convention that $\inf \emptyset = +\infty$. Clearly enough, ν_- is a stopping time with respect to the filtration $(\mathcal{F}_n)$.

For $\xi \in \mathbb{C}$ such that $\mathrm{Re}(\xi) = 0$, the independence and the i.i.d. property of the variables (X_n) yield the identity

$$\mathbb{E}\left(\sum_{n\ge 0} u^n e^{-\xi S_n}\right) = \sum_{n\ge 0} u^n \left(\mathbb{E}\left[e^{-\xi X_1}\right]\right)^n = \frac{1}{1 - u\mathbb{E}(e^{-\xi X_1})},$$

by separating the ν_- first terms of the previous sum, one gets

$$\mathbb{E}\left(\sum_{n\ge 0} u^n e^{-\xi S_n}\right) = \mathbb{E}\left(\sum_{0\le n<\nu_-} u^n e^{-\xi S_n}\right) + \mathbb{E}\left(u^{\nu_-} e^{-\xi S_{\nu_-}} \sum_{n\ge 0} u^n e^{-\xi (S_{n+\nu_-} - S_{\nu_-})}\right).$$

If $\nu_- = +\infty$, the variable $u^{\nu_-}\exp(-\xi S_{\nu_-})$ is, of course, zero (since $|u| < 1$). Lemma 2.3 shows that, conditionally on the event $\{\nu_- < +\infty\}$, the variables (ν_-, S_{ν_-}) and the sequence $(S_{n+\nu_-} - S_{\nu_-})$ are independent and the sequence $(S_{n+\nu_-} - S_{\nu_-})$ has the same distribution as (S_n). The last term of the last equality is therefore given by the expression

$$\mathbb{E}\left(u^{\nu_-} e^{-\xi S_{\nu_-}}\right) \mathbb{E}\left(\sum_{n\ge 0} u^n e^{-\xi S_n}\right),$$

one obtains the identity

$$\frac{1}{1 - u\mathbb{E}(e^{-\xi X_1})} = \frac{\mathbb{E}\left(\sum_{0\le n<\nu_-} u^n e^{-\xi S_n}\right)}{1 - \mathbb{E}(u^{\nu_-} e^{-\xi S_{\nu_-}})}. \tag{2.7}$$

By setting

$$\phi_-(u, \xi) = \frac{1}{1 - \mathbb{E}(u^{\nu_-} e^{-\xi S_{\nu_-}})},$$

for $\mathrm{Re}(\xi) \le 0$, since $S_{\nu_-} \le 0$ and $\nu_- \ge 1$, it is clear that

$$0 < 1 - |u| \le \left|1 - \mathbb{E}\left(u^{\nu_-} e^{-\xi S_{\nu_-}}\right)\right| \le 1 + |u|.$$

Therefore, the function $\phi_-(u, \cdot)$ and its inverse are continuous bounded functions on the half-plane $\{\mathrm{Re}(\xi) \le 0\}$ and holomorphic on $\{\mathrm{Re}(\xi) < 0\}$. Equation (2.7) suggests the definition

$$\phi_+(u, \xi) = \mathbb{E}\left(\sum_{0\le n<\nu_-} u^n e^{-\xi S_n}\right),$$

this function can be expressed as

$$\phi_+(u,\xi) = \sum_{n\geq 0} \mathbb{E}\left(u^n e^{-\xi S_n} 1_{\{S_1>0,\ldots,S_{n-1}>0,S_n>0\}}\right).$$

It has to be proved that ϕ_+ has the required properties on $\{\xi : \mathrm{Re}(\xi) \geq 0\}$. The vector $(X_1, X_2, \ldots, X_n)$ having the same distribution as $(X_n, X_{n-1}, \ldots, X_1)$, the identity

$$\left(S_n, 1_{\{S_1>0,\ldots,S_{n-1}>0,S_n>0\}}\right) \stackrel{\text{dist.}}{=} \left(S_n, 1_{\{S_n>S_1,S_n>S_{n-1},\ldots,S_n>0\}}\right)$$

has been obtained. The function $\phi_+(u,\cdot)$ can be written as

$$\phi_+(u,\xi) = \sum_{n\geq 0} \mathbb{E}\left(u^n e^{-\xi S_n} 1_{\{S_n>S_{n-1},\ldots,S_n>S_1,S_n>0\}}\right).$$

If ν_+ is the first time of a *record* for the random walk (S_n),

$$\nu_+ = \inf\{k > 0 : S_k > S_{k-1}, S_k > S_{k-2}, \ldots, S_k > S_1, S_k > S_0 = 0\},$$

the initial condition $S_0 = 0$ implies that ν_+ is also the hitting time of $\mathbb{R}_+ - \{0\}$, i.e. $\nu_+ = \inf\{k > 0 : S_k > 0\}$. Therefore, the function ϕ_+ can be decomposed as follows

$$\phi_+(u,\xi) = 1 + \mathbb{E}\left(u^{\nu_+} e^{-\xi S_{\nu_+}} \sum_{n\geq\nu_+} u^{n-\nu_+} e^{-\xi(S_n - S_{\nu_+})} 1_{\{S_n>S_{n-1},\ldots,S_n>S_1,S_n>0\}}\right). \tag{2.8}$$

The sum in Equation (2.8) concerns only the instants when the random walk hits a new record. For $n > \nu_+$, the inequality $S_n > S_{\nu_+}$ implies that $S_n > S_k$ for any $k \leq \nu_+$. The instants of records for (S_n) after time $n = \nu_+$ are also the instants of records for the shifted random walk $(S_{n+\nu_+} - S_{\nu_+})$, hence

$$\phi_+(u,\xi) = 1 + \mathbb{E}\left(u^{\nu_+} e^{-\xi S_{\nu_+}} \sum_{n\geq\nu_+} u^{n-\nu_+} e^{-\xi(S_n - S_{\nu_+})} \right.$$
$$\left. 1_{\{S_n - S_{\nu_+} > S_{n-1} - S_{\nu_+},\ldots,S_n - S_{\nu_+} > S_{\nu_+ +1} - S_{\nu_+}, S_n - S_{\nu_+} > 0\}}\right).$$

By using again Lemma 2.3, one gets

$$\begin{aligned}\phi_+(u,\xi) &= 1 + \mathbb{E}\left(u^{\nu_+} e^{-\xi S_{\nu_+}}\right) \mathbb{E}\left(\sum_{n\geq 0} u^n e^{-\xi S_n} 1_{\{S_n>S_{n-1},\ldots,S_n>S_1,S_n>0\}}\right)\\ &= 1 + \mathbb{E}\left(u^{\nu_+} e^{-\xi S_{\nu_+}}\right) \phi_+(u,\xi).\end{aligned}$$

The representation of $\phi_+(u,\cdot)$ is thus analogous to that of $\phi_-(u,\cdot)$,

$$\phi_+(u,\xi) = \frac{1}{1 - \mathbb{E}(u^{\nu_+} e^{-\xi S_{\nu_+}})},$$

in particular this function and its inverse are holomorphic on $\{\mathrm{Re}(\xi) > 0\}$ and continuous, bounded on $\{\mathrm{Re}(\xi) \geq 0\}$. Since $S_{\nu_+} > 0$ when $\nu_+ < +\infty$, one gets the convergence

$$\lim_{\mathrm{Re}(\xi) \to +\infty} \phi_+(u, \xi) = 1.$$

The existence of functions satisfying Conditions a) and b) is thus established.

Uniqueness. If (ψ_+, ψ_-) are functions satisfying Conditions a) and b), Relationship (2.6) gives, for $|u| < 1$ and $\mathrm{Re}(\xi) = 0$,

$$\frac{\psi_+(u, \xi)}{\phi_+(u, \xi)} = \frac{\phi_-(u, \xi)}{\psi_-(u, \xi)}.$$

The function H defined by

$$H(\xi) = \frac{\psi_+(u, \xi)}{\phi_+(u, \xi)}, \quad \text{if } \mathrm{Re}(\xi) \geq 0 \text{ and} \quad H(\xi) = \frac{\phi_-(u, \xi)}{\psi_-(u, \xi)},$$

if $\mathrm{Re}(\xi) \leq 0$, is holomorphic on $\mathbb{C} - \{\mathrm{Re}(\xi) = 0\}$ and continuous on $\mathbb{C}$, therefore holomorphic on the whole complex plane $\mathbb{C}$ (by using Morera's Theorem for example, see Rudin [**45**]). This function being bounded, by Liouville's Theorem, one concludes that H is a constant function,

$$\lim_{\mathrm{Re}(\xi) \to +\infty} H(\xi) = \lim_{\mathrm{Re}(\xi) \to +\infty} \frac{\psi_+(u, \xi)}{\phi_+(u, \xi)} = 1,$$

therefore $H \equiv 1$, hence $\psi_+ \equiv \phi_+$ and $\psi_- \equiv \phi_-$. The theorem is proved. □

In the proof of the above theorem, the following proposition has been established.

PROPOSITION 2.5. *For $|u| < 1$, the functions $\phi_+(u, \cdot)$ and $\phi_-(u, \cdot)$ of the above theorem can be expressed as*

$$\phi_+(u, \xi) = \frac{1}{1 - \mathbb{E}(u^{\nu_+} e^{-\xi S_{\nu_+}})} = \mathbb{E}\left(\sum_{0 \leq n < \nu_-} u^n e^{-\xi S_n} \right),$$

for $\mathrm{Re}(\xi) \geq 0$, *and*

$$\phi_-(u, \xi) = \frac{1}{1 - \mathbb{E}(u^{\nu_-} e^{-\xi S_{\nu_-}})},$$

for $\mathrm{Re}(\xi) \leq 0$, *with*

$$\nu_+ = \inf\{k > 0 : S_k > 0\} \quad \text{and} \quad \nu_- = \inf\{k > 0 : S_k \leq 0\}.$$

Consequently, for $|u| < 1$ and $\mathrm{Re}(\xi) = 0$, the Wiener-Hopf factorization can be written as

$$\frac{1}{1 - u\mathbb{E}(e^{-\xi X_1})} = \frac{1}{1 - \mathbb{E}(u^{\nu_+} e^{-\xi S_{\nu_+}})} \times \frac{1}{1 - \mathbb{E}(u^{\nu_-} e^{-\xi S_{\nu_-}})}. \tag{2.9}$$

The decomposition has been obtained with ϕ_+ and ϕ_- or, probabilistically, with ν_+ and ν_-. This decomposition is not completely symmetrical. The instants when the random walk reaches 0 have been arbitrarily associated with "$-$", which implies that $S_{\nu_-} \leq 0$ if ν_- is finite and $S_{\nu_+} > 0$ if ν_+ is finite, the condition

$$\lim_{\mathrm{Re}(\xi)\to+\infty} \phi_+(u,\xi) = 1$$

is satisfied. This difference does not play a role when the distribution of X_1 does not have mass on points, in this case the decomposition is symmetrical.

Spitzer's Lemma. The previous proposition gives a probabilistic expression for the functions ϕ_+ and ϕ_- of Wiener-Hopf decomposition. This factorization can also be obtained in the following way : For $|u| < 1$ and $\xi \in \mathbb{C}$ such that $\mathrm{Re}(\xi) = 0$,

$$\begin{aligned}\frac{1}{1 - u\mathbb{E}[e^{-\xi X_1}]} &= \exp\left[-\log\left(1 - u\mathbb{E}[\exp(-\xi X_1)]\right)\right] \\ &= \exp\left(\sum_1^{+\infty} \frac{u^n}{n}\mathbb{E}[\exp(-\xi X_1)]^n\right) = \exp\left(\sum_1^{+\infty} \frac{u^n}{n}\mathbb{E}[\exp(-\xi S_n)]\right),\end{aligned}$$

consequently,

$$\begin{aligned}\frac{1}{1 - u\mathbb{E}[e^{-\xi X_1}]} &= \exp\left(\sum_1^{+\infty} \frac{u^n}{n}\mathbb{E}\left[\exp(-\xi S_n)1_{\{S_n>0\}}\right]\right) \\ &\qquad\times \exp\left(\sum_1^{+\infty} \frac{u^n}{n}\mathbb{E}\left[\exp(-\xi S_n)1_{\{S_n\leq 0\}}\right]\right).\end{aligned}$$

If ψ_+ and ψ_- are respectively defined as the two functions of the right hand side of the previous identity, they also satisfy the conditions of Theorem 2.4. This is another probabilistic expression for the functions ϕ_+ and ϕ_-. For the problems considered here, this approach is not useful : The expression of $\mathbb{E}(\exp(-\xi S_n)1_{\{S_n>0\}})$ is not, in general, easy to obtain. Conversely, if the decomposition is known, this formulation is not really informative concerning the random walk. Note that the representation of Proposition 2.5 gives the distributions of (ν_+, S_{ν_+}) and (ν_-, S_{ν_-}).

2.3 Applications to the GI/GI/1 Queue

The distribution of the stationary waiting time of a $GI/GI/1$ queue is the distribution of the maximum of a random walk (Proposition 2.1). This distribution is now expressed with the factors of the Wiener-Hopf factorization.

PROPOSITION 2.6. *If $W_0 = 0$, the distribution of the sequence (W_n) of waiting times of customers of a $GI/GI/1$ queue satisfies, for $|u| < 1$ and* $\mathrm{Re}(\xi) \geq 0$,

$$(1-u)\sum_{n\geq 0} u^n \mathbb{E}\left(e^{-\xi W_n}\right) = \frac{\phi_+(u,\xi)}{\phi_+(u,0)}, \tag{2.10}$$

where $\phi_+(u,\cdot)$ is the function of the Wiener-Hopf factorization for the random walk associated with $X = \sigma_1 - \tau_1$. In particular, if $\rho < 1$, the Laplace transform of the stationary waiting time is given by

$$\mathbb{E}\left(e^{-\xi W}\right) = \lim_{u\to 1}\frac{\phi_+(u,\xi)}{\phi_+(u,0)},$$

for $\mathrm{Re}(\xi) \geq 0$.

PROOF. The sequence (W_n) defined by Identity (2.1) is a functional of the random walk (S_n). It can be decomposed in the same way as before as

$$\sum_{n\geq 0} u^n e^{-\xi W_n} = \sum_{0\leq n<\nu_-} u^n e^{-\xi W_n} + u^{\nu_-}\sum_{n\geq 0} u^n e^{-\xi W_{\nu_-+n}}. \tag{2.11}$$

For $n < \nu_-$, by definition of ν_-, $S_k > 0$ for all $k \leq n$ and, according to Relationship (2.1)

$$W_0 = 0 = S_0, \quad W_1 = (X_1)^+ = S_1, \quad W_2 = (S_1 + X_2)^+ = S_2,$$
$$W_n = (S_{n-1} + X_n)^+ = S_n,$$

for $n < \nu_-$. Moreover,

$$W_{\nu_-} = (S_{\nu_- -1} + X_{\nu_-})^+ = (S_{\nu_-})^+ = 0,$$

the customer with index ν_- is the first customer after the initial customer with index 0 who does not wait. The first term of the right hand side of Identity (2.11) can be written as

$$\sum_{0\leq n<\nu_-} u^n e^{-\xi S_n}.$$

The sequence (W_{ν_-+n}) satisfies $W_{\nu_-} = 0$ and for $n \geq 0$,

$$W_{\nu_-+n+1} = (W_{\nu_-+n} + X_{\nu_-+n+1})^+,$$

this is simply Relationship (2.1) for the shifted sequence (X_{ν_-+n}). The sequence (W_{ν_-+n}) is the sequence (W_n) associated with (X_{ν_-+n}). Lemma 2.3 shows that (W_{ν_-+n}) has the same distribution as (W_n) and is independent of ν. Taking the expectation in Relationship (2.11) yields

$$\mathbb{E}\left(\sum_{n\geq 0} u^n e^{-\xi W_n}\right) = \mathbb{E}\left(\sum_{0\leq n<\nu_-} u^n e^{-\xi S_n}\right) + \mathbb{E}(u^{\nu_-})\,\mathbb{E}\left(\sum_{n\geq 0} u^n e^{-\xi W_n}\right),$$

consequently, one gets

$$\mathbb{E}\left(\sum_{n\geq 0} u^n e^{-\xi W_n}\right) = \frac{\mathbb{E}\left(\sum_{0\leq n<\nu_-} u^n e^{-\xi S_n}\right)}{1-\mathbb{E}(u^{\nu_-})}.$$

The representation of ϕ_+ in Proposition 2.5 yields

$$\phi_+(u,0) = \frac{1-\mathbb{E}(u^{\nu_-})}{1-u},$$

and thus Relationship (2.10).

If $\rho < 1$, Proposition 2.1 shows that the sequence (W_n) converges in distribution to W,

$$\lim_{n\to+\infty} \mathbb{E}\left(e^{-\xi W_n}\right) = \mathbb{E}\left(e^{-\xi W}\right),$$

for ξ such that $\mathrm{Re}(\xi) \geq 0$, consequently

$$\lim_{u\to 1}(1-u)\sum_{n\geq 0} u^n \mathbb{E}\left(e^{-\xi W_n}\right) = \mathbb{E}\left(e^{-\xi W}\right).$$

The proposition is proved. □

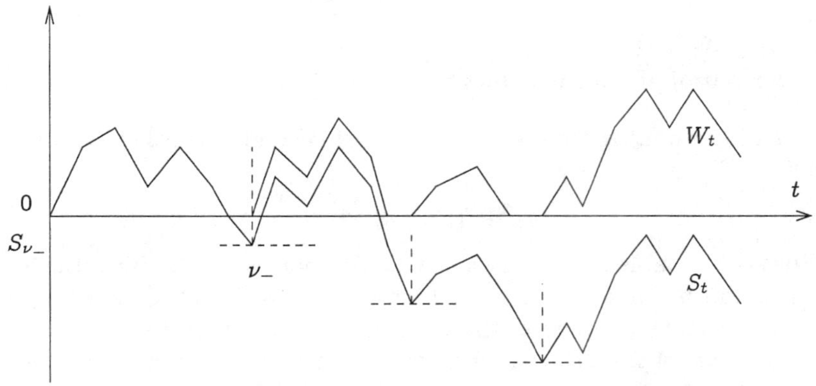

FIG. 2.1. The Sequences (W_n) and (S_n)

Once the Wiener-Hopf factorization for $\sigma - \tau$ is known, Proposition 2.6 gives the distribution of the stationary waiting time. The distributions of other variables for this queue can also be obtained in the same way.

2.3.1 Busy Periods

During the proof of Proposition 2.5 it has been shown that if ν_- is finite, ν_- is the index of the first customer after 0 who does not wait. The variable t_{ν_-} is therefore the duration of the first occupation cycle of the queue (busy period) and at $t = t_{\nu_-}$ a new cycle begins. The total amount of work processed during the first busy period is $\sigma_1 + \cdots + \sigma_\nu$ and the quantity

$$-S_{\nu_-} = t_{\nu_-} - (\sigma_1 + \cdots + \sigma_\nu)$$

is the duration of the vacation of the server during the first cycle.

If $\rho < 1$, busy periods are $\mathbb{P}$-almost surely finite. The second occupation cycle is a functional of the sequence $(\sigma_{n+\nu_+}, \tau_{n+\nu_+})$ which is independent of the sequence $(\sigma_{n\wedge\nu_-}, \tau_{n\wedge\nu_-})$ according to Lemma 2.3. The occupation cycles are therefore i.i.d. Therefore, Proposition 2.5 implies the following proposition

PROPOSITION 2.7. *If $\rho < 1$ the Laplace transform of the duration I of a period of vacation of the server is given by*

$$\mathbb{E}\left(e^{-\xi I}\right) = \lim_{u\to 1} 1 - \frac{1}{\phi_-(u,-\xi)},$$

for $\mathrm{Re}(\xi) \geq 0$, *and the generating function of the number ν_- of customers served during a busy period is*

$$\mathbb{E}\left(u^{\nu_-}\right) = 1 - \frac{1}{\phi_-(u,0)}.$$

2.3.2 The Number of Customers

If $\rho < 1$ and the waiting time W_0 of the initial customer has the stationary distribution W, the identity

$$W \stackrel{\text{dist.}}{=} (W + X_0)^+$$

shows that the variable W_1 has the same distribution as W_0. By induction on $n \in \mathbb{N}$, the waiting time W_n of the nth customer has the same distribution as W_0. The variable Q_n denotes the number of customers in the queue just before the arrival of the nth customer. If the nth customer finds less than k customers when it arrives, the service discipline being FIFO, the customer arrived at t_{n-k} has not yet leaved the queue, hence

$$\begin{aligned}\mathbb{P}(Q_n \geq k) &= \mathbb{P}(t_{n-k} + W_{n-k} + \sigma_{n-k+1} > t_n)\\ &= \mathbb{P}\left(W_{n-k} + \sigma_{n-k+1} - \tau_{n-k+1} > \sum_{i=n-k+2}^{n} \tau_i\right).\end{aligned}$$

The variables W_{n-k}, σ_{n-k+1}, (τ_i), $n-k+1 \leq i \leq n$ being independent, one gets

$$\mathbb{P}(Q_n \geq k) = \mathbb{P}\left(W_0 + \sigma_0 - \tau_0 > \sum_{i=1}^{k-1} \tau_i\right).$$

If $k \geq 2$, the variable $t_{k-1} = \tau_1 + \cdots + \tau_{k-1}$ is $\mathbb{P}$-a.s. strictly positive, consequently,

$$\mathbb{P}(Q_n \geq k) = \mathbb{P}\left((W_0 + X_0)^+ > t_{k-1}\right) = \mathbb{P}(W > t_{k-1}),$$

according to Relationship (2.2). The last identity shows that the variable Q_n converges in distribution when n tends to infinity, hence

$$\mathbb{P}(Q \geq k) = \mathbb{P}(W > t_{k-1}),$$

where Q is the stationary distribution of (Q_n). This equation is also true for $k = 1$ since $\{Q \geq 1\} = \{W > 0\}$.

PROPOSITION 2.8. *If $\rho < 1$ and Q is the number of customers in the $GI/GI/1$ at equilibrium just before the arrival of a customer, for $k \geq 1$,*

$$\mathbb{P}(Q \geq k) = \mathbb{P}(W > t_{k-1}),$$

where t_{k-1} is the sum of $k-1$ i.i.d random variables with the same distribution as τ_0 and W is a random variable independent of t_{k-1} having the stationary distribution of the waiting time.

The distribution at equilibrium of the number of customers seen by an arriving customer can be obtained from the distribution of the stationary waiting time.

2.3.3 The Virtual Waiting Time of the Queue

The *virtual waiting time* $V(t)$ of the queue at time t is the sum of all the residual services at that time. It is the time necessary to empty the queue if no other customer arrives after time t. (See also the corresponding section in Chapter 12.) The virtual waiting time just before the arrival of a customer is therefore the waiting time of the customer, for $n \in \mathbb{N}$, $V(t_n^-) = W_n$; W_n is the workload seen by the nth customer and $V(t)$ is the load seen at time t by an external observer. Proposition 12.5 page 339 of the chapter on the $G/G/1$ queue gives the relation between the distributions of the stationary waiting time W of a customer and the virtual waiting time V, for $\mathrm{Re}(\xi) \geq 0$,

$$\mathbb{E}\left(e^{-\xi V}\right) = 1 - \lambda\mathbb{E}(\sigma) + \lambda\mathbb{E}\left(e^{-\xi W}\frac{1-e^{-\xi\sigma}}{\xi}\right).$$

According to the independence of the waiting time and the service time of a customer for the FIFO discipline, Takàcs' Formula can be written as

$$\mathbb{E}\left(e^{-\xi V}\right) = 1 - \lambda\mathbb{E}(\sigma) + \lambda\mathbb{E}\left(e^{-\xi W}\right)\frac{1-\mathbb{E}(e^{-\xi\sigma})}{\xi}, \tag{2.12}$$

for $\mathrm{Re}(\xi) \geq 0$. The distribution of the virtual waiting time at equilibrium is directly expressed with the distribution of the stationary waiting time.

2.3.4 Representations of W

The condition $\rho < 1$ is assumed to be true in the following.

1. The representation of the function ϕ_+ of Proposition 2.5, for $|u| < 1$ and $\text{Re}(\xi) \geq 0$,
$$\phi_+(u,\xi) = \mathbb{E}\left(\sum_{0\leq n<\nu_-} u^n e^{-\xi S_n}\right),$$
and Theorem 2.6 yield,
$$\mathbb{E}\left(e^{-\xi W}\right) = \lim_{u\nearrow 1} \mathbb{E}\left(\sum_{0\leq n<\nu_-} u^n e^{-\xi W_n}\right) \frac{1-u}{1-\mathbb{E}(u^{\nu_-})};$$
since $W_n = S_n$ for $0 \leq n < \nu_-$,
$$\mathbb{E}\left(e^{-\xi W}\right) = \frac{\mathbb{E}\left(\sum_{0\leq n<\nu_-} e^{-\xi W_n}\right)}{\mathbb{E}(\nu_-)}.$$
The variable ν_- is the hitting time of 0 for the Markov chain (W_n) therefore, the above equation is nothing but the classical representation of the invariant probability of a Markov chain.
2. Another expression for the function ϕ_+ is given by Proposition 2.5,
$$\phi_+(u,\xi) = \frac{1}{1-\mathbb{E}(u^{\nu_+} e^{-\xi S_{\nu_+}})},$$
Proposition 2.6 shows that for $\text{Re}(\xi) > 0$,
$$\mathbb{E}\left(e^{-\xi W}\right) = \lim_{u\to 1} \frac{1-\mathbb{E}(u^{\nu_+})}{1-\mathbb{E}(u^{\nu_+} e^{-\xi S_{\nu_+}})} = \frac{1-\mathbb{P}(\nu_+ < +\infty)}{1-\mathbb{E}(e^{-\xi S_{\nu_+}} 1_{\{\nu_+<+\infty\}})},$$
in particular, since $S_{\nu_+} > 0$ when ν_+ is finite, by letting $\text{Re}(\xi)$ go to infinity, one gets
$$\mathbb{P}(W = 0) = \mathbb{P}(\nu_+ = +\infty). \tag{2.13}$$
If $\alpha = \mathbb{P}(\nu_+ < +\infty)$,
$$\mathbb{E}\left(e^{-\xi W}\right) = \frac{1-\alpha}{1-\alpha\mathbb{E}(e^{-\xi S_{\nu_+}} \mid \nu_+ < +\infty)}$$
$$= \sum_{n=0}^{+\infty} \alpha^n (1-\alpha)\mathbb{E}\left(e^{-\xi S_{\nu_+}} \mid \nu_+ < +\infty\right)^n,$$
the next proposition has therefore been proved.

PROPOSITION 2.9. *If (Z_i) is an i.i.d. sequence of random variables with the same distribution as S_{ν_+} conditioned on the event $\{\nu_+ < +\infty\}$, the stationary waiting time W has the same distribution as $Z_1 + \cdots + Z_G$, where G is geometrically distributed with parameter $\mathbb{P}(\nu_+ < +\infty)$ independent of the sequence (Z_i).*

The proposition can be obtained directly. Proposition 2.1 shows that W has the same distribution as the maximum of the random walk (S_n) associated with the random variable $\sigma - \tau$.

If the random walk goes above 0, i.e. on the event $\{\nu_+ < \infty\}$, starting from $t = \nu_+$ at the position S_{ν_+}, the random walk restarts, independently of the past (Lemma 2.3). From this position, if the random walk goes above 0, the maximum is at least the sum of two independent random variables with the same distribution as S_{ν_+} conditioned on the event $\{\nu_+ < +\infty\}$, and so on. (See Figure 2.2 below.) This maximum can thus be written as a finite sum of i.i.d random variables with the same distribution as S_{ν_+} conditioned on $\{\nu_+ < \infty\}$. The number of terms in this sum has a geometric distribution with parameter $\mathbb{P}(\nu_+ < +\infty)$. The proposition is proved.

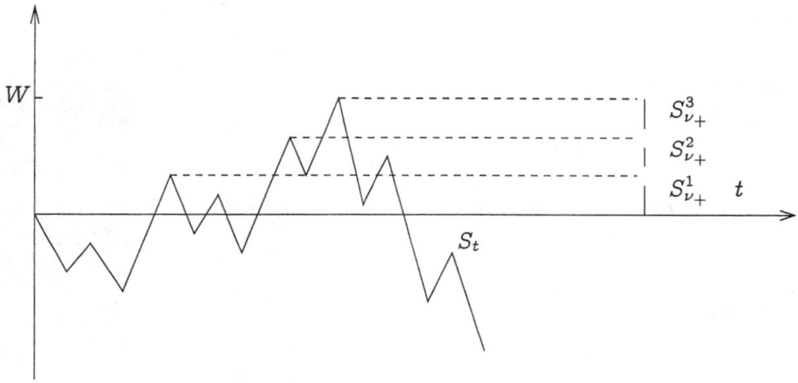

FIG. 2.2. Decomposition of the Maximum of a Random Walk

2.4 GI/M/1 and M/GI/1 Queues

In this section the variable $X = \sigma - \tau$ is such that one of the two variables τ or σ has an exponential distribution. The Wiener-Hopf factorization applies easily in these cases.

If (ϕ_+, ϕ_-) are the solutions of the Wiener-Hopf factorization for the variable X, for $|u| < 1$, denote

$$(2.14) \quad \psi_+(u,\xi) = \frac{\phi_-(u,-\xi)}{\phi_-(u,-\infty)}, \quad \text{and} \quad \psi_-(u,\xi) = \phi_+(u,-\xi)\phi_-(u,-\infty),$$

with

$$\phi_-(u,-\infty) = \lim_{\text{Re}(\xi)\to-\infty} \phi_-(u,\xi) = \frac{1}{1-\mathbb{E}(u^{\nu_-}1_{\{S_{\nu_-}=0\}})}.$$

The functions (ψ_+, ψ_-) are clearly solutions of the factorization associated with the variable $-X$. The factorization for the $M/GI/1$ queue can be obtained from the factorization of the $GI/M/1$ queue. The distribution of σ is assumed to be exponentially distributed with parameter μ.

PROPOSITION 2.10. *If $X = \sigma - \tau$ and the distribution of σ is exponential with parameter μ, the Wiener-Hopf factorization of X is given by*

$$\phi_+(u, \xi) = \frac{\mu + \xi}{b(u) + \xi}, \quad \textit{and} \quad \phi_-(u, \xi) = \frac{b(u) + \xi}{\mu + \xi - u\mu\mathbb{E}(e^{\xi\tau})},$$

for $|u| < 1$ and $b(u)$ is the unique non-negative root of

$$x - \mu\left(1 - u\mathbb{E}\left(e^{-x\tau}\right)\right) = 0.$$

In the neighborhood of $u = 1$, $b(u)$ has the following behavior:

a) If $\rho < 1$, then $\lim_{u \nearrow 1} b(u) = \beta$, where β is the unique positive root of the equation

$$x - \mu\left(1 - \mathbb{E}\left(e^{-x\tau}\right)\right) = 0.$$

b) When $\rho > 1$, then $\lim_{u \nearrow 1} b(u) = 0$ and

$$\lim_{u \nearrow 1} \frac{b(u)}{(1 - u)} = \frac{1}{\mathbb{E}(\sigma) - \mathbb{E}(\tau)}.$$

PROOF. In this case, for $\mathrm{Re}(\xi) = 0$, the Fourier transform $\mathbb{E}(\exp(-\xi X))$ is given by

$$\mu\mathbb{E}(\exp(\xi\tau))/(\mu + \xi),$$

thus for u such that $|u| < 1$,

$$\frac{1}{1 - u\mathbb{E}(e^{-\xi X_1})} = \frac{\mu + \xi}{\mu + \xi - u\mu\mathbb{E}(e^{\xi\tau})}.$$

If $f_u(\xi) = \mu + \xi - u\mu\mathbb{E}(\exp(\xi\tau))$, for u on the circle

$$C_\mu = \{\xi : |\xi + \mu| = \mu\} \subset \{\xi : \mathrm{Re}(\xi) \geq 0\},$$

the inequality

$$|u\mu\mathbb{E}(e^{\xi\tau})| \leq |u\mu| < |\mu| = |\xi + \mu|$$

holds. Rouché's Theorem (see Rudin [45] page 242 for example) shows that f_u has only one root, of multiplicity 1 inside C_μ. Moreover, any root ξ of f_u in the left half-plane satisfies $|x + \mu| = |u\mu\mathbb{E}(\exp(\xi\tau))| < \mu$. Such a root is thus inside the circle C_μ. The function f_u has a unique root $-b(u)$ in the left half-plane. This root is real, $b(u)$ is the solution of the equation

$$g_u(x) \stackrel{\text{def.}}{=} \mu\left(1 - u\mathbb{E}\left(e^{-x\tau}\right)\right) = x$$

on $\mathbb{R}_+$ (see Figure 2.3 and the proof for g_1 below).

The identity

$$\frac{1}{1 - u\mathbb{E}(e^{-\xi X_1})} = \frac{\mu + \xi}{b(u) + \xi} \times \frac{b(u) + \xi}{\mu + \xi - u\mu\mathbb{E}(e^{\xi\tau})}$$

gives the Wiener-Hopf factorization in this case. Indeed, by setting

$$\phi_+(u,\xi) = \frac{\mu+\xi}{b(u)+\xi} \qquad \text{and} \qquad \phi_-(u,\xi) = \frac{b(u)+\xi}{\mu+\xi-u\mu\mathbb{E}(e^{\xi\tau})},$$

for $|u|<1$ and $\mathrm{Re}(\xi)=0$, the functions ϕ_+ and ϕ_- clearly satisfy condition a) and b) of Theorem 2.4.

If $\rho<1$, since $g_1(0)=0$ and $g_1'(0)=\mu\mathbb{E}(\tau)=\mathbb{E}(\tau)/\mathbb{E}(\sigma)>1$ and that g_1 converges to μ at infinity, there exists another fixed point β which is positive. The function g_1 being concave, this is the only positive fixed point. For $u\in[0,1[$, the functions $x\to g_u(x)$ and $u\to g_u(x)$, $0\le u<1$ being non-increasing, one gets that the function $u\to b(u)$ is also non-increasing and the inequality $b(u)\ge\beta$ holds. The limit $\lim_{u\nearrow 1} b(u)$ is also a fixed point of g_1 greater than β, the limit is β. The first part of the proposition is proved.

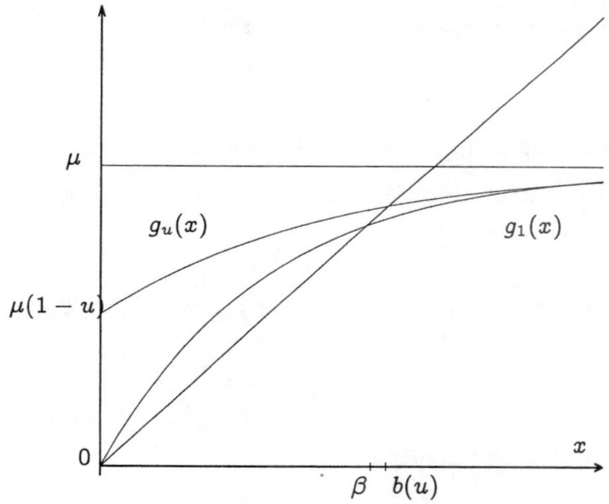

FIG. 2.3. The Functions g_u and g_1 for the Case $\rho<1$.

When $\rho>1$, the derivative of g_1 at the origin is strictly less than 1, the function g_1 being concave, 0 is the only fixed point on $\mathbb{R}_+$ of g_1, thus $\lim_{u\nearrow 1} b(u)=0$. For u such that $|u|<1$,

$$\mu - b(u) = u\mu\mathbb{E}\left(e^{-b(u)\tau}\right),$$

the elementary expansion

$$\mathbb{E}\left(e^{-b(u)\tau}\right) = 1-\mathbb{E}(\tau)b(u)+o(b(u)),$$

gives

$$\lim_{u\nearrow 1}\frac{\mu(1-u)}{b(u)} = 1-\mu\mathbb{E}(\tau).$$

The proposition is proved. □

2.4.1 The GI/M/1 Queue

In this section the distribution of services is exponential with parameter μ. The following proposition shows that the stationary waiting time of $GI/M/1$ queue conditioned to be non-zero has an exponential distribution.

PROPOSITION 2.11.
a) When $\rho < 1$, the distribution $W(dx)$ of the stationary waiting time of the $GI/M/1$ queue is given by

$$W(dx) = \frac{\beta}{\mu}\delta_0 + \left(1 - \frac{\beta}{\mu}\right)\beta e^{-\beta x}\,dx, \tag{2.15}$$

where β is the unique solution of the equation

$$\mu\left(1 - \mathbb{E}\left(e^{-\beta\tau}\right)\right) = \beta \tag{2.16}$$

in the interval $]0, \mu[$.
b) If $\rho > 1$, a busy period is infinite with probability $1 - 1/\rho$.

PROOF. The Laplace transform of W is obtained by using Propositions 2.6 and 2.10,

$$\begin{aligned}\mathbb{E}\left(e^{-\xi W}\right) &= \lim_{u\nearrow 1}\frac{\phi_+(u,\xi)}{\phi_+(u,0)} = \lim_{u\nearrow 1}\frac{\mu+\xi}{b(u)+\xi}\times\frac{b(u)}{\mu} = \frac{\mu+\xi}{\beta+\xi}\times\frac{\beta}{\mu}\\ &= \frac{\beta}{\mu} + \left(1-\frac{\beta}{\mu}\right)\frac{\beta}{\xi+\beta}\end{aligned}$$

which is Relationship (2.15) written in terms of Laplace transforms, hence a) holds. Propositions 2.7 and 2.10 give

$$\mathbb{E}\left(u^{\nu_-}\right) = 1 - \frac{\mu(1-u)}{b(u)},$$

thus $\mathbb{P}(\nu_- < +\infty) = \lim_{u\nearrow 1}\mathbb{E}(u^{\nu_-}) = 1/\rho$, b) has been proved. □

A similar result is also true for the virtual waiting time.

COROLLARY 2.12. *If $\rho < 1$, the stationary distribution of the virtual waiting time of the $GI/M/1$ queue is given by*

$$V(dx) = (1-\rho)\delta_0 + \rho\beta e^{-\beta x}\,dx,$$

where β is the solution of Equation (2.16).

PROOF. According to Takàcs' Formula (2.12),

$$\mathbb{E}\left(e^{-\xi V}\right) = 1 - \rho + \lambda\mathbb{E}\left(e^{-\xi W}\right)\mathbb{E}\left(\frac{1-e^{-\xi\sigma}}{\xi}\right),$$

the previous proposition is then used to conclude. □

PROPOSITION 2.13. *If $\rho < 1$, the stationary distribution of the number of customers Q seen by an arriving customer is a geometric distribution with parameter $1 - \beta/\mu$ where β is the unique solution of*

$$\mu\left(1 - \mathbb{E}\left(e^{-\beta\tau}\right)\right) = \beta$$

in the interval $]0, \mu[$.

PROOF. According to Proposition 2.8, for $n \geq 2$,

$$\mathbb{P}(Q \geq n) = \mathbb{P}(W > t_{n-1}),$$

W conditioned on the event $\{W > 0\}$ having an exponential distribution with parameter β, one gets

$$\mathbb{P}(Q \geq n) = \mathbb{P}(W > 0)\mathbb{P}(W > t_{n-1} \,|\, W > 0)$$
$$= \mathbb{P}(W > 0)\mathbb{E}\left(e^{-\beta t_{n-1}}\right) = \left(1 - \frac{\beta}{\mu}\right)\mathbb{E}\left(e^{-\beta\tau}\right)^{n-1},$$

according to Relationship (2.16). This last quantity is $(1-\beta/\mu)^n$. The proof is completed. □

Remark on the Conditional Distribution of W. The fact that the conditional distribution of the stationary waiting time is exponentially distributed can be seen easily with Proposition 2.9. The distribution of S_{ν_+} conditioned on that the fact the variable ν_+ is finite has to be derived. Non-negative contributions of the random walk are due to the variable σ. If the random walk goes above 0, the property of the exponential distribution shows that the value of the jump above 0 has an exponential distribution with parameter μ. Rigorously, it can be proved as follows,

$$\mathbb{P}(S_{\nu_+} > a, \nu_+ < +\infty) = \sum_{n=1}^{+\infty} \mathbb{P}(S_i \leq 0, i < n, S_n > a)$$
$$= \sum_{n=1}^{+\infty} \mathbb{P}(S_n > a \,|\, S_i \leq 0, i < n, S_n > 0)\mathbb{P}(S_i \leq 0, i < n, S_n > 0).$$

For $n \geq 1$ and $a > 0$ and the identity

$$\mathbb{P}(S_n > a \,|\, S_i \leq 0, i < n, S_n > 0) =$$
$$\mathbb{P}(\sigma_n > a - S_{n-1} + \tau_n \,|\, S_{n-1} \leq 0, \sigma_n > -S_{n-1} + \tau_n) = e^{-\mu a}$$

holds because σ is exponentially distributed, hence

$$\mathbb{P}(S_{\nu_+} > a, \nu_+ < +\infty) = e^{-\mu a}\mathbb{P}(\nu_+ < +\infty) = e^{-\mu a}.$$

The distribution of W is the distribution of the sum of G variables i.i.d. distributed as S_{ν_+} conditioned on the event $\{\nu_+ < +\infty\}$, which are exponential with parameter μ. The variable G has a geometric distribution with parameter $\alpha = \mathbb{P}(\nu_+ < +\infty)$ and is independent of these variables. By using the identity

$$\{W > 0\} \overset{\text{dist.}}{=} \left\{\sup_n S_n > 0\right\} = \{\nu_+ < +\infty\},$$

one gets that for $\xi \in \mathbb{C}$ such that $\mathrm{Re}(\xi) \geq 0$,

$$\begin{aligned}\mathbb{E}\left(e^{-\xi W} \mid W > 0\right) &= \mathbb{E}\left(e^{-\xi W} \mid \nu_+ < +\infty\right) \\ &= \sum_{n=1}^{+\infty} \alpha^{n-1}(1-\alpha)\left(\frac{\mu}{\mu+\xi}\right)^n = \frac{(1-\alpha)\mu}{(1-\alpha)\mu+\xi},\end{aligned}$$

consequently, the variable W conditioned to be non-zero has an exponential distribution with parameter $(1-\alpha)\mu$. The quantity $\alpha = \mathbb{P}(W > 0)$ has to be calculated. Relationship (2.2) yields

$$\mathbb{P}(W = 0) = \mathbb{P}(W + \sigma \leq \tau) = (1-\alpha)\mathbb{P}(\sigma \leq \tau) + \alpha\mathbb{P}(W + \sigma \leq \tau \mid W > 0),$$

hence

$$1 - \alpha = (1-\alpha)\mathbb{P}(\sigma \leq \tau) + \alpha\mathbb{P}(W + \sigma \leq \tau \mid W > 0),$$

by using that the conditional distribution of W is exponential with parameter $(1-\alpha)\mu$,

$$1 - \alpha = 1 - \mathbb{E}\left(e^{-(1-\alpha)\mu\tau}\right).$$

Therefore, $(1-\alpha)\mu$ is solution of Equation (2.16). It is either 0 or β, since $1-\alpha = \mathbb{P}(W = 0) > 0$, this implies $\alpha = 1 - \beta/\mu$. Relationship (2.15) has been obtained again.

2.4.2 The M/GI/1 Queue

The distribution of the variable τ is exponential with parameter λ.

PROPOSITION 2.14.

a) *If $\rho < 1$, the Laplace transform of the stationary waiting time of an $M/GI/1$ queue is given by Pollaczek-Khintchine's formula : for* $\mathrm{Re}(\xi) \geq 0$,

$$\mathbb{E}\left(e^{-\xi W}\right) = \frac{1-\rho}{1 - \lambda\xi^{-1}\left(1 - \mathbb{E}(e^{-\xi\sigma})\right)}. \tag{2.17}$$

b) *If $\rho > 1$, a busy period is infinite with probability β/μ, where β is solution of Equation (2.16).*

PROOF. The factorization of $\sigma - \tau$ given by ψ_+, ψ_- is obtained with Relationship (2.14) and Proposition 2.10, by replacing μ by λ and τ by σ,

$$\frac{\psi_+(u,\xi)}{\psi_+(u,0)} = \frac{\phi_-(u,-\xi)}{\phi_-(u,0)} = \frac{b(u)-\xi}{\lambda - \xi - u\lambda\mathbb{E}(e^{-\xi\sigma})} \times \frac{\lambda(1-u)}{b(u)},$$

for $|u| < 1$ and $\mathrm{Re}(\xi) \geq 0$. If $\rho < 1$,

$$\mathbb{E}\left(e^{-\xi W}\right) = \lim_{u \nearrow 1} \frac{\psi_+(u,\xi)}{\psi_+(u,0)} = \frac{(1-\rho)\xi}{\xi - \lambda(1 - \mathbb{E}(e^{-\xi\sigma}))},$$

according to the behavior of $b(u)$ in the neighborhood of 1 in this case (Proposition 2.10, recall that σ and τ are exchanged), the a) of the proposition is proved. In the same way as for the $GI/M/1$ queue, Wiener-Hopf's factorization gives the generating function of the variable ν_-,

$$\mathbb{E}(u^{\nu_-}) = 1 - \frac{b(u)}{\mu},$$

if $\rho > 1$, then $b(u) \to \beta$ as $u \to 1$, therefore $\mathbb{P}(\nu_- < +\infty) = 1 - \beta/\mu$. □

The next corollary presents a remarkable property of the $M/GI/1$ queue: At equilibrium the distribution of the workload of the queue at an arbitrary instant (the virtual waiting time) has the same distribution as the workload of the queue at the arrival of a customer (i.e. the waiting time). Section 7.6 of Chapter 7 presents a general explanation of this phenomenon.

COROLLARY 2.15. *If $\rho < 1$, the stationary virtual waiting time V has the same distribution as the stationary waiting time W.*

PROOF. Indeed, Relationship (2.17) can be written as

$$\mathbb{E}\left(e^{-\xi W}\right) = 1 - \rho + \lambda \mathbb{E}\left(e^{-\xi W}\right) \mathbb{E}\left(\frac{1 - e^{-\xi\sigma}}{\xi}\right),$$

and the right hand side is $\mathbb{E}(\exp(-\xi V))$ according to Takàcs' Formula (2.12). □

The Distribution of W. Relationship (2.17) does not have a direct probabilistic interpretation as it is the case for the $GI/M/1$ queue. This distribution can nevertheless be decomposed as follows, for $\mathrm{Re}(\xi) \geq 0$,

$$(2.18)\quad \mathbb{E}\left(e^{-\xi W}\right) = \sum_{n=0}^{+\infty} \rho^n (1-\rho) \left[\mathbb{E}\left(e^{-\xi R^\sigma}\right)\right]^n$$

$$= \sum_{n=0}^{+\infty} \rho^n (1-\rho) \mathbb{E}\left(e^{-\xi \sum_1^n R_k^\sigma}\right),$$

where R^σ is a random variable such that

$$\mathbb{E}\left(e^{-\xi R^\sigma}\right) = \frac{1 - \mathbb{E}(e^{-\xi\sigma})}{\mathbb{E}(\sigma)\xi},$$

the variable R^σ has the same distribution as the first point of the stationary renewal point process associated with σ, see Chapter 1 page 25. In particular, its density function on $\mathbb{R}_+$ is $\mathbb{P}(\sigma \geq x)/\mathbb{E}(\sigma)$. According to Relationship (2.13),

$$\mathbb{P}(W = 0) = \mathbb{P}(\nu_+ = +\infty) = 1 - \rho,$$

Proposition 2.9 and Relationship (2.18) suggest that the conditional distribution of S_{ν_+} given the event $\{\nu_+ < +\infty\}$ is the distribution of R^σ.

PROPOSITION 2.16. *If $\rho < 1$, the stationary distribution of Q the number of customers that a customer finds in the queue when it arrives is given by*

$$\begin{cases} \mathbb{P}(Q = n) &= \displaystyle\int_{]0,+\infty]} \frac{(\lambda x)^{n-1}}{(n-1)!} e^{-\lambda x}\, W(dx), \qquad n \geq 1, \\ \mathbb{P}(Q = 0) &= 1 - \rho, \end{cases} \tag{2.19}$$

where $W(dx)$ is the distribution of the stationary waiting time.

PROOF. According to Proposition 2.8, by using the fact that the arrival process $\mathcal{N}_\lambda$ is Poisson with parameter λ, for $n \geq 1$

$$\begin{aligned} \mathbb{P}(Q = n) &= \mathbb{P}(t_{n-1} < W \leq t_n) \\ &= \mathbb{P}(\mathcal{N}_\lambda(]0, W[) = n - 1) = \mathbb{E}\left[\frac{(\lambda W)^{n-1}}{(n-1)!} e^{-\lambda W} 1_{\{W>0\}}\right]. \end{aligned}$$

When $\text{Re}(\xi)$ goes to infinity in Identity (2.17), one gets $\mathbb{P}(W = 0) = 1 - \rho$ and since $\mathbb{P}(Q = 0) = \mathbb{P}(W = 0)$, the proposition is proved. □

2.5 The $\mathbf{H_1/G/1}$ queue

In this section, a typical application of the Wiener-Hopf factorization is presented. (The $M/G/1$ and $G/M/1$ queues can be analyzed without the factorization, see Chapter 7 page 179). The notation H_1 (hyper-exponential of order 1) denotes the distributions whose Laplace transform is given by

$$\mathbb{E}\left(e^{-\xi\tau}\right) = \prod_{i=1}^{N} \frac{a_i}{a_i + \xi}, \tag{2.20}$$

for $\text{Re}(\xi) \geq 0$, $N \in \mathbb{N}$ and $0 < a_1 < a_2 < \cdots < a_N$. Such a τ is the sum of N independent random variables with an exponential distribution whose parameters are given respectively by $a_1, \ldots, a_N$.

The Laplace transform of the distribution of interarrival intervals is given by Equation (2.20). To express the distribution of the stationary waiting time, the function to decompose is given by

$$\frac{1}{1 - u\mathbb{E}(e^{-\xi X_0})} = \frac{\prod_{i=1}^{N}(a_i - \xi)}{\prod_{i=1}^{N}(a_i - \xi) - u\mathbb{E}(e^{-\xi\sigma})\prod_{i=1}^{N} a_i}, \tag{2.21}$$

with $|u| < 1$ and $\text{Re}(\xi) = 0$.
If $K \geq a_N$ and $D_K = \{\xi : \text{Re}(\xi) \geq 0, |\xi - a_N| \leq K\}$; when $\xi \in \mathbb{C}$ is on the boundary of D_K, necessarily $|\xi - a_i| \geq a_i$ for any $i \leq N$, hence the inequality

$$\left|u\mathbb{E}\left(e^{-\xi\sigma}\right)\prod_{i=1}^{N} a_i\right| \leq |u| \prod_{i=1}^{N} a_i < \left|\prod_{i=1}^{N}(a_i - \xi)\right|$$

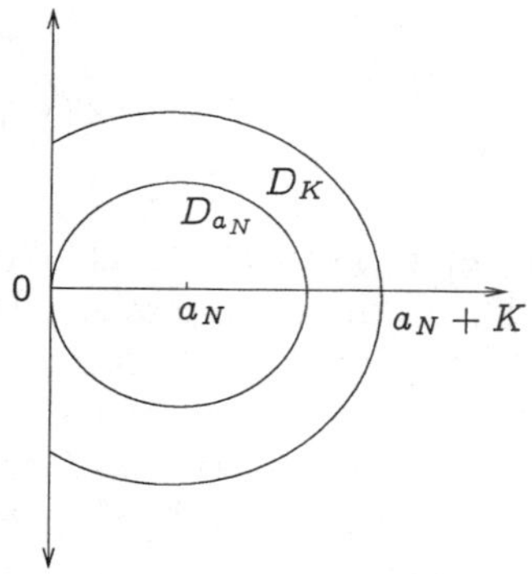

FIG. 2.4. The Contour of D_K

holds, Rouché's Theorem shows that the denominator of the right hand side of Equation (2.21) has exactly N roots $a_i(u), i = 1, \ldots, N$ in D_K. By letting K go to infinity, one gets that the right half-plane contains exactly N roots which are located in D_{a_N}.

Equation (2.21) can be rewritten as

$$\frac{1}{1 - u\mathbb{E}(e^{-\xi X_0})} = \prod_{i=1}^{N} \frac{a_i - \xi}{a_i(u) - \xi} \times \frac{\prod_{i=1}^{N}(a_i(u) - \xi)}{\prod_{i=1}^{N}(a_i - \xi) - u\mathbb{E}(e^{-\xi\sigma})\prod_{i=1}^{N} a_i},$$

for $|u| < 1$ and $\mathrm{Re}(\xi) = 0$. By setting

$$\phi_+(u, \xi) = \frac{\prod_{i=1}^{N}(a_i(u) - \xi)}{\prod_{i=1}^{N}(a_i - \xi) - u\mathbb{E}(e^{-\xi\sigma})\prod_{i=1}^{N} a_i}$$

and

$$\phi_-(u, \xi) = \prod_{i=1}^{N} \frac{a_i - \xi}{a_i(u) - \xi},$$

these two functions clearly satisfy Conditions a) and b) of Theorem 2.4. The Laplace transform of W can be expressed as

$$(2.22) \quad \mathbb{E}\left(e^{-\xi W}\right)$$

$$= \lim_{u \nearrow 1}(1 - u)\prod_{i=1}^{N} \frac{a_i}{a_i(u)} \times \frac{\prod_{i=1}^{N}(a_i(u) - \xi)}{\prod_{i=1}^{N}(a_i - \xi) - u\mathbb{E}\left(e^{-\xi\sigma}\right)\prod_{i=1}^{N} a_i}.$$

For u such that $|u| < 1$, the $a_i(u), i = 1, \ldots, N$ are solutions in the right half-plane of the denominator of Equation (2.21) and satisfy

$$f_u(\xi) = 1 - u\mathbb{E}\left(e^{-\xi(\sigma-\tau)}\right) = 1 - u\mathbb{E}\left(e^{-\xi\sigma}\right)\prod_{i=1}^{N}\frac{a_i}{a_i-\xi} = 0. \tag{2.23}$$

The function f_u is concave on the interval $]0, a_1[$, $f_u(0) = 1 - u$ and

$$\lim_{\xi \to a_1} f_u(\xi) = -\infty,$$

it has therefore a unique root $a_1(u)$ in this interval. The function $u \to f_u(\xi)$ being non-increasing for $\xi \in]0, a_1[$, the same property is true for $u \to a_1(u)$. Since $f_1(0) = 0$ and $f_1'(0) = \mathbb{E}(\sigma) - \mathbb{E}(\tau) < 0$, the fact that f_1 is concave shows that 0 is the only root of f_1 in $[0, a_1[$. The limit of $a_1(u)$ as u tends to 1 being also a root f_1 in this interval, it is necessarily 0. With a Taylor expansion in the neighborhood of 0 of Equation (2.23), one gets that $(1-u)/a_1(u)$ converges to $\mathbb{E}(\tau) - \mathbb{E}(\sigma)$ as u tends to 1.

PROPOSITION 2.17. *If the distribution of interarrival intervals is the sum of N independent exponentially distributed random variables whose parameters are respectively $0 < a_1 < a_2 < \cdots < a_N$, under the condition $\mathbb{E}(\sigma) < \mathbb{E}(\tau)$ the Laplace transform of the stationary waiting time is given by*

$$\mathbb{E}\left(e^{-\xi W}\right) = a_1(\mathbb{E}(\tau) - \mathbb{E}(\sigma))\frac{\xi \prod_{i=2}^{N} a_i(1-\xi/\alpha_i)}{\mathbb{E}(e^{-\xi\sigma})\prod_{i=1}^{N} a_i - \prod_{i=1}^{N}(a_i - \xi)}, \tag{2.24}$$

where the α_i, $i = 2, \ldots, N$ are the non-zero solutions in the right half-plane of the equation

$$\mathbb{E}\left(e^{-\xi\sigma}\right)\prod_{i=1}^{N} a_i - \prod_{i=1}^{N}(a_i - \xi) = 0.$$

PROOF. For $i \geq 2$, the set of numbers $a_i(u)$, $0 < u < 1$) being bounded, define α_i one of the limiting values when u tends to 1. By using Identity (2.22) and the behavior of $a_1(u)$ in the neighborhood of 1, one gets Relationship (2.24). Since $\mathbb{E}(\exp(-\xi W))$ is defined for any ξ such that $\mathrm{Re}(\xi) \geq 0$, the $\alpha_i, i \geq 2$ are therefore non-zero and are the roots of the denominator of (2.24) in the right half-plane. The proposition is proved. □

Notice that the stability condition plays a role only in the behavior of the root $a_1(u)$ in the neighborhood of $u = 1$.

2.6 A Probabilistic Proof of the Factorization

In this section, it is shown that the calculations of Section 2.2 have a probabilistic version. Basically, one of the key ingredients of the proof of the Wiener-Hopf factorization is the decomposition of the series

$$\phi(u, \xi) = \frac{1}{1 - u\mathbb{E}(e^{-\xi X})} = \sum_{n \geq 0} u^n \mathbb{E}\left(e^{-\xi S_n}\right)$$

at the stopping times ν_+ and ν_-, for $|u| < 1, \mathrm{Re}(\xi) = 0$. The method presented here is due, in a more general framework, to Greenwood and Pitman[GP80]. By setting

$$\psi(u,\xi) = \frac{1-u}{1-u\mathbb{E}(e^{-\xi X})},$$

for $u \in]0,1[$, the function ψ can be written as

$$\psi(u,\xi) = \sum_{n\geq 0} u^n(1-u)\mathbb{E}\left(e^{-\xi S_n}\right) = \mathbb{E}\left(e^{-\xi S_G}\right),$$

where G is a geometrically distributed random variable with parameter u, independent of the random walk (S_n). The function ψ can be decomposed in the following way

THEOREM 2.18. *If G is a geometrically distributed random variable with parameter $u \in]0,1[$, independent of the random walk (S_n), then for any $\xi \in \mathbb{C}$ such that* $\mathrm{Re}(\xi) = 0$,

$$\mathbb{E}\left(e^{-\xi S_G}\right) = \frac{1-u}{1-u\mathbb{E}(e^{-\xi X})} = \mathbb{E}\left(e^{-\xi M_G}\right)\mathbb{E}\left(e^{-\xi m_G}\right), \tag{2.25}$$

with for $n \in \mathbb{N}$,

$$M_n = \sup_{0\leq k\leq n} S_k, \text{ and } m_n = \inf_{0\leq k\leq n} S_k.$$

PROOF. The variable T denotes the first time the maximum of (S_n) before G is reached,

$$T = \inf\{k \geq 0 : S_k = M_G\},$$

by taking the variable G as above. First, the identity

$$\mathbb{E}\left(e^{-\xi S_G}\right) = \mathbb{E}\left(e^{-\xi M_G}\right)\mathbb{E}\left(e^{-\xi(S_G - M_G)}\right), \tag{2.26}$$

is established, or equivalently,

$$\mathbb{E}\left(e^{-\xi S_G}\right) = \mathbb{E}\left(e^{-\xi S_T}\right)\mathbb{E}\left(e^{-\xi(S_G - S_T)}\right).$$

If T is a stopping time, Lemma 2.3 gives directly the independence between S_T and $S_G - S_T$, and Decomposition (2.26) would be proved. The variable T is not a stopping time, the future after T must be known to decide that T is indeed the hitting time of the maximum. The independence property is nevertheless true, because G has a geometric distribution.

ASSERTION. *The variables S_T and $S_G - S_T$ are independent.*
By denoting as before,

$$\nu_+ = \inf\{k : S_k > 0\},$$

[GP80] P. Greenwood and J. Pitman, *Fluctuation identities for Lévy processes and splitting at the maximum*, Advances in Applied Probability **12** (1980), 893–902.

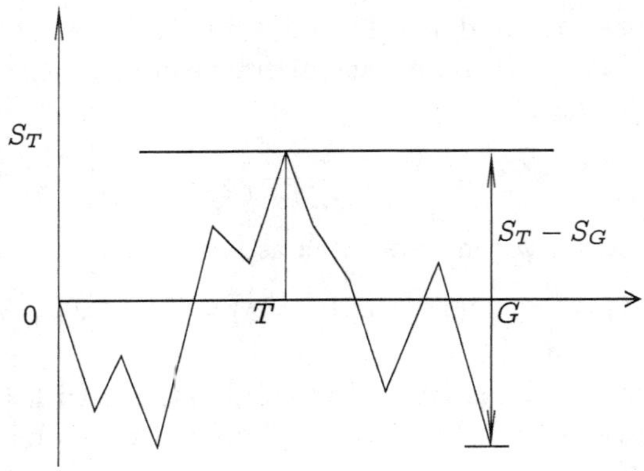

FIG. 2.5. Decomposition of the Random Walk Stopped at G

the finite sequence $\mathcal{E} = (S_k;\, 0 < k \leq \nu_+)$ is called an excursion ; this is a piece of the sample path of the random walk stopped when it becomes positive. Lemma 2.3 shows that the random walk (S_n) can be considered as a set of excursions

$$[\mathcal{E}_i : i \geq 0] = \left[\left(\sum_{p=1}^{k} X_p^i;\, 0 < k \leq \nu_+^i\right) : i \geq 0\right]$$

sticked together and i.i.d. with the same distribution as $\mathcal{E}$. It is possible that there is only a finite number of such excursions, the last one corresponding to an infinite busy period. When $\mathbb{P}(\nu_+ = +\infty) > 0$, this is the case with probability 1. If $(\widetilde{G}_i)$ is an i.i.d. sequence of geometrically distributed random variables with parameter u, for each $i \in \mathbb{N}$, if $\widetilde{G}_i < \nu_+^i$, the excursion $\mathcal{E}_i$ is painted red. Otherwise, if $\widetilde{G}_i \geq \nu_+^i$, the excursion is left unchanged. The variable $\widetilde{\tau}$ denotes the index of the first red excursion. The independence of $(\widetilde{G}_i)$ and (S_n) shows that, conditionally on the value of τ, the excursions $\mathcal{E}_i$, $0 \leq i \leq \widetilde{\tau} - 1$ are independent of the excursions $\mathcal{E}_i$, $i \geq \widetilde{\tau}$. An excursion is positive only at the last instant of its trajectory at time ν_+. Thus, the successive maxima of the random walk occur at the last instants of the excursions. By setting

$$\widetilde{G} = \nu_+^0 + \nu_+^1 + \cdots + \nu_+^{\widetilde{\tau}-1} + G_{\widetilde{\tau}},$$

since $G_{\widetilde{\tau}} < \nu_+^{\widetilde{\tau}}$, the maximum of the random walk before $\widetilde{G}$ is reached before $\nu_+^0 + \nu_+^1 + \cdots + \nu_+^{\widetilde{\tau}-1}$, consequently

$$S_{\nu_+^0 + \nu_+^1 + \cdots + \nu_+^{\widetilde{\tau}-1}} = M_{\nu_+^0 + \nu_+^1 + \cdots + \nu_+^{\widetilde{\tau}-1}} = M_{\widetilde{G}}.$$

In particular, $M_{\widetilde{G}}$ is a functional of the excursions $\mathcal{E}_i$, $0 \leq i \leq \widetilde{\tau} - 1$ and

$$S_{\widetilde{G}} - M_{\widetilde{G}} = X_1^\tau + X_2^\tau + \cdots + X_{\widetilde{G}_\tau}^\tau,$$

depends only of the excursion $\mathcal{E}_{\widetilde{\tau}}$. Therefore, the variables $S_{\widetilde{G}} - M_{\widetilde{G}}$ and $M_{\widetilde{G}}$ are independent. To prove the assertion, it is sufficient to show that the variable $\widetilde{G}$ has the same distribution as G and is independent of the random walk (S_n).

The relationship

$$\mathbb{P}\left(G - \nu_+^1 \geq n \,\middle|\, G \geq \nu_+^1\right) = u^n,$$

shows that, conditionally on the event $\{G \geq \nu_+^1\}$, the variable G can be expressed as $\nu_+^1 + G_1$, where G_1 is a geometrically distributed random variable independent of ν_+^1 and consequently of the random variable. By repeating again this argument, conditionally on the event $\{G_1 \geq \nu_+^2\}$, the identity $G_1 = \nu_+^2 + G_2$ holds, with G_2 independent of (S_n), and so on. In this way, one gets a sequence of i.i.d. geometrically distributed random variables (G_n) independent of (S_n), such that $G = \nu_+^1 + \cdots + \nu_1^{\tau-1} + G_\tau$, where τ is the first index i for which $G_i < \nu_+^i$. The variable G has the same distribution as $\widetilde{G}$. The variable $\widetilde{G}$ is thus geometrically distributed with parameter u and independent of the random walk. Relationship (2.26) is established.

The proof of Relationship (2.25), or that $S_G - M_G$ has the same distribution as m_G is done in the following way: The identity

$$S_G - M_G = \inf_{0 \leq k \leq G} \left(X_{k+1} + X_{k+2} + \cdots + X_G\right),$$

and the independence of G and the random walk give the relation

$$(X_G, X_{G-1}, \ldots, X_1) \stackrel{\text{dist.}}{=} (X_1, X_2, \ldots, X_G),$$

thus

$$S_G - M_G \stackrel{\text{dist.}}{=} \inf_{0 \leq k \leq G} (X_1 + X_2 + \cdots + X_k) = m_G,$$

the theorem is proved. □

By denoting $\psi_+(u,\xi) = \mathbb{E}(\exp(-\xi M_G))$ and $\psi_-(u,\xi) = \mathbb{E}(\exp(-\xi m_G))$, since $M_G \geq 0$ and $m_G \leq 0$, it is easily checked that the functions ψ_+ and ψ_- are the solutions of the Wiener-Hopf factorization. The uniqueness of Theorem 2.4 shows therefore that for $\mathrm{Re}(\xi) \geq 0$,

$$\phi_+(u,\xi) = \frac{\mathbb{E}\left(e^{-\xi M_G}\right)}{\mathbb{P}(M_G = 0)},$$

and for $\mathrm{Re}(\xi) \leq 0$,

$$\phi_-(u,\xi) = \mathbb{E}\left(e^{-\xi m_G}\right) \frac{1 - \mathbb{E}(u^{\nu_+})}{1-u}.$$

The identity $\{M_G = 0\} = \{G < \nu_+\}$ gives $\mathbb{P}(M_G = 0) = 1 - \mathbb{E}(u^{\nu_+})$, and from Relationship (2.9), one gets

$$\frac{1 - \mathbb{E}(u^{\nu_+})}{1-u} = \frac{1}{1 - \mathbb{E}(u^{\nu_-})}.$$

The following proposition has therefore been proved.

PROPOSITION 2.19. *If G is a geometrically distributed random variable with parameter $u \in]0,1[$, independent of the random walk, the functions ϕ_+, ϕ_- of the Wiener-Hopf factorization can be expressed as*

$$\phi_+(u,\xi) = \frac{\mathbb{E}\left(e^{-\xi M_G}\right)}{1-\mathbb{E}(u^{\nu_+})}, \qquad \phi_-(u,\xi) = \frac{\mathbb{E}\left(e^{-\xi m_G}\right)}{1-\mathbb{E}(u^{\nu_-})},$$

where, for $n \in \mathbb{N}$,

$$M_n = \sup_{0\leq k\leq n} S_k \quad \text{and} \quad m_n = \inf_{0\leq k\leq n} S_k.$$

3. Limit Theorems for $GI/GI/1$ Queues

Contents

3.1 Introduction

The Wiener-Hopf factorization gives a method to get the Laplace transform of the stationary waiting time of a $GI/GI/1$ FIFO queue. In practice, see Section 2.5 of Chapter 2, this method consists in locating the poles and the roots of a function in the complex plane. In practice, it is not always easy to get the number of poles and zeros of a complex function in general, and a fortiori to locate them. For this reason, getting some simple qualitative results for the $GI/GI/1$ FIFO queue may turn out to be quite difficult.

Since the description of the "normal" situation is difficult, it is natural to study the waiting time when one of the parameters is close to some critical value. This is the subject of this chapter. In the first part, the tail distribution of the stationary waiting time is analyzed. Under some assumptions, it is shown that this tail distribution has an exponential decay, i.e. for x sufficiently large

$$\mathbb{P}(W \geq x) \sim Ce^{-\gamma x},$$

where C and γ are some constants. The exponential decrease factor γ is the non negative solution of a fixed point equation and therefore can be easily obtained numerically. The change of probability distribution (a Girsanov's formula) is the main probabilistic tool of this section. The second part studies the $GI/GI/1$ queue when the load of the queue is just below 1, i.e. near the saturation. It is shown that the stationary waiting time properly rescaled converges in distribution to an exponentially distributed random variable.

3.2 The Biased Random Walk

As in the previous chapter the sequence (X_n) of i.i.d. random variables is defined on a probability space $(\Omega, \mathcal{F}, P)$, it is assumed that X_1 is integrable. The sequence $(\mathcal{F}_n) = (\sigma\langle X_1, \dots, X_n\rangle)$ denotes the filtration associated with (X_n) and the σ-field $\mathcal{F}$ is simply $\vee_n \mathcal{F}_n$, all possible events for this sequence. The associated random walk is denoted by $(S_n) = (X_1 + \cdots + X_n)$. Two elementary results will be useful in the sequel.

PROPOSITION 3.1. *If $\mathbb{E}(X_1) > 0$, then for any $a > 0$, the hitting time T_a of the interval $[a, +\infty[$,*

$$T_a = \inf\{n : S_n \geq a\},$$

is $\mathbb{P}$-almost surely finite and integrable.

PROOF. Because of the assumption $\mathbb{E}(X_1) > 0$, one can fix some constant $K > 0$ such that $\mathbb{E}(X_1 \wedge K) > 0$. Since the random walk associated to the i.i.d. sequence $(X_n \wedge K)$ is below the initial random walk, the corresponding hitting time is larger than T_a. It is therefore sufficient to prove the proposition for a random walk with upward jumps bounded by K.

The sequence $(S_n - n\mathbb{E}(X_1))$ is clearly a martingale and for $N \geq 0$, optional stopping Theorem gives therefore the identity

$$\mathbb{E}\left[S_{T_a \wedge N} - (T_a \wedge N)\mathbb{E}(X_1)\right] = 0,$$

consequently,

$$\begin{aligned}\mathbb{E}(T_a \wedge n)\mathbb{E}(X_1) &= \mathbb{E}(S_{T_a \wedge N}) \\ &= \mathbb{E}\left(S_{T_a \wedge N} - S_{(T_a-1)^+ \wedge N}\right) + \mathbb{E}\left(S_{(T_a-1)^+ \wedge N}\right) \leq a + K.\end{aligned}$$

From monotone convergence Theorem one gets the inequality

$$\mathbb{E}(T_a) \leq (a + K)/\mathbb{E}(X_1).$$

The proposition is proved. □

The next proposition on stopped random walks is classical, it generalizes the identity $\mathbb{E}(S_n) = n\mathbb{E}(X_1)$.

PROPOSITION 3.2 (Wald's Formula). *If τ is an integrable stopping time and X_1 is integrable, then $S_\tau = X_1 + \cdots + X_\tau$ is integrable and*

$$\mathbb{E}(S_\tau) = \mathbb{E}(X_1)\mathbb{E}(\tau). \tag{3.1}$$

PROOF. In the same way as before, for $N \in \mathbb{N}$, optional stopping Theorem applied to the martingale

$$\left(\sum_1^n |X_i| - n\mathbb{E}(|X_1|)\right)$$

and the stopping time $\tau \wedge N$ shows that

$$\mathbb{E}(|X_1| + |X_2| + \cdots + X_{\tau\wedge N}) = \mathbb{E}(\tau \wedge N)\mathbb{E}(|X_1|) \leq \mathbb{E}(\tau)\mathbb{E}(|X_1|) < +\infty.$$

Monotone convergence Theorem shows that the variable $|X_1| + \cdots + |X_\tau|$ is integrable.

Optional stopping Theorem for the martingale $(S_n - n\mathbb{E}(X_1))$ implies that $\mathbb{E}(S_{\tau\wedge N}) = \mathbb{E}(\tau \wedge N)\mathbb{E}(X_1)$. The relationship $|S_{\tau\wedge N}| \leq |X_1| + \cdots + |X_\tau|$ and Lebesgue's Theorem give the convergence of the sequence $(\mathbb{E}(S_{\tau\wedge N}))$ to $\mathbb{E}(S_\tau)$. By monotone convergence Theorem the sequence $(\mathbb{E}(\tau \wedge N))$ converges to $\mathbb{E}(\tau)$, Identity (3.1) is thus proved. □

LEMMA 3.3. *If $\mathbb{E}(X_1) < 0$ and if there exists $\theta > 0$ such that*

$$1 < \mathbb{E}\left(e^{\theta X_1}\right) < +\infty,$$

there exists a unique $\gamma \in]0, \theta[$ such that $\mathbb{E}(\exp(\gamma X_1)) = 1$.

PROOF. For $y \in [0, \theta[$, there exists some constant A such that

$$|x|^2 \exp(yx) \leq A + \exp(\theta x)$$

for $x \in \mathbb{R}$, in particular

$$\mathbb{E}\left(|X_1|^2 \exp(yX_1)\right) \leq A + \mathbb{E}(\exp(\theta X_1)) < +\infty.$$

The function $y \to f(y) = \mathbb{E}(\exp(yX_1))$ is therefore strictly convex on $[0, \theta[$ because its second derivative is finite and positive. By assumption, $f(0) = 1$, $f'(0) = \mathbb{E}(X_1) < 0$ and $f(\theta) > 1$, therefore there is a unique $\gamma > 0$ such that $f(\gamma) = 1$. □

Change of Probability Measure and Positive Martingales

In the rest of this section the conditions of the previous lemma are assumed, the quantity γ denotes the unique positive solution of the equation in x, $\mathbb{E}(\exp(xX_1)) = 1$.

PROPOSITION 3.4 (Girsanov's Formula). *If the assumptions of Lemma 3.3 are satisfied, there is a unique probability distribution $\mathbb{Q}$ on $(\Omega, \mathcal{F})$ such that for $n \geq 0$ and any non-negative, $\mathcal{F}_n$-measurable function Y,*

$$\mathbb{E}_{\mathbb{Q}}(Y) \stackrel{def.}{=} \int_\Omega Y\, d\mathbb{Q} = \int_\Omega Y e^{\gamma S_n}\, d\mathbb{P}.$$

For the probability distribution $\mathbb{Q}$, the sequence (S_n) is also a random walk and the distribution of X_1 for $\mathbb{Q}$ has the characteristic function

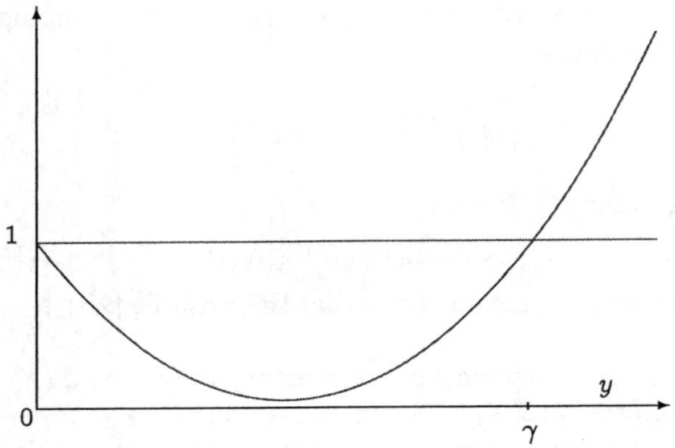

FIG. 3.1. The Function $y \to \mathbb{E}(\exp(yX_1))$

$$\mathbb{E}_{\mathbb{Q}}\left(e^{-\xi X_1}\right) = \mathbb{E}\left(e^{-(\xi-\gamma)X_1}\right),$$

for $\xi \in \mathbb{C}$ *such that* $\mathrm{Re}(\xi) = 0$. *The variable* X_1 *is integrable for the probability distribution* $\mathbb{Q}$ *and* $\mathbb{E}_{\mathbb{Q}}(X_1) > 0$.

If τ *is a stopping time and* Y *a non-negative* $\mathcal{F}_\tau$*-measurable random variable, the identity*

$$\int_\Omega Y 1_{\{\tau<+\infty\}}\, d\mathbb{P} = \int_\Omega Y e^{-\gamma S_\tau} 1_{\{\tau<+\infty\}}\, d\mathbb{Q} \tag{3.2}$$

holds.

PROOF. It is easily checked that $(Z_n) = (\exp(\gamma S_n))$ is a martingale. Indeed, for $n \geq 1$, by independence of X_{n+1} and $\mathcal{F}_n$,

$$\mathbb{E}\left(e^{\gamma S_{n+1}} \mid \mathcal{F}_n\right) = e^{\gamma S_n}\mathbb{E}\left(e^{\gamma X_{n+1}}\right) = e^{\gamma S_n}.$$

For any $A \in \mathcal{F}_n$, define

$$\mathbb{Q}_n(A) = \int_A e^{\gamma S_n}\, d\mathbb{P},$$

note that $\mathbb{Q}_n$ is a probability measure since

$$\mathbb{Q}_n(\Omega) = \mathbb{E}[\exp(\gamma S_n)] = (\mathbb{E}[\exp(\gamma X_1)])^n = 1.$$

The probability measures $(\mathbb{Q}_n)$ are compatible: i.e. if $p \leq n$, then the identity $\mathbb{Q}_n(A) = \mathbb{Q}_p(A)$ holds for any $A \in \mathcal{F}_p \subset \mathcal{F}_n$ since

$$\mathbb{Q}_n(A) = \mathbb{E}\left(1_A e^{\gamma S_n}\right) = \mathbb{E}\left[1_A \mathbb{E}\left(e^{\gamma S_n} \mid \mathcal{F}_p\right)\right]$$

since A is $\mathcal{F}_p$-measurable, the martingale property of (Z_n) gives

$$\mathbb{Q}_n(A) = \mathbb{E}\left(1_A e^{\gamma S_p}\right) = \mathbb{Q}_p(A).$$

The sequence of probability measures $(\mathbb{Q}_n)$ is thus compatible. If Ω is the canonical probability space associated with the i.i.d. sequence (X_n), Daniell-Kolmogorov's Theorem (see Rogers and Williams [44] for example) shows that there exists a unique probability measure $\mathbb{Q}$ on $\mathcal{F}$ such that the equality $\mathbb{Q} = \mathbb{Q}_n$ holds on the σ-field $\mathcal{F}_n$.

The next assertion to prove is about the distribution of (S_n) for $\mathbb{Q}$. If $\xi \in \mathbb{C}$ is such that $\text{Re}(\xi) = 0$, for $n \geq 1$ the definition of $\mathbb{Q}$ shows that

$$\mathbb{E}_{\mathbb{Q}}\left[e^{-\xi S_n}\right] = \mathbb{E}\left[e^{-(\xi-\gamma)S_n}\right] = \left(\mathbb{E}\left[e^{-(\xi-\gamma)X_1}\right]\right)^n.$$

Therefore, the variable S_n is the sum of n i.i.d. variables and the Laplace transform of their common distribution is given by $\mathbb{E}(\exp(-(\xi-\gamma)X_1))$. In other words, for the probability measure $\mathbb{Q}$ the sequence (S_n) is a random walk.

Since there is some $\theta > 0$ such that $1 < \mathbb{E}(\exp(\theta X_1)) < +\infty$, the constant γ is strictly less that θ, therefore there exists some constant $A > 0$ such that $|x|\exp(\gamma x) \leq A + \exp(\theta x)$ for any $x \in \mathbb{R}$. By definition of $\mathbb{Q}$,

$$\mathbb{E}_{Q}(|X_1|) = \mathbb{E}\left(|X_1|e^{\gamma X_1}\right),$$

thus, the variable X_1 is integrable with respect to $\mathbb{Q}$.

If τ is a stopping time and Y a non-negative $\mathcal{F}_\tau$-measurable random variable,

$$\int_\Omega Y 1_{\{\tau<+\infty\}}\, d\mathbb{P} = \sum_{i=0}^{+\infty} \int_\Omega Y 1_{\{\tau=i\}}\, d\mathbb{P},$$

for $i \in \mathbb{N}$,

$$\int_\Omega Y 1_{\{\tau=i\}}\, d\mathbb{P} = \int_\Omega Y e^{-\gamma S_i} 1_{\{\tau=i\}} e^{\gamma S_i}\, d\mathbb{P}.$$

The variable Y being $\mathcal{F}_\tau$-measurable, $Ye^{-\gamma S_i}1_{\{\tau=i\}}$ is $\mathcal{F}_i$-measurable. According to the definition of $\mathbb{Q}$,

$$\int_\Omega Y 1_{\{\tau=i\}}\, d\mathbb{P} = \int_\Omega Y e^{-\gamma S_\tau} 1_{\{\tau=i\}}\, d\mathbb{Q},$$

by summing up this last identity, one gets Relationship (3.2).

If $f(y) = \mathbb{E}(\exp(yX_1)$, since $f(0) = f(\gamma) = 1$ and $f'(0) < 0$, the derivative of f is 0 at some point in $]0,\gamma[$, the function f being strictly convex on $]0,\theta[$ (see the proof of Lemma 3.3), one gets that

$$0 < f'(\gamma) = \mathbb{E}\left(X_1 e^{\gamma X_1}\right) = \mathbb{E}_{\mathbb{Q}}(X_1).$$

The proposition is proved. □

If, for the probability measure $\mathbb{P}$ the random walk (S_n) has a negative drift $\mathbb{E}(X_1)$, for the probability measure $\mathbb{Q}$, this drift is positive (which explains the name biased random walk). This phenomenon can be explained as follows : The probability measure $\mathbb{Q}$ is absolutely continuous with respect to $\mathbb{P}$,

$$d\mathbb{Q} = \exp(\gamma S_t)\, d\mathbb{P}$$

on the σ-field $\mathcal{F}_t$. Because of the coefficient $\exp(\gamma S_t)$ the probability measure $\mathbb{Q}$ puts a larger weight on the sample paths of the random walk for which S_t is very large. This sample paths above 0 are therefore more likely for the probability distribution $\mathbb{Q}$. The property of positive drift is important as it will be seen later.

This change of probability measure can be generalized to a family of nonnegative martingales.

A Family of Biased Random Walks

In the same way as in the proof of Proposition 3.4, if

$$0 \leq \theta < \theta_c = \sup\left\{\theta : \mathbb{E}\left(e^{\theta X_1}\right) < +\infty\right\},$$

the function $\phi(\theta)$ is defined by

$$\phi(\theta) = \mathbb{E}\left(e^{\theta X_1}\right),$$

the sequence $(\exp(\theta S_n)/\phi(\theta)^n)$ is a martingale. It is therefore possible to define a probability measure $\mathbb{Q}_\theta$ on Ω such that for all $n \geq 0$,

$$d\mathbb{Q}_\theta = \frac{e^{\theta S_n}}{\phi(\theta)^n}\, d\mathbb{P} \text{ on the } \sigma\text{-field } \mathcal{F}_n.$$

For the probability measure $\mathbb{Q}_\theta$, the sequence (S_n) is also a random walk. The characteristic function of the distribution of the jumps of this random walk is given by

$$\mathbb{E}_{\mathbb{Q}_\theta}\left(e^{-\xi X_1}\right) = \frac{1}{\phi(\theta)}\mathbb{E}\left(e^{-(\xi-\theta)X_1}\right),$$

in particular, the expected value of a jump is

$$\mathbb{E}_{\mathbb{Q}_\theta}(X_1) = \mathbb{E}\left(X_1 e^{\theta X_1}\right)/\mathbb{E}\left(e^{\theta X_1}\right).$$

As before, if τ is a stopping time and Y a non-negative random variable $\mathcal{F}_\tau$-measurable then,

$$\int_\Omega Y 1_{\{\tau<+\infty\}}\, d\mathbb{P} = \int_\Omega Y e^{-\theta S_\tau}\phi(\theta)^\tau 1_{\{\tau<+\infty\}}\, d\mathbb{Q}_\theta.$$

As θ varies, the random walk for $\mathbb{Q}_\theta$ is concentrated on different samples paths of the original random walk. When $\theta = 0$, this is the original random walk.

The probability measures $\mathbb{Q}_\theta$ are used to derive large deviations results, to estimate the probability of the event $\{S_n \sim \alpha n\}$ with $\alpha \neq \mathbb{E}(X_1)$ for example. To get this kind of result, one has to find a θ such that the random walk for $\mathbb{Q}_\theta$ grows at rate α, i.e. $\mathbb{E}_{\mathbb{Q}_\theta}(X_1) = \alpha$. Theorem 5.20 page 136 on the large deviations of the $M/M/1$ queue is proved in this way. Notice that if X_1 is $\mathbb{P}$-almost surely bounded by a constant C, the maximal value of $\mathbb{E}_{\mathbb{Q}_\theta}(X_1)$ cannot be greater than C. In the rest of this chapter only the parameter γ is used.

In the more delicate framework of diffusion processes, the relation between the probability measures associated with exponential martingales are Cameron-Martin's Formula or Girsanov's Formula. For example if $(B(t))$ is the standard Brownian motion, for $\alpha \in \mathbb{R}$, a change of probability measure is associated with the exponential martingale

$$\left(\exp\left(\alpha B(t) - \alpha^2 t/2\right)\right).$$

For this new probability measure, it is easily checked that $(B(t))$ is a Brownian motion with drift α. This result has numerous applications, the distribution of some of the functionals of the Brownian motion with drift can be derived from the distribution of the same functional for the standard Brownian motion. (See Rogers and Williams [**43**] or Revuz and Yor [**42**].)

3.3 The Tail Distribution of W

The case of $GI/GI/1$ queue is considered from now on. The random variable X_n is $\sigma_n - \tau_n$, where (τ_n) is the sequence of interarrival intervals and (σ_n) the sequence of services. These sequences are i.i.d., independent and the condition $\mathbb{E}(\sigma_1) < \mathbb{E}(\tau_1)$ holds. In Chapter 2 it has been shown that under this condition, the sequence of waiting times of customers of the $GI/GI/1$ queue is an ergodic Markov chain. Moreover, the stationary waiting time W has the same distribution as M, the maximum of the random walk (S_n). The next theorem presents an asymptotic equivalent of the tail distribution of W. The first results in this domain are due to Kingman[Kin70].

THEOREM 3.5. *If the conditions*

a) $\mathbb{E}(X_1) < 0$ *where* $X_1 = \sigma_1 - \tau_1$*;*
b) there exists $\theta > 0$ *such that*

$$1 < \mathbb{E}\left(e^{\theta X_1}\right) < +\infty,$$

are satisfied, then the inequality

$$\mathbb{P}(W \geq a) = \mathbb{P}\left(\max_{n\geq 0} S_n \geq a\right) \leq e^{-\gamma a} \tag{3.3}$$

holds for $a \geq 0$*, where* γ *in the unique positive solution of the equation*

$$\mathbb{E}(\exp(\gamma X_1)) = 1.$$

Moreover, if the distribution of the variable σ *is non-lattice (See the definition page 25), then the distribution of* W *has a tail distribution with an exponential decay,*

$$\lim_{a\to+\infty} e^{\gamma a}\mathbb{P}(W \geq a) = C_1 = \frac{\mathbb{P}(\nu_+ = +\infty)}{\gamma \mathbb{E}(X_1 e^{\gamma X_1})\mathbb{E}\left(\nu_+ e^{\gamma S_{\nu_+}} 1_{\{\nu_+ < +\infty\}}\right)}, \tag{3.4}$$

[Kin70] J.F.C. Kingman, *Inequalities in the theory of queues*, Journal of the Royal Statistical Society B **32** (1970), 102–110.

where $C_1 > 0$ and ν_+ is the hitting time of $\mathbb{R}_+$ by the random walk.

$$\nu_+ = \inf\{n : S_n > 0\}.$$

As it will be seen in the proof, the constant C_1 of Equation (3.4) is well defined. Its expression can be obtained with the joint distribution of (ν_+, S_{ν_+}). In any case, the important parameter is γ which can be easily obtained by numerical methods.

PROOF. For $a > 0$, T_a denotes the hitting time of $[a, +\infty[$ by the random walk,

$$T_a = \inf\{n : S_n \geq a\}.$$

If $M = \sup\{S_n : n \geq 0\}$, Identity (3.2) of Proposition 3.4 yields

$$\mathbb{P}(W \geq a) = \mathbb{P}(M \geq a) = \mathbb{P}(T_a < +\infty) = \int e^{-\gamma S_{T_a}} 1_{\{T_a<+\infty\}}\, d\mathbb{Q},$$

$$\mathbb{P}(W \geq a) = e^{-\gamma a}\mathbb{E}_{\mathbb{Q}}\left(e^{-\gamma(S_{T_a}-a)} 1_{\{T_a<+\infty\}}\right), \tag{3.5}$$

since $S_{T_a} \geq a$, one gets

$$\mathbb{P}(W \geq a) \leq e^{-\gamma a},$$

and therefore Inequality (3.3) holds. Since $\gamma < \theta$ and $\mathbb{E}(\exp(\theta X_1) < +\infty$, the variable $X_1 \exp(\gamma X_1)$ is integrable. Formula (3.2) of Proposition 3.4 gives the inequality $\mathbb{E}_{\mathbb{Q}}(X_1) > 0$ and, according to the law of large numbers, $\mathbb{Q}$-almost surely,

$$\lim_{n\to+\infty} \frac{S_n}{n} = \mathbb{E}_{\mathbb{Q}}(X_1) > 0.$$

Consequently, $\mathbb{Q}$-almost surely, the sequence (S_n) converges to infinity. In particular, if $a \geq 0$, the stopping time T_a is $\mathbb{Q}$-almost surely finite. Identity (3.5) can thus be rewritten as

$$\mathbb{P}(W \geq a) = e^{-\gamma a}\mathbb{E}_{\mathbb{Q}}\left(e^{-\gamma(S_{T_a}-a)}\right) \tag{3.6}$$

For $n \geq 0$, the sequence (ν_+^n) is defined by induction by $\nu_+^0 = 0$ and

$$\nu_+^{n+1} = \inf\{n \geq \nu_+^n : S_n - S_{\nu_+^n} > 0\},$$

assuming, as usual, that $\inf \emptyset = +\infty$, in particular

$$\nu_+^1 = \inf\{n : S_n > 0\} = \nu_+.$$

Since the random walk converges $\mathbb{Q}$-almost surely to infinity, the variables ν_+^n are $\mathbb{Q}$-a.s. finite. The sequence $(S_{\nu_+^n})$ is by definition non-decreasing, Lemma 2.3 page 33 shows that for the probability $\mathbb{Q}$ the increments

$$\left(S_{\nu_+^{n+1}} - S_{\nu_+^n}\right)$$

are i.i.d. with the same distribution as S_{ν_+}. The sequence $(S_{\nu_+^n})$ is therefore a renewal process associated with the distribution of S_{ν_+}. The assumptions of renewal Theorem (Proposition 1.28 page 26) are satisfied:

– Integrability. Since $\mathbb{E}_{\mathbb{Q}}(X_1) > 0$, Lemma 3.1 shows that the variable ν_+ is integrable for the probability measure $\mathbb{Q}$. From Wald's Formula (3.1) one gets that S_{ν_+} is also integrable for $\mathbb{Q}$.
– Non-lattice jumps. The non-negative jumps of the random walk are due to the distribution of the random variable σ. The assumption on σ implies that S_{ν_+} is also non-lattice.

The sequence $(S_{\nu_+^n})$ is the sequence of the successive records of the random walk. The first time the random walk (S_n) is above the level a is necessarily at one of these instants, i.e. T_a is one of the ν_+^p for $p \geq 1$.

On the event $\{T_a < +\infty\}$, the quantity S_{T_a} is the first point of the renewal process $(S_{\nu_+^n})$ after a. According to renewal Theorem, as a tends to infinity, the variable $S_{T_a} - a$ converges in distribution (for the probability distribution $\mathbb{Q}$) to the random variable F_{ν_+} distributed as the first point of the stationary renewal process associated with the distribution of S_{ν_+}. For $x \geq 0$,

$$\mathbb{Q}(F_{\nu_+} \geq x) = \frac{1}{\mathbb{E}_{\mathbb{Q}}(S_{\nu_+})} \int_x^{+\infty} \mathbb{Q}(S_{\nu_+} > y)\, dy.$$

From Equation (3.6) one gets the convergence

$$\lim_{a \to +\infty} e^{\gamma a} \mathbb{P}(W \geq a) = \mathbb{E}_{\mathbb{Q}}\left(e^{-\gamma F_{\nu_+}}\right).$$

The right hand side has to be calculated. The expression of the distribution of F_{ν_+} shows the identity

$$\mathbb{E}_{\mathbb{Q}}\left(e^{-\gamma F_{\nu_+}}\right) = \frac{1 - \mathbb{E}_{\mathbb{Q}}\left(e^{-\gamma S_{\nu_+}}\right)}{\gamma \mathbb{E}_{\mathbb{Q}}(S_{\nu_+})}.$$

Wald's Identity $\mathbb{E}_{\mathbb{Q}}(S_{\nu_+}) = \mathbb{E}_{\mathbb{Q}}(\nu_+)\mathbb{E}_{\mathbb{Q}}(X_1)$ and Relationship (3.2) imply

$$\mathbb{E}_{\mathbb{Q}}\left(e^{-\gamma F_{\nu_+}}\right) = \frac{1 - \mathbb{E}\left(e^{-\gamma S_{\nu_+}} e^{\gamma S_{\nu_+}} 1_{\{\nu_+ < +\infty\}}\right)}{\gamma \mathbb{E}(X_1 e^{\gamma X_1})\, \mathbb{E}\left(\nu_+ e^{\gamma S_{\nu_+}} 1_{\{\nu_+ < +\infty\}}\right)},$$

hence

$$\mathbb{E}_{\mathbb{Q}}\left(e^{-\gamma F_{\nu_+}}\right) = \frac{\mathbb{P}(\nu_+ = +\infty)}{\gamma \mathbb{E}(X_1 e^{\gamma X_1})\, \mathbb{E}\left(\nu_+ e^{\gamma S_{\nu_+}} 1_{\{\nu_+ < +\infty\}}\right)},$$

and Identity (3.4) is proved. □

This result is related to the following phenomenon: If T_a is the time to cross level a, when a tends to infinity, the random walk conditioned to cross level a behaves, between 0 and T_a, as a random walk associated with the distribution of X_1 for $\mathbb{Q}$. In particular this random walk has a positive drift. Section 3.6 proves this property.

Application to the M/G/1 and G/M/1 Queues

Throughout this section the mean interarrival time [resp. service time] is $1/\lambda$ [resp. $1/\mu$].

The $G/M/1$ Queue. The equation $\mathbb{E}[\exp(\gamma(\sigma - \tau))] = 1$ can be written as

$$\gamma = \mu\,(1 - \mathbb{E}[\exp(-\gamma\tau)]),$$

in this case γ is the constant β of Identity (2.16) page 46 of Proposition 2.11. The asymptotic result on the tail distribution is trivial since for $x > 0$, $\mathbb{P}(W \geq x) = (1 - \beta/\mu)\exp(-\beta x)$.

The $M/G/1$ Queue. Assuming that $\mathbb{E}(\exp(\theta\sigma)) < +\infty$ for some $\theta > 0$, the rate γ of exponential decay of the tail distribution is given by the positive solution of the fixed point equation

$$\gamma = \lambda\,(\mathbb{E}\,(e^{\gamma\sigma}) - 1)\,.$$

These results suppose that σ has an exponential moment. The next proposition gives the order of magnitude of the tail distribution of W when σ has a tail distribution with a polynomial decay.

PROPOSITION 3.6. *If the arrival process is Poisson and there exist a, $b > 0$ and $q > 1$ such that*

$$\frac{a}{x^q} \leq \mathbb{P}(\sigma \geq x) \leq \frac{b}{x^q},$$

for x sufficiently large, there exist A, $B > 0$ such that

$$\frac{A}{x^{q-1}} \leq \mathbb{P}(W \geq x) \leq \frac{B}{x^{q-1}}.$$

PROOF. Proposition 2.9 and the remark of page 49 shows that W can be represented as $Z_1 + \cdots + Z_G$, where G is geometrically distributed random variable with parameter ρ, independent of the i.i.d. sequence (Z_i) whose common distribution has the density $\mathbb{P}(\sigma \geq x)/\mathbb{E}(\sigma)$, $x \geq 0$ (this is the distribution of S_{ν_+}). The tail distribution of Z_1 has also a polynomial decay:

$$\mathbb{P}(Z_1 \geq x) = \int_x^{+\infty} \frac{\mathbb{P}(\sigma \geq y)}{\mathbb{E}(\sigma)}\,dy,$$

for x sufficiently large,

$$\int_x^{+\infty} \frac{a}{y^q\mathbb{E}(\sigma)}\,dy \leq \mathbb{P}(Z_1 \geq x) \leq \int_x^{+\infty} \frac{b}{y^q\mathbb{E}(\sigma)}\,dy = \frac{b}{(q-1)\mathbb{E}(\sigma)}\frac{1}{x^{q-1}},$$

hence there exist a_1 and $b_1 > 0$ such that for x large enough,

$$\frac{a_1}{x^{q-1}} \leq \mathbb{P}(Z_1 \geq x) \leq \frac{b_1}{x^{q-1}}.$$

According to the representation of the distribution of W, for $x > 0$,

$$\mathbb{P}(W \geq x) = \sum_{n=1}^{+\infty} \rho^n(1-\rho)\mathbb{P}\left(\sum_{i=1}^{n} Z_i \geq x\right),$$

in particular

$$\mathbb{P}(W \geq x) \geq \mathbb{P}(G \geq 1, Z_1 \geq x) = \rho \mathbb{P}(Z_1 \geq x) \geq \frac{\rho a_1}{x^{q-1}}.$$

For $n \geq 1$, the elementary inequality

$$\mathbb{P}\left(\sum_{i=1}^{n} Z_i \geq x\right) \leq \sum_{i=1}^{n} \mathbb{P}\left(Z_i \geq \frac{x}{n}\right) = n\mathbb{P}\left(Z_1 \geq \frac{x}{n}\right) \leq \frac{b_1 n^q}{x^{q-1}},$$

shows the existence of a finite constant $B > 0$ such that $\mathbb{P}(W \geq x) \leq B/x^{q-1}$. The proposition is proved. □

REMARK.

Queueing systems associated with distributions having a non-exponential tail play a quite important role in the modeling of communication networks. They are used, for example, to describe the variability of data transfers in communication links. See Boxma and Dumas [BD98] for a survey and further references.

3.4 The Maximum of a Busy Period

In this section, the tail distribution of the maximum of a busy period is analyzed. The busy periods corresponding to large values of the maximum is described in Section 3.6.

DEFINITION 9. *Condition (C) is satisfied if*

1. $\mathbb{E}(X_1) < 0$ *where* $X_1 = \sigma - \tau$ *and the distribution of* σ *is spread-out (see the definition page 25);*
2. *there exists* $\theta > 0$ *such that* $1 < \mathbb{E}(\exp(\theta\sigma)) < +\infty$.

Throughout the rest of the chapter, γ denotes the unique positive solution of the equation $\mathbb{E}(\exp(\gamma X_1)) = 1$ (Lemma 3.3). If $\mathbb{E}(X_1) < 0$, the stopping time $\nu_- = \inf\{n > 0 : S_n \leq 0\}$ is $\mathbb{P}$-almost surely finite, the random walk (S_n) has a positive excursion on the interval $\{0, \ldots, \nu_- - 1\}$. Using the results of Chapter 2, if a customer begins a busy cycle at $t = 0$, ν_- is the index of the first customer who finds the queue empty after $t = 0$. The main result of this section is the next theorem.

THEOREM 3.7. *Under Condition (C) the maximum of the random walk* (S_n) *on a positive excursion satisfies the relation*

[BD98] O.J. Boxma and V. Dumas, *Fluid queues with long-tailed activity period distributions*, Computer Communications **21** (1998), 1509–1529, Special issue on "Stochastic Analysis and Optimization of Communication Systems".

$$\lim_{a\to+\infty} e^{\gamma a}\mathbb{P}\left(\max_{0\le k<\nu_-} S_k \ge a\right) = \frac{\mathbb{P}(\nu_+ = +\infty)\left(1-\mathbb{E}\left(e^{\gamma S_{\nu_-}}\right)\right)}{\gamma\mathbb{E}\left(X_1 e^{\gamma X_1}\right)\mathbb{E}\left(\nu_+ e^{\gamma S_{\nu_+}} 1_{\{\nu_+<+\infty\}}\right)},$$

where

$$\nu_+ = \inf\{n \ge : S_n > 0\} \quad \textit{and} \quad \nu_- = \inf\{n > 0 : S_n \le 0\}.$$

PROOF. If $n \ge 0$, define $M_n = \max_{0\le k\le n} S_k$. For $a > 0$,

$$\mathbb{P}(M_{\nu_-} < a \le M_\infty) = \mathbb{P}(M_{\nu_-} < a,\ T_a < +\infty),$$

where T_a is, as before, the hitting time of the interval $[a, +\infty[$ by the random walk. Notice that the event

$$\begin{aligned}\{\nu_- < +\infty,\ M_{\nu_-} < a,\ T_a < +\infty\} &= \{M_{\nu_-} < a \le M_\infty\}\\ &= \{M_{\nu_-} < a,\ \nu_- < T_a < +\infty\}\end{aligned}$$

is $\mathcal{F}_{T_a}$-measurable, Identity (3.2) implies that

$$\mathbb{P}(M_{\nu_-} < a \le M_\infty) = \mathbb{E}_\mathbb{Q}\left(e^{-\gamma S_{T_a}} 1_{\{\nu_-<+\infty,\ M_{\nu_-}<a,\ T_a<+\infty\}}\right).$$

Since $\mathbb{E}_\mathbb{Q}(X_1) > 0$, the variable T_a is $\mathbb{Q}$-almost surely finite (Proposition 3.1), therefore,

$$\mathbb{P}(M_{\nu_-} < a \le M_\infty) = e^{-\gamma a}\mathbb{E}_\mathbb{Q}\left(e^{-\gamma(S_{T_a}-a)} 1_{\{\nu_-<+\infty,\ M_{\nu_-}<a\}}\right).$$

From Lebesgue's Theorem, one gets that there exist $a_0 > 0$ and $k_0 \ge 0$ such that, if $a \ge a_0$ and $k \ge k_0$, then

$$\mathbb{E}_\mathbb{Q}\left(\left|1_{\{\nu_-<+\infty,\ M_{\nu_-}<a\}} - 1_{\{\nu_-<+\infty\}}\right|\right) \le \varepsilon \quad\text{and}\quad \mathbb{E}_\mathbb{Q}\left(1_{\{k\le\nu_-<+\infty\}}\right) \le \varepsilon,$$

consequently, for $a \ge a_0$ and $k \ge k_0$ the inequality

$$\left|\mathbb{E}_\mathbb{Q}\left(e^{-\gamma(S_{T_a}-a)} 1_{\{\nu_-<+\infty,\ M_{\nu_-}<a\}}\right) - \mathbb{E}_\mathbb{Q}\left(e^{-\gamma(S_{T_a}-a)} 1_{\{\nu_-<k\}}\right)\right| \le 2\varepsilon$$

is satisfied. The event $\{\nu_- < k\}$ depends only on the variables $X_1, \dots, X_k$, Proposition 1.28 page 26 shows that the variable $S_{T_a} - a$ converges in distribution and is asymptotically independent of the first points of the renewal process, and therefore of the variables $X_1, \dots, X_k$, thus

$$\lim_{a\to+\infty}\mathbb{E}_\mathbb{Q}\left(e^{-\gamma(S_{T_a}-a)} 1_{\{\nu_-<k\}}\right) = \mathbb{E}_\mathbb{Q}\left(e^{-\gamma F_{\nu_+}}\right)\mathbb{E}_\mathbb{Q}\left(1_{\{\nu_-<k\}}\right).$$

The variable F_{ν_+} has the distribution of the first point of the stationary renewal process associated with S_{ν_+} for the probability measure $\mathbb{Q}$. Notice one must check that the distribution of the variable S_{ν_+} is spread-out if the distribution of σ is. By using the fact that the variable ν_+ is not bounded, this is straightforward. Finally one gets,

$$\lim_{a\to+\infty} e^{\gamma a}\mathbb{P}(M_{\nu_-} < a \le M_\infty) = \mathbb{E}_\mathbb{Q}\left(e^{-\gamma F_{\nu_+}}\right)\mathbb{E}_\mathbb{Q}\left(1_{\{\nu_-<+\infty\}}\right),$$

and according to the proof of Theorem 3.5,

$$\lim_{a\to+\infty} e^{\gamma a}\mathbb{P}(M_\infty \geq a) = \mathbb{E}_{\mathbb{Q}}\left[e^{-\gamma F_{\nu_+}}\right],$$

hence,

$$\begin{aligned}\lim_{a\to+\infty} e^{\gamma a}\mathbb{P}(M_{\nu_-} \geq a) &= \mathbb{E}_{\mathbb{Q}}\left[e^{-\gamma F_{\nu_+}}\right]\left(1 - \mathbb{E}_{\mathbb{Q}}\left[1_{\{\nu_-<+\infty\}}\right]\right) \\ &= \frac{\mathbb{P}(\nu_+ = +\infty)\left(1-\mathbb{E}\left[e^{\gamma S_{\nu_-}}\right]\right)}{\gamma\mathbb{E}[X_1 e^{\gamma X_1}]\,\mathbb{E}\left[\nu_+ e^{\gamma S_{\nu_+}} 1_{\{\nu_+<+\infty\}}\right]}.\end{aligned}$$

The theorem is proved. □

Section 3.6 describes more precisely the random walk conditioned to hit a very high level.

3.5 The GI/GI/1 Queue near Saturation

In this section the $GI/GI/1$ queue is analyzed when $\mathbb{E}(X_1) = \mathbb{E}(\sigma_1 - \tau_1)$ is negative but tends to 0. Proposition 2.2 shows that when $\mathbb{E}(\sigma_1) = \mathbb{E}(\tau_1)$, the waiting time of the nth customer grows at rate $\sqrt{n}$. If W_{X_1} denotes the associated stationary waiting time, the variable W_{X_1} should converge in distribution to infinity as $\mathbb{E}(X_1)$ tends to 0. The main result of this section is the convergence in distribution of the variable W_{X_1} properly renormalized. The first results in this domain are also due to Kingman[Kin65].

In the following, (X_1^n) is an i.i.d. sequence of non-degenerated, square integrable random variables such that

1. for $n \geq 0$, $\mathbb{E}(X_1^n) = -1/K_n < 0$;
2. $\lim_{n\to+\infty} \mathbb{E}(X_1^n) = 0$;
3. if $\eta_n = \sqrt{\text{var}(X_1^n)}$, then $\lim_{n\to+\infty} \eta_n = \eta > 0$.

For a fixed $n \geq 0$, the sequence (X_i^n) is i.i.d. and (W_i^n) is the sequence of corresponding waiting times,

$$W_0^n = 0, \quad W_i^n = \left(W_{i-1}^n + X_i^n\right)^+,$$

for $i \geq 1$. If $M_i^n = -W_i^n\mathbb{E}(X_1^n)/\eta_n$, since $\mathbb{E}(X_1^n) < 0$, the above equations can be rewritten as

$$M_0^n = 0, \quad M_i^n = \left(M_{i-1}^n - \frac{\mathbb{E}(X_1^n)}{\eta_n}X_i^n\right)^+,$$

[Kin65] J.F.C. Kingman, *The heavy traffic approximation in the theory of queues*, Proc. Symp. on Congestion theory (Chapel Hill), Univ. of North Carolina Press, 1965, pp. 137–169.

for $i \geq 1$. After this scaling on the space variable, the time scale is changed in the following way, if $M^n(t) = M^n_{\lfloor tK_n^2 \rfloor}$ for $t \geq 0$, then the process $(M^n(t))$ is approximately a diffusion.

THEOREM 3.8. *Under the above assumptions, for $T > 0$, when n tends to infinity, the process $(M^n(t); 0 \leq t \leq T)$ converges in distribution to the process $(R(t); 0 \leq t \leq T)$ which is the Brownian motion with drift $(B(t) - t/\eta)$ reflected at 0, i.e.*

$$R(t) = \sup_{0 \leq s \leq t} \left[B(t) - t/\eta - (B(s) - s/\eta) \right], \tag{3.7}$$

where $(B(t))$ is the standard Brownian motion starting from 0.

Appendix D page 375 presents definitions and results concerning reflected processes.

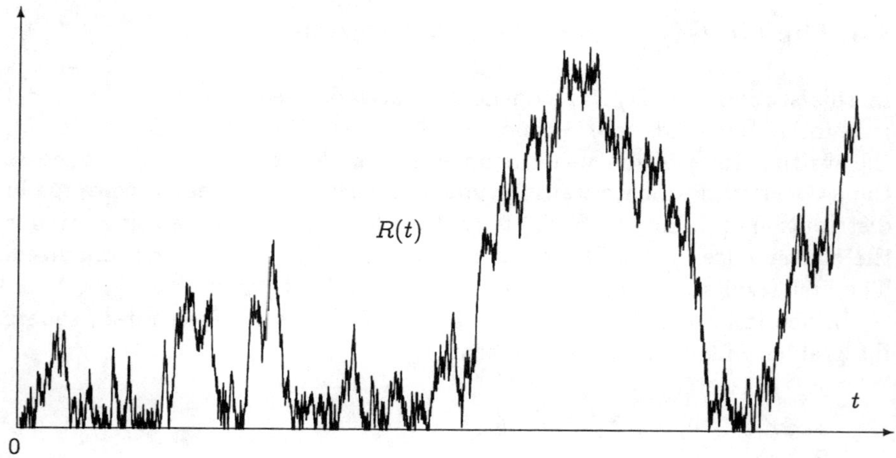

FIG. 3.2. A Sample Path of $(R(t))$

PROOF. By iterating the recurrence relation defining the waiting times (W_i^n), one gets the well known identity (cf. Relationship 2.3 page 30),

$$W_i^n = \sup_{1 \leq k \leq i} \left(\sum_{j=k}^{i} X_j^n \right) \vee 0,$$

since $\mathbb{E}(X_1^n) < 0$,

$$M^n(t) = \sup_{1\leq k\leq \lfloor tK_n^2\rfloor}\left(\frac{1}{\eta_n K_n}\sum_{j=k}^{\lfloor tK_n^2\rfloor} X_j^n\right)\vee 0$$

$$(3.8)\qquad = \sup_{1\leq k\leq \lfloor tK_n^2\rfloor}\frac{1}{\eta_n}\left(\frac{1}{K_n}\sum_{j=k}^{\lfloor tK_n^2\rfloor}(X_j^n - \mathbb{E}(X_1^n)) - \frac{(\lfloor tK_n^2\rfloor - k)}{K_n^2}\right)\vee 0.$$

According to Donsker's Theorem, see Billingsley [7], for $T > 0$, when n tends to infinity, on the interval $[0,T]$ the process

$$(D^n(t)) \stackrel{\text{def.}}{=} \left(\frac{1}{\eta_n K_n}\sum_{1}^{\lfloor tK_n^2\rfloor}(X_i^n - \mathbb{E}(X_1^n)) - \frac{\lfloor tK_n^2\rfloor}{K_n^2}\right)$$

converges in distribution to the Brownian motion with drift

$$(B(t) - t/\eta\,;\, 0\leq t\leq T),$$

where $(B(t))$ is the standard Brownian motion starting from 0. Identity (3.8) shows that $(M^n(t))$ can be expressed as a functional of $(D^n(t))$ in the following way :

$$M^n(t) = \sup_{0\leq s\leq t}(D^n(t) - D^n(s)).$$

On the space of functions on $[0,T]$, continuous on the right and with left limits endowed with the topology of uniform convergence, the functional defined by

$$f \longrightarrow \psi(f),\quad \psi(f)(t) = \sup_{0\leq s\leq t}(f(t) - f(s)),$$

is clearly continuous. (See Appendix C or Billingsley [7]). As a consequence, when n tends to infinity, the process $(M^n(t);\, 0\leq t\leq T)$ converges in distribution to $(R(t);\, 0\leq t\leq T)$ defined by Equation (3.7). □

For $t\geq 0$,

$$R(t) = \sup_{0\leq s\leq t}(B(t) - B(s) - (t-s)/\eta) = \sup_{0\leq s\leq t}(B(t) - B(t-s) - s/\eta),$$

since $(B(t) - B(t-s);\, 0\leq s\leq t)$ has the same distribution as a Brownian motion on $[0,t]$, the variable $R(t)$ has the same distribution as

$$\sup\{B(s) - s/\eta : 0\leq s\leq t\},$$

therefore, it converges in distribution to $R = \sup(B(t) - t/\eta : t\geq 0)$. A classical result on the scaling of Brownian motion shows that $(tB(1/t))$ is also a Brownian motion starting from 0. In particular, $\mathbb{P}$-almost surely, $tB(1/t)$ converges to 0 as t goes to 0, thus $B(t)/t$ converges also to 0 when t goes to infinity, hence $\mathbb{P}$-a.s.

$$\lim_{t\to+\infty} B(t) - t = -\infty.$$

The variable R is therefore almost surely finite.

The distribution of R is the invariant measure of the process $(R(t))$. Therefore, this suggests that the variables $(W^n\mathbb{E}(X_1^n)/\eta_n)$ should converge in distribution to this invariant measure. Another way of stating this result is to say that the diagram below commutes,

$$\begin{array}{ccc} (M^n(t)) & \xrightarrow{t\to+\infty} & W^n\,\mathbb{E}(X_1^n)/\eta_n \\ {\scriptstyle n\to+\infty}\big\downarrow & & {\scriptstyle n\to+\infty}\big\downarrow \\ (R(t)) & \xrightarrow{t\to+\infty} & R. \end{array}$$

The following lemma gives the distribution of R.

LEMMA 3.9. *If $(B(t))$ is a standard Brownian motion starting from 0, the distribution of the random variable $R = \sup\{B(t)-t : t\geq 0\}$ is exponentially distributed with parameter 2.*

PROOF. It is easy to check that $(\exp(2(B(t)-t)))$ is a martingale. For $a>0$, the variable $H_a = \inf\{t\geq 0 : B(t)-t\geq a\}$ is a stopping time with respect to the natural filtration of Brownian motion $(B(t))$. Doob's optional stopping Theorem shows that, for $t\geq 0$,

$$\mathbb{E}\left[\exp(2(B(H_a\wedge t)-H_a\wedge t))\right] = 1,$$

since $\lim_{t\to+\infty} B(t)-t = -\infty$, one gets $\mathbb{P}$-a.s.

$$\lim_{t\to+\infty} \exp(2(B(H_a\wedge t)-H_a\wedge t)) = e^{2a}1_{\{H_a<+\infty\}}.$$

The positive variable $\exp(2(B(H_a\wedge t))-H_a\wedge t))$ being bounded by $\exp(2a)$, from Lebesgue's Theorem, by letting t go to infinity one obtains

$$\mathbb{E}\left(e^{2a}1_{\{H_a<+\infty\}}\right) = 1,$$

hence $\mathbb{P}(H_a<+\infty) = \mathbb{P}(R\geq a) = \exp(-2a)$. The lemma is proved. □

PROPOSITION 3.10. *If W^n is a random variable distributed as the stationary waiting time associated with the random variable X_1^n, under Assumptions 1, 2 and 3 page 69, the variable*

$$\frac{-\mathbb{E}(X_1^n)}{\eta}W^n$$

converges in distribution to an exponentially distributed random variable with parameter 2.

This results shows that if $\rho = \mathbb{E}(\sigma)/\mathbb{E}(\tau)$ is very close to 1, the stationary waiting time of the $GI/GI/1$ queue is approximately $\lambda\eta^2 V/(1-\rho)$ where V is an exponentially distributed random variable with parameter 2.

PROOF. According to Relationship (3.8), the variable

$$\frac{-\mathbb{E}(X_1^n)}{\sqrt{\operatorname{var}(X_1^n)}}\,W^n$$

has the same distribution as

$$\sup_{t\geq 0}\frac{1}{\eta_n}\left(\sum_{j=1}^{\lfloor tK_n^2\rfloor}-X_j^n\mathbb{E}(X_1^n)\right)=\sup_{t\geq 0}\frac{1}{\eta_n}\left(\frac{1}{K_n}\sum_{j=1}^{\lfloor tK_n^2\rfloor}(X_j^n-\mathbb{E}(X_1^n))-\frac{\lfloor tK_n^2\rfloor}{K_n^2}\right).$$

Theorem 3.8 shows that for a fixed T, the convergence

$$\sup_{0\leq t\leq T}\frac{1}{\eta_n}\left(\frac{1}{K_n}\sum_{j=1}^{\lfloor tK_n^2\rfloor}(X_j^n-\mathbb{E}(X_1^n))-\frac{\lfloor tK_n^2\rfloor}{K_n^2}\right)\overset{\text{dist.}}{\longrightarrow}\sup_{0\leq t\leq T}\left(B(t)-\frac{t}{\eta}\right),$$

holds as n tends to infinity. The main result of the proposition is that the convergence also holds for $T=+\infty$. In other words, the residual terms

$$\sup_{k\geq\lfloor TK_n^2\rfloor}\frac{1}{\eta_n}\left(\sum_{j=1}^{k}-X_j^n\mathbb{E}(X_1^n)\right)\text{ and }\sup_{t\geq T}\left(B(t)-\frac{t}{\eta}\right),$$

do not contribute really to the maximum on the whole real half-line. This is clearly true for the second term if T is sufficiently large. For the first term, this is also true if it is uniformly negative in n as T is large.

A martingale inequality is used to prove this assertion. The martingale presented here can also be used to prove the law of large numbers (see Durrett [**17**] for such a proof and Asmussen [**3**] for its use in this framework). For a fixed n and $k\geq 0$, if $Y_j^n=-X_j^n\mathbb{E}(X_1^n)+\mathbb{E}(X_1^n)^2$,

$$S_k=Y_1^n+Y_2^n+\cdots+Y_k^n,$$

and $\mathcal{G}_k$ is the σ-field generated by the variables $S_k,Y_{k+1}^n,Y_{k+2}^n,\ldots$ The filtration $(\mathcal{G}_k)$ is non-increasing and S_k is $\mathcal{G}_k$-measurable.

Assertion. The sequence (S_k/k) is a backwards martingale for $(\mathcal{G}_k)$, i.e. if $k\geq 1$,

$$\mathbb{E}(\,S_{k-1}/(k-1)\mid\mathcal{G}_k)=S_k/k.$$

Since $S_{k-1}=S_k-Y_k^n$, it is sufficient to express $\mathbb{E}(Y_k^n/\mathcal{G}_k)$. By symmetry, $\mathbb{E}(Y_j^n\mid\mathcal{G}_k)=\mathbb{E}(Y_k^n\mid\mathcal{G}_k)$ holds for any $j\leq k$. The sum S_k of the Y_j^n for $j\leq k$ is $\mathcal{G}_k$-measurable, hence $\mathbb{E}(Y_k^n\mid\mathcal{G}_k)=S_k/k$, therefore,

$$\mathbb{E}\left(\frac{S_{k-1}}{k-1}\,\middle|\,\mathcal{G}_k\right)=\frac{S_k-S_k/k}{k-1}=\frac{S_k}{k}.$$

The assertion is proved. The sequence $\left((S_n/n)^2\right)$ is therefore a backwards sub-martingale, Doob's inequality implies therefore the identity

$$\mathbb{P}\left(\sup_{j\geq k}\frac{|S_j|}{j}>a\right)\leq\frac{1}{a^2}\frac{\mathbb{E}(S_k^2)}{k^2}. \tag{3.9}$$

Going back to W^n, one has

$$\mathbb{P}\left(\sup_{k\geq\lfloor TK_n^2\rfloor}\sum_{j=1}^{k}-X_j^n\mathbb{E}(X_1^n)\geq 0\right)=\mathbb{P}\left(\sup_{k\geq\lfloor TK_n^2\rfloor}\frac{1}{k}\sum_{j=1}^{k}Y_j^n\geq\mathbb{E}(X_1^n)^2\right)$$

and Inequality (3.9) shows that

$$\mathbb{P}\left(\sup_{k\geq \lfloor TK_n^2\rfloor}\sum_{j=1}^{k} -X_j^n\mathbb{E}(X_1^n)\geq 0\right)$$
$$\leq \frac{1}{\mathbb{E}(X_1^n)^4}\frac{\mathbb{E}(X_1^n)^2\operatorname{var}(X_1^n)}{\lfloor TK_n^2\rfloor} = \frac{K_n^2}{\lfloor TK_n^2\rfloor}\operatorname{var}(X_1^n).$$

The sequence $(\operatorname{var}(X_1^n))$ converges to η^2 and $(K_n^2/\lfloor TK_n^2\rfloor)$ to $1/T$. Thus, for $\varepsilon>0$, there exists T_0 such that, for $T\geq T_0$,

$$\mathbb{P}\left(\sup_{k\geq \lfloor TK_n^2\rfloor}\sum_{j=1}^{k} -X_j^n\mathbb{E}(X_1^n)\geq 0\right)\leq \varepsilon, \quad \text{for all } n\in\mathbb{N}, \text{ and}$$
$$\mathbb{P}\left(\sup_{s\geq T}(B(t)-t/\eta)\right)\leq\varepsilon.$$

One obtains that, for $a\geq 0$,

$$\left|\mathbb{P}\left(-\frac{\mathbb{E}(X_1^n)}{\sqrt{\operatorname{var}(X_1^n)}}W_n\geq a\right)-\mathbb{P}\left(\sup_{0\leq t}(B(t)-t)\geq a\right)\right|$$
$$\leq 2\varepsilon+\left|\mathbb{P}\left(\sup_{0\leq t\leq T}\sum_{j=1}^{\lfloor tK_n^2\rfloor} -X_j^n\mathbb{E}(X_1^n)\geq a\right)-\mathbb{P}\left(\sup_{0\leq t\leq T}(B(t)-t)\geq a\right)\right|,$$

and this last term converges to 0 as n tends to infinity. The convergence in distribution is therefore proved.

It has to be shown that the variable

$$X = 1/\eta \sup_{t\geq 0}(B(t)-t/\eta)$$

has an exponential distribution with parameter 2. The scaling property of Brownian motion implies that $(\eta B(t/\eta^2))$ is also a Brownian motion. One gets that X has the same distribution as

$$\sup_{t\geq 0}\left(B(t/\eta^2)-t/\eta^2\right)=\sup_{t\geq 0}(B(t)-t).$$

Lemma 3.9 concludes the proof. □

The convergence of the sequence (W_n) properly renormalized is achieved mainly with a martingale inequality. If this technical point is important, the crucial part of the proof relies on the explicit representation of the variable W_n as a functional of the random walk associated with X_1^n.

3.6 The Random Walk Conditioned to Hit Level a

The notations and definitions of Section 3.3 are used. The random walk conditioned to cross the level a for a large is described in this section. This conditional random walk is crucial (but somewhat hidden) for the results obtained in Sections 3.3 and 3.4.

THEOREM 3.11. *If* $T_a = \inf\{n : S_n \geq a\}$ *and Condition (C) page 67 is satisfied,*

1. *conditionally on the event* $\{T_a < +\infty\}$, *as* a *tends to infinity, the sequence* $(S_{n\wedge T_a})$ *converges in distribution to the distribution of the stopped random walk* $(S_{n\wedge T_a})$ *for the probability measure* $\mathbb{Q}$.
2. *The conditional distribution of the sequence* (S_{n+T_a}) *given the event* $\{T_a < +\infty\}$ *is the distribution of the random walk* (S_n) *starting from* S_{T_a} *for the initial probability* $\mathbb{P}$.

This result shows that the conditional sample path of a random walk with a negative drift and a large maximum can be decomposed into two pieces: The first part is a random walk with a positive drift $\mathbb{E}_\mathbb{Q}(X_1)$ and the second is a sample path going to 0 of the unbiased random walk (thus with a negative drift) starting above a.

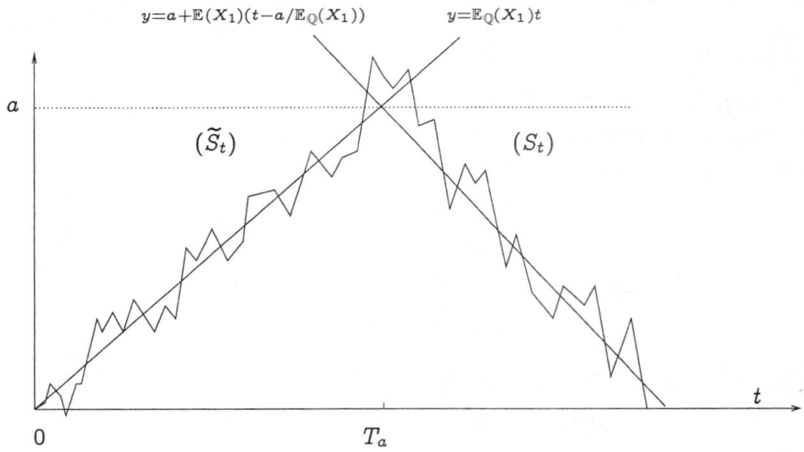

FIG. 3.3. The Random Walk Conditioned to Hit a

PROOF. Assertion 2 of the theorem is just the strong Markov property of the random walk (Lemma 2.3 page 33).

To prove a) of the theorem, it is sufficient to prove that if $K \in \mathbb{N}$ and $f : \mathbb{R}^\mathbb{N} \to \mathbb{R}_+$ is a bounded Borelian function depending only of the K first coordinates, the relation

$$\lim_{a\to+\infty} \mathbb{E}\left[f((S_{n\wedge T_a}))\,|\,T_a < +\infty\right] = \mathbb{E}_\mathbb{Q}(f(S_n))$$

holds. Identity (3.2) yields, T_a being $\mathbb{Q}$-a.s. finite, that

$$\mathbb{E}\left[f((S_{n\wedge T_a}))1_{\{T_a<+\infty\}}\right] = \mathbb{E}_{\mathbb{Q}}\left[f((S_{n\wedge T_a}))e^{-\gamma S_{T_a}}\right].$$

Proposition 1.28 shows that, as a tends to infinity, under the probability measure $\mathbb{Q}$ the vector $(S_{T_a}-a,(X_i,0\leq i\leq K))$ converges in distribution to the distribution of the vector $(F_{\nu_+},(X_i,0\leq i\leq K))$, where F_{ν_+} and $X_i, 0\leq i\leq K$ are independent and F_{ν_+} is distributed as the first point of the stationary renewal process associated with the variable S_{ν_+}. The last identity gives the relation

$$\mathbb{E}(f((S_{n\wedge T_a}))\,|\,T_a<+\infty) = \frac{\mathbb{E}_{\mathbb{Q}}\left(f(S_{n\wedge T_a})e^{-\gamma(S_{T_a}-a)}\right)}{\mathbb{E}_{\mathbb{Q}}\left(e^{-\gamma(S_{T_a}-a)}\right)},$$

and, by letting a go to infinity, one gets

$$\lim_{a\to+\infty}\mathbb{E}[f((S_{n\wedge T_a}))\,|\,T_a<+\infty] = \frac{\mathbb{E}_{\mathbb{Q}}[f((S_n))]\,\mathbb{E}_{\mathbb{Q}}\left[e^{-\gamma F_{\nu_+}}\right]}{\mathbb{E}_{\mathbb{Q}}\left[e^{-\gamma F_{\nu_+}}\right]} = \mathbb{E}_{\mathbb{Q}}[f((S_n))].$$

The theorem is proved. □

Conditioning the random walk to hit a very large level transforms it, in the ascending part, into a random walk with a positive drift. The next result shows that forbidding the return to 0 transforms the random walk in a Markov chain.

PROPOSITION 3.12. *If X_1 is a variable such that $\mathbb{E}(X_1)>0$, the random walk (S_n) conditioned to never return to 0 is a Markov chain (M_n), its transition probability $r(x,dy)$ is given by*

$$r(x,dy) = \frac{\mathbb{P}_y(\widetilde{\nu}_-=+\infty)}{\mathbb{P}_x(\widetilde{\nu}_-=+\infty)}\,\mathbb{P}(x+X_1\in dy),$$

with $\mathbb{P}_x(\cdot)=\mathbb{P}(\cdot\,|\,S_0=x)$ and

$$\widetilde{\nu}_- = \inf\{k\geq 0 : M_0+S_k\leq 0\}.$$

The above identity defines a probability distribution since, for $x>0$,

$$\mathbb{P}_x(\widetilde{\nu}_-=+\infty) = \mathbb{E}\left[\mathbb{P}_{x+X_1}(\widetilde{\nu}_-=+\infty)\right],$$

thus, $r(x,\mathbb{R}) = \mathbb{E}[\mathbb{P}_{x+X_1}(\widetilde{\nu}_-=+\infty)]/\mathbb{P}_x(\widetilde{\nu}_-=+\infty) = 1$. Note that, of course, $r(x,]-\infty,0])=0$ for $x>0$.

PROOF. The Markov property is a direct consequence of Proposition B.7 page 365 (this proposition is shown for a continuous time process, but it can easily be adapted to the discrete time case). If f is a bounded Borelian function on $\mathbb{R}_+$ and $x>0$,

$$\mathbb{E}\left(f(x+X_1)1_{\{T_0=+\infty\}}\right) = \int_{]0,+\infty[} f(y)\mathbb{P}_y(T_0=+\infty)\,\mathbb{P}(x+X_1\in dy),$$

consequently,

$$\begin{aligned}\int f(y)\, r(x, dy) &= \mathbb{E}(f(x + X_1) \,|\, T_0 = +\infty) \\ &= \int_{]0,+\infty[} f(y) \frac{\mathbb{P}_y(T_0 = +\infty)}{\mathbb{P}_x(T_0 = +\infty)} \mathbb{P}(x + X_1 \in dy).\end{aligned}$$

This completes the proof of the proposition. □

4. Stochastic Networks and Reversibility

Contents

4.1 Introduction

This chapter investigates the equilibrium behavior of some queueing networks. The state of these networks is described by a Markov jump process

$$(X(t)) = \big(X_i(t); 1 \leq i \leq N\big)$$

with values in $\mathbb{N}^N$, for $i \leq N$, $X_i(t)$ is the number of customers at the ith node of the network. The results concerning Poisson point processes are recalled in Chapter 1 and in Appendix B for the Markov jump processes.

The dynamic of these networks can be easily described: Customers arrive at one of the queues of the network, where they wait before being served then, they go to another queue or leave the network. With these transitions, for a fixed $t \geq 0$, the variables $X_i(t)$ are very likely to be correlated. Nevertheless, if arrival processes are Poisson and services are exponentially distributed, some of these queueing networks have the remarkable property of having an invariant measure with a product form. This implies that, at equilibrium, for a fixed t, the variables $X_i(t)$ are independent.

This result on the invariant measure with a product form is related to some reversibility properties of the Markov processes. The main results concerning reversibility are recalled in Section 1. As a consequence, Burke's Theorem on the Poisson property of the departure process of an $M/M/1$ queue at equilibrium is proved. Kelly [31] and Whittle [54] are good references on this subject.

4.2 Reversibility of Markov Processes

In the following, $(X(t))$ is a Markov process in a countable state space $\mathcal{S}$, continuous on the right with a limit on the left everywhere (càdlàg to use French notation, see Definition 40 page 359). The quantity $X(t-)$ denotes the left limit of X at t. This process is assumed to be homogeneous, irreducible and its Q-matrix is denoted by $Q = (q_{ij})$. When it exists, the invariant measure $(\pi(i); i \in \mathcal{S})$ of the Markov process satisfies the equilibrium equations, *the global balance equations*,

$$\pi(i)\left(\sum_{j\neq i} q_{ij}\right) = \sum_{j\neq i} \pi(j) q_{ji}, \tag{4.1}$$

for any $i \in \mathcal{S}$. See Section B.2 page 362.

4.2.1 The Reversibility Property

The next proposition shows that reversing the time for a *stationary* Markov process preserves the Markov property and the time-homogeneity of the transitions. Recall that a Markov process $(X(t))$ is stationary if the distribution of $X(0)$ is the invariant probability distribution of the process.

PROPOSITION 4.1. *If $(X(t))$ is a stationary Markov jump process, for $T > 0$, the reversed process*

$$\big(X^*(t)\big) = \big(X\left((T-t)-\right)\big)$$

is a Markov process defined on time interval $[0,T]$ with a Q-matrix defined by $Q^* = (q^*_{ij}; i,j \in \mathcal{S})$, with

$$q^*_{ij} = q_{ji}\frac{\pi(j)}{\pi(i)}, \tag{4.2}$$

for i, $j \in \mathcal{S}$, where $(\pi(i); i \in \mathcal{S})$ is the invariant probability of the Markov process $(X(t))$.

With a regularization on the left with "$-$", such a process $\big(X^*(t); 0 \leq t \leq T\big)$ is indeed continuous on the right. This regularization does not change the distribution of the process since for $t \geq 0$, $\mathbb{P}\big(X(t) = X(t-)\big) = 1$. (See Appendix B page 359 on the jump processes.) The variables $X(t)$ and $X(t-)$ have the same distribution. More generally, for $0 \leq t_1 \leq t_2 \leq \cdots \leq t_n$, the vectors $\big(X(t_1),\ldots,X(t_n)\big)$ and $\big(X(t_1-),\ldots,X(t_n-)\big)$ have the same distribution.

PROOF. It is sufficient to prove that $\big(X(T-t);0 \le t \le T\big)$ is a Markov process, the same property is also true for the regularized version since they have the same distribution. For n-tuples $0 \le t_1 \le t_2 \le \cdots \le t_n \le T$ and $i_1,\dots,i_n \in \mathcal{S}$,

$$(4.3)\quad \mathbb{P}\big(X(T-t_n)=i_n \mid X(T-t_{n-1})=i_{n-1},\dots,X(T-t_1)=i_1\big) = \frac{\mathbb{P}\big(X(T-t_n)=i_n,\dots,X(T-t_1)=i_1\big)}{\mathbb{P}\big(X(T-t_{n-1})=i_{n-1},\dots,X(T-t_1)=i_1\big)}.$$

The numerator of the right hand side of this equality can be written as

$$\mathbb{P}\big(X(T-t_1)=i_1 \mid X(T-t_n)=i_n,\dots,X(T-t_2)=i_2\big) \times \mathbb{P}\big(X(T-t_n)=i_n,\dots,X(T-t_2)=i_2\big),$$

by the Markov property of $(X(t))$, this quantity is

$$\mathbb{P}\big(X(T-t_1)=i_1 \mid X(T-t_2)=i_2\big)\mathbb{P}\big(X(T-t_n)=i_n,\dots,X(T-t_2)=i_2\big).$$

The right hand side of Relationship (4.3) is

$$\frac{\mathbb{P}\big(X(T-t_n)=i_n,\dots,X(T-t_2)=i_2\big)}{\mathbb{P}\big(X(T-t_{n-1})=i_{n-1},\dots,X(T-t_2)=i_2\big)} = \mathbb{P}\big(X(T-t_n)=i_n \mid X(T-t_{n-1})=i_{n-1},\dots,X(T-t_2)=i_2\big).$$

By induction, one finally gets

$$\mathbb{P}\left(X(T-t_n)=i_n \mid X(T-t_{n-1})=i_{n-1},\dots,X(T-t_1)=i_1\right) = \mathbb{P}\big(X(T-t_n)=i_n \mid X(T-t_{n-1})=i_{n-1}\big),$$

and therefore the Markov property of the process $(X(T-t);0\le t\le T)$.

The left continuity of $(X(t-))$ yields

$$\lim_{s_n\searrow t_n,\dots,s_1\searrow t_1} \mathbb{P}\big(X(T-s_n)=i_n,\dots,X(T-s_1)=i_1\big) = \mathbb{P}\big(X^*(t_n)=i_n,\dots,X^*(t_1)=i_1\big).$$

Consequently, the process $(X^*(t);0\le t\le T)$ is Markov because this property holds for process $(X(T-t);0\le t\le T)$.

The transitions of this Markov process are easy to derive. For $0<t<T$, $0<h\le T-t$ and $i,j\in\mathcal{S}$,

$$\mathbb{P}\big(X^*(t+h)=i \mid X^*(t)=j\big) = \frac{\mathbb{P}\big(X((T-t-h)-)=i, X((T-t)-)=j\big)}{\mathbb{P}(X((T-t)-)=j)} = \frac{\lim_{s\nearrow T-t}\mathbb{P}(X(s-h)=i,X(s)=j)}{\lim_{s\nearrow T-t}\mathbb{P}(X(s)=j)},$$

the process $(X(t))$ being stationary (see Section 10.4 of Chapter 10 page 297),

$$\mathbb{P}\big(X(s-h)=i,X(s)=j\big)=\mathbb{P}\big(X(0)=i,X(h)=j\big),$$

one gets

$$\mathbb{P}\big(X^*(t+h) = i \mid X^*(t) = j\big) = \frac{\mathbb{P}\big(X(0) = i, X(h) = j\big)}{\mathbb{P}\big(X(0) = j\big)}$$
$$= \mathbb{P}\big(X(h) = j \mid X(0) = i\big)\frac{\pi(i)}{\pi(j)}.$$

Dividing by h and letting h go to 0 show that the reversed process is indeed time homogeneous and that its Q-matrix is given by the matrix Q^*. □

REMARK.
It is important to note that the process $(X^*(t))$ is defined as the time-reversed version of a *stationary* version of the Markov process with Q-matrix Q. In particular, a Markov process with Q-matrix Q^* and starting from some fixed initial state $x \in \mathcal{S}$ *cannot* be interpreted a priori as a time-reversed version of some Markov process with Q-matrix Q. For a non-stationary Markov process with Q-matrix Q, the Markov property of the time-reversed process holds without the stationarity assumption (see the proof of the Markovian property above). But the time homogeneity of the transitions is a consequence of the stationarity assumption. Time-reversed versions of non-stationary Markov processes are considered by Nagasawa[Nag64] and Chung and Walsh[CW69].

DEFINITION 10. *A stochastic process $(X(t))$ is reversible if, for any $T > 0$ and $n \in \mathbb{N}$ and every sequence $0 \leq t_1 \leq t_2 \leq \cdots \leq t_n \leq T$, the identity in distribution*

$$(X(t_1), X(t_2), \ldots, X(t_n)) \stackrel{dist.}{=} (X^*(t_1), X^*(t_2), \ldots, X^*(t_n))$$

holds.

Reversibility is the equality in distribution of the process $(X(t); 0 \leq t \leq T)$ and the reversed process $(X^*(t); 0 \leq t \leq T)$ for all $T > 0$.

PROPOSITION 4.2. *A reversible process $(X(t))$ is stationary.*

PROOF. As before, just to ease the notation, the $-$ of $X(t-)$ is removed. If $0 \leq t_1 \leq t_2 \leq \cdots \leq t_n$, $t \in \mathbb{R}$ and $T > t_n$, according to the reversibility property

$$\big(X(T-t_1), X(T-t_2), \ldots, X(T-t_n)\big) \stackrel{\text{dist.}}{=} (X(t_1), X(t_2), \ldots, X(t_n)),$$

consequently, for $t \in \mathbb{R}$,

$$\big[X(T+t-(t+t_1)), X(T+t-(t+t_2)), \ldots, X(T+t-(t+t_n))\big]$$
$$\stackrel{\text{dist.}}{=} \big[(X(t+t_1), X(t+t_2), \ldots, X(t+t_n)\big].$$

[Nag64] M. Nagasawa, *Time reversions of Markov processes*, Nagoya Mathematical Journal **24** (1964), 177–204.

[CW69] K.L. Chung and J.B. Walsh, *To reverse a Markov process*, Acta Mathematica **123** (1969), 225–251.

Hence the identity

$$\big(X(t_1), X(t_2), \ldots, X(t_n)\big) \stackrel{\text{dist.}}{=} \big(X(t+t_1), X(t+t_2), \ldots, X(t+t_n)\big),$$

holds. The process $(X(t))$ is therefore stationary. □

If the invariant probability distribution is known, the reversibility property can be checked very easily with the $\mathcal{Q}$-matrix (provided that the state space is not too complicated of course).

THEOREM 4.3. *A Markov process is reversible if and only if it is stationary and there exists a positive sequence $(\pi(i); i \in \mathcal{S})$ such that*

$$\pi(i)q_{ij} = \pi(j)q_{ji}, \tag{4.4}$$

for i, $j \in \mathcal{S}$ and $\sum_{i\in\mathcal{S}} \pi(i) = 1$.

PROOF. If $(X(t))$ is reversible, according to the last proposition, this process is necessarily stationary. Its invariant measure is denoted by $(\pi(i); i \in \mathcal{S})$. For i, $j \in \mathcal{S}, i \neq j$, the identity

$$\mathbb{P}(X(0) = i, X(t) = j) = \mathbb{P}(X(0) = j, X(t) = i)$$

holds because of the reversibility, hence

$$\pi(i)\mathbb{P}(X(t) = j \mid X(0) = i) = \pi(j)\mathbb{P}(X(t) = i \mid X(0) = j).$$

Dividing by t and letting t go to 0 give Equation (4.4).

Conversely if $(X(t))$ is a Markov process satisfying Relationship (4.4), by summing up Relationship (4.4) on $j \in \mathcal{S}$, one obtains that $(\pi(i))$ is the invariant probability of the jump process. If the distribution of $X(0)$ is $(\pi(i); i \in \mathcal{S})$, according to Equations (4.2) and (4.4), the process $(X(t))$ and the reversed process are càdlàg Markov processes with the same $\mathcal{Q}$-matrix and same initial distribution (the stationary distribution). Proposition B.2 page 361 shows that the processes $(X(t))$ and $(X^*(t))$ have the same distribution, therefore $(X(t))$ is reversible. □

An Operator Point of View. The reversibility property can be translated into a functional analysis framework through some properties of the associated functional operators. A sketch of the correspondence between these properties is presented.

If $(X(t))$ is a reversible Markov process with $\mathcal{Q}$-matrix Q and invariant probability distribution π, for $t \geq 0$ the operator P_t is defined by

$$P_t(f)(i) = \mathbb{E}_i[f(X(t))], \quad i \in \mathcal{S},$$

for a bounded real function f on $\mathcal{S}$, where, as usual, $\mathbb{E}_i(\cdot) = \mathbb{E}(\cdot\,|X(0) = i)$ and $\mathbb{E}_\pi$ is the expected value when the distribution of $X(0)$ is π. The Markov property of $(X(t))$ implies that these operators have the semi-group property, for s and $t \in \mathbb{R}_+$,

$$P_{t+s} = P_t \circ P_s.$$

See Appendix B for the main definitions and results on Markov processes. The scalar product of $L^2(\pi)$ is denoted by $<\cdot,\cdot>_\pi$:

$$<f,g>_\pi = \sum_{i\in\mathcal{S}} \pi(i)f(i)g(i), \quad f,g \in L^2(\pi).$$

For $t \geq 0$, Markov property, stationarity and reversibility (with $T = 2t$) used successively give, for f and $g \in L^2(\pi)$,

$$\begin{aligned}<f,P_t(g)>_\pi &= \sum_{i\in\mathcal{S}} \pi(i)f(i)\mathbb{E}_i[g(X(t))]\\ &= \mathbb{E}_\pi\left[f(X(0))g(X(t))\right] = \mathbb{E}_\pi\left[f(X(t))g(X(2t))\right]\\ &= \mathbb{E}_\pi\left[f(X(t))g(X(0))\right],\end{aligned}$$

therefore $<f,P_t(g)>_\pi = <P_t(f),g>_\pi$.

Thus, reversibility implies that the operators $(P_t,\, t \geq 0)$ are self-adjoint.

If f and g are non-negative functions on $\mathcal{S}$, if

$$Q(f)(i) = \sum_{j\in\mathcal{S}} q_{ij} f(j),$$

Relationship (4.4) gives

$$<f,Q(g)>_\pi = \sum_{i,j\in\mathcal{S}} \pi(i)q_{ij}f(i)g(j) = \sum_{i,j\in\mathcal{S}} \pi(j)q_{ji}f(i)g(j) = <Q(f),g>_\pi,$$

the infinitesimal generator Q is also a self-adjoint operator on an appropriate function space. It is not difficult to see that this property is in fact equivalent to reversibility.

Birth and Death Processes. A birth and death process is a Markov jump process on $\mathbb{N}$ such that q_{ij} is 0 except if $|i-j| \leq 1$ (jumps on the neighbors only). The Q-matrix has an invariant probability distribution if the equations

$$(q_{i,i+1} + q_{i,i-1})\,\pi(i) = q_{i+1,i}\,\pi(i+1) + q_{i-1,i}\,\pi(i-1),$$

for $i \in \mathbb{N}$, have a positive solution with a finite sum. They can be rewritten as

$$q_{i,i-1}\,\pi(i) - q_{i-1,i}\,\pi(i-1) = q_{i+1,i}\,\pi(i+1) - q_{i,i+1}\,\pi(i).$$

The left hand side is thus constant with i and, since it is 0 for $i = 0$, one obtains

$$q_{i+1,i}\,\pi(i+1) = q_{i,i+1}\,\pi(i),$$

which is the reversibility criterion seen before. Consequently, if a birth and death process has an invariant probability measure, it is necessarily reversible (see remark page 362).

The $M/M/K$ Queue. This queue has K servers, customers arrive according to a Poisson process with parameter λ. At most K of them can be served simultaneously. Services are exponentially distributed with parameter μ. In this case, the equilibrium equations are

$$\begin{aligned}(i+1)\mu\pi(i+1) &= \lambda\pi(i), & 0 \leq i < K,\\ K\mu\pi(i+1) &= \lambda\pi(i), & i \geq K.\end{aligned}$$

This gives

$$\pi(n) = C\frac{1}{n!}\left(\frac{\lambda}{\mu}\right)^n, \qquad n \leq K,$$
$$\pi(n) = C\frac{1}{K!}\left(\frac{\lambda}{\mu}\right)^K\left(\frac{\lambda}{K\mu}\right)^{n-K}, \qquad n \geq K,$$

it is a probability distribution if $\lambda/K\mu < 1$ and C is the normalizing constant. Under this condition, the number of customers in the $M/M/K$ at equilibrium is a reversible process.

In particular the invariant distribution of the $M/M/1$ queue is geometrically distributed with parameter $\rho = \lambda/\mu$ if $\rho < 1$ and, for the $M/M/\infty$ queue, it is a Poisson distribution with parameter ρ.

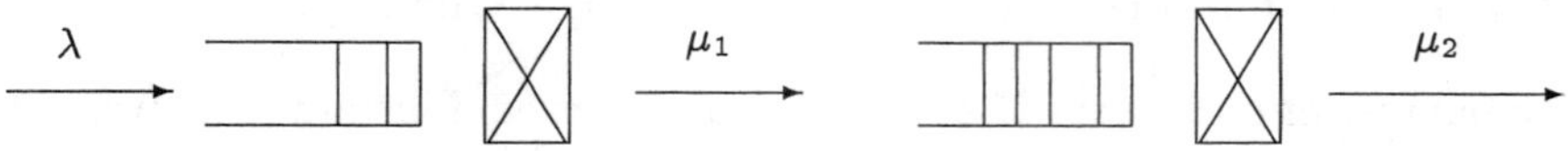

FIG. 4.1. Tandem Queues.

A Counter-example: Two $M/M/1$ Queues in Tandem. This queueing network is a sequence of two FIFO queues with exponentially distributed services. The first queue receives a Poisson arrival process and when a customer has been served in this queue, it goes to the second queue before leaving definitely the network. See Figure 4.1. The number of customers in each queue $(L_1(t), L_2(t))$ is a Markov process, the non-zero elements outside the main diagonal of the corresponding Q-matrix Q are given by

$$q_{(i,j),(i+1,j)} = \lambda, \qquad i, j \in \mathbb{N},$$
$$q_{(i,j),(i-1,j+1)} = \mu_1, \qquad i \geq 1, j \in \mathbb{N},$$
$$q_{(i,j),(i,j-1)} = \mu_2, \qquad i \in \mathbb{N}, j \geq 1.$$

The process cannot be reversible since the transition from $(1,0)$ to $(0,1)$ has intensity μ_1 and the opposite transition cannot occur. Thus, the identity

$$\pi((1,0))q_{(1,0)(0,1)} = \pi((0,1))q_{(0,1)(1,0)}$$

cannot be satisfied. (Recall that *all* the elements of the sequence $(\pi(i))$ have to be positive.)

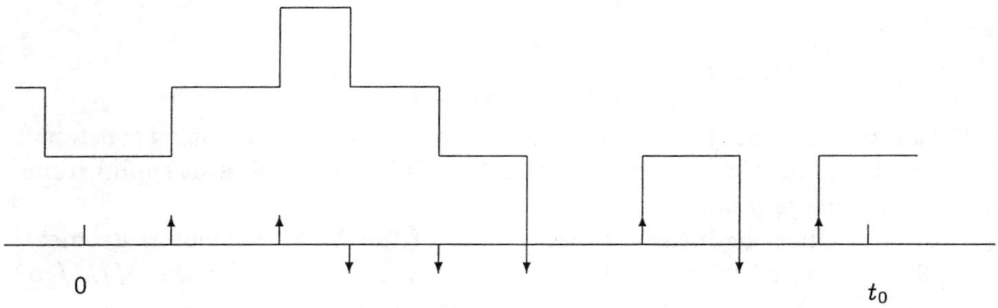

FIG. 4.2. The Number of Customers of the $M/M/1$ Queue

4.2.2 The Departure Process of the M/M/1 Queue

The number of customers of an $M/M/1$ queue at equilibrium is a reversible Markov process. The following important result is a simple consequence of this fact.

PROPOSITION 4.4 (Burke). *The departure process of an $M/M/1$ queue at equilibrium is a Poisson process and the number of customers at $t = t_0$ is independent of the departures before time $t = t_0$.*

PROOF. The number of customers of this queue at time $t \geq 0$ at equilibrium is denoted by $L(t)$. The process $(L(t))$ being reversible, for $t_0 > 0$ the processes $(L(t); 0 \leq t \leq t_0)$ and $(L((t_0 - t)-); 0 \leq t \leq t_0)$ have the same distribution. In particular, the processes of the instants where the jumps of $(L(t); 0 \leq t \leq t_0)$ and $(L((t_0 - t)-); 0 \leq t \leq t_0)$ are positive have the same distribution. The instants where the jumps are positive for the process $(L((t_0 - t)-); 0 \leq t \leq t_0)$ are precisely the departure instants between 0 and t_0. For the initial process, these instants are arrival times of customers, which form a Poisson process. Consequently, the arrival process restricted to the interval $[0, t_0]$ has the same distribution as the departure process restricted on the same interval. Thus the departure process is Poisson.

The Poisson property for the arrivals shows that the number of customers at time $t = 0$ is independent of the arrivals between 0 and t_0. For the reversed process, this statement can be translated as follows: The number of customers at time t_0 is independent of departures before t_0. The proposition is proved. □

Two $M/M1$ Queues in Tandem. With the same notation as before for this example, if $\rho_i = \lambda/\mu_i$, for $i = 1$ and 2. At equilibrium, the arrival process at the second queue is the departure process of the first one. Consequently, the arrival process at the second queue is Poisson, hence at equilibrium the number of customers at the second queue is geometrically distributed. The properties of this small queueing network are presented in the next proposition. These results are generalized later for more general queueing networks.

PROPOSITION 4.5. *If* $\lambda < \mu_1 \wedge \mu_2$, *the number of customers* $L_1(t)$ *and* $L_2(t)$ *in each file of the tandem queues at equilibrium are independent with a geometric distribution, i.e. for* $t \in \mathbb{R}$,

$$\mathbb{P}\left(L_1(t) = n_1, L_2(t) = n_2\right) = (1-\rho_1)\rho_1^{n_1}(1-\rho_2)\rho_2^{n_2}.$$

PROOF. If the first queue is at equilibrium, its number of customers $L_1(t)$ is independent of departures before t. The number of customers in the second queue at time t is precisely a functional of the departure process of the first file before t. At equilibrium, the random variables $L_1(t)$ and $L_2(t)$ are therefore independent. □

This proposition can be generalized to the case of N $M/M/1$ queues if they are located on a tree where customers can only go downward (the root of the tree is at the top).

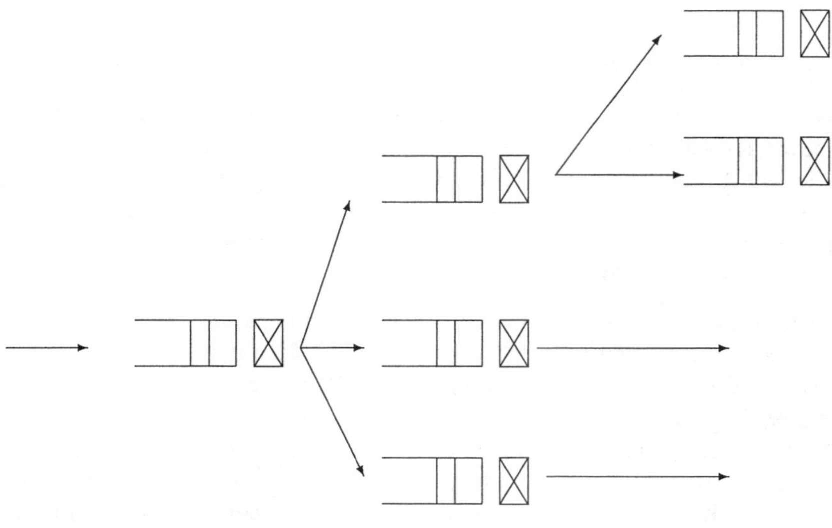

FIG. 4.3. Tree Network.

4.2.3 Truncation of Reversible Processes

The reversibility property is invariant with respect to some transformations of the $\mathcal{Q}$-matrix.

PROPOSITION 4.6. *If $Q = (q_{ij})$ is the $\mathcal{Q}$-matrix of a reversible Markov process and if $\mathcal{A}$ is a subset of $\mathcal{S}$, $\alpha \geq 0$ and $\widetilde{Q} = (\widetilde{q}_{ij})$ the matrix defined by*

$$\begin{cases} \widetilde{q}_{ij} = \alpha q_{ij}, & i \in \mathcal{A},\ j \in \mathcal{S} - \mathcal{A}, \\ \widetilde{q}_{ij} = q_{ij}, & \text{otherwise for } i \neq j, \end{cases}$$

The Markov jump process associated with $\widetilde{Q}$ has the invariant probability distribution defined by $\widetilde{\pi} = (\widetilde{\pi}_i; i \in \mathcal{S})$ given by

$$\begin{cases} \widetilde{\pi}_i = C\pi(i), & i \in \mathcal{A}, \\ \widetilde{\pi}_i = C\alpha\pi(i), & i \in \mathcal{S} - \mathcal{A}, \end{cases}$$

with the normalization constant

$$C = \frac{1}{\pi(\mathcal{A}) + \alpha\pi(\mathcal{S} - \mathcal{A})}.$$

Furthermore, this process is also reversible.

PROOF. The verification is straightforward. □

Queues with Finite Capacity. If α is 0 in the previous proposition, the modification turns out to suppress some transitions outside $\mathcal{A}$. In the case of queues, if $\mathcal{A} = \{0, 1, \ldots, C\}$, this is equivalent to reject customers when the queue has C customers. If an infinite capacity queue is reversible (i.e. the number of customers is reversible), the same property holds for this queue when the capacity is finite.

Application to Loss Networks. These networks are classical to study communication networks. See Kelly[Kel91]. The terminology of this domain is used to describe them. The network has N nodes, vertices of a graph $\mathcal{G}$. A *route* r connecting a node x to a node y is a finite sequence of nodes $r = (x_1, \ldots, x_k)$, such that $x_1 = x$, $x_k = y$ and, for $i = 1, \ldots, k-1$, x_i and x_{i+1} are neighbors in the graph, i.e. $(x_i, x_{i+1}) \in r$. Figure 4.4 presents a network with 6 nodes and 2 routes, the first one connecting x_1 to x_4 and the other x_2 to x_3. The set of routes of the network is denoted by $\mathcal{R}$.

The arrivals of calls requesting route $r \in \mathcal{R}$ is assumed to be Poisson with parameter λ_r and the service distribution is exponentially distributed with parameter μ_r. If $X_r(t)$ is the number of calls in progress on route r at time t, the vector $X(t) = (X_r(t); r \in \mathcal{R})$ is a Markov jump process describing the network. If the number of servers (i.e. the maximum number of simultaneous communications on a link of the network) is infinite, for $r \in \mathcal{R}$, the process

[Kel91] F.P. Kelly, *Loss networks*, Annals of Applied Probability **1** (1991), no. 3, 319–378.

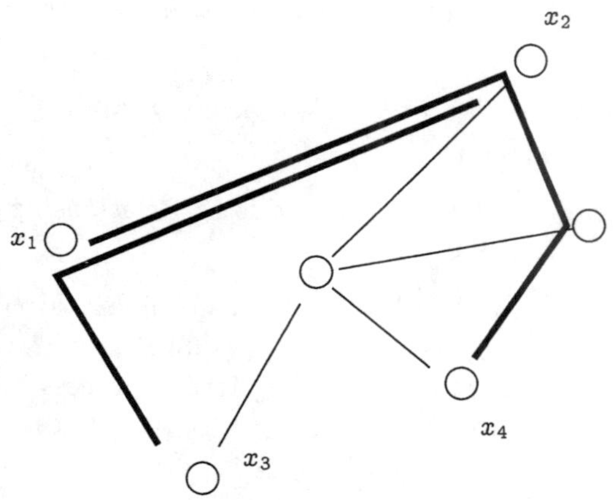

FIG. 4.4. Routes in a Loss Network.

$(X_r(t))$ is the number of customers of an $M/M/\infty$ queue, with arrival rate λ_r and service rate μ_r. The invariant probability measure of this process is Poisson with parameter λ_r/μ_r and at equilibrium $(X_r(t))$ is reversible. The processes $(X_r(t)), r \in \mathcal{R}$ are clearly independent ; the process $(X(t))$ is consequently reversible, and its invariant measure $\tilde{\pi}$ is given by

$$\tilde{\pi}(n) = \prod_{r\in\mathcal{R}} \frac{(\lambda_r/\mu_r)^{n_r}}{n_r!} e^{-\lambda_r/\mu_r},$$

for $n = (n_r) \in \mathbb{N}^{|\mathcal{R}|}$ ($|\mathcal{R}|$ is the cardinality of $\mathcal{R}$).

Of course, the capacity of these networks is, in practice, finite. For i and $j \le N$, C_{ij} denotes the maximum number of simultaneous calls using the link connecting nodes x_i and x_j (C_{ij} is the capacity of the link). This amounts to say that the transitions outside the set

$$\mathcal{S} = \left\{ n = (n_r) \in \mathbb{N}^{|\mathcal{R}|} : \sum_{r\in\mathcal{R},(x_i,x_j)\in r} n_r \le C_{ij}, \qquad \forall i,j \le k \right\}$$

are not allowed. According to the above proposition, the invariant probability measure $\tilde{\pi}_{\mathcal{S}}$ of the associated Markov jump process is the probability $\tilde{\pi}$ restricted to $\mathcal{S}$, i.e. for $n \in \mathcal{S}$,

$$\tilde{\pi}_{\mathcal{S}}(n) = C_{\mathcal{S}} \prod_{r\in\mathcal{R}} \frac{(\lambda_r/\mu_r)^{n_r}}{n_r!}, \quad \text{where} \quad C_{\mathcal{S}} = \sum_{n\in\mathcal{S}} \prod_{r\in\mathcal{R}} \frac{(\lambda_r/\mu_r)^{n_r}}{n_r!}$$

is the normalizing constant.

Another Criterion of Reversibility. This section presents a criterion due to Kolmogorov. It has the advantage of not using the explicit expression for the invariant probability.

PROPOSITION 4.7. *A Markov process on $\mathcal{S}$ with an invariant probability distribution is reversible if and only if*

$$q_{i_1 i_2} q_{i_2 i_3} \cdots q_{i_{n-1} i_n} q_{i_n i_1} = q_{i_1 i_n} q_{i_n i_{n-1}} \cdots q_{i_3 i_2} q_{i_2 i_1}, \tag{4.5}$$

for all $n \in \mathbb{N}$ and any sequence $i_1, \ldots, i_n$ of $\mathcal{S}$.

PROOF. If the Markov process is reversible, by using Relationship (4.4) repeatedly, one gets for $i_1, \ldots, i_n \in \mathcal{S}$ and $k \leq n$,

$$\pi(i_1) q_{i_1 i_2} q_{i_2 i_3} \cdots q_{i_{k-1} i_k} = \pi(i_k) q_{i_k i_{k-1}} q_{i_{k-1} i_{k-2}} \cdots q_{i_3 i_2} q_{i_2 i_1},$$

then if $i_{n+1} = i_1$, since $\pi(i_1) > 0$, Relationship (4.4) follows.

Conversely, if $i_0 \in \mathcal{S}$ is fixed, for $i \in \mathcal{S}$, the Markov process being irreducible, there exist $i_1, \ldots, i_n$ such that $q_{i i_n} q_{i_n i_{n-1}} \cdots q_{i_1 i_0} > 0$, set

$$\pi(i) = \frac{q_{i_0 i_1} q_{i_1 i_2} \cdots q_{i_n i}}{q_{i i_n} q_{i_n i_{n-1}} \cdots q_{i_1 i_0}}.$$

If $j_1, \ldots, j_p$ is another sequence such that $q_{i j_p} q_{j_p j_{p-1}} \cdots q_{j_1 i_0} > 0$, Equation (4.5) shows the identity

$$\frac{q_{i_0 i_1} q_{i_1 i_2} \cdots q_{i_n i} q_{i j_n} q_{j_n j_{n-1}} \cdots q_{j_1 i_0}}{q_{i_0 j_1} q_{j_1 j_2} \cdots q_{j_n i} q_{i i_n} q_{i_n i_{n-1}} \cdots q_{i_1 i_0}} = 1,$$

or

$$\frac{q_{i_0 i_1} q_{i_1 i_2} \cdots q_{i_n i}}{q_{i i_n} q_{i_n i_{n-1}} \cdots q_{i_1 i_0}} = \frac{q_{i_0 j_1} q_{j_1 j_2} \cdots q_{j_n i}}{q_{i j_n} q_{j_n j_{n-1}} \cdots q_{j_1 i_0}}.$$

Thus the definition of $\pi(i)$ does not depend on the choice of the particular sequence (i_k). If $i, j \in \mathcal{S}$ are such that $q_{ji} > 0$, then

$$\pi(i) q_{ij} = q_{ji} \frac{q_{i_0 i_1} q_{i_1 i_2} \cdots q_{i_n i} q_{ij}}{q_{ji} q_{i i_n} q_{i_n i_{n-1}} \cdots q_{i_1 i_0}} = q_{ji} \pi(j).$$

Hence, if $q_{ji} > 0$ or $q_{ij} > 0$, Equations (4.4) are satisfied for this sequence $(\pi(i))$ which is necessarily the invariant distribution. The Markov process is reversible. □

4.3 Local Balance Equations

Reversibility is a strong property of symmetry for a Markov process. It is equivalent to the fact that the Q-matrix Q is a self-adjoint operator on $L^2(\pi)$, the space of sequences with a finite second moment with respect to the invariant measure π of Q. To show that a Markov process is reversible, a candidate for the invariant probability measure has to be found so that Equations (4.4) are satisfied.

Markov processes are not, in general, reversible (cf. the tandem queues example). Nevertheless, a weaker property, quasi-reversibility, is satisfied in some important cases. For these processes, the adjoint operator of Q on $L^2(\pi)$ is not anymore Q but in the same class of operators as Q. In the same way as for reversibility, one must check some identities for the invariant measure.

As a consequence, this property characterizes some of the departures of the networks considered in this chapter.

PROPOSITION 4.8. *If there exist a positive sequence* $(\pi(i); i \in \mathcal{S})$ *and, for any element* i_0 *of* $\mathcal{S}$*, a partition* $(\mathcal{A}_j^{i_0}; j \in I)$ *of* $\mathcal{S} - \{i_0\}$ *such that if the sequence* $(\widetilde{q}_{ij}; i, j \in \mathcal{S})$ *is defined by*

$$\pi(i) q_{ij} = \pi(j) \widetilde{q}_{ji}, \quad i, j \in \mathcal{S}, \tag{4.6}$$

the equations

$$\sum_{k \in \mathcal{A}_j^{i_0}} \widetilde{q}_{i_0 k} = \sum_{k \in \mathcal{A}_j^{i_0}} q_{i_0 k}, \quad j \in I, \tag{4.7}$$

hold, then the sequence $(\pi(i); i \in \mathcal{S})$ *is the invariant measure of the Markov process associated with the Q-matrix Q. If the sum of the sequence* $(\pi(i))$ *is finite, the matrix* $\widetilde{Q} = (\widetilde{q}_{ij}; i, j \in \mathcal{S})$ *is the Q-matrix of the reversed process.*

PROOF. Equation (4.6) gives

$$\sum_{j \in \mathcal{S} - \{i_0\}} \pi(j) q_{j i_0} = \pi(i_0) \sum_{j \in \mathcal{S} - \{i_0\}} \widetilde{q}_{i_0 j},$$

and from Identity (4.7), the relationship

$$\sum_{j \in \mathcal{S} - \{i_0\}} \widetilde{q}_{i_0 j} = \sum_{l \in I} \sum_{j \in \mathcal{A}_l^{i_0}} \widetilde{q}_{i_0 j} = \sum_{l \in I} \sum_{j \in \mathcal{A}_l^{i_0}} q_{i_0 j} = \sum_{j \in \mathcal{S} - \{i_0\}} q_{i_0 j}$$

follows. If these two equations are combined, *global balance* Equations (4.1) for $(\pi(i); i \in \mathcal{S})$ are obtained. The result on the reversed process is a direct consequence of Proposition 4.1. The proposition is proved. □

REMARKS.

1. If, for $j \neq i$, $\mathcal{A}_j^i = \{j\}$, the Markov process satisfying the above proposition with these partitions is reversible.
2. Under the assumptions of the proposition, for $j \in I$, by summing on $\mathcal{A}_j$ Equation (4.6), one gets the *local balance* equations for $(\pi(i))$

   $$\pi(i) \sum_{k \in \mathcal{A}_j^i} q_{ik} = \sum_{k \in \mathcal{A}_j^i} \pi(k) q_{ki}, \quad i \in \mathcal{S}, j \in I,$$

   which are more detailed equations than Equations (4.1). At equilibrium, the stationary flow from i to $\mathcal{A}_j^i$ is the same as the stationary flow in reversed direction.
3. According to Relationship (4.6), if $q_{i_0 k} = 0$ and $q_{k i_0} = 0$, the same identity is true for $\widetilde{q}$, in particular if

   $$N_0 = \{k : q_{i_0 k} = 0 \text{ and } q_{k i_0} = 0\},$$

   Relationship (4.7) holds, i.e.

$$\sum_{j \in N_0} \widetilde{q}_{i_0 j} = \sum_{j \in N_0} q_{i_0 j}.$$

Consequently, it is sufficient to find a partition of $\mathcal{S} \setminus (N_0 \cup \{i_0\}($ to prove that π is the invariant distribution of the jump process.

COROLLARY 4.9. *The sequence* $(\pi(i); i \in \mathcal{S})$ *is an invariant measure for the Markov process associated with the* Q*-matrix* Q *if and only if the sequence* $(\widetilde{q}_{ij}; i, j \in \mathcal{S})$ *defined by (4.6) satisfies*

$$\sum_k \widetilde{q}_{ik} = \sum_k q_{ik}, \quad i \in \mathcal{S}.$$

PROOF. If the sequence $(\mathcal{A}_j^i; j \in I)$ is reduced to the single set $\mathcal{S} - \{i\}$, Equations (4.6) and (4.7) are equivalent to global balance Relationship (4.1) which is always satisfied by the invariant measure. □

4.4 Queueing Networks with Product Form

4.4.1 Jackson Networks

A Jackson network is a set of N FIFO queues which works as follows: for $1 \leq i \leq N$, the ith queue (also called the ith node of the network) has an exponential service time with parameter μ_i and the arrival process of external customers at queue i is Poisson with parameter λ_i. After being served at queue i, a customer goes to queue j with probability p_{ij} (with $p_{ii} = 0$) or leaves definitively the network with probability $1 - (p_{i1} + \cdots + p_{iN})$. The matrix $R = (r_{ij}; i, j = 0, \ldots, N)$ is defined as follows, for $i \neq 0$ and $j \neq 0$, $r_{ij} = p_{ij}$, $r_{i0} = 1 - p_{i1} - p_{i2} - \cdots - p_{iN}$ and $r_{00} = 1$, so that R is a Markovian matrix.

The Markov jump process associated with this network has the state space $\mathcal{S} = \mathbb{N}^N$. For $1 \leq i \leq N$, e_i denotes the ith vector of the canonical basis of $\mathbb{R}^N$, $e_i = (1_{\{j=i\}}; 1 \leq j \leq N)$, the Q-matrix Q of this Markov process is, for i and $j \leq N$,

$$\begin{aligned} q_{n,n+e_i-e_j} &= \mu_j r_{ji}, && n_j>0, \\ q_{n,n-e_j} &= \mu_j r_{j0}, && n_j > 0, \\ q_{n,n+e_i} &= \lambda_i. \end{aligned}$$

The Markovian matrix $R = (r_{ij}; i, j = 0, \ldots N)$ is assumed to have a unique absorbing point 0, i.e. if (Y_n) is a Markov chain with transitions defined by R, almost surely, (Y_n) is constant equal to 0 after some time.

Furthermore, it is assumed that $r_{ii} = 0$ for any $1 \leq i \leq N$ (a customer does not feedback after its service). If this condition is not fulfilled, it is enough to replace μ_i by $\mu_i/(1 - r_{ii})$ and r_{ij} by $r_{ij}/(1 - r_{ii})$ for $j \neq i$ in the associated Q-matrix to be in that setting.

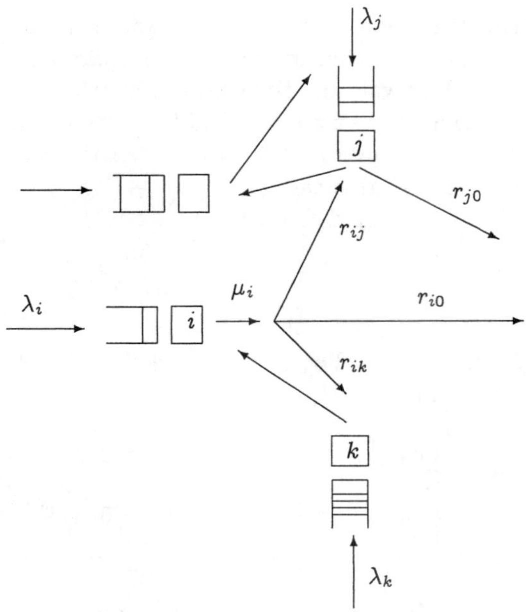

FIG. 4.5. A Jackson Network

LEMMA 4.10. *There exists a positive vector* $(\bar{\lambda}_i;\ 1 \le i \le N)$ *such that the traffic equations hold,*

$$\bar{\lambda}_i = \lambda_i + \sum_{j=1}^{N} \bar{\lambda}_j p_{ji}. \tag{4.8}$$

For $1 \le i \le N$, *the quantity* $\bar{\rho}_i$ *will denote* $\bar{\lambda}_i/\mu_i$.

PROOF. If $\lambda = \sum_1^N \lambda_i$ and $\alpha = (\lambda_i/\lambda; 1 \le i \le N)$ is the initial distribution of the Markov chain associated with the transition matrix R, since this Markov chain is transient on $\{1, \dots, N\}$ and the state 0 absorbing, the number of visits N_i in the state $i = 1, \dots, N$ is integrable and

$$\mathbb{E}(N_i) = \mathbb{P}(Y_0 = i) + \sum_{j=1}^{N} \sum_{n=0}^{+\infty} \mathbb{P}(Y_n = j, Y_{n+1} = i).$$

The Markov property gives the identity

$$\lambda \mathbb{E}(N_i) = \lambda_i + \sum_{j=1}^{N} \sum_{n=0}^{+\infty} \lambda \mathbb{P}(Y_n = j) r_{ji} = \lambda_i + \sum_{j=1}^{N} \lambda \mathbb{E}(N_j) r_{ji},$$

the vector $(\lambda \mathbb{E}(N_i); 1 \le i \le N)$ is solution of System (4.8). □

It means that, if the network is at equilibrium, $\bar{\lambda}_i$ is the intensity of the total flow of customers going through node i, this quantity is the sum of the intensity λ_i of external arrivals and the intensity of other arrivals from nodes of the network. Equations (4.8) are equivalent to the fact that, at equilibrium, the total arrival rate at a node is equal to its departure rate. The average load arriving at node i is in this case $\bar{\rho}_i$.

The vector $(\pi(n); n \in \mathcal{S})$ is defined by

$$\pi(n) = \prod_1^N \bar{\rho}_i^{n_i}(1 - \bar{\rho}_i),$$

for $n = (n_i\,; 1 \leq i \leq N) \in \mathcal{S}$. The matrix $\widetilde{Q}$, defined by Relationship (4.6) for this vector is, for i and $j \leq N$,

$$(4.9) \qquad \begin{cases} \widetilde{q}_{n,n+e_i-e_j} = \dfrac{\bar{\lambda}_i}{\bar{\rho}_j} r_{ij}, & n_j > 0, \\ \widetilde{q}_{n,n-e_j} = \dfrac{\lambda_j}{\bar{\rho}_j}, & n_j > 0, \\ \widetilde{q}_{n,n+e_i} = \bar{\lambda}_i r_{i0}. \end{cases}$$

For $1 \leq j \leq N$, if $\mathcal{D}_j = \{e_i - e_j/i \neq j\} \cup \{-e_j\}$ and $\mathcal{A}_0 = \{e_i/1 \leq i \leq N\}$, $n + \mathcal{D}_j$ is the set of all the possible states from state $n \in \mathcal{S}$ after a departure from queue j (if $n_j > 0$) and $n + \mathcal{A}_0$ is the state of the network after an external arrival. Traffic Equations (4.8) give the relations

$$(4.10) \qquad \begin{cases} \sum_{m\in\mathcal{D}_j} q_{n,n+m} = \mu_j r_{j0} + \sum_i \mu_j r_{ji} = \mu_j, \\ \displaystyle\sum_{m\in\mathcal{D}_j} \widetilde{q}_{n,n+m} = \frac{\lambda_j + \sum_i \bar{\lambda}_i p_{ij}}{\rho_j} = \frac{\bar{\lambda}_j}{\bar{\rho}_j} = \mu_j, \end{cases}$$

and

$$\sum_{m\in\mathcal{A}_0} q_{n,n+m} = \sum_{i=1}^N \lambda_i,$$

$$\sum_{m\in\mathcal{A}_0} \widetilde{q}_{n,n+m} = \sum_{i=1}^N \bar{\lambda}_i r_{i0} = \sum_{i=1}^N \bar{\lambda}_i \left(1 - \sum_{j=1}^N p_{ij}\right)$$
$$= \sum_{j=1}^N \left(\bar{\lambda}_j - \sum_{i=1}^N \bar{\lambda}_i p_{ij}\right) = \sum_{i=1}^N \lambda_i = \sum_{m\in\mathcal{A}_0} q_{n,n+m}.$$

The only possible states starting from n are necessarily in one of the sets $n + \mathcal{D}_j$ for $j = 1\ldots, N$ or in $n + \mathcal{A}_0$, thus Proposition 4.8 shows that the next theorem holds.

THEOREM 4.11. *If the solutions* $(\bar{\lambda}_i\,;\,1 \le i \le N)$ *of the equations*

$$\bar{\lambda}_i = \lambda_i + \sum_{j=1}^{N} \bar{\lambda}_j p_{ji},$$

satisfy the condition $\bar{\rho}_i = \bar{\lambda}_i/\mu_i < 1$ *for* $i = 1, \ldots, N$, *the Markov process associated to the Jackson network is ergodic and its invariant probability measure is given by the product formula*

$$\pi(n) = \prod_{1}^{N} \bar{\rho}_i^{n_i}(1 - \bar{\rho}_i),$$

for $n = (n_i\,;\,1 \le i \le N) \in \mathcal{S}$.

The number of customers of the Jackson network at equilibrium is equivalent to a set of N *independent queues delivering service rates* $(\mu_i\,;\,1 \le i \le N)$ *and with Poisson arrival processes with intensities* $(\lambda_i\,;\,1 \le i \le N)$.

At equilibrium, the state of the network at time t *is independent of the departure process before* t. *The departure processes out of the network from any of the queues of the network are independent Poisson processes with respective intensities* $(\bar{\lambda}_i r_{i0}\,;\,1 \le i \le N)$.

Notice that the Poisson property is not for all departures from a queue, but only for departures of customers who leave definitively the network. In general this total departure process is not Poisson. See Walrand [**51**].

PROOF. Only the last part of the theorem has to be proved. By using the fact that the matrix $\widetilde{Q}$ is the Q-matrix of the reversed process. According to Equation (4.9), this Q-matrix $\widetilde{Q}$ corresponds also to a Jackson network of N queues with the parameters $(\widetilde{\lambda}_i)$, $(\widetilde{\mu}_i)$, $(\widetilde{p}_{ij})$ given by

$$\widetilde{\lambda}_i = \bar{\lambda}_i r_{i0}, \quad \widetilde{\mu}_i = \mu_i, \quad \widetilde{p}_{ij} = \frac{\bar{\lambda}_j}{\bar{\lambda}_i} r_{ji}, \quad \widetilde{p}_{i0} = \frac{\lambda_i}{\bar{\lambda}_i},$$

for $i, j = 1, \ldots, N$. The ith queue of this network receives external arrivals with intensity $\bar{\lambda}_i r_{i0}$ and exponentially distributed services with parameter μ_i (Equation (4.10)). The reversed process being a Jackson network, the external arrival process at each queue is Poisson and the state of the network at time t is independent of the arrival process after t. For the original process, since arrivals of the reversed process are departures from the original network, this proves the last statement of the theorem. □

For this network it is possible to exhibit an invariant positive measure solution of global balance equations. Since the associated Markov process is irreducible, if this measure has a finite sum, the Markov process is necessarily ergodic. In general, ergodicity is not proved in such a way. The classical method to prove this kind of results is to show that if the initial state of the Markov process is "very large" then it drifts toward small states in average (see Chapter 8 for precise statements). The ergodicity conditions being established, the next step is usually to try to get some information on the equilibrium measure. This is very difficult in general, the case of Jackson networks is very uncommon from this point of view.

Proposition 9.6 page 256 gives a "natural" proof (i.e. without using the knowledge of the product formula) of the ergodicity of a Jackson network under the condition $\max_i \bar{\rho}_i < 1$.

4.4.2 Gordon-Newel Networks

A Gordon-Newel network is a Jackson network but without external arrivals and departures from the network. It is a closed network of N queues in which M customers move from one queue to another. For $i = 1, \dots, N$, the ith queue delivers an exponentially distributed service with parameter μ_i. When a customer leaves this queue, it goes to queue j with probability p_{ij}. The Q-matrix Q of the associated Markov process is given by

$$q_{n,n+e_i-e_j} = \mu_j p_{ji}, \quad n_j > 0, \quad i, j \leq N.$$

The matrix $P = (p_{ij}\,;\, i, j = 1, \dots, N)$ is the transition matrix of an irreducible Markov chain, it is therefore ergodic since the state space is finite. Define $(\nu_i\,;\, i = 1, \dots, N)$ the invariant distribution of this Markov chain, the equilibrium equations are given by

$$\nu_i = \sum_1^N \nu_j p_{ji}, \quad i = 1, \dots, N. \tag{4.11}$$

For the closed network, these equilibrium equations are the analogue of traffic Equations (4.8), in the same way the quantities ρ_i are defined by

$$\rho_i = \frac{\nu_i}{\mu_i}, \quad i = 1, \dots, N.$$

The Markov jump process describing the number of customers at each of the nodes of the network has the state space

$$\mathcal{S} = \left\{ n = (n_i\,;\, 1 \leq i \leq N) : \sum_1^N n_i = M \right\}.$$

The irreducibility of the matrix P on $\{1, \dots, N\}$ implies the same property for the Markov process on $\mathcal{S}$. As before, take

$$\pi(n) = \prod_1^N \rho_i^{n_i}, \quad n = (n_i\,;\, 1 \leq i \leq N) \in \mathcal{S},$$

and the corresponding matrix $\widetilde{Q}$ defined by Equation (4.6) is

$$\widetilde{q}_{n,n+e_i-e_j} = \frac{\nu_i}{\rho_j} p_{ij}, \quad n_j > 0, \quad i, j \leq N.$$

With the same notations as in the last proof, Equations (4.11) give that, for $n = (n_i\,;\, 1 \le i \le N) \in \mathcal{S}$ and $j = 1, \dots, N$ such that $n_j > 0$, then

$$(4.12) \qquad \begin{cases} \displaystyle\sum_{m \in \mathcal{D}_j} q_{n,n+m} = \sum_i \mu_j p_{ji} = \mu_j, \\ \displaystyle\sum_{m \in \mathcal{D}_j} \widetilde{q}_{n,n+m} = \frac{\sum_i \nu_i p_{ij}}{\rho_j} = \frac{\nu_j}{\rho_j} = \mu_j. \end{cases}$$

Starting from $n \in \mathcal{S}$, the state of the Markov process is necessarily in one of the subsets $n + \mathcal{D}_j$, $j = 1, \dots, N$. Proposition 4.8 and Equations (4.12) give the following theorem for the Gordon-Newel network.

THEOREM 4.12. *The closed queueing network with the routing matrix (p_{ij}) described above has the invariant probability distribution*

$$\pi(n) = \frac{1}{K} \prod_1^N \rho_i^{n_i}, \qquad n = (n_i\,;\, 1 \le i \le N) \in \mathcal{S},$$

where, for $i = 1, \dots, N$, $\rho_i = \nu_i / \mu_i$ where (ν_i) satisfy the system of equations

$$\nu_i = \sum_1^N \nu_j p_{ji},$$

and K is the normalizing constant.

The normalizing constant K is not (in general) easy to get, mainly because of the size of the state space and complex combinatorial relations defining the state space. Recursive procedures can express numerically this constant K, see Walrand [**51**].

4.4.3 Multi-class Networks

To begin with, the network is a single server queue where the arrival process is the superposition of J Poisson processes with respective parameters $\lambda_1, \dots, \lambda_J$. For $j = 1, \dots, J$, a customer from jth flow is said of class j. When it arrives, if n customers are already in the queue, this customer is located at lth position in the queue, $l = 1, \dots, n+1$, with probability $\gamma(l, n)$, where $\gamma(1, n) + \gamma(2, n) + \cdots + \gamma(n+1, n) = 1$. If n customers are present in the queue, the customer in position l receives an exponentially distributed service with parameter $\phi(n)\delta(l, n)$ for $l = 1, \dots, n$, with the relation $\delta(1, n) + \cdots + \delta(n, n) = 1$. The server works therefore at rate $\phi(n)$. The function $n \to \phi(n)$ is assumed to be positive when $n > 0$.

The queue can be described by a Markov jump process whose state space is given by

$$\mathcal{S} = (\{1, \dots, J\})^{(\mathbb{N})} = \{c = (c_i\,;\, 1 \le i \le n) : n \in \mathbb{N}, c_i \in \{1, \dots, J\}\},$$

the vector 0 corresponds to the empty queue. If $c = (c_i\,;\, 1 \le i \le n) \in \mathcal{S}$, $|c| = n$ denotes the length of c and, for $i = 1, \dots, n$, c_i is the class of ith customer in the queue. The transitions of the Markov process on $\mathcal{S}$ have to be defined. The insert operator (A_j^i) and delete operator (S_i) are defined by

$$\begin{aligned} A_i^j(c) &= (c_1, c_2, \ldots, c_{i-1}, j, c_i, c_{i+1}, \ldots, c_n), \quad i = 1, \ldots, n+1 \quad j = 1, \ldots, J, \\ S_i(c) &= (c_1, c_2, \ldots, c_{i-1}, c_{i+1}, \ldots, c_n), \qquad i = 1, \ldots, n, \end{aligned}$$

for $c = (c_i\,;\, 1 \leq i \leq n) \in \mathcal{S}$. The operator A_j^i inserts a j at ith place and operator S_i suppresses the element at ith place. The Q-matrix Q can be expressed as follows

$$\begin{aligned} q_{c,A_i^j(c)} &= \lambda_j \gamma(i, |c|), \qquad i = 1, \ldots, |c|+1, \quad j = 1, \ldots, J, \\ q_{c,S_i(c)} &= \phi(|c|)\delta(i, |c|), \qquad i = 1, \ldots, |c|. \end{aligned}$$

Notice the service received by a customer or the place it occupies in the queue does not depend on its class.

Examples.

The $M/M/K$ FIFO Queue. There is only one class of customers for this queue. An arriving customer is placed at the end of the queue, and only the K first customers receive an exponentially distributed service, for $n \in \mathbb{N}$,

$$\begin{aligned} \phi(n) &= n \wedge K, \\ \delta(l, n) &= \begin{cases} 1/(n \wedge K), & 1 \leq l \leq n \wedge K, \\ 0, & l > n \wedge K, \end{cases} \\ \gamma(l, n) &= \begin{cases} 0, & l = 1, \ldots, n, \\ 1, & l = n+1. \end{cases} \end{aligned}$$

The $M/M/K$ LIFO Queue. The functions ϕ and γ do not change, only the K last customers are served,

$$\delta(l, n) = \begin{cases} 1/(n \wedge K), & (n-K)^+ < l \leq n,\ n \neq 0, \\ 0, & l \leq (n-K)^+. \end{cases}$$

The $M/M/1$ Processor-Sharing Queue. The variable δ is changed as follows

$$\delta(l, n) = \frac{1}{n}, \quad l = 1, \ldots, n.$$

Going back to the general case, an invariant measure for this Markov process is given by

$$\pi(c) = \prod_{l=1}^{|c|} \frac{\lambda_{c_l}}{\phi(l)}, \quad c \in \mathcal{S}.$$

Indeed, using Definition (4.6), the corresponding matrix $\widetilde{Q}$ for π is

$$\begin{aligned} \widetilde{q}_{c,A_i^j(c)} &= \lambda_j \delta(i, |c|+1), \qquad i = 1, \ldots, |c|+1, \\ \widetilde{q}_{c,S_i(c)} &= \phi(|c|)\gamma(i, |c|-1), \qquad i = 1, \ldots, |c|, \quad j = 1, \ldots, J. \end{aligned}$$

For $j = 1, \ldots, J$ fixed, it is easily verified that

$$\sum_{i=1}^{|c|+1} \widetilde{q}_{c,A_i^j(c)} = \lambda_j = \sum_{i=1}^{|c|+1} q_{c,A_i^j(c)} \text{ and } \sum_{i=1}^{|c|} \widetilde{q}_{c,S_i(c)} = \phi(|c|) = \sum_{i=1}^{|c|} q_{c,S_i(c)}.$$

Starting from $c \in \mathcal{S}$, the only possible states for the matrices Q and $\widetilde{Q}$ are $A_i^j(c)$ for $j = 1, \ldots, J$ and $i = 1, \ldots, |c| + 1$ and $S_i(c)$ for $i = 1, \ldots, |c|$. The two identities above and Proposition 4.8 show that π is the invariant measure of the Markov process.

PROPOSITION 4.13. *The invariant distribution of the multi-class single server queue is given by*

$$\pi(c) = \prod_{l=1}^{|c|} \frac{\lambda_{c_l}}{\phi(l)}, \quad c \in \mathcal{S},$$

provided that $K = \sum_{c \in \mathcal{S}} \pi(c)$ *is finite. For* $j \in \{1, \ldots, J\}$*, the departure process of class* j *customers is Poisson with parameter* λ_j*.*

PROOF. The Q-matrix $\widetilde{Q}$ of the reversed process also corresponds to a multi-class single server queue. For this queue, the arrival process of class j customers is Poisson with parameter λ_j, which implies the Poisson property for departures of class j customers from the original queue. □

At equilibrium, the reversed process is a multi-class single server queue with δ and γ exchanged. The detailed description of the queue given above is not necessary to study the total number of customers in the queue. Indeed, the total number of customers is a Markov process, in fact a birth and death process whose rates are given by

$$q_{i,\,i+1} = \sum_{i=1}^{J} \lambda_i \text{ for } i \geq 0 \text{ and } q_{i,\,i-1} = \phi(i), \text{ for } i > 0.$$

The invariant distribution can thus be expressed quite easily. This detailed description will be nevertheless necessary to study networks of such queues.

Now a network with N multi-class single server queues described above is considered, the external arrivals are given by J Poisson processes. For $1 \leq j \leq J$, the jth flow has intensity λ_j, a customer of this flow arrives at queue $f_1^j \in \{1, \ldots, N\}$ then, after being served, it goes to queue f_2^j, and so on, until it leaves the network when it leaves queue $f_{n_j}^j$. The jth flow travels along the route $f_1^j, f_2^j, \ldots, f_{n_j}^j$ in the network. If $1 \leq k \leq N$, the functions ϕ, δ and γ associated with node k are defined by ϕ_k, δ_k and γ_k. Figure 4.6 gives an example of a network with two routes $(1, 2, 3)$ and $(4, 3, 2, 4)$. Notice the difference with Jackson networks where routes in the network are possibly random. The state of ith queue, for $i \leq N$, is described by the vector

$$c_i = ((c_{i1}, s_{i1}), (c_{i2}, s_{i2}), \ldots, (c_{im_i}, s_{im_i}))$$

in the state space

$$\mathcal{S}_i = \left(\{(j, s) : f_s^j = i\}\right)^{(\mathbb{N})}$$

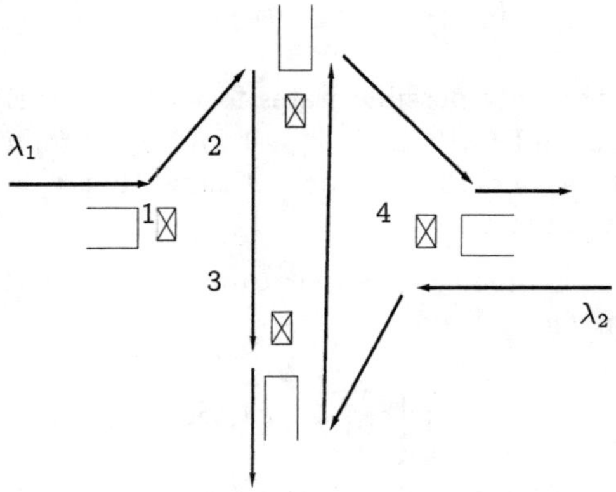

FIG. 4.6. A Multi-Class Network with Fixed Routes.

where c_{ik} is the class of the customer at kth position in the queue and s_{ik} is the number of queues already visited by this customer. As before, $|c_i| = m_i$ denotes the length of the vector c_i.

For the Markovian description of the network, the detailed description given by the vectors $(s_{ik}\,;\, 1 \leq k \leq m_i)$ is necessary only if a queue occurs more than once in one of the routes.

It is easily seen that $C(t) = (c_i(t)\,;\, 1 \leq i \leq N)$ is an irreducible Markov jump process for the state space $\mathcal{S} = \mathcal{S}_1 \times \mathcal{S}_2 \times \cdots \times \mathcal{S}_N$. If the state of the network is $C = (c_i)$, from this state, the possible transitions are given by:

- A customer of class j arrives (with intensity λ_j) in the network at queue $f = f_1^j$, the element $(j, 1)$ is inserted at position l of this queue with probability $\gamma_f(l, |c_f|)$.
- A customer of class j at the stage $s < n_j$ of its route, at position l in the queue $f = f_s^j$, leaves this queue and goes to position m of the queue $g = f_{s+1}^j$ with intensity
$$\phi_f\left(|c_f|\right) \delta_f\left(l, |c_f|\right) \gamma_g(m, |c_g|)\,,$$
the element $(j, s+1)$ is then at the position m of c_g.
- A customer of class j at stage n_j of its route, at position l of the queue $f = f_{n_j}^j$, completes its service and leaves the network with intensity
$$\phi_f\left(|c_f|\right) \delta_f\left(l, |c_f|\right).$$

For $i \in \{1, \ldots, N\}$, the measure π is defined by

$$\text{(4.13)} \qquad \pi_i(c_i) = \prod_{k=1}^{|c_i|} \frac{\lambda_{c_{ik}}}{\phi_i(k)}, \text{ and for } C = (c_i) \in \mathcal{S}, \quad \pi(C) = \prod_{i=1}^{N} \pi_i(c_i),$$

π is a natural candidate for the invariant measure of the Markov process.

The matrix $\widetilde{Q}$ associated with π is clearly the Q-matrix of a multi-class network with J classes of customers arriving according to Poisson processes with parameters (λ_j). The jth flow has the route $f^j_{n_j}, f^j_{n_j-1}, \ldots, f^j_1$. The transitions of the Q-matrix $\widetilde{Q}$ are described as follows:

- A customer of jth flow arrives with intensity λ_j in the network at the queue $f = f^j_{n_j}$, the element $(j, 1)$ is inserted in lth position of this queue with probability $\delta_f(l, |c_f| + 1)$.
- A customer of the class j at stage $s < n_j$ of its route, at position l of the queue $f = f^j_{n_j-s}$, leaves to go to position m of the queue $g = f^j_{n_j-s-1}$ with intensity
$$\phi_f\left(|c_f|\right) \gamma_f\left(l, |c_f| - 1\right) \delta_g\left(m, |c_g| + 1\right),$$
the element $(j, s+1)$ is at position m of c_g.
- A customer of class j at stage n_j of its route, at position l of the queue $f = f^j_1$, leaves the network with intensity
$$\phi_f\left(|c_f|\right) \gamma_f\left(l, |c_f| - 1\right).$$

Theorem 4.14. *The invariant probability distribution of the multi-class queueing network with fixed routes (also called Kelly's network) is given by product Formula (4.13), provided that it has a finite sum. The departure process of each class of customers is a Poisson process.*

Proof. If the initial state is $C \in \mathcal{C}$, a customer of class j arrives in the network at rate λ_j for Q and for $\widetilde{Q}$. A departure of the queue f occurs at rate $\phi(|c_f|)$ for Q and $\widetilde{Q}$. A partition of possible states starting from C satisfying the conditions of Proposition 4.8 has been constructed. The proposition is proved. □

5. The M/M/1 Queue

Contents

In this chapter the explicit distributions of the $M/M/1$ queue and the related limit results are investigated. For the moment, only the equilibrium behavior of this queue has been considered (the reversibility property for example in Chapter 4). Here, transient distributions, hitting times or rate of convergence to equilibrium are analyzed. The associated renormalized processes are introduced in this simple case, a functional law of large numbers as well as a functional central limit theorem for these processes are also proved. This study gives a simple example of a more general procedure introduced in Chapter 9 to study general queueing networks. A large deviations result concludes this chapter, it gives an estimation of the probability that the sample path of the process of the number of customers follows a very unlikely path.

The $M/M/1$ queue can also be seen as a continuous time version of the simple random walk on $\mathbb{Z}$ whose jumps are either 1 or 1 with a reflection at 0. For this reason, this queue is a classical stochastic process of probability theory. The analysis of this queue is also an occasion to appreciate the full power of elementary stochastic calculus: exponential martingales together with some changes of probability measures give a very detailed description of the behavior of this queue. This approach shows also that the usual and cumbersome calculations of transient distributions at a given time are no more necessary to get a full picture of this stochastic process (even for the rate of convergence).

5.1 Introduction

In the following, $\mathcal{N}_\xi(dx)$ denotes a Poisson point process on $\mathbb{R}$ with parameter $\xi \in \mathbb{R}_+$. All Poisson point processes used in this chapter are assumed to be defined on the same probability space and independent. The notation of Chapter 1 is used: If I is an interval of $\mathbb{R}_+$, the quantity $\mathcal{N}_\xi(I)$ is the number of points of the point process $\mathcal{N}_\xi$ in this interval.

The arrival rate of customers is λ and the service rate is μ, $\mathcal{N}_\lambda$ and $\mathcal{N}_\mu$ are the corresponding Poisson point processes and $\mathcal{F} = (\mathcal{F}_t)$ is the associated filtration,

$$\mathcal{F}_t = \sigma \left\langle \mathcal{N}_\lambda(]0,s]), \mathcal{N}_\mu(]0,s]), s \leq t \right\rangle,$$

for $t \geq 0$. If $L(t)$ is the number of customers of the queue at time t, $(L(t))$ is a jump process on $\mathbb{N}$ whose Q-matrix is given by $Q = (q_{ij})$, with

$$q_{i\,i+1} = \lambda,\ i \in \mathbb{N}; \quad q_{i\,i-1} = \mu,\ i > 0; \quad q_{ij} = 0, \text{ when } |i-j| > 1.$$

The properties of this queue at equilibrium seen in Chapter 2 and Chapter 4 are recalled in the next proposition.

PROPOSITION 5.1. *If $\lambda < \mu$, the Markov process $(L(t))$ is ergodic, the number L of customers at equilibrium is geometrically distributed with parameter $\rho = \lambda/\mu$ and the stationary distribution of the waiting time W of a customer is given by*

$$\mathbb{P}(W > x) = \rho e^{-(\mu-\lambda)x}, \qquad x \geq 0.$$

At equilibrium, $(L(t))$ is a reversible process and the departure process is Poisson with parameter λ.

The distribution of the process $(L(t))$ can also be described as the solution of a stochastic differential equation.

PROPOSITION 5.2. *The Markov process with Q-matrix Q starting from $x \in \mathbb{N}$ has the same distribution as the unique solution of the stochastic differential equation*

$$dL(t) = \mathcal{N}_\lambda(dt) - 1_{\{L(t-)>0\}}\mathcal{N}_\mu(dt), \tag{5.1}$$

with $L(0) = x$, or

$$L(t) = x + \mathcal{N}_\lambda(]0,t]) - \int_{]0,t]} 1_{\{L(s-)>0\}}\mathcal{N}_\mu(ds).$$

The process

$$(M(t)) = \left(L(t) - x - \lambda t + \mu \int_0^t 1_{\{L(s)>0\}}\, ds\right)$$

is a martingale with respect to the filtration $\mathcal{F} = (\mathcal{F}_t)$.

Proposition A.11 page 356 in Appendix A shows that such stochastic differential equations have a unique solution which is right continuous and have left limits (abbreviated as càdlàg) almost surely. If $(X(t))$ is some càdlàg process, the notation $dX(t)$ is for the limit

$$\lim_{s \nearrow t} X(t) - X(s).$$

PROOF. Equation (5.1) is quite natural, the process $(L(t))$ jumps by $+1$ at the point of the Poisson process $\mathcal{N}_\lambda$ with parameter λ. If $L(t-) > 0$, the residual service σ of the customer at the head of the queue is exponentially distributed with parameter μ, the residual service σ can be described so that $t + \sigma$ is the first point after t of the point process $\mathcal{N}_\mu$.

Rigorously, to establish the correspondence between differential Equation (5.1) and the Markov process with Q-matrix Q and initial state x, Proposition B.6 page 365 is used. It is sufficient to show that if $(L(t))$ is the solution of stochastic differential Equation (5.1), then for any function f the process

$$\left(f(L(t)) - f(x) - \int_0^t Q(f)(L(s))\, ds \right)$$

is a local martingale. Recall that, for $x \in \mathbb{N}$,

$$Q(f)(x) = \sum_{i \neq x} q_{xi}(f(i) - f(x)).$$

The differential expression for $(L(t))$ gives, for $t \geq 0$,

$$\begin{aligned} df(L(t)) = {} & \left[f(L(t-)+1) - f(L(t-))\right] \mathcal{N}_\lambda(dt) \\ & + 1_{\{L(t-)>0\}} \left[f(L(t-)-1) - f(L(t-))\right] \mathcal{N}_\mu(dt). \end{aligned} \tag{5.2}$$

Since, for $t \geq 0$,

$$\begin{aligned} \int_0^t Q(f)(L(s))\, ds = {} & \lambda \int_0^t \left[f(L(s)+1) - f(L(s))\right] ds \\ & + \mu \int_0^t 1_{\{L(s)>0\}} \left[f(L(s)-1) - f(L(s))\right] ds, \end{aligned}$$

by integrating differential Equation (5.2), one gets

$$\begin{aligned} & f(L(t)) - f(x) - \int_0^t Q(f)(L(s))\, ds \\ & \quad = \int_0^t \left[f(L(s-)+1) - f(L(s-))\right] (\mathcal{N}_\lambda(ds) - \lambda\, ds) \\ & \quad + \int_0^t 1_{\{L(s-)>0\}} \left[f(L(s-)-1) - f(L(s-))\right] (\mathcal{N}_\mu(ds) - \mu\, ds). \end{aligned}$$

Proposition A.9 page 355 shows that the right hand side of the above expression is a local martingale. The distributions of $(L(t))$ and the Markov process with Q-matrix Q and initial state x are therefore identical. The last part of the proposition is obtained by taking $f(y) = y$ for $y \in \mathbb{N}$. □

DEFINITION 11. *The* free process *is the process* $(Z(t))$ *defined by*

$$Z(t) = \mathcal{N}_\lambda(]0,t]) - \mathcal{N}_\mu(]0,t]).$$

If $L(t-) > 0$, in the neighborhood of t, the process $(L(t))$ behaves locally as the free process, i.e. $dL(t) = dZ(t)$. Equation (5.1) can be rewritten as

$$\begin{aligned} dL(t) &= \mathcal{N}_\lambda(dt) - \mathcal{N}_\mu(dt) + 1_{\{L(t-)=0\}}\,\mathcal{N}_\mu(dt) \\ &= dZ(t) + 1_{\{L(t-)=0\}}\,\mathcal{N}_\mu(dt), \\ L(t) &= L(0) + Z(t) + \int_0^t 1_{\{L(s-)=0\}}\,\mathcal{N}_\mu(ds), \end{aligned}$$

the expression $1_{\{L(t-)=0\}}\,\mathcal{N}_\mu(dt)$ compensates the negative part of the free process $(Z(t))$ (see Appendix D on reflected processes).

Several Markov processes are considered in this chapter, the notations $\mathbb{E}_x$, $\mathbb{P}_x$ are *only* for the Markov process $(L(t))$ starting from $x \in \mathbb{N}$.

5.2 Exponential Martingales

The following proposition is the analogue of Proposition 3.4 page 59 for the continuous time process $(Z(t))$.

PROPOSITION 5.3. *For $u \in \mathbb{R}$, the process*

$$(H(t)) = \left(u^{Z(t)} \exp\left[\lambda t(1-u) + \mu t(1-1/u)\right]\right),$$

is a martingale. There exists a unique probability measure $\mathbb{P}^u$ such that if Y is positive and $\mathcal{F}_t$-measurable random variable

$$\mathbb{E}^u(Y) = \int Y\, d\mathbb{P}^u = \int Y\, H(t)\, d\mathbb{P},$$

and, conversely

$$\mathbb{E}(Y) = \int Y\, H(t)^{-1}\, d\mathbb{P}^u = \int Y\, \frac{\exp\left[-\lambda t(1-u) - \mu t(1-1/u)\right]}{u^{Z(t)}}\, d\mathbb{P}^u.$$

For the probability distribution $\mathbb{P}^u$, $\mathcal{N}_\lambda$ and $\mathcal{N}_\mu$ are independent Poisson processes with respective parameters λu and μ/u.

The change of probability induced by the martingale $(H(t))$ preserves the Poisson property but changes the parameter of Poisson processes. This situation is very similar to the case of random walks seen in Section 3.2.

PROOF. The processes $\mathcal{N}_\lambda$ and $\mathcal{N}_\mu$ having independent increments, the free process $(Z(t))$ has also independent increments. The martingale property of $(H(t))$ is a direct consequence of this fact: If $s \leq t$,

$$\begin{aligned}\mathbb{E}(H(t) \mid \mathcal{F}_s) &= H(s)e^{(t-s)(\lambda(1-u)+\mu(1-1/u))}\mathbb{E}\left(u^{Z(t)-Z(s)} \,\middle|\, \mathcal{F}_s\right)\\ &= H(s)e^{(t-s)(\lambda(1-u)+\mu(1-1/u))}\mathbb{E}\left(u^{Z(t)-Z(s)}\right) = H(s).\end{aligned}$$

To show the existence and uniqueness $\mathbb{P}^u$, Daniell-Kolmogorov's Theorem (see Rogers and Williams [**44**]) is again used. It is sufficient to check the compatibility property : if $s \leq t$ and $A \in \mathcal{F}_s$,

$$\int_A H(t)\, d\mathbb{P} = \int_A H(s)\, d\mathbb{P}.$$

This equation is a direct consequence of the martingale property of $(H(t))$, indeed,

$$\int_A H(t)\, d\mathbb{P} = \int_A \mathbb{E}(H(t) \mid \mathcal{F}_s)\, d\mathbb{P},$$

because $A \in \mathcal{F}_s$, hence

$$\int_A H(t)\, d\mathbb{P} = \int_A H(s)\, d\mathbb{P}.$$

Now, the distribution of the couple $(\mathcal{N}_\lambda, \mathcal{N}_\mu)$ for the probability measure $\mathbb{P}^u$ has to be identified. In this situation, the Laplace transform of point processes is a convenient tool (see Section 1 of Chapter 1). If f and g are non-negative Borelian functions on $\mathbb{R}_+$ with support in $[0,t]$, for $t \geq 0$, the random variables

$$\int f(s)\, \mathcal{N}_\lambda(ds), \quad \text{and} \int g(s)\, \mathcal{N}_\mu(ds),$$

are clearly $\mathcal{F}_t$-measurable. By definition of $\mathbb{P}^u$,

$$\begin{aligned}\mathbb{E}^u &\left[\exp\left(-\int f(s)\, \mathcal{N}_\lambda(ds)\right) \exp\left(-\int g(s)\, \mathcal{N}_\mu(ds)\right)\right]\\ &= \mathbb{E}\left[u^{\mathcal{N}_\lambda(]0,t])} e^{\lambda t(1-u)} \exp\left(-\int f(s)\, \mathcal{N}_\lambda(ds)\right)\right.\\ &\qquad \left.\times (1/u)^{\mathcal{N}_\mu(]0,t])}\, e^{\mu t(1-1/u)} \exp\left(-\int g(s)\, \mathcal{N}_\mu(ds)\right)\right].\end{aligned}$$

From the independence of $\mathcal{N}_\lambda$ and $\mathcal{N}_\mu$ for the probability measure $\mathbb{P}$, one gets that this last term is given by

$$\begin{aligned}\mathbb{E}&\left[u^{\mathcal{N}_\lambda(]0,t])} e^{\lambda t(1-u)} \exp\left(-\int f(s)\, \mathcal{N}_\lambda(ds)\right)\right]\\ &\qquad \times \mathbb{E}\left[(1/u)^{\mathcal{N}_\mu(]0,t])}\, e^{\mu t(1-1/u)} \exp\left(-\int g(s)\, \mathcal{N}_\mu(ds)\right)\right].\end{aligned}$$

In particular, when $g = 0$, this yields

$$\mathbb{E}^u\left[\exp\left(-\int f(s)\,\mathcal{N}_\lambda(ds)\right)\right] = \mathbb{E}\left[u^{\mathcal{N}_\lambda(]0,t])}e^{\lambda t(1-u)}\exp\left(-\int f(s)\,\mathcal{N}_\lambda(ds)\right)\right],$$

for any Borelian function f with compact support on $\mathbb{R}_+$. The identity

$$\mathbb{E}^u\left[\exp\left(-\int f(s)\,\mathcal{N}_\lambda(ds)\right)\exp\left(-\int g(s)\,\mathcal{N}_\mu(ds)\right)\right] = \mathbb{E}^u\left[\exp\left(-\int f(s)\,\mathcal{N}_\lambda(ds)\right)\right)\mathbb{E}^u\left(\exp\left(-\int g(s)\,\mathcal{N}_\mu(ds)\right)\right].$$

is thus valid for all non-negative Borelian functions f and g with compact support on $\mathbb{R}$,

Therefore, the point processes $\mathcal{N}_\lambda$ and $\mathcal{N}_\mu$ are independent for $\mathbb{P}^u$. The last calculation shows that, if f is a non-negative Borelian function with support in $[0,t]$ then, for the probability distribution $\mathbb{P}^u$, the Laplace transform of the point process $\mathcal{N}_\lambda$ at f is given by, see Section 1.1 page 3,

$$\mathbb{E}\left[u^{\mathcal{N}_\lambda(]0,t])}e^{\lambda t(1-u)}\exp\left(-\int f(s)\,\mathcal{N}_\lambda(ds)\right)\right],$$

which is equal to

$$e^{\lambda t(1-u)}\mathbb{E}\left[\exp\left(-\int_0^t (f(s)-\log u)\,\mathcal{N}_\lambda(ds)\right)\right].$$

The expression of the Laplace of a Poisson point process seen in Proposition 1.5 page 6 gives the relationship

$$\begin{aligned}\mathbb{E}^u\left[\exp\left(-\int f(s)\,\mathcal{N}_\lambda(ds)\right)\right] &= e^{\lambda t(1-u)}\exp\left(-\lambda\int_0^t\left[1-\exp(-(f(s)-\log u))\right]ds\right)\\ &= \exp\left(-\lambda u\int\left[1-\exp(-f(s))\right]ds\right).\end{aligned}$$

For the probability distribution $\mathbb{P}^u$, the point process $\mathcal{N}_\lambda$ is a Poisson point process with parameter λu. The corresponding result for $\mathcal{N}_\mu$ is established by replacing u with $1/u$. □

REMARKS.

1. By using Corollary B.5 page 364, the martingale property of $(H(t))$ can be shown analytically. One has to remark that the mapping
$$(t, y) \to u^y \exp\left[(\lambda(1-u) + \mu(1-1/u))t\right]$$
is a space-time harmonic function for the Q-matrix of the free process.
2. If $u = \sqrt{\mu/\lambda}$, $(Z(t))$ is a symmetrical process for the probability measure $\mathbb{P}^u$. It has the same distribution as the difference of two independent Poisson processes with the same parameter $\sqrt{\lambda\mu}$. The corresponding exponential martingale is
$$\left[\frac{1}{\sqrt{\rho}^{\,Z(t)}} \exp\left[\left(\sqrt{\lambda}-\sqrt{\mu}\right)^2 t\right]\right].$$
3. If $u = \mu/\lambda$, for the probability measure $\mathbb{P}^u$, $\mathcal{N}_\lambda$ and $\mathcal{N}_\mu$ are independent Poisson processes whose respective parameters are given by μ and λ. The parameters of the Poisson processes are thus exchanged in this case. The exponential martingale is $(1/\rho^{Z(t)})$.

5.3 Hitting Times : Downward

In this section, the distributions of some of the hitting times associated with the Markov process $(L(t))$ are analyzed. The knowledge of these distributions allows a very detailed description of the transient behavior of the Markov process.

DEFINITION 12. *For $b \in \mathbb{N}$, T_b is the hitting time of b for the Markov process $(L(t))$, i.e.*
$$T_b = \inf\{s > 0 : L(s) = b\},$$
with the convention that $\inf\{\emptyset\} = +\infty$.

If $L(0) = 1$, T_0 is the time necessary to empty the $M/M/1$ queue which starts with one customer ; T_0 is the duration of a busy period of this queue.

PROPOSITION 5.4. *If $a \geq b$, The Laplace transform of the hitting time of b starting from a is given by*
$$\mathbb{E}_a\left(e^{-xT_b}\right) = \left(\mathbb{E}_1\left(e^{-xT_0}\right)\right)^{a-b},$$
for $x \geq 0$, with

$$\mathbb{E}_1\left(e^{-xT_0}\right) = \frac{\lambda+\mu+x-\sqrt{(\lambda+\mu+x)^2-4\lambda\mu}}{2\lambda}. \tag{5.3}$$

When $\lambda \leq \mu$, the hitting time T_0 has a finite exponential moment of order $(\sqrt{\lambda}-\sqrt{\mu})^2$ and

$$\mathbb{E}_1\left(e^{(\sqrt{\lambda}-\sqrt{\mu})^2 T_0}\right) \leq \sqrt{\frac{\mu}{\lambda}}. \tag{5.4}$$

If $\lambda \geq \mu$ and $L(0) = 1$, the variable T_0 is finite with probability μ/λ.

PROOF. If $a > b$ and $L(0) = a$, the hitting time of b can be decomposed as the sum of two terms: The hitting time of $a-1$ and the hitting time of b starting from $a-1$. The strong Markov property of $(L(t))$ (Proposition B.1 page 360) shows that these two hitting times are independent: Conditionally on $\mathcal{F}_{T_{a-1}}$, the process $(L(T_{a-1}+t))$ depends only of the value $L(T_{a-1}) = a-1$, and is therefore independent of $\mathcal{F}_{T_{a-1}}$ and in particular of T_{a-1}. By induction, it is easily seen that T_b can be decomposed as the sum of $b-a$ independent stopping times $(\tau_i; b+1 \leq i \leq a)$, the variable τ_i being the hitting time of $i-1$ starting from i. Since the process $(L(t))$ is identical (in distribution) to the free process $(Z(t))$ as long as 0 is not reached, the transitions of $(L(t))$ are space homogeneous until T_b. In particular, for $b+1 \leq i \leq a$ the distribution of τ_i does not depend on i,

$$\mathbb{E}_a\left(e^{-xT_b}\right) = \mathbb{E}_1\left(e^{-xT_0}\right)^{a-b}.$$

If $L(0) = 1$, the condition at 0 of differential Equation (5.1) does not play a role until the hitting time of 0, $L(t)$ is given by $1 + Z(t)$ until that moment. Consequently, the variable T_0 is the hitting time of -1 for the free process $(Z(t))$. The variable T_0 being a stopping time, the martingale property of $(H(t))$ and the optional stopping Theorem give

$$\mathbb{E}\left[u^{Z(t\wedge T_0)}\exp\left(\left(\lambda(1-u)+\mu(1-1/u)\right)(t\wedge T_0)\right)\right] = 1. \tag{5.5}$$

If $x \in \mathbb{R}$, the equation

$$\lambda(1-u) + \mu(1-1/u) = -x$$

has two solutions

$$\frac{\lambda+\mu+x \pm \sqrt{(\lambda+\mu+x)^2 - 4\lambda\mu}}{2\lambda}$$

if $(\lambda+\mu+x)^2 - 4\lambda\mu \geq 0$, which is true if $x \geq -\left(\sqrt{\lambda}-\sqrt{\mu}\right)^2$. It is easy to show that, if $x \geq 0$, the quantity

$$u_x = \frac{\lambda+\mu+x - \sqrt{(\lambda+\mu+x)^2 - 4\lambda\mu}}{2\lambda}$$

is non-negative and bounded by 1. By taking $u = u_x$ in Relationship (5.5), Lebesgue's Theorem shows that

$$1 = \lim_{t\to+\infty} \mathbb{E}\left(u_x^{Z(t\wedge T_0)}e^{-x\, t\wedge T_0}\right) = \mathbb{E}\left(u_x^{Z(T_0)}e^{-xT_0}\right) = \mathbb{E}\left(u_x^{-1}e^{-xT_0}\right).$$

Relationship (5.3) is proved.

If $\lambda < \mu$, the quantity $x = -(\sqrt{\lambda}-\sqrt{\mu})^2$ satisfies the relations

$$(\lambda+\mu+x)^2 - 4\lambda\mu = 0$$

and $u_x = \sqrt{\mu/\lambda} > 1$; by taking $u = u_x$ in Equation (5.5), this yields

$$\mathbb{E}_1\left(u_x^{-1}e^{\left(\sqrt{\lambda}-\sqrt{\mu}\right)^2 t\wedge T_0}\right) \leq \mathbb{E}_1\left(u_x^{Z(t\wedge T_0)}e^{\left(\sqrt{\lambda}-\sqrt{\mu}\right)^2 t\wedge T_0}\right) = 1,$$

and by letting t go to infinity, monotone convergence Theorem implies the inequality for the exponential moment of T_0 of order $(\sqrt{\lambda}-\sqrt{\mu})^2$.

If $\lambda \geq \mu$, for $x > 0$,

$$\mathbb{E}_1\left(e^{-xT_0}\right) = \mathbb{E}_1\left(e^{-xT_0} 1_{\{T_0<+\infty\}}\right),$$

this gives the identity

$$\mathbb{P}_1(T_0 < +\infty) = \lim_{x \searrow 0} \mathbb{E}_1\left(e^{-xT_0}\right) = \frac{\mu}{\lambda},$$

according to Relationship (5.3). The proposition is proved. ☐

The last part of the proposition has been already shown in the case of the $GI/M/1$ queue (see b) of Proposition 2.11 page 46).

Relationship (5.4) can be also be obtained analytically. The right hand side of Identity (5.3) is a continuous function in the right half-plane

$$\left\{\mathrm{Re}(x) \geq -(\sqrt{\lambda}-\sqrt{\mu})^2\right\}$$

and holomorphic inside. Monotone convergence Theorem shows that

$$\sqrt{\frac{\mu}{\lambda}} = \lim_{x \searrow -(\sqrt{\lambda}-\sqrt{\mu})^2} \mathbb{E}_0\left(e^{-xT_0}\right) = \mathbb{E}_0\left(e^{(\sqrt{\lambda}-\sqrt{\mu})^2 T_0}\right).$$

Inequality (5.4) is in fact an identity.

If $\lambda < \mu$, by taking the derivatives of order 1 and 2 of Equation (5.3) with respect to $x = 0$, one gets the expected value and the variance of T_0,

$$\mathbb{E}_1(T_0) = \frac{1}{\mu-\lambda}, \quad \text{and} \quad \mathrm{var}_1(T_0) = \frac{\mu+\lambda}{(\mu-\lambda)^3}.$$

In the last proof, it has been shown that if $L(0) = n$, T_0 can be written as the sum of n independent random variables with the same distribution as T_0 starting from 1. Consequently, the central limit Theorem proves the proposition.

Proposition 5.5. *If $\lambda < \mu$, if $L(0) = n$, then the convergence*

$$\lim_{n \to +\infty} \frac{T_0}{n} = \frac{1}{\mu-\lambda},$$

holds in distribution. The hitting time of 0 for the $M/M/1$ queue starting from n is asymptotically linear in n. Moreover, the variable

$$\frac{T_0 - n/(\mu-\lambda)}{\sqrt{n(\mu+\lambda)/(\mu-\lambda)^3}}$$

converges in distribution to a centered Gaussian random variable with variance 1 as n tends to infinity.

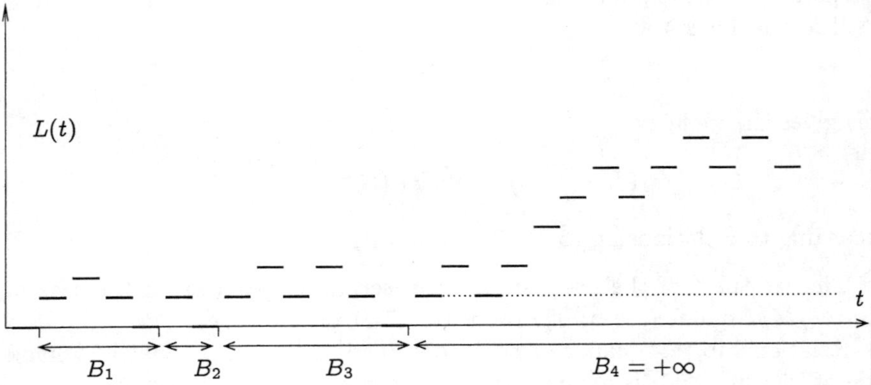

FIG. 5.1. A Sample Path of $(L(t))$ when $\lambda > \mu$

The Transient Case

This part is a description of sample paths of the $M/M/1$ queue when it is transient, i.e. when $\lambda > \mu$. If $\rho = \lambda/\mu > 1$, according to Proposition 5.4, a busy period is infinite with probability $1 - 1/\rho > 0$. In this case, a sample path $(L(t))$ can be described easily : It consists of G busy periods, the first $G-1$ busy periods are finite and the last one is infinite. The strong Markov property of $(L(t))$ shows that the number of busy periods G is a geometrically distributed random variable with parameter

$$\mu/\lambda = \mathbb{P}(T_0 < +\infty).$$

In Figure 5.1, the B_i's are the durations of the busy periods, the fourth one is infinite. The distribution of a finite busy period is given by the proposition.

PROPOSITION 5.6 (The Conditional Distribution of a Finite Busy Period). *In the transient case, when $\lambda > \mu$, the conditional distribution of the process $(L(t \wedge T_0))$ given the event $\{T_0 < +\infty\}$ has the distribution as the process $(L(t \wedge T_0))$ when the parameters λ and μ are exchanged, i.e. with arrival rate μ and service rate λ.*

PROOF. If f is a bounded measurable function on the set $F(\mathbb{R}_+, \mathbb{N})$ of integer valued functions on $\mathbb{R}_+$ endowed with the topology of simple convergence, for $a \geq 0$ and $x \in \mathbb{N}$, the identity

$$\mathbb{E}_x\left(\rho^{-Z(T_0\wedge a)} f(L(T_0 \wedge \cdot\,))1_{\{T_0<a\}}\right) = \mathbb{E}_x^{\rho^{-1}}\left(f(L(T_0 \wedge \cdot\,))1_{\{T_0<a\}}\right)$$

is obtained with the formula of change of probability in Proposition 5.3 when $u = 1/\rho$, by denoting $L(T_0 \wedge \cdot\,)$ the stopped process $(L(T_0 \wedge t))$. Since, for $t \leq T_0$, $L(t) = Z(t) + x$ when $L(0) = x$, one gets the relation

$$\mathbb{E}_x\left(\rho^{x-L(T_0\wedge a)}f(L(T_0\wedge\cdot))1_{\{T_0<a\}}\right)=\mathbb{E}_x^{\rho^{-1}}\left(f(L(T_0\wedge\cdot))1_{\{T_0<a\}}\right).$$

For the probability distribution $\mathbb{P}_x^{\rho^{-1}}$, λ and μ are exchanged and consequently T_0 is $\mathbb{P}_x^{\rho^{-1}}$-almost surely finite. Moreover, ρ being greater than 1, by letting a go to infinity and by using Lebesgue's Theorem, the last identity becomes

$$\mathbb{E}_x\left[\rho^x f(L(T_0\wedge\cdot))1_{\{T_0<+\infty\}}\right]=\mathbb{E}_x^{\rho^{-1}}\left[f(L(T_0\wedge\cdot))\right],$$

by taking $f\equiv 1$, this yields $\mathbb{P}_x(T_0<+\infty)=\rho^{-x}$, consequently,

$$\mathbb{E}_x\left[f(L(T_0\wedge\cdot))\mid T_0<+\infty\right]=\mathbb{E}_x^{\rho^{-1}}\left[f(L(T_0\wedge\cdot))\right],$$

the proposition is proved. □

The next proposition gives the distribution of $(L(t))$ on the last busy period which is infinite. It will conclude the description of the transient case.

Proposition 5.7. *Under the condition $\rho=\lambda/\mu>1$, conditionally on the event $\{T_0=+\infty\}$, the process $(L(t))$ is a birth and death process with a Q-matrix $R=(r_{ij})$ given by*

$$r_{ii+1}=\mu\frac{\rho^{i+1}-1}{\rho^i-1}\quad\text{and}\quad r_{ii-1}=\lambda\frac{\rho^{i-1}-1}{\rho^i-1},$$

for $i\geq 1$.

Proof. Proposition B.7 page 365 of Appendix B on Markov jump processes shows the Markov property of the conditioned process, its Q-matrix $R=(r_{ij})$ can be expressed with the Q-matrix $Q=(q_{ij})$ of $(L(t))$ in the following way

$$r_{ij}=\frac{\mathbb{P}_j(T_0=+\infty)}{\mathbb{P}_i(T_0=+\infty)}q_{ij},$$

for $i\neq j$. According to Proposition 5.3, $\mathbb{P}_1(T_0<+\infty)=\mu/\lambda$ and the strong Markov property and spatial homogeneity give

$$\begin{aligned}\mathbb{P}_n(T_0<+\infty)&=\mathbb{P}_n(T_{n-1}<+\infty)\mathbb{P}_{n-1}(T_0<+\infty)\\&=\prod_{i=1}^n\mathbb{P}_i(T_{i-1}<+\infty)=\left(\frac{\mu}{\lambda}\right)^n.\end{aligned}$$

Notice that this identity can also be obtained by using the fact that $\left(1/\rho^{Z(t)}\right)$ is a bounded martingale. The proposition is proved. □

5.4 Convergence to Equilibrium

The exponential moment of the duration of a busy period obtained in Proposition 5.4 gives an estimate for the convergence rate to equilibrium of this queue. The notation $\|\cdot\|_{tv}$ is for the total variation norm on measures, i.e. if μ is some measure on $\mathbb{N}$, then

$$\|\mu\|_{tv} = \frac{1}{2}\sum_{x\in\mathbb{N}} |\mu(x)|.$$

See Appendix C.

PROPOSITION 5.8. *If G_ρ is the geometric distribution on $\mathbb{N}$ with parameter $\rho = \lambda/\mu < 1$, then, for $x \in \mathbb{N}$ and $t \geq 0$, the distance to equilibrium of the distribution of $L(t)$ satisfies the inequality*

$$\|\mathbb{P}_x(L(t)\in\cdot\,) - G_\rho\|_{tv} \\ = \frac{1}{2}\sum_{n=0}^{+\infty}\left|\mathbb{P}_x(L(t)=n) - \rho^n(1-\rho)\right| \leq \left(\sqrt{\frac{\mu}{\lambda}}^{\,x} + 1\right) e^{-(\sqrt{\lambda}-\sqrt{\mu})^2 t},$$

where $\mathbb{P}_x(L(t)\in\cdot\,)$ is the distribution of the variable $L(t)$ for $\mathbb{P}_x$. The quantity $(\sqrt{\lambda}-\sqrt{\mu})^2$ is the exact exponential convergence rate to equilibrium, that is,

$$\lim_{t\to+\infty}\frac{1}{t}\log\|\mathbb{P}_0(L(t)\in\cdot\,) - G_\rho\|_{tv} = -\left(\sqrt{\lambda}-\sqrt{\mu}\right)^2.$$

It is important to note that the upper bound for the distance to equilibrium has two terms. The first one is connected with the initial state, this is the spatial component. The second term, the time component, gives the exponential decay with respect to time.

PROOF. A coupling is used to prove the first inequality. By denoting $(\widetilde{L}(t))$, the process at equilibrium, i.e. a Markov process with Q-matrix Q and such that, for $y \in \mathbb{N}$, $\mathbb{P}(\widetilde{L}(0) = y) = \rho^y(1-\rho)$. The process $(L(t))$ starting from $x \in \mathbb{N}$ behaves independently of the process $(\widetilde{L}(t))$ as long as the two processes do not meet. After they meet, $(L(t))$ follows the same sample path as the process $(\widetilde{L}(t))$. If the process at equilibrium is forgotten, it is easily checked that the process $(L(t))$ constructed in that way has the same distribution as the number of customers of an $M/M/1$ queue with x initial customers: A Markov process with Q-matrix Q and starting from x. If H is the meeting time of the two processes, then

$$\begin{aligned}\|\mathbb{P}_x(L(t)\in\cdot\,) - G_\rho\|_{tv} &= \sup_{A\subset\mathbb{N}}|\mathbb{P}_x(L(t)\in A) - \mathbb{P}(\widetilde{L}(t)\in A)| \\ &= \sup_{A\subset\mathbb{N}}|\mathbb{E}_x(1_{\{L(t)\in A\}} - 1_{\{\widetilde{L}(t)\in A\}})| \\ &\leq \sup_{A\subset\mathbb{N}}\mathbb{E}_x\left(\left|1_{\{L(t)\in A\}} - 1_{\{\widetilde{L}(t)\in A\}}\right|\right) \leq \mathbb{P}_x(H>t)\end{aligned}$$

because of the inequality

$$\left|1_{\{L(t)\in A\}} - 1_{\{\widetilde{L}(t)\in A\}}\right| \leq 1_{\left\{L(t)\neq\widetilde{L}(t)\right\}} = 1_{\{H>t\}}.$$

If $L(0) = x > \widetilde{L}(0)$, the jumps of each process being 1 or -1, when $(L(t))$ hits 0, the process $(L(t))$ has already crossed the sample path of the process at equilibrium. The coupling time of these two processes is therefore less than the hitting time of 0 by $(L(t))$. Otherwise, if $L(0) < \widetilde{L}(0)$, the hitting time of 0 of the process at equilibrium is larger than the coupling time of the two processes.

Using Relationship (5.4), one gets

$$\mathbb{E}_x\left(e^{(\sqrt{\lambda}-\sqrt{\mu})^2 H}\right) \leq \mathbb{E}_x\left(e^{(\sqrt{\lambda}-\sqrt{\mu})^2 T_0}\right)\mathbb{P}(\widetilde{L}(0)\leq x) \\ + \mathbb{E}\left(\mathbb{E}_{\widetilde{L}(0)}\left(e^{(\sqrt{\lambda}-\sqrt{\mu})^2 T_0}\right)1_{\{\widetilde{L}(0)>x\}}\right).$$

Consequently,

$$\mathbb{E}_x\left(e^{(\sqrt{\lambda}-\sqrt{\mu})^2 H}\right) \leq \sqrt{\frac{\mu}{\lambda}}^{x}\mathbb{P}(\widetilde{L}(0)\leq x) + \sum_{y=x+1}^{+\infty}\sqrt{\frac{\mu}{\lambda}}^{y}\mathbb{P}(\widetilde{L}(0)=y)$$

$$= \sqrt{\frac{\mu}{\lambda}}^{x}(1-\rho^{x+1}) + \sum_{y=x+1}^{+\infty}\sqrt{\frac{\mu}{\lambda}}^{y}\rho^y(1-\rho) \leq \sqrt{\frac{\mu}{\lambda}}^{x} + \sqrt{\frac{\lambda}{\mu}}^{x+1},$$

and, by using Chebishov's Inequality,

$$\|\mathbb{P}_x(L(t)\in\cdot\,) - G_\rho\|_{tv} \leq \mathbb{E}_x\left(e^{(\sqrt{\lambda}-\sqrt{\mu})^2 H}\right)e^{-(\sqrt{\lambda}-\sqrt{\mu})^2 t},$$

hence

$$\|\mathbb{P}_x(L(t)\in\cdot\,) - G_\rho\|_{tv} \leq \left(\sqrt{\frac{\mu}{\lambda}}^{x}+1\right)e^{-(\sqrt{\lambda}-\sqrt{\mu})^2 t}.$$

The first inequality of the proposition is proved. One deduces

$$\limsup_{t\to+\infty}\frac{1}{t}\log\|\mathbb{P}_0(L(t)\in\cdot\,) - G_\rho\|_{tv} \leq -\left(\sqrt{\lambda}-\sqrt{\mu}\right)^2,$$

and, since

$$\|\mathbb{P}_0(L(t)\in\cdot\,) - G_\rho\|_{tv} \geq |\mathbb{P}_0(L(t)=0)-(1-\rho)|,$$

to establish the last assertion, it is sufficient to show that

$$\liminf_{t\to+\infty}\frac{1}{t}\log|\mathbb{P}_0(L(t)=0)-(1-\rho)| \geq -\left(\sqrt{\lambda}-\sqrt{\mu}\right)^2.$$

Another coupling is used where $(\widetilde{L}(t))$ is, as before, the process at equilibrium and $(L(t))$ the process starting from 0. Contrary to the previous coupling, the two processes use the same Poisson processes $\mathcal{N}_\lambda$ and $\mathcal{N}_\mu$. Consequently, the initial condition $L(0) = 0$ shows that the process at equilibrium is always above $(L(t))$, i.e. $L(t) \leq \widetilde{L}(t)$ for any t. Moreover, on the intervals where $L(t) > 0$, the difference $\widetilde{L}(t) - L(t)$ remains constant.

The variable $\widetilde{T}_0$, the hitting time of 0 by $(\widetilde{L}(t))$, it is the first time the two processes meet. Indeed, at $\widetilde{T}_0$ the two processes are equal to 0. Since the difference $\widetilde{L}(t) - L(t)$ decreases only when $L(t) = 0$, this implies that it can be 0 for the first time only when $\widetilde{L}(t) = 0$, hence at $\widetilde{T}_0$. Therefore, the inequality

$$1_{\{L(t)=0\}} - 1_{\{\widetilde{L}(t)=0\}} \geq 1_{\{\widetilde{T}_0>t,\, \widetilde{L}(t)=1\}}$$

holds. Indeed, the left hand side of this inequality is non-negative. If the right hand side is 1, then $\widetilde{T}_0 > t$, or equivalently the two processes have not met before time t, hence $L(t) < \widetilde{L}(t)$. Since $\widetilde{L}(t) = 1$, the quantity $L(t)$ is necessarily 0. The left hand side is also 1 in this case. The inequality is established. One gets the relation

$$\begin{aligned} |\mathbb{P}_0(L(t)=0) - (1-\rho)| &= \mathbb{P}_0(L(t)=0) - \mathbb{P}_0\left(\widetilde{L}(t)=0\right) \\ &\geq \mathbb{P}\left(\widetilde{L}(t)=1, \widetilde{T}_0 > t\right). \end{aligned}$$

The process $(\widetilde{L}(t))$ being reversible (Proposition 5.1), this last quantity is

$$\mathbb{P}(\widetilde{L}(0)=1, \widetilde{T}_0 > t) = \rho(1-\rho)\mathbb{P}_1(T_0 > t).$$

The above inequality becomes

$$|\mathbb{P}_0(L(t)=0) - (1-\rho)| \geq \rho(1-\rho)\mathbb{P}_1(T_0 > t).$$

It is therefore sufficient to study the tail distribution of T_0. The next lemma concludes the proof of the proposition. □

LEMMA 5.9. *Under the condition* $\rho < 1$,

$$\lim_{t\to+\infty} \frac{1}{t} \log \mathbb{P}_1(T_0 > t) = -\left(\sqrt{\lambda} - \sqrt{\mu}\right)^2. \tag{5.6}$$

PROOF. As before, T_0 is interpreted as the hitting time of -1 by the free process $(Z(t))$. If $v > 1/\sqrt{\rho}$, then $\lambda v - \mu/v > 0$. Thus, for the probability distribution $\mathbb{P}^v$, the variable $(Z(t))$ has a positive drift and according to Proposition 5.4, the inequality $\mathbb{P}^v(T_0 = +\infty) > 0$ holds. By using the formula of Proposition 5.3 for the change of probability distribution, one gets

$$\begin{aligned} \mathbb{P}(T_0 > t) &= e^{-(\lambda(1-v)+\mu(1-1/v))t}\, \mathbb{E}^v\left(v^{-Z(t)} 1_{\{T_0>t\}}\right) \\ &\geq e^{-(\lambda(1-v)+\mu(1-1/v))t}\, \mathbb{E}^v\left(v^{-Z(t)} 1_{\{T_0=+\infty\}}\right). \end{aligned} \tag{5.7}$$

Jensen's inequality for the probability distribution $\mathbb{P}^v$ conditioned on the event $\{T_0 = +\infty\}$ gives the relation

$$\begin{aligned} \mathbb{E}^v\left(v^{-Z(t)} 1_{\{T_0=+\infty\}}\right) &= \mathbb{E}^v\left(v^{-Z(t)} \,\middle|\, T_0 = +\infty\right) \mathbb{P}^v(T_0 = +\infty) \\ &\geq v^{-\mathbb{E}^v(Z(t)\,|\,T_0=+\infty)} \mathbb{P}^v(T_0 = +\infty). \end{aligned} \tag{5.8}$$

The exponent of v in the right hand side is given by

$$\mathbb{E}^v(Z(t)\,|\,T_0 = +\infty) = \frac{1}{\mathbb{P}^v(T_0 = +\infty)} \mathbb{E}^v\left(Z(t) 1_{\{T_0=+\infty\}}\right).$$

When t goes to infinity, the variable $Z(t)/t$ converges to $\lambda v - \mu/v$ in $L_2(\mathbb{P}^v)$ (Law of large numbers, Corollary 10.10 page 288), one deduces the relation

$$\lim_{t\to+\infty} \mathbb{E}^v \left(\frac{Z(t)}{t} \;\middle|\; T_0 = +\infty \right) = \lambda v - \frac{\mu}{v}.$$

Since

$$\lim_{t\to+\infty} \frac{1}{t} \log \mathbb{P}^v(T_0 = +\infty) = 0,$$

Identity (5.8) gives the lower bound

$$\liminf_{t\to+\infty} \frac{1}{t} \log \mathbb{E}^v \left(v^{-Z(t)} 1_{\{T_0=+\infty\}} \right) \geq -\left(\lambda v - \frac{\mu}{v} \right) \log v,$$

hence, with Relationship (5.7), one gets

$$\liminf_{t\to+\infty} \frac{1}{t} \log \mathbb{P}(T_0 > t) \geq -\left[\lambda(1-v) + \mu\left(1 - \frac{1}{v}\right)\right] - \left(\lambda v - \frac{\mu}{v} \right) \log v.$$

By letting v go to $1/\sqrt{\rho}$, finally

$$\liminf_{t\to+\infty} \frac{1}{t} \log \mathbb{P}(T_0 > t) \geq -\left(\sqrt{\lambda} - \sqrt{\mu} \right)^2.$$

With Inequality (5.4) and Chebishov's Inequality, one gets

$$\limsup_{t\to+\infty} \frac{1}{t} \log \mathbb{P}(T_0 > t) \leq -\left(\sqrt{\lambda} - \sqrt{\mu} \right)^2.$$

Relationship (5.6) is therefore established. □

The quantity $(\sqrt{\lambda} - \sqrt{\mu})^2$ is the exponential rate of convergence to equilibrium, it is also the exponential rate of the tail distribution of T_0. This result is not really surprising since T_0 can be seen as the hitting time of a non-negligible set for the invariant distribution: The neighborhood of 0.

The fact that the variable T_0 does not have an exponential moment of order greater than $(\sqrt{\lambda} - \sqrt{\mu})^2$ can be proved analytically by using Identity (5.3). The function $z \to \mathbb{E}(\exp(-zT_0))$ is holomorphic in the half-plane

$$\left\{ \mathrm{Re}(z) > -\left(\sqrt{\lambda} - \sqrt{\mu} \right)^2 \right\}$$

and $-(\sqrt{\lambda} - \sqrt{\mu})^2$ is a singular point.

5.5 Hitting Times : Upward

For $a \leq b$, the distribution of the hitting time T_b of b starting from a is investigated. This stopping time is also interpreted as the time to overflow a single server queue with finite capacity. If $C \in \mathbb{N}^*$, the number of customers $(L_C(t))$

of the $M/M/1/C$ queue satisfies the same stochastic differential equation as the $M/M/1$ queue, with an additional boundary condition,

$$dL_C(t) = 1_{\{L_C(t-)\leq C\}} \mathcal{N}_\lambda(dt) - 1_{\{L_C(t-)>0\}} \mathcal{N}_\mu(dt).$$

If both queues start with a customers, $a < C$, before time T_{C+1}, the quantities $L(t)$ and $L_C(t)$ are the same. The time T_{C+1} is the first moment a customer is rejected in the $M/M/1/C$ queue.

In the case considered in this section, the reflection of $(L(t))$ at 0 plays a role, the free process $(Z(t))$ and its martingales cannot be used directly as in Section 5.3. To get an interesting martingale to obtain the distribution of T_b, a linear combination of the two martingales associated with the free process is used. The coefficients of this combination are chosen so that the martingale property is still valid if the process visits 0. Proposition 5.3 gives two such martingales, for $x \geq 0$,

$$\phi_1(x) = \frac{\lambda + \mu + x - \sqrt{(\lambda+\mu+x)^2 - 4\lambda\mu}}{2\lambda},$$
$$\phi_2(x) = \frac{\lambda + \mu + x + \sqrt{(\lambda+\mu+x)^2 - 4\lambda\mu}}{2\lambda},$$

then for $i = 1$ and 2, $(\phi_i(x)^{Z(t)} \exp(-xt))$ is a martingale. The next construction is due, in a more general framework, to Kennedy[Ken76].

PROPOSITION 5.10 (Martingales of the Reflected Process). *For $x \geq 0$, if $\phi_1(x)$ and $\phi_2(x)$ are the solutions of the equation*

$$\lambda(1-u) + \mu\left(1 - \frac{1}{u}\right) = -x,$$

the process

$$(M(t)) = \left(\exp(-xt)\left[(1-\phi_1(x))\,\phi_2(x)^{L(t)+1} + (\phi_2(x)-1)\phi_1(x)^{L(t)+1}\right]\right),$$

is a martingale.

PROOF. Set

$$h(t,y) = \exp(-xt)\left[(1-\phi_1(x))\phi_2(x)^{y+1} + (\phi_2(x)-1)\phi_1(x)^{y+1}\right],$$

for $t \geq 0$ then $M(t) = h(t, L(t))$ and, according to Corollary B.5 page 364, it is sufficient to show that h is space-time harmonic for the Q-matrix Q, that is, for $t \geq 0$ and $y \in \mathbb{N}$,

$$\frac{\partial h}{\partial t}(t,y) + Q(h)(t,y) = 0.$$

This property is clearly true for $y > 0$ since for $i = 1$ and 2,

$$(t,y) \to \phi_i(x)^y e^{-xt}$$

[Ken76] D.P. Kennedy, *Some martingales related to cumulative sum tests and single-server queues*, Stochastic Processes and their Applications **4** (1976), 261–269.

is space-time harmonic for the $\mathcal{Q}$-matrix of the free process (See Remark 1 page 109), which coincides with Q for $y > 0$. Therefore, the above identity has only to be checked at $y = 0$,

$$e^{xt}\frac{\partial h}{\partial t}(t,0) = -x(\phi_2(x) - \phi_1(x)),$$
$$e^{xt}Q(h)(t,0) = \lambda\left(1-\phi_1(x)\right)\left(\phi_2(x)-1\right)\left(\phi_2(x)-\phi_1(x)\right),$$

The identity uses the relationship $\lambda\,(1-\phi_1(x))\,(\phi_2(x)-1) = x$, for $x \geq 0$, which is clear since $\phi_1(x)+\phi_2(x) = (\lambda+\mu+x)/\lambda$ and $\phi_1(x)\phi_2(x) = \mu/\lambda$. □

The main result of this section is the next proposition.

PROPOSITION 5.11. *If $a \leq b$, then for $x \geq 0$,*

$$\mathbb{E}_a\left(e^{-xT_b}\right) = (\rho\phi_1(x))^{b-a}\frac{1-\phi_1(x)+(1-\rho\phi_1(x))\rho^a\phi_1(x)^{2a+1}}{1-\phi_1(x)+(1-\rho\phi_1(x))\rho^b\phi_1(x)^{2b+1}}, \tag{5.9}$$

with

$$\phi_1(x) = \frac{\lambda+\mu+x-\sqrt{(\lambda+\mu+x)^2-4\lambda\mu}}{2\lambda}.$$

Moreover, when $\lambda < \mu$ and n tends to infinity, the variable $(\lambda/\mu)^n T_n$ converges in distribution to an exponentially distributed random variable with parameter $(\mu-\lambda)^2/\mu$.

PROOF. The irreducibility of the Markov process $(L(t))$ implies that the variable T_b is $\mathbb{P}_a$-almost surely finite. The martingale $(M(t))$ of Proposition 5.10 stopped at T_b is bounded and therefore converges in L_1 as t tends to infinity. One gets $\mathbb{E}(M(T_b)) = \mathbb{E}(M(0))$, consequently,

$$\mathbb{E}_a\left(e^{-xT_b}\right) = \frac{(1-\phi_1(x))\phi_2(x)^{a+1}+(\phi_2(x)-1)\phi_1(x)^{a+1}}{(1-\phi_1(x))\phi_2(x)^{b+1}+(\phi_2(x)-1)\phi_1(x)^{b+1}},$$

Identity (5.9) is proved since $\phi_1(x)\phi_2(x) = \mu/\lambda$.

For the second part of the proposition, a is assumed to be 0, the proof for the general case is similar. The Laplace transform of T_n is given by

$$\mathbb{E}_0\left(e^{-xT_n}\right) = (\rho\phi_1(x))^n\frac{1-\phi_1(x)+(1-\rho\phi_1(x))\phi_1(x)}{1-\phi_1(x)+(1-\rho\phi_1(x))\rho^n\phi_1(x)^{2n+1}}, \tag{5.10}$$

for $x \geq 0$. The derivative of ϕ_1 at 0 being $-1/(\mu-\lambda)$, condition $\lambda < \mu$ shows that for n in the neighborhood of infinity,

$$\phi_1(x\rho^n) = 1 - \frac{x}{\mu-\lambda}\rho^n + o(\rho^n),$$

that is, $\phi_1(x\rho^n)^n \sim 1$. By using this relationship in Equation (5.10), one gets

$$\mathbb{E}_0\left(e^{-x\rho^nT_n}\right) = \rho^n\frac{1-\rho+o(1}{x\rho^n/(\mu-\lambda)+(1-\rho)\rho^n+o(\rho^n)}$$
$$= \frac{1-\rho}{x/(\mu-\lambda)+1-\rho} + o(1).$$

Consequently,

$$\lim_{n\to+\infty} \mathbb{E}_0\left(e^{-x\rho^n T_n}\right) = \frac{(\mu-\lambda)^2/\mu}{(\mu-\lambda)^2/\mu + x},$$

the proposition is proved. □

If $\lambda < \mu$, the process $(L(t))$ starting from n goes very quickly to 0, within a duration of time linear with respect to n (Proposition 5.5). In the reverse direction, the time necessary to reach n starting from 0 is exponential in n, of order $(\mu/\lambda)^n$.

In the two sections devoted to hitting times, the Laplace transforms of the stopping times have been obtained. Inverting these transforms is not a simple operation, and even when it is possible, the expressions of the densities do not give directly a qualitative description of the behavior of the process (see Section 5.9).

The bound of an exponential moment of T_0 obtained, via a probabilistic argument, has given a quite precise estimation of the rate of convergence to equilibrium. As it will be seen, a description of rare events occurring for this queue is obtained from the convergence result of Proposition 5.11.

5.6 Rare Events

5.6.1 Large Busy Periods

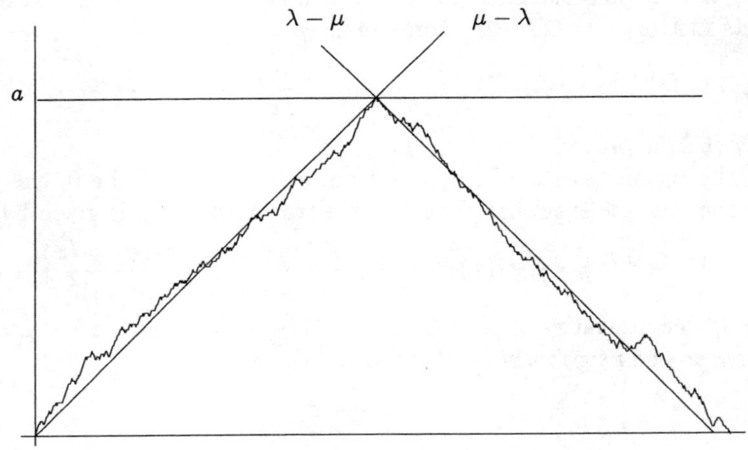

FIG. 5.2. Simulation of a Large Busy Period

If $\rho = \lambda/\mu < 1$, a sample path of the $M/M/1$ queue reaches large values on an exponential time scale, of order $(\mu/\lambda)^a$, to hit a (Proposition 5.11).

Throughout this section, it is assumed that $\rho < 1$, the process $(L(t))$ is analyzed when it hits large values, i.e. the limiting distribution of a busy period when the number of customers reaches the value a, when a goes to infinity. Section 3.6 of Chapter 3 contains a similar analysis for random walks. For a general Markov process, such a study is quite technical. The restricted framework of the $M/M/1$ queue allows an explicit description of these busy periods.

PROPOSITION 5.12. *If $\lambda < \mu$, the distribution of $(L(t \wedge T_a))$ given the event $\{T_a < T_0\}$ converges, as a tends to infinity, to the distribution of a birth and death process whose Q-matrix $R = (r_{ij})$ is given by*

$$r_{xx+1} = \mu \frac{1-\rho^{x+1}}{1-\rho^x}, \qquad r_{xx-1} = \lambda \frac{1-\rho^{x-1}}{1-\rho^x}. \tag{5.11}$$

PROOF. It is assumed that $L(0) = 1$, the general case is similar. For $a \in \mathbb{R}_+$, the variable τ denotes the stopping time $T_a \wedge T_0$. On the time interval $[0, \tau]$, the process $(L(t))$ is identical to the process $(1 + Z(t))$. If

$$\begin{aligned} F : D(\mathbb{R}_+, \mathbb{N}) &\to \mathbb{R}_+ \\ y &\to F(y), \end{aligned}$$

is a bounded measurable functional on the set of càdlàg functions (see Definition 40 page 359) from $\mathbb{R}_+$ to $\mathbb{N}$ endowed with Skorohod topology (See Appendix C or Billingsley [7]). Since $\left(1/\rho^{Z(t)}\right)$ is the exponential martingale corresponding to $u = 1/\rho$, one gets that

$$\begin{aligned} \mathbb{E}_1 \left[F(L(\tau \wedge t \wedge \cdot))\right] &= \mathbb{E}_1 \left[\rho^{Z(\tau\wedge t)} F(L(\tau \wedge t \wedge \cdot)) \frac{1}{\rho^{Z(\tau\wedge t)}}\right] \\ &= \mathbb{E}_1^{\rho^{-1}} \left[\rho^{Z(\tau\wedge t)} F(L(\tau \wedge t \wedge \cdot))\right] = \mathbb{E}_1^{\rho^{-1}} \left[\rho^{L(\tau\wedge t)-1} F(L(\tau \wedge t \wedge \cdot))\right], \end{aligned}$$

with the notations of Section 5.2. The variable τ being $\mathbb{P}_1^{\rho^{-1}}$-a.s. finite and since $\rho < 1$, Lebesgue's Theorem shows that

$$\mathbb{E}_1 \left[F(L(\tau \wedge \cdot))\right] = \mathbb{E}_1^{\rho^{-1}} \left[\rho^{L(\tau)-1} F(L(\tau \wedge \cdot))\right].$$

If $T_a(y)$ is the hitting time of a for the function y), the last identity applied to the functional

$$\begin{aligned} D(\mathbb{R}_+, \mathbb{N}) &\to \mathbb{R}_+ \\ y &\to F(y) 1_{\{T_a(y) < T_0(y)\}} \end{aligned}$$

gives the identity

$$\begin{aligned} \mathbb{E}_1 \left(F(L(\tau \wedge \cdot)) 1_{\{T_a < T_0\}}\right) &= \mathbb{E}_1^{\rho^{-1}} \left(\rho^{L(\tau)-1} F(L(\tau \wedge \cdot)) 1_{\{T_a < T_0\}}\right) \\ &= \mathbb{E}_1^{\rho^{-1}} \left(\rho^{L(T_a)-1} F(L(T_a \wedge \cdot)) 1_{\{T_a < T_0\}}\right), \end{aligned}$$

hence,

(5.12) $$\mathbb{E}_1\left[F(L(T_a\wedge\cdot))1_{\{T_a<T_0\}}\right]=\mathbb{E}_1^{\rho^{-1}}\left[\rho^{a-1}F(L(T_a\wedge\cdot))1_{\{T_a<T_0\}}\right].$$

According to Remark 3 page 109, the intensities of the two processes $\mathcal{N}_\lambda$ and $\mathcal{N}_\mu$ are exchanged for the probability measure $\mathbb{P}_1^{u_0}$ with $u_0=\rho^{-1}$. The process $(L(t))$ is transient for the probability distribution $\mathbb{P}_1^{u_0}$ if $\rho<1$, consequently $\mathbb{P}_1^{\rho^{-1}}(T_0=+\infty)>0$. By letting a go to infinity in (5.12), Lebesgue's Theorem shows that

$$\lim_{a\to+\infty}\rho^{1-a}\mathbb{E}_1\left[F(L(T_a\wedge\cdot))1_{\{T_a<T_0\}}\right]=\mathbb{E}_1^{\rho^{-1}}\left[F(L(\cdot))1_{\{T_0=+\infty\}}\right],$$

consequently,

(5.13) $$\lim_{a\to+\infty}\mathbb{E}_1\left[F(L(T_a\wedge\cdot))\mid T_a<T_0\right]=\mathbb{E}_1^{\rho^{-1}}\left[F(L(\cdot))\mid T_0=+\infty\right]$$

The right hand side is expressed with the conditional distribution of $(L(t))$ given that the hitting time of 0 is infinite. Proposition 5.7 is used to conclude. □

Optional stopping Theorem for the martingale $(\mu/\lambda)^{1+Z(t)}$ and the stopping time $T_a\wedge T_0$ gives

$$\mathbb{P}_1(T_a<T_0)=\frac{1/\rho-1}{1/\rho^a-1}=\rho^{a-1}\frac{1-\rho}{1-\rho^a},$$

which can be seen in a discrete setting as the formula for the ruin probability of players, $\lim_{a\to+\infty}\rho^{-a}\mathbb{P}_1(T_a<T_0)=(1-\rho)/\rho$.

Relationship (5.13) shows that a typical sample path of a busy period until it hits the level a is the same as the corresponding sample path of a busy period of an $M/M/1$ queue with arrival rate μ and service rate λ. This last queue is, of course, transient, in particular it reaches a quickly in a time linear with respect to a (Proposition 5.4 for the hitting time of 0 by $(a-L(t))$).

The downward part of the busy period after T_a is, because of the strong Markov property, identical to the sample path of the free process starting from a. In particular, the drift of the process on this part is $\lambda-\mu$. According to Equation (5.11), the drift of the first part for x large is $\mu-\lambda$.

5.6.2 The Excursion Process in the Neighborhood of a

To describe the instants when the number of customers of the $M/M/1$ queue is greater than a, the sequence of stopping times (R_n) and (H_n) are introduced by induction,

$$\begin{aligned}H_n&=\inf\{t>R_{n-1}:L(t)=a\},\\R_n&=\inf\{t>H_n:L(t)=0\}\\&=\inf\left\{t>R_{n-1}:L(t)=0,L(s)\geq a\text{ for some }s\in[R_{n-1},t]\right\},\end{aligned}$$

with $R_0=0$. The point process N_a associated with the non-decreasing sequence (H_n) is the point process of the hitting times of a after a visit to 0.

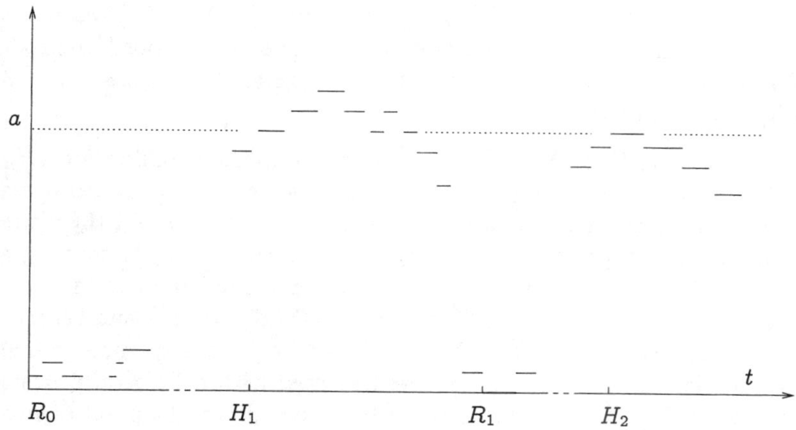

FIG. 5.3. A Sequence of Excursions of $(L(t))$ in the Neighborhood of a

The strong Markov property of $(L(t))$ shows that the sequences $(H_n - R_{n-1})$ and $(R_n - H_n)$, are i.i.d. and independent.

For $n \in \mathbb{N}$, the variable $H_n - R_{n-1}$ has the same distribution as the hitting time of a starting from 0. Proposition 5.11 shows that when a tends to infinity, the variable $\rho^a(H_n - R_{n-1})$ converges in distribution to an exponentially distributed random variable with parameter $(\mu - \lambda)^2/\mu$. According to Proposition 5.5, the time to go to 0 is very short, $\mathbb{P}$-almost surely,

$$\lim_{a \to +\infty} \rho^a(R_n - H_n) = 0,$$

since Proposition 5.5 shows that $(R_n - H_n)/a$ converges $\mathbb{P}$-a.s. to a constant as a tends to infinity. The increments of the point process

$$N_a = \sum_{n \geq 1} \delta_{\rho^a H_n},$$

converge in distribution to i.i.d. exponentially distributed random variables with parameter $(\mu - \lambda)^2/\mu$, therefore the following proposition holds.

PROPOSITION 5.13. *If $\rho < 1$, the point process*

$$N_a = \sum_{n} \delta_{\rho^a H_n}$$

converges in distribution to a Poisson process with parameter $(\mu - \lambda)^2/\mu$.

PROOF. Since the sequence $(\rho^a H_n - \rho^a H_{n-1})$ converges in distribution to an i.i.d. sequence of exponentially distributed random variables, Proposition 1.23 page 22 shows that the point process N_a converges to a Poisson point process with parameter $(\mu - \lambda)^2/\mu$. □

The process N_a does not describe all the hitting times of the level a by $(L(t))$, i.e. the instants s for which $L(s) = a$ and $L(s-) \neq a$. This point process cannot be asymptotically Poisson: If $L(t_0)$ is a, with a positive probability, $(L(t))$ visits a several times after t_0 before going to 0. See Figure 5.3 in the neighborhood of H_1.

An Informal Description. For a large a and $n \in \mathbb{N}$, in the neighborhood H_n the process $(L(t))$ behaves like $(a + Z_n(t - H_n))$, where $(Z_n(t))$ is a process with the same distribution as the free process $(Z(t))$. Just after H_n, the visits of $(L(t))$ at a, correspond to visits of $(Z_n(t))$ at 0. Since $(Z_n(t))$ converges $\mathbb{P}$-a.s. to $-\infty$ when t tends to infinity, only a finite number of such visits occur after H_n. Then, the process $(L(t))$ goes to 0 (at $t = R_n$) and returns to a after a very long time after R_n. If (Z_n) is an i.i.d. sequence of free processes, the visits of $(L(t))$ at a can be described with a Poisson process with parameter $\rho^a(\mu - \lambda)^2/\mu$ (the instants H_n), and to the nth point of this point process one associates the visits at 0 of the process $(Z_n(t))$.

The next proposition gives the distribution of the time spent at 0 by the free process.

PROPOSITION 5.14. *If $\rho < 1$, the variable*

$$S_0 = \int_0^{+\infty} 1_{\{Z(s)=0\}}\, ds$$

is exponentially distributed with parameter $\mu - \lambda$.

PROOF. The free process stays at 0 for a duration of time $E_{\lambda+\mu}$, exponentially distributed with parameter $\lambda + \mu$, before a jump at 1 [resp. -1] with probability $\lambda/(\lambda + \mu)$ [resp. $\mu/(\lambda + \mu)$]. If the jump is at 1, then $(L(t))$ visits 0 again, in this case $S_0 = E_{\lambda+\mu} + S_0'$, where S_0' is independent of $E_{\lambda+\mu}$ and with the same distribution as S_0. If the jump is at -1, with probability λ/μ, the process goes back to 0 (Proposition 5.4). One concludes that S_0 is the sum of G i.i.d. random variables, exponentially distributed with parameter $\lambda + \mu$, and G is an independent, geometrically distributed random variable with parameter $2\lambda/(\lambda + \mu)$. by taking Laplace transform, it is easily seen that the distribution of S_0 is exponential with parameter $\mu - \lambda$. □

This description of rare events as a Poisson process with a "cloud of points" around each point of the point process is quite common in a Markovian setting. Aldous [**1**] gives many examples of such situations.

5.7 Fluid Limits

Fluid limits are the results of a scaling of the number of customers of an $M/M/1$ queue. The scaling considered here consists in speeding up the time scale with the size of its initial state and in scaling the process with the same quantity. The procedure suppresses some random fluctuations around what appears to be the main trajectory of the process. For the $M/M/1$ queue the behavior of the rescaled process is very simple. This example serves also as an introduction to these questions which are considered in Chapter 9.

Initially, there are x_N customers in the queue and the sequence (x_N) is such that

$$\lim_{N\to+\infty} \frac{x_N}{N} = x \in \mathbb{R}_+.$$

The *renormalized process* $(\bar{L}_N(t))$ is defined by

$$\bar{L}_N(t) = \frac{L(Nt)}{N},$$

where $(L(t))$ is the solution of differential Equation (5.1) with initial condition x_N. Notice that $\bar{L}_N$ lives on a very rapid time scale $t \to Nt$, arrivals and services are sped up by a factor N. The scaling by $1/N$ compensates the acceleration of time.

The next lemma presents a uniform estimate of the rate of convergence of the law of large numbers for Poisson processes.

Lemma 5.15. *For ξ, $t \in \mathbb{R}_+$, ε, $\delta > 0$, there exists N_0 such that if $N \geq N_0$, then*

$$\mathbb{P}\left(\sup_{0\leq s\leq t}\left|\frac{\mathcal{N}_\xi(]0,sN])}{N} - \xi s\right| < \delta\right) \geq 1-\varepsilon.$$

Proof. Proposition 1.21 page 20 shows that the process

$$\left(\frac{\mathcal{N}_\xi(]0,Nt])}{N} - \xi t\right)$$

is martingale with respect to its natural filtration. Since it is square integrable, Doob's inequality gives the identity

$$\mathbb{P}\left(\sup_{0\leq s\leq t}\left|\frac{\mathcal{N}_\xi(]0,Ns])}{N} - \xi s\right| \geq \delta\right) \leq \frac{\mathbb{E}\left(\mathcal{N}_\xi(]0,Nt])/N - \xi t\right)^2}{\delta^2} = \frac{\xi t}{N\delta^2},$$

and thus the lemma. □

The proposition below is the main result concerning the convergence of the rescaled process $(\overline{X}(t))$.

Proposition 5.16 (Fluid Limits of the $M/M/1$ Queue). *The renormalized process $(\bar{L}_N(t))$ converges to a deterministic function, piecewise linear, i.e. $\mathbb{P}$-almost surely, the convergence of processes associated with the uniform norm on compact sets holds*

$$\lim_{N\to+\infty} \left(\bar{L}_N(t)\right) = \left(\bar{L}(t)\right) = \left(x + (\lambda-\mu)\,t\right)^+.$$

Moreover, if the sequence $\left(\sqrt{N}(x_N/N - x)\right)$ tends to 0 *and $(B(t))$ is the standard Brownian motion, then*

$$\sqrt{N}\left(\frac{\bar{L}_N(t) - \bar{L}(t)}{\sqrt{\lambda+\mu}}\right) \overset{dist.}{\longrightarrow} (B(t)), \tag{5.14}$$

on the time interval $[0, x/(\mu - \lambda)^+[$ *when* N *tends to infinity.*

PROOF. If $\lambda < \mu$, $\tau_N = T_0/N$ is the hitting time of 0 by $(\bar{L}_N(t))$. Proposition 5.5 shows that

$$\lim_{N \to +\infty} \tau_N = \frac{x}{\mu - \lambda},$$

$\mathbb{P}$-almost surely. According to Identity (5.1), until the hitting time of 0, the process $(L(t) - x_N)$ can be expressed as the difference of two Poisson process,

$$\bar{L}_N(t \wedge \tau_N) = \frac{x_N}{N} + \frac{\mathcal{N}_\lambda(]0, N(t \wedge \tau_N)]) - \mathcal{N}_\mu(]0, N(t \wedge \tau_N)])}{N}. \tag{5.15}$$

For $t < x/(\mu - \lambda)$, $\mathbb{P}$-almost surely $t < \tau_N$ after some (random) rank N_0, and consequently, Lemma 5.15 shows the convergence of the processes on the interval $[0, x/(\mu - \lambda)[$,

$$\lim_{N \to +\infty} \left(\bar{L}_N(t)\right) = \left(x + (\lambda - \mu)t\right),$$

almost surely. It remains to show that for $t \geq x/(\mu - \lambda)$, $\mathbb{P}$-a.s.

$$\lim_{N \to +\infty} \bar{L}_N(t) = 0.$$

If $(\bar{L}_N^y(t))$ is the rescaled process with the initial condition $L(0) = \lfloor yN \rfloor$ with the same Poisson processes, it has been seen that if $y + (\lambda - \mu)t > 0$, then $\mathbb{P}$-almost surely

$$\lim_{N \to +\infty} \bar{L}_N^y(t) = y + (\lambda - \mu)t.$$

The function $y \to \bar{L}_N^y(t)$ is clearly non-decreasing.

If $\varepsilon > 0$ is fixed and $y = \varepsilon + (\mu - \lambda)t$, then $y > x$. There exists N_0 such that, if $N \geq N_0$, then $x_N \leq \lfloor yN \rfloor$. The monotonicity property gives, for $N \geq N_0$,

$$\bar{L}_N(t) \leq \bar{L}_N^y(t),$$

hence, the inequality

$$\limsup_{N \to +\infty} \bar{L}_N(t) \leq \limsup_{N \to +\infty} \bar{L}_N^y(t) = y + (\lambda - \mu)t = \varepsilon$$

holds $\mathbb{P}$-a.s. This upper bound shows that for $\varepsilon > 0$, $\mathbb{P}$-a.s.

$$\limsup_{N \to +\infty} \bar{L}_N(t) \leq \varepsilon,$$

and thus, $\lim_{N \to +\infty} \bar{L}_N(t) = 0$.

The Case $\lambda > \mu$. The law of large numbers shows that the quantity

$$\inf_{s \geq 0} \left(\mathcal{N}_\lambda(]0, s]) - \mathcal{N}_\mu(]0, s])\right),$$

is $\mathbb{P}$-almost surely finite, since the generic term of the infimum converges to infinity. Since x_N tends to infinity, the right hand side of Equation (5.15) is $\mathbb{P}$-a.s. positive for N sufficiently large. Hence, $\mathbb{P}$-a.s. the variable τ_N is infinite for N sufficiently large. The process $(\bar{L}_N(t))$ can therefore be expressed as the difference of two Poisson processes, the rest of proof of this case is similar to the previous one.

For $\xi \in \mathbb{R}_+$, Donsker's Theorem for Poisson processes shows the convergence in distribution

$$\left(\frac{\mathcal{N}_\xi(]0, Nt]) - \xi Nt}{\sqrt{\xi N}}\right) \stackrel{\text{dist.}}{\longrightarrow} (B(t)),$$

as N tends to infinity. By independence of $\mathcal{N}_\lambda$ and $\mathcal{N}_\mu$, the convergence

$$\left(\frac{\mathcal{N}_\lambda(]0, Nt]) - \mathcal{N}_\mu(]0, Nt] - (\lambda - \mu)Nt}{\sqrt{(\lambda + \mu)N}}\right) \stackrel{\text{dist.}}{\longrightarrow} (B(t))$$

holds. Convergence (5.14) is then obtained with the help of Representation (5.15) of the process $(\bar{L}_N(t))$ for $t \leq \tau_N$ and the a.s. convergence of τ_N to $x/(\mu - \lambda)$. □

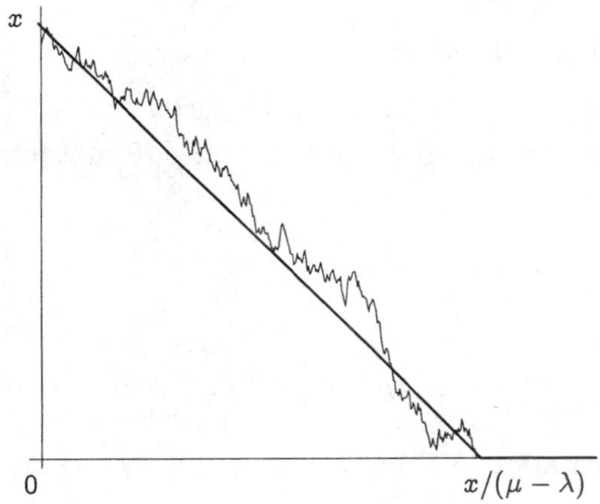

FIG. 5.4. The Renormalized Process $(\bar{L}(t))$ and the Fluctuations of $(\bar{L}_N(t))$

With the scaling defined here, the stochastic process $(L(t))$ is reduced to a deterministic drift $(\lambda - \mu)t$. The Brownian fluctuations around this trajectory vanish with such a scaling. Basically, if $L(0) \sim Nx$, then

$$L(Nt) \sim N\overline{L}(t) + \sqrt{N(\lambda + \mu)}B(t),$$

for $t \in [0, x/(\mu - \lambda)^+[$ and N large.

If $\lambda < \mu$, once the renormalized process hits 0, it remains at 0. This property is, in some sense, characteristic of ergodic Markov processes (See Proposition 9.9 page 262). It has been shown using the natural stochastic monotonicity of this queue. Intuitively, it can be argued as follows: Once the process $(L(t))$ hits 0, with the coupling argument of Proposition 5.8, it is approximately at equilibrium and it is mainly living in bounded neighborhood of 0. The time scale, linear with N, does not allow the visit of large values since an exponential time scale is necessary for this purpose (Proposition 5.11), of the order $(\mu/\lambda)^{Ny}$ to reach the value Ny. The scaling factor $1/N$ suppresses these small variations. This explains that the renormalized process is stuck at 0. A similar argument is used in the proof of Proposition 5.17.

We conclude this section by an estimate which will be useful in the analysis of large deviations for this queue.

PROPOSITION 5.17. *For ε, $\delta > 0$, there exists $N_0 \in \mathbb{N}$ such that, if $N \geq N_0$, then*

$$\inf_{|z/N-x|<\delta/2} \mathbb{P}_z\left(\sup_{0\leq s\leq t} |\bar{L}_N(s) - (x + (\lambda-\mu)\, s)^+| < \delta\right) \geq 1 - \varepsilon. \tag{5.16}$$

PROOF. For $z \in \mathbb{N}$, $\delta > 0$, denoting

$$\Delta_N(s) = \bar{L}_N(s) - (x + (\lambda-\mu)s)^+$$

and, as in the previous proof, τ_N the hitting time of 0 by the renormalized process $(\bar{L}_N(t))$,

$$\mathbb{P}_z\left(\sup_{0\leq s\leq t} |\Delta_N(s)| \geq \delta\right) \tag{5.17}$$
$$\leq \mathbb{P}_z\left(\sup_{0\leq s\leq t\wedge\tau_N} |\Delta_N(s)| \geq \delta\right) + \mathbb{P}_z\left(\sup_{\tau_N\wedge t< s\leq t} |\Delta_N(s)| \geq \delta\right).$$

According to Representation (5.15) of $\bar{L}_N(\cdot\wedge\tau_N)$, if $|z/N - x| < \delta/2$,

$$\mathbb{P}_z\left(\sup_{0\leq s\leq t\wedge\tau_N} |\Delta_N(s)| \geq \delta\right) =$$
$$\mathbb{P}_z\left(\sup_{0\leq s\leq t\wedge\tau_N}\left| z/N - x + \frac{\mathcal{N}_\lambda(]0,Ns])}{N} - \lambda s - \left(\frac{\mathcal{N}_\mu(]0,Ns]}{N} - \mu s\right)\right| \geq \delta\right)$$

and this last term is bounded by

$$\mathbb{P}\left(\sup_{0\leq s\leq t}\left|\frac{\mathcal{N}_\lambda(]0,Ns])}{N} - \lambda s\right| \geq \frac{\delta}{4}\right) + \mathbb{P}\left(\sup_{0\leq s\leq t}\left|\frac{\mathcal{N}_\mu(]0,Ns])}{N} - \mu s\right| \geq \frac{\delta}{4}\right).$$

Lemma 5.15 shows the existence of N_1 such that, if $N \geq N_1$, then

$$\sup_{|z/N-x|<\delta/2} \mathbb{P}_z\left(\sup_{0\leq s\leq t\wedge\tau_N} |\Delta_N(s)| \geq \delta\right) \leq \frac{\varepsilon}{2}.$$

The last expression of Inequality (5.17) has to be estimated. The almost sure convergence of the sequence (τ_N) to $x/(\mu-\lambda)$ gives the existence of N_2, such that if $N \geq N_2$,

$$\mathbb{P}\left(\tau_N \leq \frac{x-\delta/2}{\mu-\lambda}\right) \leq \frac{\varepsilon}{4}.$$

By considering the cases whether τ_N is greater than $(x-\delta/2)/(\mu-\lambda)$ or not, one gets the inequalities

$$\mathbb{P}_z\left(\sup_{\tau_N\wedge t<s\leq t}|\Delta_N(s)|\geq\delta\right)\leq\mathbb{P}_z\left(\sup_{\tau_N<s\leq t+\tau_N}|\Delta_N(s)|\geq\delta\right)$$
$$\leq\mathbb{P}_z\left(\sup_{\tau_N<s\leq t+\tau_N}|\bar{L}_N(s)|\geq\delta/2,\ \tau_N\geq\frac{x-\delta/2}{\mu-\lambda}\right)+\mathbb{P}_z\left(\tau_N<\frac{x-\delta/2}{\mu-\lambda}\right).$$

The strong Markov property of $(L(t))$ yields the relationship

$$\mathbb{P}_z\left(\sup_{\tau_N\wedge t<s\leq t}|\Delta_N(s)|\geq\delta\right)\leq\mathbb{P}_0\left(\sup_{0\leq s\leq t}|\bar{L}_N(s)|\geq\delta/2\right)+\frac{\varepsilon}{4}$$
$$\leq\mathbb{P}_0\left(T_{\lfloor N\delta/2\rfloor}\leq Nt\right)+\frac{\varepsilon}{4}.$$

Proposition 5.11 shows that the sequence $\left(\rho^{\lfloor N\delta\rfloor}T_{\lfloor N\delta\rfloor}\right)$ converges in distribution to an exponentially distributed random variable. In particular, $\mathbb{P}_0(T_{\lfloor N\delta/2\rfloor}\leq Nt)$ tends to 0, there exists N_3 such that if $N\geq N_3$,

$$\mathbb{P}_z\left(\sup_{\tau_N\wedge t<s\leq t}|\Delta_N(s)|\geq\delta\right)\leq\frac{\varepsilon}{2}$$

and by taking $N_0=N_1\vee N_3$, if $N\geq N_0$, Inequality (5.16) is then satisfied. □

5.8 Large Deviations

The notations of the last section are used. If $L_N(0)=z\in\mathbb{N}$, for a small $\delta>0$ and $z\in\mathbb{N}$, Proposition 5.17 shows that the event

$$\mathcal{R}^t_{N,\delta}(\phi)=\left\{\sup_{0\leq s\leq t}\left|\bar{L}_N(s)-\phi(s)\right|<\delta\right\}, \tag{5.18}$$

is asymptotically very likely when function ϕ is

$$\phi(s)=(x+(\lambda-\mu)s)^+$$

and with the condition $|z/N-x|\leq\delta/2$. Otherwise, for $\phi(s)=x+\alpha s$, with $x>0$ and $\alpha\neq\lambda-\mu$, it will be seen that this event is asymptotically negligible, in fact exponentially negligible with respect to N.

This section estimates this exponential rate of decay of the probability of the event (5.18) when ϕ is an arbitrary function. This is equivalent to study the probability that a sample path of $(L(t))$ remains within a tube of diameter δ around some unlikely path $s\to\phi(s)$. The main result is given by Theorem 5.20 page 136. The purpose of this study is to give an elementary presentation of these questions in the simple case of the $M/M/1$ queue. The main results concerning large deviations of queues in a much more general framework are presented in Shwartz and Weiss [**48**].

In this study, the following cases are successively investigated: ϕ is a linear function, then piecewise linear and then a C^1 function. The transient case being not really different from the ergodic case, only the latter case is worked out (see the remark below). Thus, throughout this section it is assumed that the relation $\rho < 1$ holds.

In the sequel, the same notation N_1 is used for different integers, this is a slight abuse of notation. It just indicates an integer greater than the previous one which satisfies an additional property.

5.8.1 The Linear Case

The function ϕ is given by

$$\phi(s) = (x + \alpha s)^+,$$

with $x \geq 0$ and $\alpha \in \mathbb{R}$. The reflected behavior of $(L(t))$ near 0 gives rise to three different cases.

1. The function ϕ is positive on the interval $[0, t]$;
2. For $s \leq t$, $\phi(s) = \alpha s$ with $\alpha > 0$, in particular $\phi(0) = 0$;
3. For $s \leq t$, $\phi(s) = 0$.

A Sample Path in $\mathbb{R}_+ - \{0\}$. In this case, $x > 0$ and $x + \alpha t > 0$. The formula for the change of probability distribution (Proposition 5.3) shows, for $u \in \mathbb{R}_+$,

$$\text{(5.19)} \quad \mathbb{P}_z\left(\sup_{0\leq s\leq t} \left|\bar{L}_N(s) - (x + \alpha s)\right| < \delta\right) \\ = \mathbb{E}_z^u\left(u^{-Z(Nt)} e^{-(\lambda N(1-u)+\mu N(1-1/u))t} \, 1_{\mathcal{R}^t_{N,\delta}(\phi)}\right).$$

By setting

$$u(\alpha) = \frac{\alpha + \sqrt{\alpha^2 + 4\lambda\mu}}{2\lambda},$$

$u(\alpha)$ is the non-negative solution of $\lambda u - \mu/u = \alpha$, when N tends to infinity, the probability of the event $\mathcal{R}^t_{N,\delta}(\phi)$ is 1 for the probability measure $\mathbb{P}_z^{u(\alpha)}$ if z is such that $|z/N - x| \leq \delta/2$ (Proposition 5.17).

Since $x > 0$, if δ is sufficiently small, the process $(L(Ns))$ is positive on the interval $[0, t]$ on the event $\mathcal{R}^t_{N,\delta}(\phi)$, and thus identical to the free process translated by z, $(z + Z(Ns))$, in particular

$$\text{(5.20)} \qquad Z(Nt) \in [\alpha Nt + xN - z - \delta N, \alpha Nt + xN - z + \delta N].$$

Identity (5.19) for $u = u(\alpha)$ gives that $\mathbb{P}_z(\mathcal{R}^t_{N,\delta}(\phi))$ is equal to

$$\mathbb{E}_z^{u(\alpha)}\left[\exp\left(-N\left[\frac{Z(Nt)}{N}\log u(\alpha)\right.\right.\right.$$
$$\left.\left.\left.+\lambda(1-u(\alpha))t+\mu\left(1-\frac{1}{u(\alpha)}\right)t\right]\right)1_{\mathcal{R}^t_{N,\delta}(\phi)}\right].$$

Let us define

$$H(x,\alpha)=\alpha\log u(\alpha)+\lambda(1-u(\alpha))+\mu\left(1-\frac{1}{u(\alpha)}\right),$$

$$H(x,\alpha)=\alpha\log\frac{\alpha+\sqrt{\alpha^2+4\lambda\mu}}{2\lambda}+\lambda+\mu-\sqrt{\alpha^2+4\lambda\mu}, \tag{5.21}$$

notice that for the moment H does not depend on x. Relationship (5.20) shows that the probability of the event $\mathcal{R}^t_{N,\delta}(\phi)$ satisfies the inequalities

$$\begin{aligned}\mathbb{P}_z^{u(\alpha)}&\left[\mathcal{R}^t_{N,\delta}(\phi)\right]\exp[-(\delta N+|xN-z|)|\log u(\alpha)|]\\ &\leq\mathbb{P}_z\left[\mathcal{R}^t_{N,\delta}(\phi)\right]e^{H(x,\alpha)Nt}\leq\\ &\qquad\mathbb{P}_z^{u(\alpha)}\left(\mathcal{R}^t_{N,\delta}(\phi)\right)\exp[(\delta N+|xN-z|)|\log u(\alpha)|],\end{aligned}$$

consequently,

$$\begin{aligned}&\frac{1}{N}\log\mathbb{P}_z\left(\mathcal{R}^t_{N,\delta}(\phi)\right)+H(x,\alpha)t\\ &\qquad\geq\frac{1}{N}\log\left[\mathbb{P}_z^{u(\alpha)}(\mathcal{R}^t_{N,\delta}(\phi))\right]-(\delta+|x-z/N|)\,|\log u(\alpha)|\end{aligned} \tag{5.22}$$

and

$$\frac{1}{N}\log\mathbb{P}_z\left(\mathcal{R}^t_{N,\delta}(\phi)\right)+H(x,\alpha)t\leq\left(\delta+\left|x-\frac{z}{N}\right|\right)|\log u(\alpha)|. \tag{5.23}$$

If $\varepsilon>0$, for δ_0 sufficiently small so that $2\delta_0|\log u(\alpha)|\leq\varepsilon/2$ and $\delta\leq\delta_{\mathrm{C}}$, according to Inequality (5.16), there exists $N_1\in\mathbb{N}$ such that, for $N\geq N_1$,

$$\inf_{|z/N-x|\leq\delta/2}\frac{1}{N}\log\mathbb{P}_z^{u(\alpha)}\left(\mathcal{R}^t_{N,\delta}(\phi)\right)\geq-\frac{\varepsilon}{2}.$$

Finally, for $N\geq N_1$ and $\delta\leq\delta_0$, Relationships (5.22) and (5.23) give the inequalities

$$\begin{cases}\displaystyle\sup_{|z/N-x|<\delta}\frac{1}{N}\log\mathbb{P}_z\left(\mathcal{R}^t_{N,\delta}(\phi)\right)+H(x,\alpha)t & \leq\varepsilon,\\ \displaystyle\inf_{|z/N-x|<\delta/2}\frac{1}{N}\log\mathbb{P}_z\left(\mathcal{R}^t_{N,\delta}(\phi)\right)+H(x,\alpha)t & \geq-\varepsilon,\end{cases} \tag{5.24}$$

and the estimation of the probability of the event $\mathcal{R}^t_{N,\delta}(\phi)$ is completed for this case. The other cases are more delicate because of the reflection at 0.

The Problem of the Boundary $x = 0$. Here the function is given by $\phi(s) = \alpha s$, for $s \leq t$ and $\alpha > 0$. In this case, the free process cannot be used because of the reflection at 0. The function H is continued at $x = 0$ by defining $H(0, \alpha)$ by Relationship (5.21), in particular $H(x, \alpha) = H(0, \alpha)$, for any $x \in \mathbb{R}_+$ if $\alpha > 0$.

If $\eta_0 = 2\delta/\alpha$ (with δ sufficiently small so that $\eta_0 \leq t$), then

$$\mathcal{R}^t_{N,\delta}(\phi) = \left\{ \mathcal{R}^{\eta_0}_{N,\delta}(\phi), \sup_{\eta_0 \leq s \leq t} \left|\bar{L}_N(s) - \alpha s\right| < \delta \right\},$$

and the Markov property of $(L(t))$ shows, denoting by $\phi_1(s) = \alpha(\eta_0 + s)$

$$\mathbb{P}_z \left[\mathcal{R}^t_{N,\delta}(\phi)\right] = \mathbb{E}_z \left[1_{\mathcal{R}^{\eta_0}_{N,\delta}(\phi)} \mathbb{P}_{L(N\eta_0)}(\mathcal{R}^{t-\eta_0}_{N,\delta}(\phi_1))\right], \tag{5.25}$$

hence the inequality

$$\mathbb{P}_z \left[\mathcal{R}^t_{N,\delta}(\phi)\right] \leq \mathbb{E}_z \left[1_{\{|\bar{L}_N(\eta_0) - \alpha\eta_0| < \delta\}} \mathbb{P}_{L(N\eta_0)}(\mathcal{R}^{t-\eta_0}_{N,\delta}(\phi_1))\right].$$

According to Inequality (5.24), for δ sufficiently small, there exists N_1 such that if $N \geq N_1$,

$$\frac{1}{N} \log \mathbb{P}_{\bar{L}_N(\eta_0)} \left[\mathcal{R}^{t-\eta_0}_{N,\delta}(\phi_1)\right] + H(0, \alpha)(t - \eta_0) \leq \frac{\varepsilon}{2},$$

on the event $\{|\bar{L}_N(\eta_0) - \alpha\eta_0| < \delta\}$; consequently Relationship (5.25) shows that for δ sufficiently small, less than $\alpha\varepsilon/(4H(0, \alpha))$, there exists N_1 such that if $N \geq N_1$,

$$\sup_{z/N \leq \delta} \frac{1}{N} \log \mathbb{P}_z \left[\mathcal{R}^t_{N,\delta}(\phi)\right] + H(0, \alpha)t \leq \varepsilon, \tag{5.26}$$

the upper bound is proved.

A lower bound for the probability of these events. Clearly

$$\mathbb{P}_z \left[\mathcal{R}^t_{N,\delta}(\phi)\right] \geq \mathbb{P}_z \left[\mathcal{R}^t_{N,\delta}(\phi) \cap \{Z(s) \geq 0, \forall s \geq 0\}\right],$$

with the change of probability measure formula of Proposition 5.3, this last term can be expressed as

$$\mathbb{E}^{u(\alpha)}_z \left[\exp\left[-Nt(\lambda[1 - u(\alpha)] + \mu[1 - 1/u(\alpha)]) - Z(Nt) \log u(\alpha)\right] \times 1_{\{\mathcal{R}^t_{N,\delta}(\phi), Z(s) \geq 0, \forall s \geq 0\}}\right].$$

Since $(Z(t))$ and $(L(t))$ are identical as long as Z is positive,

$$\exp[(\delta N + z)|\log u(\alpha)| + NH(0, \alpha)t]\, \mathbb{P}_z \left[\mathcal{R}^t_{N,\delta}(\phi)\right] \geq \mathbb{P}^{u(\alpha)}_z \left[\mathcal{R}^t_{N,\delta}(\phi), Z(s) \geq 0, \forall s \geq 0\right].$$

By using Inequality (5.16), it is easy to show that the expression

$$\inf_{z/N \leq \delta/2} \mathbb{P}^{u(\alpha)}_z \left(\mathcal{R}^t_{N,\delta}(\phi), Z(s) \geq 0, \forall s \geq 0\right)$$

converges to

$$\mathbb{P}^{u(\alpha)}(Z(s) \geq 0, \forall s \geq 0) = \mathbb{P}^{u(\alpha)}(T_0 = +\infty) = 1 - \frac{\mu}{\lambda u^2(\alpha)} = \frac{\alpha}{\lambda u(\alpha)} > 0$$

when N tends to infinity. From Inequality (5.26), one gets that for δ sufficiently small, there exists N_1 such that $N \geq N_1$,

$$\begin{cases} \sup\limits_{z/N \leq \delta} \dfrac{1}{N} \log \mathbb{P}_z\left(\mathcal{R}^t_{N,\delta}(\phi)\right) + H(0,\alpha)t & \leq \varepsilon, \\ \inf\limits_{z/N \leq \delta/2} \dfrac{1}{N} \log \mathbb{P}_z\left(\mathcal{R}^t_{N,\delta}(\phi)\right) + H(0,\alpha)t & \geq -\varepsilon. \end{cases} \tag{5.27}$$

The analysis of this case is completed.

The Flat Trajectory $x = 0$ and $\alpha = 0$. Function ϕ is identically 0 on the interval $[0, t]$, the event $\mathcal{R}^t_{N,\delta}(\phi)$ is therefore very likely since the queue is stable. By setting

$$H(0,0) = 0, \tag{5.28}$$

Inequalities (5.27) are also satisfied according to Proposition 5.17.

Similar results can be proved in the same way by replacing the event $\mathcal{R}^t_{N,\delta}(\phi)$ by the event

$$\left\{\left|\bar{L}_N(t) - \phi(t)\right| \leq \delta\right\} = \left\{\left|\bar{L}_N(t) - (x + \alpha t)^+\right| \leq \delta\right\}.$$

For this event, the process $(\bar{L}_N(s))$ is forced to be close to ϕ only at 0 and t. Inequalities (5.24), (5.27) and (5.28) give therefore the proposition.

PROPOSITION 5.18. *If $\rho < 1$, $\alpha \in \mathbb{R}$, x, $t \in \mathbb{R}_+$ and $\phi(s) = (x + \alpha s)^+$, for $\varepsilon > 0$ and δ sufficiently small, there exists N_0 such that if $N \geq N_0$, then*

$$\inf_{|z/N - x| < \delta/2} \frac{1}{N} \log \mathbb{P}_z\left(\sup_{0 \leq s \leq t} \left|\bar{L}_N(s) - \phi(s)\right| < \delta\right) + H(x,\alpha)t \geq -\varepsilon,$$

$$\sup_{|z/N - x| < \delta} \frac{1}{N} \log \mathbb{P}_z\left(\sup_{0 \leq s \leq t} \left|\bar{L}_N(s) - \phi(s)\right| < \delta\right) + H(x,\alpha)t \leq \varepsilon$$

and

$$\inf_{|z/N - x| < \delta/2} \frac{1}{N} \log \mathbb{P}_z\left(\left|\bar{L}_N(t) - \phi(t)\right| < \delta\right) + H(x,\alpha)t \geq -\varepsilon,$$

$$\sup_{|z/N - x| < \delta} \frac{1}{N} \log \mathbb{P}_z\left(\left|\bar{L}_N(t) - \phi(t)\right| < \delta\right) + H(x,\alpha)t \leq \varepsilon,$$

where $H(x, \alpha)$ is defined by

$$H(x,\alpha) = \alpha \log \frac{\alpha + \sqrt{\alpha^2 + 4\lambda\mu}}{2\lambda} + \lambda + \mu - \sqrt{\alpha^2 + 4\lambda\mu},$$

if $(x, \alpha) \neq (0,0)$ and $H(0,0) = 0$.

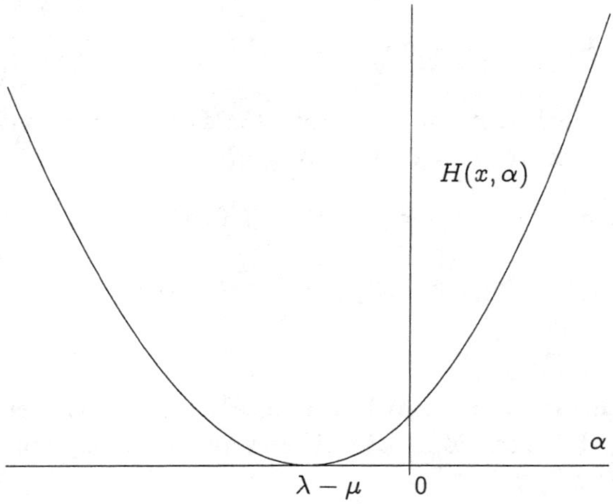

FIG. 5.5. The Function $\alpha \to H(x, \alpha)$ for $x > 0$

The probability of the event $\mathcal{R}_{N,\delta}^{t}(\phi)$ for which the whole sample path of $(\bar{L}_N(s), s \leq t)$ is close to ϕ converges exponentially fast to 0 at rate $H(x, \alpha)Nt$ when N tends to infinity. It is remarkable that this exponential decay is the same for the probability that the process is close to ϕ only at the last point t. This is a consequence of the formula for the change of probability measure. If the sample path is constrained at the origin and at t to be close to some specified points, the sample path between must be a random walk, whose fluid limit is therefore just a straight line.

5.8.2 The Piecewise Linear Case

The last result is generalized to the case of a piecewise linear function ϕ

$$\phi(s) = \begin{cases} \phi_1(s) = x + \alpha_1 s, & \text{for } s \leq v, \\ \phi_2(s - v), & \text{for } v < s \leq t, \end{cases}$$

where $\phi_2(s) = y + \alpha_2 s$ with $y = \phi_1(v) = x + \alpha_1 v$. The functions $\phi_i, i = 1, 2$ are assumed to be such that ϕ is non-negative on the interval $[0, t]$.

From Definition (5.18), the Markov property of $(L(t))$ yields

$$\begin{aligned} &\mathbb{P}_z\left(\mathcal{R}_{N,\delta}^{t}(\phi)\right) \\ &\quad = \mathbb{P}_z\left[\sup_{0 \leq s \leq v} \left|\bar{L}_N(s) - \phi_1(s)\right| < \delta, \sup_{v \leq s \leq t} \left|\bar{L}_N(s) - \phi_2(s - v)\right| < \delta\right] \\ &\quad = \mathbb{E}_z\left[1_{\mathcal{R}_{N,\delta}^{v}(\phi_1)} \mathbb{P}_{L(Nv)}\left(\mathcal{R}_{N,\delta}^{t-v}(\phi_2)\right)\right]. \end{aligned}$$

Proposition 5.18 shows that for $\varepsilon > 0$ and δ sufficiently small, there exists $N_1 \in \mathbb{N}$ such that, if $N \geq N_1$,

$$\mathbb{P}_{L(Nv)}\left(\mathcal{R}_{N,\delta}^{t-v}(\phi_2)\right) \leq \exp\left[N(\varepsilon/2 - H(y, \alpha_2)(t - v))\right],$$

on the event $\{|L(Nv)/N - y| \leq \delta\}$, and

$$\sup_{|z/N-x|\leq\delta} \mathbb{P}_z\left[\mathcal{R}^v_{N,\delta}(\phi_1)\right] \leq \exp\left[N(\varepsilon/2 - H(x,\alpha_1)v)\right].$$

Since $\mathcal{R}^v_{N,\delta}(\phi_1)$ is a subset of $\{|L(Nv)/N - y| \leq \delta\}$, for $N \geq N_1$, the relation

$$\sup_{|z/N-x|\leq\delta} \frac{1}{N}\log\mathbb{P}_z\left[\mathcal{R}^t_{N,\delta}(\phi)\right] + H(x,\alpha_1)v + H(y,\alpha_2)(t-v) \leq \varepsilon$$

is established. By definition of H,

$$H(x,\alpha_1)v + H(y,\alpha_2)(t-v) = \int_0^t H(\phi(s),\phi'(s))\,ds,$$

hence, the last inequality can be rewritten as

$$\sup_{|z/N-x|\leq\delta} \frac{1}{N}\log\mathbb{P}_z\left[\mathcal{R}^t_{N,\delta}(\phi)\right] + \int_0^t H(\phi(s),\phi'(s))\,ds \leq \varepsilon.$$

The analogue inequality for the lower bound

$$\inf_{|z/N-x|\leq\delta/4} \frac{1}{N}\log\mathbb{P}_z\left[\mathcal{R}^t_{N,\delta}(\phi)\right] + \int_0^t H(\phi(s),\phi'(s))\,ds \geq -\varepsilon,$$

is obtained in the same way by using the relation

$$\mathbb{P}_z\left[\mathcal{R}^t_{N,\delta}(\phi)\right] \geq \mathbb{E}_z\left(1_{\mathcal{R}^v_{N,\delta/2}(\phi_1)}\mathbb{P}_{L(Nv)}\left(\mathcal{R}^{t-v}_{N,\delta}(\phi_2)\right)\right)$$

and Proposition 5.18.

As before, these results remain unchanged if the event $\mathcal{R}^t_{N,\delta}(\phi)$ is replaced by the event

$$\left\{\left|\bar{L}_N(v) - \phi(v)\right| < \delta, \left|\bar{L}_N(t) - \phi(t)\right| < \delta\right\}.$$

The next proposition is obtained by induction on the number of the segments of the piecewise function ϕ,

PROPOSITION 5.19. *If $\rho < 1$, for x and $t \in \mathbb{R}_+$, if ϕ is a non-negative, continuous and linear function on the intervals $[v_i, v_{i+1}[$, $i = 0,\ldots,n-1$, with $0 = v_0 \leq v_1 \leq \cdots \leq v_n = t$ and $\phi(0) = x$, then for $\varepsilon > 0$ and $\delta > 0$ sufficiently small, there exist $N_1 \in \mathbb{N}$ and $\delta_1 > 0$ such that if $N \geq N_0$, then*

$$\sup_{|z/N-x|<\delta_1}\left|\frac{1}{N}\log\mathbb{P}_z\left(\sup_{0\leq s\leq t}\left|\bar{L}_N(s) - \phi(s)\right| < \delta\right) + \int_0^t H(\phi(s),\phi'(s))\,ds\right| \leq \varepsilon,$$

and

$$\sup_{|z/N-x|<\delta_1}\left|\frac{1}{N}\log\mathbb{P}_z\left(\sup_{0\leq i\leq n}\left|\bar{L}_N(v_i) - \phi(v_i)\right| < \delta\right) + \int_0^t H(\phi(s),\phi'(s))\,ds\right| \leq \varepsilon, \tag{5.29}$$

for the function $H(\cdot,\cdot)$ defined by

$$H(y,\alpha) = \alpha \log \frac{\alpha + \sqrt{\alpha^2 + 4\lambda\mu}}{2\lambda} + \lambda + \mu - \sqrt{\alpha^2 + 4\lambda\mu},$$

if $(y,\alpha) \neq (0,0)$ *and* $H(0,0) = 0$.

5.8.3 Generalization

Now ϕ is a positive C^1-function on $\mathbb{R}_+$. The large deviations result is given by the theorem.

THEOREM 5.20. *If* $\rho < 1$ *and* x, $t \in \mathbb{R}_+$, $\phi : \mathbb{R}_+ \to \mathbb{R}_+$ *is a* C^1*-function, such that* $\phi(0) = x$, *for* $\varepsilon > 0$ *and* $\delta > 0$ *sufficiently small, there exist* $N_0 \in \mathbb{N}$ *and* δ_1 *such that if* $N \geq N_0$, *then for any* $z \in \mathbb{R}$ *such that* $|z/N - x| < \delta_1$, *the inequality*

$$\left| \frac{1}{N} \log \mathbb{P}_z \left(\sup_{0 \leq s \leq t} \left| \bar{L}_N(s) - \phi(s) \right| < \delta \right) + \int_0^t H(\phi(s), \phi'(s))\, ds \right| \leq \varepsilon,$$

holds, where H *is the* rate function *defined by*

$$H(x,\alpha) = \alpha \log \frac{\alpha + \sqrt{\alpha^2 + 4\lambda\mu}}{2\lambda} + \lambda + \mu - \sqrt{\alpha^2 + 4\lambda\mu}, \tag{5.30}$$

if $(x,\alpha) \neq (0,0)$ *and* $H(0,0) = 0$.

PROOF. According to what has been proved up to now, the theorem is true for piecewise linear functions. The function ϕ is approximated by such functions (ϕ_n). The main difficulty to use this approximation is the discontinuity of rate function (5.30) at $(0,0)$,

$$\lim_{\alpha \to 0, \alpha > 0} H(0,\alpha) \neq H(0,0).$$

The estimation of

$$\int_0^t H(\phi(s), \phi'(s))\, ds \quad \text{by} \quad \int_0^t H(\phi_n(s), \phi_n'(s))\, ds,$$

is not a priori true. It is easily seen that the rate function is nevertheless H semi-continuous, that is,

$$\lim_{\substack{x \to x_0, x \leq x_0, \\ \alpha \to \alpha_0, \alpha \leq \alpha_0}} H(x,\alpha) = H(x_0,\alpha_0),$$

moreover,

$$\liminf_{x \to x_0, \alpha \to \alpha_0} H(x,\alpha) = H(x_0,\alpha_0). \tag{5.31}$$

These properties are in fact sufficient to prove the large deviations result of the theorem.

Lower Bound. The function ϕ being C^1, for all $\eta > 0$, it is possible to have a function ϕ_η, piecewise linear on $[0,t]$ and such that

$$\phi_\eta(s) \leq \phi(s), \quad \phi'_\eta(s) \leq \phi'(s), \quad 0 \leq s \leq t,$$

and

$$\sup_{0\leq s\leq t} |\phi_\eta(s) - \phi(s)| < \eta, \quad \sup_{0\leq s\leq t} |\phi'_\eta(s) - \phi'(s)| < \eta.$$

The lower semi-continuity of H gives

$$\lim_{\eta\to 0} H(\phi_\eta(s), \phi'_\eta(s)) = H(\phi(s), \phi'(s)), \quad s \in [0,t],$$

and, from Lebesgue's Theorem, one gets

$$\lim_{\eta\to 0} \int_0^t H(\phi_\eta(s), \phi'_\eta(s))\, ds = \int_0^t H(\phi(s), \phi'(s))\, ds.$$

For $\varepsilon > 0$, there exists $\eta_0 > 0$ such that if $\eta \leq \eta_0$, then

$$\left| \int_0^t H(\phi_\eta(s), \phi'_\eta(s))\, ds - \int_0^t H(\phi(s), \phi'(s))\, ds \right| \leq \frac{\varepsilon}{2}.$$

For $\delta > 0$,

$$\mathbb{P}_z\left(\sup_{0\leq s\leq t} \left|\bar{L}_N(s) - \phi(s)\right| < \delta + \eta \right) \geq \mathbb{P}_z\left(\sup_{0\leq s\leq t} \left|\bar{L}_N(s) - \phi_\eta(s)\right| < \delta \right),$$

if $\eta \leq \eta_0$ is fixed and $\delta > 0$ is chosen sufficiently small, there exist (Proposition 5.19) $N_0 \in \mathbb{N}$ and δ_1 such that for $N \geq N_0$,

$$\inf_{|z/N-x|<\delta_1} \frac{1}{N} \log \mathbb{P}_z\left(\sup_{0\leq s\leq t} \left|\bar{L}_N(s) - \phi_\eta(s)\right| < \delta \right) \geq -\frac{\varepsilon}{2} - \int_0^t H(\phi_\eta(s), \phi'_\eta(s))\, ds,$$

hence

$$\inf_{|z/N-x|<\delta_1} \frac{1}{N} \log \mathbb{P}_z\left(\sup_{0\leq s\leq t} \left|\bar{L}_N(s) - \phi(s)\right| < \delta + \eta \right) \geq -\varepsilon - \int_0^t H(\phi(s), \phi'(s))\, ds. \tag{5.32}$$

Upper Bound. For $n \geq 1$, ϕ_n denotes the piecewise linear function

$$\phi_n(s) = \phi\left(\frac{k}{n}\right) + (ns - k)\left[\phi\left(\frac{k+1}{n}\right) - \phi\left(\frac{k}{n}\right)\right],$$

if $\lfloor sn \rfloor = k$. Function ϕ_n coincides with ϕ at the points k/n, $k \leq \lfloor tn \rfloor$. The uniform continuity property of ϕ and ϕ' shows that this function converges uniformly to ϕ and its right derivative converges uniformly to ϕ'.

The rate function H being non-negative, Fatou's Lemma and the lower semi-continuity property (5.31) give

$$\int_0^t \liminf_{n\to+\infty} H(\phi_n(s),\phi_n'(s))\,ds \le \liminf_{n\to+\infty} \int_0^t H(\phi_n(s),\phi_n'(s))\,ds,$$

$$\int_0^t H(\phi(s),\phi'(s))\,ds \le \liminf_{n\to+\infty} \int_0^t H(\phi_n(s),\phi_n'(s))\,ds.$$

Thus, there exists n_0 such that for $n \ge n_0$,

$$\int_0^t H(\phi(s),\phi'(s))\,ds - \frac{\varepsilon}{2} \le \int_0^t H(\phi_n(s),\phi_n'(s))\,ds.$$

The inequality

$$\begin{aligned}\mathbb{P}_z\left(\sup_{0\le s\le t}\left|\bar{L}_N(s)-\phi(s)\right|<\delta\right)& \\ &\le \mathbb{P}_z\left(\sup_{0\le k\le \lfloor n_0 t\rfloor}\left|\bar{L}_N(k/n_0)-\phi(k/n_0)\right|<\delta\right)\\ &= \mathbb{P}_z\left(\sup_{0\le k\le \lfloor n_0 t\rfloor}\left|\bar{L}_N(k/n_0)-\phi_{n_0}(k/n_0)\right|<\delta\right),\end{aligned}$$

and Identity (5.29) of Proposition 5.19 applied to the function ϕ_{n_0} show that, for $\varepsilon > 0$ δ sufficiently small, there exist $N_0 \in \mathbb{N}$ and $\delta_1 > 0$ such that if $N \ge N_0$, then

$$\begin{aligned}\sup_{|z/N-x|<\delta_1} \frac{1}{N}\log \mathbb{P}_z\left(\sup_{0\le k\le \lfloor n_0 t\rfloor}\left|\bar{L}_N(k/n_0)-\phi(k/n_0)\right|<\delta\right)& \\ &\le \frac{\varepsilon}{2} - \int_0^t H(\phi_{n_0}(s),\phi_{n_0}'(s))\,ds.\end{aligned}$$

The inequality

$$\begin{aligned}\sup_{|z/N-x|<\delta_1} \frac{1}{N}\log \mathbb{P}_z\left(\sup_{0\le k\le \lfloor n_0 t\rfloor}\left|\bar{L}_N(k/n_0)-\phi(k/n_0)\right|<\delta\right)& \\ &\le \varepsilon - \int_0^t H(\phi(s),\phi'(s))\,ds\end{aligned}$$

has been proved, finally

$$\sup_{|z/N-x|<\delta_1} \frac{1}{N}\log \mathbb{P}_z\left(\sup_{0\le s\le t}\left|\bar{L}_N(s)-\phi(s)\right|<\delta\right) \le \varepsilon - \int_0^t H(\phi(s),\phi'(s))\,ds.$$

This inequality and Relationship (5.32) finish the proof of the theorem. □

This theorem can be generalized to the case where the function ϕ is absolutely continuous with respect to Lebesgue's measure. (In this case the derivative of ϕ exists almost everywhere.)

REMARK.

If the proof of the last theorem is somewhat technical, the main ideas are quite simple. The change of probability is, of course, the basic ingredient of the proof. Looking at the sample paths of $(L(t))$ which follow $t \to x+\alpha t$ is reduced to a change of probability measure so that the drift of $(L(t))$ is α. The second important point is the discontinuity at 0. The rate function $(x,\alpha) \to H(x,\alpha)$ is not continuous but has a crucial property: The lower semi-continuity (5.31). This property is used to approximate an arbitrary functions by piecewise linear functions. See Dembo and Zeitouni [**16**] for an introduction to large deviations.

5.9 Appendix

The Density of a Busy Period

The density of the duration of the busy period is expressed by inverting Laplace transform (5.3). The expressions of the densities, when they are known, use generally special functions (Bessel functions in this case) which are not really easy to handle to get some asymptotic results. In general limit theorems are easier to obtain via probabilistic arguments.

PROPOSITION 5.21. *If $\lambda \leq \mu$ and $L(0) = 1$, the density of T_0 is given by*

$$\sqrt{\frac{\mu}{\lambda}}\,\frac{I_1\left(2\sqrt{\lambda\mu t}\right)}{t}\,e^{-(\lambda+\mu)t}, \qquad t > 0,$$

where I_1 is the modified Bessel function of order 1,

$$I_1(t) = \sum_{k=0}^{+\infty} \frac{(t/2)^{2k+1}}{k!(k+1)!}, \qquad t \in \mathbb{R}.$$

PROOF. By using the expansion of $\sqrt{1+x}$ for $|x| < 1$,

$$\sqrt{1+x} = \sum_{k=0}^{+\infty} C_{1/2}^k x^k,$$

with

$$C_{1/2}^k = \frac{1/2(1/2-1)(1/2-2)\dots(1/2-k+1)}{k!},$$

with $C_{1/2}^0 = 1$, one gets that

$$\frac{1-\sqrt{1-x^2}}{x} = \sum_{k\geq 0} \frac{2k!}{k!(k+1)!}(x/2)^{2k+1},$$

consequently,

$$\mathbb{E}_1\left(e^{-xT_0}\right) = \sqrt{\frac{\mu}{\lambda}}\,\frac{1-\sqrt{1-4\lambda\mu/(\lambda+\mu+x)^2}}{2\sqrt{\lambda\mu}/(\lambda+\mu+x)}$$

$$= \sum_{k\geq 0}\frac{2k!}{k!(k+1)!}\left(\frac{\lambda}{\lambda+\mu}\right)^k\left(\frac{\mu}{\lambda+\mu}\right)^{k+1}\left(\frac{\lambda+\mu}{\lambda+\mu+x}\right)^{2k+1}. \tag{5.33}$$

If $k \in \mathbb{N}$ and $x \in \mathbb{R}$,

$$\left(\frac{\lambda+\mu}{\lambda+\mu+x}\right)^{2k+1}$$

is the Laplace transform in x of the sum of $2k+1$ independent random variables with an exponential distribution with parameter $\lambda+\mu$, therefore this is the Laplace transform in x of the function

$$(\lambda+\mu)^{2k+1}\frac{t^{2k}}{(2k)!}e^{-(\lambda+\mu)t},$$

on $\mathbb{R}_+$. Hence T_0 has the density

$$\sum_{k\geq 0}\frac{1}{k!(k+1)!}\mu(\lambda\mu)^k t^{2k}e^{-(\lambda+\mu)t}$$

$$= \sqrt{\frac{\mu}{\lambda}}\sum_{k\geq 0}\frac{1}{k!(k+1)!}(\sqrt{\lambda\mu})^{2k+1}t^{2k}e^{-(\lambda+\mu)t} = \sqrt{\frac{\mu}{\lambda}}\,\frac{I_1\left(2\sqrt{\lambda\mu}t\right)}{t}e^{-(\lambda+\mu)t}$$

on $\mathbb{R}_+$. The proposition is proved. □

Relationship (5.33) has a simple interpretation. If $p \in [0,1]$ and $q = 1-p$, the expression

$$\frac{2k!}{k!(k+1)!}p^k q^{k+1}$$

is the probability that a simple random walk starting from 1 hits 0 for the first time at $t = 2k+1$ (see Feller [**21**]). The jumps of the Markov process $(L(t))$ between 0 and T_0 occur at the points of Poisson process with parameter $\lambda+\mu$. A jump of this process is $+1$ with probability $\lambda/(\lambda+\mu)$ and -1 otherwise. The general term of the series (5.33) is thus the Laplace transform of the duration of a busy period of the $M/M/1$ queue with k customers.

6. The M/M/∞ Queue

Contents

6.1 Introduction

The queue with an infinite number of servers is, with the $M/M/1$ queue, a basic model. It plays a crucial role in most of the stochastic models of communication networks. If the $M/M/1$ queue is the basic element of a Jackson network (see Section 4.4.1 page 92), the $M/M/\infty$ queue is the basic element of loss networks (see Example 4.2.3 page 88) used to represent communication networks. As it is the case for the $M/M/1$ queue, the stochastic processes describing this queue are not only important in the analysis of queueing systems but also in various areas such as statistical physics or theoretical computer science.

The behavior of this queue at equilibrium has been studied in Chapter 4 and 7. In this chapter, the transient distributions, the hitting times, the fluid limits and a finite capacity model of this queue are investigated. If the plan of the study bears some similarity with the analysis of the $M/M/1$ queue, the properties of these two queues are significantly different.

1. The total service rate being potentially infinite (recall that when n customers are presents, the total service rate is proportional to n), the $M/M/\infty$ queue is always stable. The tail distribution $\pi([n,+\infty[)$ of the number of customers at equilibrium satisfies the inequality
$$\sum_{k\geq n}\frac{\rho^k}{k!}e^{-\rho}\leq\frac{\rho^n}{n!}\sim\frac{1}{\sqrt{2\pi}}n^{-n-\frac{1}{2}}e^{n(\log\rho+1)},$$
where ρ is the average workload of the queue (see the definition below). Recall that the corresponding tail distribution of the $M/M/1$ queue is only geometrically decreasing. The number of customers of this queue is very stable, i.e. it is often close to 0. The limit results for the distributions of the hitting times illustrate this property (Propositions 6.8 and 6.10).
2. The number of customers, when it is positive, is not anymore a random walk. Therefore, and this complicates the analysis, the associated fundamental positive martingales are not simple exponential martingales as for the $M/M/1$ queue (Proposition 6.4).
3. The scaling procedure used for this queue is the following: The arrival rate is scaled by a factor N and the number of customers by a factor $1/N$. With this scaling, the process of the number of customers converges in distribution to a fluid limit which is a deterministic process. The fluid limit is not, as for the $M/M/1$ queue, a piecewise linear function but a function converging exponentially fast to ρ (Theorem 6.13). In particular if the fluid limit starts at ρ, it remains constant equal to ρ (similarly, for a stable $M/M/1$ queue, a fluid limit starting at 0 remains at 0)
4. The stochastic fluctuation around the fluid limit is not anymore a Brownian motion but an ergodic diffusion related to Ornstein-Uhlenbeck process (Theorem 6.14).

The results are again obtained via elementary stochastic calculus, thereby showing how powerful the probabilistic methods can be. Stochastic differential equations and positive martingales together with elementary calculus give a very detailed description of the behavior of this queue.

Definitions and Notations. As in Chapter 5 for the $M/M/1$ queue, $\mathcal{N}_\xi(dx)$ denotes a Poisson point process on $\mathbb{R}$ with parameter $\xi\in\mathbb{R}_+$. With the same notation as in Chapter 1, when I is a Borelian subset of $\mathbb{R}_+$, the quantity $\mathcal{N}_\xi(I)$ is the number of points of the point process $\mathcal{N}_\xi$ in the set I. An upper index $\mathcal{N}_\xi^i, i\in\mathbb{N}$ is added when an i.i.d. sequence of Poisson point processes with parameter ξ is required.

The arrival process $\mathcal{N}_\lambda$ is Poisson with rate λ, services are exponentially distributed with parameter μ and $\rho=\lambda/\mu$ is the load of the queue. An i.i.d. sequence of Poisson processes $(\mathcal{N}_\mu^i)$ is fixed. The filtration $\mathcal{F}=(\mathcal{F}_t)$ is defined by
$$\mathcal{F}_t=\sigma\left\langle\mathcal{N}_\lambda(]0,s]),\mathcal{N}_\mu^i(]0,s]):i\in\mathbb{N},s\leq t\right\rangle,$$

for $t \geq 0$. The process of the number of customers $(L(t))$ is a Markov jump process whose Q-matrix $Q = (q_{ij})$ is given by

$$q_{i\,i+1} = \lambda, \quad i \in \mathbb{N}, q_{i\,i-1} = i\mu, \quad i \in \mathbb{N}, q_{ij} = 0, \text{ when } |i-j| > 1.$$

The transition rates are not bounded, contrary to the case of the $M/M/1$ queue. Nevertheless, the transition rates of the upward jumps are bounded, which implies that there is no problem with finite time explosion of the process (see Appendix B). If f is a function on $\mathbb{N}$, the Q-matrix seen as a functional operator can be expressed as

$$Q(f)(n) = \lambda[f(n+1) - f(n)] + n\mu[f(n-1) - f(n)],$$

for $n \in \mathbb{N}$. Proposition 7.2 and the reversibility property of ergodic birth and death processes (see Chapter 4) give the proposition.

PROPOSITION 6.1. *The invariant distribution of the number of customers is Poisson with parameter $\rho = \lambda/\mu$. At equilibrium the process $(L(t))$ of the number of customers in the queue is reversible and the departure process is Poisson with parameter λ. Moreover, for $t \in \mathbb{R}_+$, the total variation distance to equilibrium at time t (see Appendix C) satisfies the inequality*

$$\|\mathbb{P}_0(L(t) \in \cdot\,) - \mathrm{Pois}(\rho)\|_{tv} \leq 1 - e^{-\rho \exp(-\mu t)},$$

where $\mathrm{Pois}(\rho)$ is the Poisson distribution on $\mathbb{N}$ with parameter ρ.

The process $(L(t))$ can also be seen as the solution of a stochastic differential equation. This formulation is useful to derive the main properties of the fluid limits associated with this process.

PROPOSITION 6.2. *The Markov process with Q-matrix Q starting from $x \in \mathbb{N}$ has the same distribution as the unique solution of the stochastic differential equation*

$$dL(t) = \mathcal{N}_\lambda(dt) - \sum_{i=1}^{L(t-)} \mathcal{N}_\mu^i(dt), \tag{6.1}$$

with $L(0) = x$ or, equivalently,

$$L(t) = x + \mathcal{N}_\lambda(]0,t]) - \sum_{i=1}^{+\infty} \int_{]0,t]} 1_{\{L(s-)\geq i\}} \mathcal{N}_\mu^i(ds). \tag{6.2}$$

The process

$$(M(t)) = \left(L(t) - L(0) - \lambda t + \mu \int_0^s L(s)\, ds \right)$$

is a square integrable martingale whose increasing process is given by

$$\langle M \rangle (t) = \lambda t + \mu \int_0^t L(s)\, ds.$$

The arrival process is given by $\mathcal{N}_\lambda$. For $i \geq 1$, the process $\mathcal{N}_\mu^i$ has the following interpretation : If at least i customers are in the queue at time t, the distance to the next point of $\mathcal{N}_\mu^i$ at the right of t is the residual service time of the ith customer.

PROOF. Proposition A.11 page 356 in Appendix A gives the existence and uniqueness of the solution $(L(t))$ of the differential Equation (6.1). To establish the correspondence between this solution and the Markov process with Q-matrix Q, Proposition B.6 page 365 shows that it is sufficient to prove that, if f is a function on $\mathbb{N}$, the process

$$\left(f(L(t)) - f(L(0)) - \int_0^t Q(f)(L(s))\,ds\right)$$

is a local martingale. Since the process $(L(t))$ has, almost surely, jumps of size 1, it is easily checked that

$$\begin{aligned} df(L(t)) = \lim_{s\nearrow t} f(L(t)) - f(L(s)) \\ &= \left[f(L(t-)+1) - f(L(t-))\right] \mathcal{N}_\lambda(dt) \\ &\quad + \left[f(L(t-)-1) - f(L(t-))\right] \sum_{i=1}^{L(t-)} \mathcal{N}_\mu^i(dt). \end{aligned}$$

This shows that $f(L(t))$ can be represented as

$$\begin{aligned} f(L(t)) &= f(L(0)) + R(t) + \lambda \int_0^t \left[f(L(s-)+1) - f(L(s-))\right] ds \\ &\quad + \mu \int_0^t \left[f(L(s-)-1) - f(L(s-))\right] L(s-)\,ds \\ &= f(L(0)) + R(t) + \int_0^t Q(f)(L(s-))\,ds, \\ f(L(t)) &= f(L(0)) + R(t) + \int_0^t Q(f)(L(s))\,ds, \end{aligned}$$

where $(R(t))$ is the local martingale defined by,

$$\begin{aligned} R(t) &= \int_{]0,t]} \left[f(L(s-)+1) - f(L(s-))\right] \left(\mathcal{N}_\lambda(ds) - \lambda\,ds\right) \\ &\quad + \sum_{i=1}^{+\infty} \int_{]0,t]} 1_{\{L(s-)\geq i\}} \left[(f(L(s-)-1) - f(L(s-))\right] \left(\mathcal{N}_\mu^i(ds) - \mu\,ds\right). \end{aligned} \tag{6.3}$$

The last identity shows that the process $(L(t))$ is indeed a solution of the martingale problem associated with the Q-matrix Q, that is, for any function f on $\mathbb{N}$, the process

$$\left(f(L(t)) - f(L(0)) - \int_0^t Q(f)(L(s))\,ds\right),$$

is a local martingale. Thus, the process $(L(t))$ has the same distribution as the Markov process associated with $\mathcal{Q}$-matrix Q and starting from x. By taking $f(x) = x$, one gets that

$$\left(f(L(t)) - f(L(0)) - \int_0^t Q(f)(L(s))\,ds\right) = \left(L(t) - L(0) - \lambda t - \mu \int_0^t L(s)\,ds\right) = (M(t)),$$

is a local martingale. For $s \leq t \in \mathbb{R}_+$, the variable $L(s)$ is upper bounded by $N_\lambda(]0,t]) + L(0)$, hence the variable $\sup\{M(s)\colon s \leq t\}$ is integrable. The process $(M(t))$ is therefore a martingale according to Proposition A.7. Expression (6.3) of this martingale shows that it can be written as a stochastic integral with respect to the martingales associated with the Poisson processes $\mathcal{N}_\lambda$ and $(\mathcal{N}_\mu^i)$,

$$M(t) = \mathcal{N}_\lambda(]0,t]) - \lambda t - \sum_{i=1}^{+\infty} \int_{]0,t]} 1_{\{L(s-)\geq i\}} \left(\mathcal{N}_\mu^i(ds) - \mu\,ds\right).$$

The variable $M(t)^2$ is the sum of

$$(\mathcal{N}_\lambda(]0,t]) - \lambda t)^2 + \sum_{i=1}^{+\infty} \left(\int_{]0,t]} 1_{\{L(s-)\geq i\}} \left(\mathcal{N}_\mu^i(ds) - \mu\,ds\right)\right)^2, \tag{6.4}$$

and

$$2\int_{]0,t]} 1_{\{L(s-)\geq i\}} (\mathcal{N}_\mu^i(ds) - \mu\,ds) \int_{]0,t]} 1_{\{L(s-)\geq j\}} (\mathcal{N}_\mu^j(ds) - \mu\,ds), \tag{6.5}$$

for $i \neq j \in \mathbb{N}$, and

$$-\,2\,(\mathcal{N}_\lambda(]0,t]) - \lambda t) \int_{]0,t]} 1_{\{L(s-)\geq i\}} \left(\mathcal{N}_\mu^i(ds) - \mu\,ds\right), \quad i \in \mathbb{N}. \tag{6.6}$$

Since indicator functions are bounded by 1, it is easily seen that all these random variables are integrable. Proposition A.10 page 356 shows that the processes defined by Equations (6.5) and (6.6) are martingales. The increasing process of $(M(t))$ is therefore the process that compensates Expression (6.4) so that it becomes a martingale.

For $\xi \in \mathbb{R}_+$ and h a function on $\mathbb{N}$, the increasing process of the martingale

$$\left(\int_{]0,t]} h(L(s-))\,[\mathcal{N}_\xi(\omega, ds) - \xi\,ds]\right)$$

is, according to Proposition A.9 page 355,

$$\left(\xi \int_0^t h^2(L(s))\,ds\right).$$

Therefore, the process

$$\left(M(t)^2 - \lambda t - \mu \int_0^t L(s)\, ds\right)$$

is a martingale. The proposition is proved. □

6.2 Positive Martingales

The fundamental martingales defined in this section are, for the $M/M/\infty$ queue, the analogue of the exponential martingales for the $M/M/1$ queue. These martingales play a central role, in particular to construct other martingales used to express distributions of hitting times.

LEMMA 6.3. *For $c \in \mathbb{R}$, the function*

$$h_c : (t,x) \to \left(1 + ce^{\mu t}\right)^x e^{-\rho c \exp(\mu t)}, \tag{6.7}$$

is space-time harmonic with respect to the Q-matrix Q of $(L(t))$, that is,

$$\frac{\partial h_c}{\partial t}(t,x) + Q(h_c)(t,x) = 0,$$

for any $t \in \mathbb{R}_+$ and $x \in \mathbb{N}$, with the convention that

$$Q(h_c)(t,x) = Q(h_c(t,\cdot))(x).$$

PROOF. For $t \in \mathbb{R}_+$ and $x \in \mathbb{N}$,

$$\begin{aligned} \frac{\partial h_c}{\partial t}(t,x) &= e^{-\rho c \exp(\mu t)} \left[\mu x c e^{\mu t} \left(1 + ce^{\mu t}\right)^{x-1} - \lambda c e^{\mu t} \left(1 + e^{\mu t}\right)^x\right] \\ &= \mu x e^{-\rho c \exp(\mu t)} \left[\left(1 + ce^{\mu t}\right)^x - \left(1 + ce^{\mu t}\right)^{x-1}\right] \\ &\qquad - \lambda c e^{-\rho c \exp(\mu t)} \left[\left(1 + e^{\mu t}\right)^{x+1} - \left(1 + e^{\mu t}\right)^x\right] \\ &= -\, Q(h_c)(t,x). \end{aligned}$$

The function h_c is space-time harmonic with respect to the Q-matrix Q. □

Corollary B.5 page 364 gives a probabilistic translation of the above lemma.

PROPOSITION 6.4. *For $c \in \mathbb{R}$ and $L(0) = x \in \mathbb{N}$, the process*

$$(\mathcal{E}_c(t)) \overset{def.}{=} \left(\left(1 + ce^{\mu t}\right)^{L(t)} e^{-\rho c \exp(\mu t)}\right)$$

is a martingale.

This is a Doléans-Dade's exponential martingale (see Fricker *et al.*[FRT99]). The general framework of positive martingales associated with a stochastic process is presented in Rogers and Williams [**43**] (in the section on exponential martingales).

[FRT99] C. Fricker, Ph. Robert, and D. Tibi, *On the rates of convergence of Erlang's model*, Journal of Applied Probability **36** (1999), no. 4, 1167–1184.

PROOF. The function $t \to \partial h_c/\partial t$ being continuous, Corollary B.5 shows that the process $(h_c(t, L(t)) = (\mathcal{E}_c(t))$ is a local martingale. Since

$$L(s) \leq x + \mathcal{N}_\lambda(]0,t])$$

for $s \leq t \in \mathbb{R}_+$, the variable $\sup\{\mathcal{E}_c(s) : s \leq t\}$ is integrable, hence the process $(\mathcal{E}_c(t))$ is a martingale (Proposition A.7 page 354). □

The martingale property gives the identity $\mathbb{E}_x(\mathcal{E}_c(t)) = \mathbb{E}_x(\mathcal{E}_c(0))$, that is

$$\mathbb{E}_x\left[\left(1+ce^{\mu t}\right)^{L(t)} e^{-\rho c \exp(\mu t)}\right] = (1+c)^x e^{-\rho c},$$

for t, $c \in \mathbb{R}_+$ and $x \in \mathbb{N}$. With a change of variables, one gets for $u \in [0,1]$,

$$\mathbb{E}_x\left(u^{L(t)}\right) = e^{\rho(1-\exp(-\mu t))(u-1)}\left(1+(u-1)e^{-\mu t}\right)^x. \tag{6.8}$$

The variable $L(t)$ has the same distribution as the sum of two independent random variables whose distributions are respectively a Poisson distribution with parameter $\rho(1-\exp(-\mu t))$ and a binomial distribution with parameters x and $\exp(-\mu t)$. The interpretation of this result is simple: The Poisson distribution is the number of customers arrived after $t = 0$ and still in the queue. The binomial distribution is the distribution of the number of the x initial customers who have not completed their services by time t. This result is also proved, in a different way, in Section 7.2 of Chapter 7.

The variable c can be chosen arbitrarily, the derivatives of $\mathcal{E}(t)$ with respect to c give a family of martingales. A classical result (see Chihara [**11**] for example) shows that, for $w \geq 0$ and $x \in \mathbb{N}$,

$$C(w,x) = (1+w)^x e^{-\rho w} = \sum_{n=0}^{+\infty} p_n^\rho(x)\frac{w^n}{n!},$$

where $p_n^\rho(x)$ is a polynomial of degree n with respect to x. The (p_n^ρ) are Poisson-Charlier polynomials. This is one of the classical families of orthogonal polynomials. The process $(C[c\exp(\mu t), L(t)])$ being a martingale,

$$\mathbb{E}[C(ce^{\mu t}, L(t)) \mid \mathcal{F}_s] = C(ce^{\mu s}, L(s)), \qquad s \leq t,$$

by identifying the coefficients in c of this identity, one gets the corollary.

COROLLARY 6.5. *For $n \geq 0$, if $p_n^\rho(x)$ is the Poisson-Charlier polynomial with degree n, the process*

$$\left[p_n^\rho(L(t))\exp(n\mu t)\right],$$

is a martingale. In particular, by taking successively $n = 1$ and $n = 2$, the processes

$$\left(e^{\mu t}(L(t)-\rho)\right) \text{ and } \left(e^{2\mu t}\left[(L(t)-\rho)^2 - L(t)\right]\right) \tag{6.9}$$

are martingales.

A similar result holds for Brownian motion $(B(t))$. For $c \in \mathbb{R}$, the process $\left(\exp\left(cB(t) - c^2t/2\right)\right)$ is a martingale, the exponential martingale of Brownian motion. The decomposition

$$H(c,x) = e^{cx-c^2/2} = \sum_{n=0}^{+\infty} \frac{c^n}{n!} h_n(x),$$

where the $(h_n(x))$ are Hermite polynomials, shows in the same way that, for $n \geq 0$, the process

$$\left(t^{n/2} h_n\left(B(t)/\sqrt{t}\right)\right)$$

is a martingale. The cases $n = 1$ and $n = 2$ correspond to the classical martingales $(B(t))$ and $(B(t)^2 - t)$.

Contrary to the case of the $M/M/1$ queue, the martingale $(\mathcal{E}_c(t))$ does not give directly results on the distributions of hitting times. The expressions $c\exp(\mu t)$ and $L(t)$ are not separated in the expression of the martingale. To overcome this difficulty, it is natural to integrate this martingale with respect to some absolutely continuous measure $f(c)\,dc$. A change of variables isolates the term $\exp(\mu t)$ so that the variables t and $L(t)$ are separated. This is the general method used to construct the martingales of the proposition. A much more general result concerning the set of *all* harmonic functions for birth and death processes is presented in Doob *et al.* [DSW60].

Proposition 6.6. *For $\alpha > 0$, $t \in \mathbb{R}_+$ and $L(0) = x \in \mathbb{N}$, if*

$$I_\alpha(t) = e^{-\alpha\mu t} \int_0^{+\infty} (1+y)^{L(t)} y^{\alpha-1} e^{-\rho y}\, dy, \tag{6.10}$$

$$J_\alpha(t) = e^{-\alpha\mu t} \int_0^{1} (1-y)^{L(t)} y^{\alpha-1} e^{\rho y}\, dy,$$

and $T_0 = \inf\{t > 0 : L(t) = 0\}$, the processes $(I_\alpha(t))$ and $(J_\alpha(t \wedge T_0))$ are martingales.

The time variable t is separated from the space variable $(L(t))$ for these martingales. Moreover, the martingale $(J_\alpha(t))$ is bounded as a function of the variable $L(t)$.

Proof. For $s \leq t$, Expression (6.8) of the distribution of $L(s)$ shows that it is bounded by $x + Y$ where Y is a Poisson random variable with parameter $\rho(1 - \exp(-\mu s))$. It is then easily checked that the variable

$$\int_0^{+\infty} (1+y)^{L(s)} y^{\alpha-1} e^{-\rho y}\, dy$$

[DSW60] J.L. Doob, J.L. Snell, and R.E. Williamson, *Application of boundary theory to sums of independent random variables.*, Contributions to probability and statistics, Stanford Univ. Press, Stanford, Calif., 1960, pp. 182–197.

has a finite expected value when $\alpha > 0$, in particular it is $\mathbb{P}$-almost surely finite for any $s \le t$. The process defined by Equation (6.10) is therefore $\mathbb{P}$-almost surely finite and integrable. The process $(J_\alpha(t))$ is obviously well defined

$$J_\alpha(t) \le e^{-\alpha\mu t} \int_0^1 y^{\alpha-1} e^{\rho y}\, dy < +\infty.$$

For $0 \le s \le t$ and if Y is some non-negative $\mathcal{F}_s$-measurable random variable, the martingale property of $(\mathcal{E}_c(t))$ gives the relation,

$$\mathbb{E}[Y\mathcal{E}_c(t)] = \mathbb{E}\left[Y\mathbb{E}(\mathcal{E}_c(t) \mid \mathcal{F}_s)\right] = \mathbb{E}[Y\mathcal{E}_c(s)].$$

If this identity is integrated on $\mathbb{R}_+$ with respect to the measure $c^{\alpha-1}\, dc$ on $\mathbb{R}_+$, one gets

$$\int_0^{+\infty} \mathbb{E}\left[Y(1+ce^{\mu t})^{L(t)}\right] e^{-\rho c \exp(\mu t)} c^{\alpha-1}\, dc$$
$$= \int_0^{+\infty} \mathbb{E}\left[Y(1+ce^{\mu s})^{L(s)}\right] e^{-\rho c \exp(\mu s)} c^{\alpha-1}\, dc.$$

Fubini's formula and a change of variables give the identity

$$\mathbb{E}\left[Ye^{-\alpha\mu t}\int_0^{+\infty} (1+y)^{L(t)} e^{-\rho y} y^{\alpha-1}\, dy\right]$$
$$= \mathbb{E}\left[Ye^{-\alpha\mu s}\int_0^{+\infty} (1+y)^{L(s)} e^{-\rho y} y^{\alpha-1}\, dy\right]$$

for any $\mathcal{F}_s$-measurable function Y, therefore $\mathbb{P}$-almost surely,

$$\mathbb{E}(I_\alpha(t) \mid \mathcal{F}_s) = I_\alpha(s).$$

The process $(I_\alpha(t))$ is a martingale. The first part of the proposition is proved.

For $c \in [0,1]$ and $\alpha > 0$, by using Definition (6.7) of h_c, the process

$$(h_{-c}[t, L(t)]) = \left(\left(1 - ce^{\mu t}\right)^{L(t)} e^{\rho c \exp(\mu t)}\right)$$

is a martingale. As before, this process is integrated with respect to the measure $c^{\alpha-1}\, dc$, but on the interval $[0, \exp(-\mu t)]$, i.e. if

$$g(t,x) \stackrel{\text{def.}}{=} \int_0^{\exp(-\mu t)} h_{-c}(t,x) c^{\alpha-1}\, dc$$
$$= \int_0^{\exp(-\mu t)} \left(1 - ce^{\mu t}\right)^x e^{\rho c \exp(\mu t)} c^{\alpha-1}\, dc.$$

It will be shown that the process $(g[t, L(t)])$ is indeed a martingale. Since integration bounds depends on time, the above proof of the martingale property of the integrated process cannot be used directly. Nevertheless, it is remarkable that the martingale property still holds if the process is stopped at time T_0.

The process

$$\left(g\left(t, L(t)\right) - \int_0^t \left[\frac{\partial g}{\partial t} + Q(g)\right](s, L(s))\, ds\right)$$

being a local martingale (Proposition B.4 page 364), Doob's optional stopping Theorem shows that

$$\left(g\left(t \wedge T_0, L(t \wedge T_0)\right) - \int_0^{t\wedge T_0} \left[\frac{\partial g}{\partial t} + Q(g)\right](s, L(s))\, ds\right) \tag{6.11}$$

is also martingale. If $s < T_0$, the partial derivative of g with respect to t is given by

$$\begin{aligned}\frac{\partial g}{\partial t}(s, L(s)) &= -\mu e^{-\mu s} h_{-\exp(-\mu s)}(s, L(s)) e^{-(\alpha-1)\mu s} \\ &\quad + \int_0^{\exp(-\mu s)} \frac{\partial h_{-c}}{\partial t}(s, L(s)) c^{\alpha-1}\, dc \\ &= \int_0^{\exp(-\mu s)} \frac{\partial h_{-c}}{\partial t}(s, L(s)) c^{\alpha-1}\, dc,\end{aligned}$$

since $s < T_0$ implies $L(s) > 0$, and therefore $h_{-\exp(-\mu s)}(s, L(s)) = 0$. On the other hand, it is easy to check that the identity

$$Q(g)(s, L(s)) = \int_0^{\exp(-\mu s)} Q(h_{-c})(s, L(s)) c^{\alpha-1}\, dc$$

holds. Finally, for $s < T_0$,

$$\begin{aligned}&\left[\frac{\partial g}{\partial t} + Q(g)\right](s, L(s)) \\ &\qquad = \int_0^{\exp(-\mu s)} \left[\frac{\partial h_{-c}}{\partial t} + Q(h_{-c})\right](s, L(s)) c^{\alpha-1}\, dc = 0,\end{aligned}$$

since h_{-c} is a space-time harmonic function for the Q-matrix Q by Lemma 6.3.

The process defined by Expression (6.11) being a local martingale, the same property holds for the process $(g(t \wedge T_0, L(t \wedge T_0)))$ which is, after a change of variables, the process $(J_\alpha(t \wedge T_0))$. The proposition is proved since a bounded local martingale is a martingale (see Proposition A.7 page 354). □

6.3 Hitting Times : Downward

The notation for hitting times is the same as for the $M/M/1$ queue.

DEFINITION 13. *The variable T_a is the hitting time of the integer a for the Markov process $(L(t))$, i.e.*

$$T_a = \inf\{t > 0 : L(t) = a\}.$$

The process $(L(t))$ being ergodic, the variable T_a is $\mathbb{P}$-almost surely finite for all $a \in \mathbb{N}$. As for the $M/M/1$ queue, the study is divided in two parts depending on the fact that the point to reach is below or above the initial point.

PROPOSITION 6.7. *For $\alpha > 0$ and $a \geq b \geq 0$,*

$$\mathbb{E}_a\left(e^{-\alpha T_b}\right) = \frac{B_a(\alpha)}{B_b(\alpha)}, \tag{6.12}$$

with, for $y \geq 0$,

$$B_y(\alpha) = \int_0^1 (1-c)^y\, c^{\alpha/\mu - 1} e^{\rho c}\, dc.$$

The result is due to Takàcs[Tak56] for the expression of the Laplace transform of a busy period (hitting time of 0 starting from 1). The above result has been obtained by Guillemin and Simonian[GS95] via analytical methods.

PROOF. The martingale $(J_\alpha(t \wedge T_0))$ of Proposition 6.6 is bounded, hence uniformly integrable. Therefore,

$$\mathbb{E}_a\left(J_\alpha(0)\right) = \mathbb{E}_a\left(J_\alpha(T_b \wedge T_0)\right) = \mathbb{E}_a\left(J_\alpha(T_b)\right)$$

and Identity (6.12) can be deduced. □

For $y \in \mathbb{N}$, the function $\alpha \to B_y(\alpha)$ is divergent at 0. The left hand side of Identity (6.12) is clearly continuous at $\alpha = 0$ and the limit is 1. Continuity of the right hand side of Identity (6.12) at 0 can be easily checked via some elementary calculus.

Asymptotic Results for the Hitting Time of 0. When the process $(L(t))$ starts from n large, after an exponential time with expected value $1/(n\mu+\lambda)$, it goes to $n+1$ with probability $\lambda/(n\mu+\lambda)$, otherwise it goes to $n-1$. Therefore, if n is large, the next jump is very likely to occur at $n-1$. With this approximation, the mean hitting time of 0 is of the order

$$\sum_{i=1}^{n} \frac{1}{i\mu+\lambda} \sim \frac{\log n}{\mu}.$$

This non-rigorous description implies that the process $(L(t))$ returns very quickly to 0. The next proposition shows that this heuristic reasoning is correct.

[Tak56] L. Takàcs, *On a probability problem arising in the theory of counters*, Proceedings of the Cambridge Philosophical Society **52** (1956), 488–498.

[GS95] F. Guillemin and A. Simonian, *Transient characteristics of an $M/M/\infty$ system*, Advances in Applied Probability **27** (1995), no. 3, 862–888.

PROPOSITION 6.8. *If $L(0) = n \in \mathbb{N}$, the hitting time of 0 is of the order $\log n$. More precisely, the random variable $T_0/\log n$ converges in probability to the constant $1/\mu$ as n tends to infinity.*

PROOF. The above proposition shows that, for $\alpha > 0$,

$$\phi_n(\alpha) = \mathbb{E}_n\left(e^{-\alpha\mu T_0}\right) = \int_0^1 (1-c)^n\, c^{\alpha-1}e^{\rho c}\,dc \Big/ \int_0^1 c^{\alpha-1}e^{\rho c}\,dc.$$

If $\alpha_n = \alpha/\log n$, the Laplace transform of $\mu T_0/\log n$ at α is given by

$$\phi_n(\alpha_n) = \alpha_n \int_0^1 (1-c)^n\, c^{\alpha_n-1}e^{\rho c}\,dc \Big/ \alpha_n \int_0^1 c^{\alpha_n-1}e^{\rho c}\,dc.$$

By expanding the exponential function, the expression of the denominator can be expressed as

$$\alpha_n \int_0^1 c^{\alpha_n-1}e^{\rho c}\,dc = 1 + \alpha_n \sum_{k=1}^{+\infty} \frac{\rho^k}{(k+\alpha_n)k!},$$

hence,

$$\lim_{n\to+\infty} \alpha_n \int_0^1 c^{\alpha_n-1}e^{\rho c}\,dc = 1.$$

It is thus sufficient to study the convergence of

$$\begin{aligned}
&\alpha_n \int_0^1 (1-c)^n\, c^{\alpha_n-1}e^{\rho c}\,dc \\
&\qquad = \alpha_n \int_0^1 (1-c)^n\, c^{\alpha_n-1}\,dc + \alpha_n \sum_{k=1}^{+\infty} \frac{\rho^k}{k!} \int_0^1 (1-c)^n\, c^{\alpha_n+k-1}\,dc.
\end{aligned}$$

The series is bounded by $\exp(\rho)-1$ and the first term of the right hand side of this identity converges clearly to 0 as n tends to infinity. One gets that $\phi_n(\alpha_n)$ is equivalent to

$$\alpha_n \int_0^1 (1-c)^n\, c^{\alpha_n-1}\,dc.$$

The last integral can be expressed as $B(n+1,\alpha_n)$ where B is a Beta function. Beta functions can represented with Gamma function Γ as

$$B(x,y) \stackrel{\text{def.}}{=} \int_0^1 c^{x-1}(1-c)^{y-1}\,dc = \frac{\Gamma(x)\Gamma(y)}{\Gamma(x+y)}, \text{ for } x \text{ and } y > 0,$$

with

$$\Gamma(x) = \int_0^{+\infty} c^{x-1}e^{-c}\,dc\ .$$

Gamma function satisfies $\Gamma(n) = (n-1)!$ for $n \in \mathbb{N}$ and

$$\lim_{x\to 0} x\Gamma(x) = 1 \quad \text{and} \quad \Gamma(x) \sim x^{x-1}\sqrt{2\pi x}e^{-x},$$

when x tends to infinity, according to Stirling's formula, (see Whittaker and Watson [**53**] for a presentation of general properties of these functions). The representation of $\phi_n(\alpha_n)$ with the help of Gamma functions gives an equivalent as n tends to infinity,

$$\phi_{n-1}(\alpha_{n-1}) \sim \alpha_{n-1}\frac{\Gamma(\alpha_{n-1})\Gamma(n)}{\Gamma(\alpha_{n-1}+n)}.$$

One gets the relationships

$$\begin{aligned}\phi_{n-1}(\alpha_{n-1}) &\sim \frac{n^{n-1}e^{-n}\sqrt{2\pi n}}{(\alpha_{n-1}+n)^{\alpha_{n-1}+n-1}\exp(-(\alpha_{n-1}+n))\sqrt{2\pi(n+\alpha_{n-1})}}\\ &\sim \left(1-\frac{\alpha_{n-1}}{\alpha_{n-1}+n}\right)^{n-1}(\alpha_{n-1}+n)^{-\alpha_{n-1}}\\ &\sim \exp\left(-(n-1)\frac{\alpha_{n-1}}{\alpha_{n-1}+n}-\frac{\alpha}{\log(n-1)}\log(\alpha_{n-1}+n)\right).\end{aligned}$$

This last term converges to $\exp(-\alpha)$ as n tends to infinity, hence

$$\lim_{n\to+\infty}\mathbb{E}_n\left[\exp\left(-\alpha\mu T_0/\log n\right)\right]=\exp(-\alpha).$$

The proposition is proved. □

6.4 Hitting Times : Upward

PROPOSITION 6.9. *For $\alpha>0$ and $0\le a\le b$,*

$$\mathbb{E}_a\left(e^{-\alpha T_b}\right)=\frac{\Gamma_a(\alpha)}{\Gamma_b(\alpha)}, \tag{6.13}$$

where, for $y\ge 0$,

$$\Gamma_y(\alpha)=\int_0^{+\infty}(1+c/\rho)^y\,c^{\alpha/\mu-1}e^{-c}\,dc.$$

The function $\alpha\to\Gamma_0(\alpha\mu)$ is the usual Gamma function seen above. The Laplace transform (6.13) of T_b starting from a is the integral representation of a formula due to Takàcs [**50**] in the case of the $G/M/\infty$ queue.

PROOF. The martingale

$$(I_\alpha(t))=\left(e^{-\alpha\mu t}\int_0^{+\infty}(1+x)^{L(t)}x^{\alpha-1}e^{-\rho x}\,dx\right)$$

of Proposition 6.6 stopped at T_b is bounded and therefore uniformly integrable, hence

$$\mathbb{E}_a(I_\alpha(T_b))=\mathbb{E}_a(I_\alpha(0)).$$

Since $L(T_b)=b$, one gets

$$\mathbb{E}_a\left(e^{-\alpha\mu T_b}\right)\int_0^{+\infty}(1+c)^b c^{\alpha-1}e^{-\rho c}\,dc=\int_0^{+\infty}(1+c)^a c^{\alpha-1}e^{-\rho c}\,dc,$$

Formula (6.13) is established. □

Rare Events. The next proposition shows that it takes a very long time to reach a large n.

PROPOSITION 6.10. *If $L(0)=x$, as n tends to infinity, the variable*

$$\rho^n T_n/(n-1)!$$

converges in distribution to an exponentially distributed random variable with parameter $\mu\exp(-\rho)$.

PROOF. Identity (6.13) shows that it is sufficient to prove the convergence for $x=0$. According to Relationship (6.13) of Proposition 6.9, for $\alpha>0$, the Laplace transform of T_n at $\alpha\mu$ is given by

$$\mathbb{E}_0\left(e^{-\alpha\mu T_n}\right)=\int_0^{+\infty}c^{\alpha-1}e^{-c}\,dc\Big/\int_0^{+\infty}(1+c/\rho)^n\,c^{\alpha-1}e^{-c}\,dc.$$

By expanding $(1+c/\rho)^n$, the denominator of this expression is

$$\sum_{k=0}^{n}\binom{n}{k}\frac{1}{\rho^k}\Gamma(\alpha+k),$$

and, by setting $\phi_n(\alpha)=1/\mathbb{E}_0\left(e^{-\alpha\mu T_n}\right)$, one gets

$$\phi_n(\alpha)=\sum_{k=0}^{n}\binom{n}{k}\frac{1}{\rho^k}\frac{\Gamma(\alpha+k)}{\Gamma(\alpha)}.$$

Gamma function satisfies the classical relation $\Gamma(x+1)=x\Gamma(x)$ for $x>0$ (which can easily be obtained with integration by parts), hence

$$\phi_n(\alpha)=1+\alpha\sum_{k=1}^{n}\binom{n}{k}\frac{1}{\rho^k}\prod_{i=1}^{k-1}(\alpha+i). \tag{6.14}$$

If $\alpha_n=\alpha\rho^n/(n-1)!$, the last equation gives the relationship

$$\phi_n(\alpha_n)=1+\alpha\sum_{k=1}^{n}\frac{n}{k}\frac{\rho^{n-k}}{(n-k)!}\frac{\prod_{i=1}^{k-1}(\alpha_n+i)}{(k-1)!},$$

$$\phi_n(\alpha_n)=1+\alpha\sum_{k=0}^{n-1}(k+1)\frac{\rho^k}{k!}\frac{n}{(k+1)(n-k)}\prod_{i=1}^{n-k-1}(1+\alpha_n/i). \tag{6.15}$$

For any $0\le k\le n-1$, the inequality

$$1\le\prod_{i=1}^{n-k-1}(1+\alpha_n/i)\le(1+\alpha_n)^n,$$

clearly holds and its right hand side converges to 1 as n tends to infinity. Consequently, the kth term of Series (6.15) converges to $\rho^k/k!$ when n tends to infinity. Moreover, for $0 \leq k \leq n-1$, since the quantity $n/(k+1)(n-k)$ is bounded by 1, one gets

$$\frac{n}{(k+1)(n-k)} \prod_{i=1}^{n-k-1} (1+\alpha_n) \leq (1+\alpha_n)^n,$$

this last expression being bounded with respect to n, Lebesgue's Theorem applied to the series of Identity (6.15) (seen as an integral over $\mathbb{N}$ with respect to the finite measure with density $((k+1)\rho^k/k!)$) shows that

$$\lim_{n\to+\infty} \phi_n(\alpha_n) = 1 + \alpha e^{\rho}.$$

The Laplace transform of T_n verifies

$$\lim_{n\to+\infty} \mathbb{E}_0\left[\exp\left(-\alpha\mu\rho^n T_n/(n-1)!\right)\right] = 1/(1+\alpha e^{\rho}) = e^{-\rho}/(e^{-\rho}+\alpha),$$

the convergence in distribution is therefore proved. □

Starting from a large n, the process $(L(t))$ goes down to 0 very quickly, of the order $\log n$ (Proposition 6.8) and, at the opposite, starting from 0 this process reaches the level n very slowly, after a duration of time of the order $(n-1)!/\rho^n$.

This last result is the analogue, for the $M/M/\infty$ queue, of Proposition 5.11 for the $M/M/1$ queue. In both cases, the result is the same: The distribution of the hitting time T_A of a rare event A scaled by a factor $\pi(A)$ is very close to an exponential distribution. This kind of result can be established formally for general reversible processes (see Keilson [**30**]), the Markov processes considered for the $M/M/1$ and $M/M/\infty$ queues fit, of course, in that framework. The book by Aldous [**1**] shows that this type convergence is likely to be valid under more general assumptions: Reversibility is not a crucial property in this respect.

Rare Events

The hitting time of a large n by the process $L(t))$ is of the order $(n-1)!/\rho^n$ and to reach $n+1$, an additional factor n/ρ is required. This suggests the following result : The process reaches n and then goes back very quickly to 0. In particular, very likely, it does not visit $n+1$. Notice here the difference with the $M/M/1$ queue. For this queue, when the value n is reached, the process stays some time in the neighborhood of n before going down to 0. In particular, the value $n+1$ can be reached before the next visit to 0.

This phenomenon can also be seen as follows, if $\alpha_n = \rho^n/(n-1)!$ and $\xi > 0$, the strong Markov property of $(L(t))$ gives the relation

$$\begin{aligned}(6.16)\quad &\mathbb{E}_0\left[\exp(-\xi\alpha_{n+1}T_{n+1})\right] \\ &\qquad = \mathbb{E}_0\left[\exp(-\xi\rho\alpha_n T_n/n)\right]\mathbb{E}_n\left[\exp(-\xi\alpha_{n+1}T_{n+1})\right].\end{aligned}$$

It is easily shown that $\mathbb{E}_0(\exp(-\xi\rho\alpha_n T_n/n))$ converges to 1 as n tends to infinity : For $n_0 \in \mathbb{N}$ and $n \geq n_0$

$$\mathbb{E}_0\left[\exp(-\xi\rho\alpha_n T_n/n)\right] \geq \mathbb{E}_0\left[\exp(-\xi\rho\alpha_n T_n/n_0)\right],$$

when n converges to infinity, one gets

$$\begin{aligned}\liminf_{n\to+\infty} \mathbb{E}_0\left[\exp(-\xi\rho\alpha_n T_n/n)\right] &\geq \liminf_{n\to+\infty} \mathbb{E}_0\left[\exp(-\xi\rho\alpha_n T_n/n_0)\right] \\ &= \mu e^{-\rho}/(\mu e^{-\rho} + \xi\rho/n_0).\end{aligned}$$

The convergence is obtained by letting n_0 go to infinity, Relationship (6.16) shows that the variable $\alpha_{n+1}T_{n+1}$ starting from $L(0) = n$ converges also in distribution to an exponentially distributed random variable with parameter $\mu\exp(-\rho)$. This can be summarized as follows: in the path from 0 to $n+1$ the last step, from n to $n+1$ is the most significant part.

The time to return to 0 being small, the successive hitting times of n form, approximately, a renewal process associated with the exponential distribution with parameter $\mu\exp(-\rho)\rho^n/(n-1)!$ The point process associated with the hitting times of n should be asymptotically Poisson. A similar result holds for the $M/M/1$ queue (Proposition 5.13 page 123), but only for a subset of the hitting times.

PROPOSITION 6.11. *If $(\tau_{i,n})$ is the sequence of the successive hitting times of $n \geq 0$ for the process $(L(t))$,*

$$\{\tau_{i,n}, i \in \mathbb{N}\} = \{t \in \mathbb{R}_+ : L(t) = n, L(t-) \neq n\},$$

then the point process

$$\sum_{i=1}^{+\infty} \delta_{t_{i,n}} = \sum_{i=1}^{+\infty} \delta_{\rho^n \tau_{i,n}/(n-1)!},$$

converges in distribution to a Poisson process with parameter $\mu\exp(-\rho)$ as n tends to infinity.

PROOF. The above arguments contain essentially the proof. The strong Markov property of $(L(t))$ shows that the sequence $(t_{i+1,n} - t_{i,n})$ is i.i.d. If $\alpha_n = \rho^n/(n-1)!$, $L(0) = n$ and

$$U_n = \inf\{t > 0 : L(t) = n, L(t-) \neq n\},$$

the variable $\alpha_n U_n$ has the same distribution as $t_{i+1,n} - t_{i,n}$. Proposition 1.23 page 22 shows that it is sufficient to prove that the variable $\alpha_n U_n$ starting from $L(0) = n$ converges in distribution to an exponentially distributed random variable with parameter $\mu\exp(-\rho)$.

On the event $\{T_0 < U_n, L(0) = n\}$ the variable $\alpha_n U_n$ can be decomposed as the sum of two independent random variables. Their respective distribution is given by the distribution of $\alpha_n T_0$ with $L(0) = n$ and the distribution of $\alpha_n T_n$ with $L(0) = 0$. The first variable converges in distribution to 0 (Proposition 6.8) and the distribution of the second variable converges to an exponentially distributed random variable with parameter $\mu\exp(-\rho)$ (Proposition 6.10). It is therefore sufficient to prove that the probability of the event $\{T_0 < U_n\}$ converges to 1 as n gets large.

Starting from n, in order that the instant U_n occurs before the hitting time of 0, either the process hits $n+1$ before 0, or it hits n after having visited $n-1$, hence

$$\mathbb{P}_n(T_0 \geq U_n) \leq \mathbb{P}_n(T_{n+1} \leq T_0) + \mathbb{P}_{n-1}(T_n \leq T_0). \tag{6.17}$$

For $\varepsilon > 0$, the sequence $(\mathbb{P}_n(T_{n+1} \leq T_0))$ verifies the inequalities

$$\begin{aligned}\mathbb{P}_n(T_{n+1} \leq T_0) &= \mathbb{P}_n(\alpha_{n+1}T_{n+1} \leq \alpha_n T_0)\\ &\leq \mathbb{P}_n(\alpha_{n+1}T_0 > \varepsilon) + \mathbb{P}_n(\alpha_{n+1}T_{n+1} \leq \varepsilon)\\ &= \mathbb{P}_n\left(T_0/\log n > \varepsilon/(\alpha_{n+1}\log n)\right) + \mathbb{P}_n(\alpha_{n+1}T_{n+1} \leq \varepsilon).\end{aligned}$$

Since the random variable $T_0/\log n$ converges to a constant and the sequence $(\varepsilon/\alpha_{n+1}\log n)$ converges to infinity the first term of the right hand side of the last inequality converges to 0 as n gets large. The convergence in distribution of the variable $\alpha_{n+1}T_{n+1}$ starting from $L(0) = n$ to an exponentially distributed random variable with parameter $\mu\exp(-\rho)$ gives the convergence

$$\limsup_{n\to+\infty} \mathbb{P}_n(T_{n+1} \leq T_0) \leq (1 - e^{-\varepsilon\mu\exp(-\rho)})/\mu e^{-\rho}.$$

By letting ε go to 0, one gets that the sequence $(\mathbb{P}_n(T_{n+1} \leq T_0))$ converges to 0 as n tends to infinity. The proof is concluded by using Inequality (6.17). □

6.5 Fluid Limits

The scaling considered for the process $(L(t))$ increases the arrival rate by a factor $N \in \mathbb{N}$ and the process itself is scaled by the factor $1/N$. This scaling has been introduced by Kelly[Kel86] to study the convergence of invariant measures of loss networks (see also Kelly[Kel91] on this subject).

Thus the arrivals are a Poisson point process with intensity λN, represented by $(\mathcal{N}_\lambda(]0, tN]))$. The associated process of the number of customers for this queue is denoted by $(L_N(t))$ and

$$\bar{L}_N(t) = \frac{L_N(t)}{N}$$

is the renormalized process. The initial value is deterministic, it is an integer $L_N(0)$ such that

$$\lim_{N\to+\infty} \frac{L_N(0)}{N} = x \in \mathbb{R}_+.$$

[Kel86] F.P. Kelly, *Blocking probabilities in large circuit-switched networks*, Advances in Applied Probability **18** (1986), 473–505.

[Kel91] F.P. Kelly, *Loss networks*, Annals of Applied Probability **1** (1991), no. 3, 319–378.

Integral Equation (6.2) can be written as

$$L_N(t) = L_N(0) + \mathcal{N}_\lambda(]0, Nt]) - \sum_{i=1}^{+\infty} \int_0^t 1_{\{L_N(s-)\geq i\}} \mathcal{N}_\mu^i(ds). \tag{6.18}$$

LEMMA 6.12. *For $N \in \mathbb{N}$, the process*

$$(M_N(t)) = \left(\bar{L}_N(t) - \bar{L}_N(0) - \lambda t + \mu \int_0^t \bar{L}_N(s)\, ds\right)$$

is a martingale such that

$$\mathbb{E}\left(M_N^2(t)\right) \leq \frac{1}{N}\left[\mu \bar{L}_N(0) + \lambda\left(1 + \frac{\mu}{2}t\right)\right] t,$$

for $t \in \mathbb{R}_+$ and $N \in \mathbb{N}$.

PROOF. Proposition 6.2 shows that

$$\left(L_N(t) - L_N(0) - \lambda N t + \mu \int_0^t L_N(s)\, ds\right)$$

is a martingale whose increasing process is given by

$$\left(\lambda N t + \mu \int_0^t L_N(s)\, ds\right),$$

hence,

$$\mathbb{E}(M_N^2(t)) = \frac{1}{N}\left(\lambda t + \mu \int_0^t \mathbb{E}(\bar{L}_N(s))\, ds\right) \leq \frac{1}{N}\left[\mu \bar{L}_N(0) t + \lambda\left(1 + \frac{\mu}{2}t\right) t\right],$$

since for all $s \in \mathbb{R}_+$, $L_N(s) \leq L_N(0) + \mathcal{N}_\lambda(]0, Ns])$. □

The last lemma shows that $(\bar{L}_N(t))$ satisfies the identity

$$\bar{L}_N(t) = \bar{L}_N(0) + M_N(t) + \lambda t - \mu \int_0^t \bar{L}_N(s)\, ds. \tag{6.19}$$

The martingale $(M_N(t))$ becomes negligible, of the order $1/\sqrt{N}$ as N tends to infinity since

$$\mathbb{E}(M_N(t)) \leq \sqrt{\mathbb{E}(M_N(t)^2)} \leq \frac{C(t)}{\sqrt{N}},$$

for some $C(t)$ independent of N. Assume that the sequence of processes $(\bar{L}_N(t))$ has a limiting value $(y(t))$ for the topology on the processes induced by the topology of the uniform convergence on compacts sets (in particular $y(0) = x$) and that the martingale $(M_N(t))$ really vanishes. The function $(y(t))$ should verify the equation

$$y(t) = x + \lambda t - \mu \int_0^t y(s)\, ds,$$

or $y'(t) = \lambda - \mu y(t)$ with $y(0) = x$, that is, $y(t) = \rho + (x - \rho)\exp(-\mu t)$. If there is a limiting value, it is therefore unique. The theorem below establishes this convergence rigorously. Note that the phenomenon observed for the scaling of $M/M/1$ queue is still valid: The scaling of the sample paths suppresses the stochastic perturbation (the martingale) around a deterministic trajectory.

THEOREM 6.13 (Functional law of large numbers). *For x, $t \in \mathbb{R}_+$ and*

$$x(t) \stackrel{def.}{=} \rho + (x-\rho)e^{-\mu t},$$

the random variable

$$\sup_{0\leq s\leq t}\left|\frac{L_N(s)}{N} - x(s)\right|$$

converges in distribution to 0 in L_1 as N tends to infinity. In particular, for $\varepsilon > 0$ and $t \geq 0$,

$$\lim_{N\to+\infty} \mathbb{P}\left(\sup_{0\leq s\leq t}\left|\frac{L_N(s)}{N} - x(s)\right| > \varepsilon\right) = 0. \tag{6.20}$$

PROOF. The function $t \to x(t)$ satisfies the equation

$$x(t) = x + \lambda t - \mu\int_0^t x(s)\,ds.$$

By setting $Z_N(t) = \bar{L}_N(t) - x(t)$, Equation (6.19) gives

$$Z_N(t) = \left(\bar{L}_N(0) - x\right) + M_N(t) - \mu\int_0^t Z_N(s)\,ds,$$

therefore,

$$\sup_{0\leq s\leq t}|Z_N(s)| \leq \left|\bar{L}_N(0) - x\right| + \sup_{0\leq s\leq t}|M_N(s)| + \mu\int_0^t \sup_{0\leq u\leq s}|Z_N(u)|\,ds. \tag{6.21}$$

By using successively Cauchy-Schwartz's inequality, Doob's inequality and the last lemma, one gets the inequality

$$\begin{aligned}\left(\mathbb{E}\left[\sup_{0\leq s\leq t}|M_N(s)|\right]\right)^2 &\leq \mathbb{E}\left(\sup_{0\leq s\leq t}|M_N(s)|^2\right)\\ &\leq 4\mathbb{E}(M_N^2(t)) \leq \frac{4}{N}\left(\mu\bar{L}_N(0) + \lambda\left(1+\frac{\mu}{2}t\right)\right)t \leq \frac{C}{N},\end{aligned}$$

where C is a constant independent of N. Therefore, if the function f_N is defined by

$$f_N(t) = \mathbb{E}\left(\sup_{0\leq s\leq t}|Z_N(s)|\right),$$

Inequality (6.21) shows that, for $u \leq t$, the relationship

$$f_N(u) \leq \left|\bar{L}_N(0) - x\right| + \sqrt{C/N} + \mu\int_0^u f_N(s)\,ds$$

holds. By applying Gronwall's Lemma 6.24 page 174, one gets

$$f_N(t) \leq \left(\left|\bar{L}_N(0) - x\right| + \sqrt{C/N}\right)e^{\mu t},$$

hence $f_N(t)$ converges to 0 as N tends to infinity. The L_1-convergence is established. □

6.6 A Functional Central Limit Theorem

The notations of the last section are used. The stochastic fluctuations of $(L_N(t))$ around the deterministic function $(x(t))$,

$$\left(\sqrt{N}\left[\bar{L}_N(t) - x(t)\right]\right) = \left(\frac{L_N(t) - Nx(t)}{\sqrt{N}}\right), \tag{6.22}$$

are analyzed when the initial values are assumed to satisfy the relation

$$\lim_{N\to+\infty} \frac{L_N(0) - Nx}{\sqrt{N}} = v,$$

where $v \in \mathbb{R}$. The process defined by Equation (6.22) has clearly the Markov property. The following theorem is due, in a more general case, to Borovkov[Bor67]. It describes the limit in distribution of the process (6.22).

THEOREM 6.14. *The process*

$$\left(\frac{L_N(t) - Nx(t)}{\sqrt{N}}\right)$$

converges in distribution, as N tends to infinity, to a process $(X(t))$ with independent (non-homogeneous) increments given by

$$X(t) = ve^{-\mu t} + \int_0^t e^{-\mu(t-s)}\sqrt{h(s)}\, dB(s),$$

for $t \geq 0$, where $(B(t))$ is a standard Brownian motion, and h is a function defined by

$$h(t) = 2\lambda + \mu(x-\rho)e^{-\mu t}.$$

When $x = \rho$, then $x(s) = \rho$ for all $s \in \mathbb{R}_+$ and the limit $(X(t))$ is an Ornstein-Uhlenbeck process,

$$X(t) = ve^{-\mu t} + \sqrt{2\lambda}\int_0^t e^{-\mu(t-s)}\, dB(s),$$

or, equivalently, $(X(t))$ satisfies the stochastic differential equation

$$X(0) = v, \qquad dX(t) = -\mu X(t)\, dt + \sqrt{2\lambda}\, dB(t).$$

Section 6.8.1 page 174 recalls the definition of Ornstein-Uhlenbeck processes.

[Bor67] A.A. Borovkov, *Limit laws for queueing processes in multichannel systems*, Sibirsk. Mat. Ž. **8** (1967), 983–1004.

PROOF. By setting

$$G_N(t) = \frac{L_N(t) - Nx(t)}{\sqrt{N}},$$

Lemma 6.12 shows that the process

$$(Y_N(t)) \stackrel{\text{def.}}{=} \left(G_N(t) - G_N(0) + \mu \int_0^t G_N(s)\, ds\right) \tag{6.23}$$

is a square integrable martingale whose increasing process is given by

$$\langle Y_N \rangle (t) = \lambda t + \mu \int_0^t \bar{L}_N(s)\, ds.$$

According to Theorem 6.13, the increasing process $(<Y_N>(t))$ converges in probability to $(\gamma(t))$ defined by

$$\gamma(t) = \lambda t + \mu \int_0^t x(s)\, ds = 2\lambda t + (x - \rho)(1 - e^{-\mu t}),$$

for $t \geq 0$. One obtains, with Theorem C.10 page 373, that the process $(Y_N(t))$ converges in distribution to $(B(\gamma(t))$, where $(B(t))$ is the standard Brownian motion.

The tightness of the sequence of processes $(G_N(t))$ can now be proved. The modulus of continuity w_N of $(G_N(t))$ is defined by

$$w_N(\delta) = \sup_{u,v \leq t, |u-v| \leq \delta} |G_N(u) - G_N(v)|.$$

It satisfies the inequality

$$\begin{aligned} w_N(\delta) &\leq \sup_{u,v \leq t, |u-v| \leq \delta} \left(|Y_N(u) - Y_N(v)| + \mu \int_u^v |G_N(s)|\, ds \right) \\ &\leq \mu\delta \sup_{u \leq t} |G_N(u)| + \sup_{u,v \leq t, |u-v| \leq \delta} |Y_N(u) - Y_N(v)|. \end{aligned} \tag{6.24}$$

For s and $t \in \mathbb{R}_+$ such that $s \leq t$, the definition of $(Y_N(t))$ gives that

$$|G_N(s)| \leq |Y_N(s)| + |G_N(0)| + \mu \int_0^s |G_N(u)|\, du,$$

$$\sup_{0 \leq u \leq s} |G_N(u)| \leq \sup_{0 \leq u \leq s} |Y_N(u)| + |G_N(0)| + \mu \int_0^s \sup_{0 \leq v \leq u} |G_N(v)|\, du.$$

By using Gronwall's Lemma, one gets the inequality

$$\sup_{0 \leq u \leq s} |G_N(u)| \leq \left(\sup_{0 \leq u \leq s} |Y_N(u)| + |G_N(0)| \right) e^{\mu s}, \tag{6.25}$$

for $s \leq t$. Since the sequence of processes $(Y_N(t))$ converges in distribution, with the continuity argument used in the proof of Proposition 2.2 page 32, it is easily seen that the sequence of random variables

$$\left(\sup_{0\le u\le t}|Y_N(u)|\right)$$

converges in distribution. Moreover, since the sequence $(G_N(0))$ converges to v, for $\eta>0$, there exist $K>0$ and N_0 such that if $N\ge N_0$, then

$$\mathbb{P}\left(\left[\sup_{0\le u\le t}|Y_N(u)|+|G_N(0)|\right]e^{\mu t}\ge K\right)\le\frac{\eta}{2},$$

Relationship (6.25) gives the upper bound

$$\mathbb{P}\left(\sup_{0\le u\le t}|G_N(u)|\ge K\right)\le\frac{\eta}{2}. \tag{6.26}$$

The sequence of processes $(Y_N(t))$ being tight, for η and $\varepsilon>0$, there exist δ_0 and N_1 such that, for $N\ge N_1$, the modulus of continuity of $(Y_N(t))$ satisfies the inequality

$$\mathbb{P}\left(\sup_{u,v\le t,|u-v|\le\delta_0}|Y_N(u)-Y_N(v)|\ge\frac{\varepsilon}{2}\right)\le\frac{\eta}{2}. \tag{6.27}$$

Relationship (6.24) shows that the inequality

$$\mathbb{P}(w_N(\delta)\ge\varepsilon)\le\mathbb{P}\left(\sup_{u,v\le t,|u-v|\le\delta}|Y_N(u)-Y_N(v)|\ge\frac{\varepsilon}{2}\right)$$
$$+\,\mathbb{P}\left(\delta\mu\sup_{0\le u\le t}|G_N(u)|\ge\frac{\varepsilon}{2}\right),$$

holds when $\delta\le\delta_0\wedge\varepsilon/(2\mu K)$ and $N\ge N_0\vee N_1$. According to Relationships (6.27) and (6.26), the two expressions of the right hand side of the last inequality are bounded by $\eta/2$. For ε and $\eta>0$ there exist δ and N_2 such that the inequality $\mathbb{P}(w_N(\delta)\ge\varepsilon)\le\eta$ is valid for $N\ge N_2$.

Theorem C.9 page 373 shows that the sequence of processes $(G_N(t))$ is tight for the Skorohod topology on the càdlàg functions on $[0,t]$ (see Definition 40 page 359) and that any limiting point is almost surely a continuous process. If $(X(t))$ is a limiting point of this sequence, Relationship (6.23) shows that $\mathbb{P}$-a.s. for any $s\le t$,

$$B(\gamma(s))=X(s)-v+\mu\int_0^s X(u)\,du. \tag{6.28}$$

The process $(B[\gamma(t)])$ has the same distribution as the process

$$\left(\int_0^t\sqrt{\gamma'(s)}\,dB_1(s)\right),$$

where $(B_1(t))$ is a standard Brownian motion (this is a continuous martingale with the same increasing process). Equation (6.28) can be rewritten as a stochastic differential equation

$$dX(s) = \sqrt{\gamma'(s)}\, dB_1(s) - \mu X(s)\, ds. \tag{6.29}$$

It is well known (see Rogers and Williams [**43**] for example) that there is only one solution such an equation. The sequence $(G_N(t))$ has therefore only one limiting point therefore, it converges in distribution.

The limit can be expressed as

$$X(t) = v + \int_0^t \sqrt{\gamma'(s)}\, dB(s) - \mu \int_0^t X(s)\, ds,$$

and it is easily seen that

$$X(t) = ve^{-\mu t} + \int_0^t e^{-\mu(t-s)} \sqrt{\gamma'(s)}\, dB(s)$$

is a solution of the above equation. In the case where $x = \rho$, then $\gamma(t) = 2\lambda t$, Relationship (6.29) shows that $(X(t))$ is indeed the solution of the stochastic differential equation

$$dX(t) = \sqrt{2\lambda}\, dB(t) - \mu X(t)\, dt.$$

The theorem is proved. □

Basically, the last theorem describes the renormalized process $(\overline{L}_N(t))$ of the number of customers in a $M/M/\infty$ queue as a deterministic differential equation

$$dx(t) = -\mu x(t)\, dt,$$

perturbated by a stochastic noise $(W(t))$. More precisely, it can be represented as the solution of a stochastic differential equation of the form

$$d\overline{L}_N(t) = \varepsilon\, dW(t) - \mu \overline{L}_N(t)\, dt.$$

where $\varepsilon = 1/\sqrt{N}$.

A systematic treatment of perturbated deterministic differential equations is presented in Freidlin and Wentzell [**23**].

A Related Model. The parking of a super-market : The same probabilistic assumptions are taken. Places in the queue are numbered. When a customer arrives, it chooses the empty place with the smallest number. The total number of customers is the same as for the $M/M/\infty$ queue, this model has an additional feature, a spatial component. For this model, several characteristics are interesting such as the largest occupied number or the number of empty places between the head of the queue and the last customer. This model has been analyzed, via scaling, by Aldous[Ald86].

[Ald86] D. Aldous, *Some interesting processes arising as heavy traffic limits in an $M/M/\infty$ storage process*, Stochastic Processes and their Applications **22** (1986), 291–313.

6.7 The M/M/N/N Queue

This queue is a truncated version of the $M/M/\infty$ queue, if λ is the arrival rate, μ the service rate and $\rho = \lambda/\mu$, the process of the number of customers at equilibrium is a reversible process. Its invariant measure π_N is given by

$$\pi_N(k) = C^{-1}\frac{\rho^k}{k!}, \qquad 0 \le k \le N, \tag{6.30}$$

where C is the normalization constant $C = 1 + \rho + \rho^2/2 + \cdots + \rho^N/N!$

This is the basic model of a link in a communication network where N calls can be accommodated simultaneously. The duration of a call is exponentially distributed with parameter μ. When N calls are present, any new request is rejected. A loss network is just a set of such links (see Section 4.2.3 page 88). In this setting, it is interesting to estimate the probability that an arriving call is rejected, or to get the distribution of the time between two rejected calls, etc...

This queue is analyzed when the arrival rate is λN, in particular it is proportional to the capacity N. The process of the number of customers is denoted by $(L_N^*(t))$, it satisfies the stochastic differential equation

$$L_N^*(t) = 1_{\{L_N^*(t-)<N\}}\mathcal{N}_{\lambda N}(dt) - \sum_{i=1}^{L_N^*(t-)} \mathcal{N}_\mu^i(dt).$$

In the same way, the hitting time of $a \in \{0, \dots, N\}$ is denoted by T_a^*,

$$T_a^* = \inf\{t > 0 : L_N^*(t) = a\}.$$

The values of the renormalized process $(\bar{L}_N^*(t))$ defined by

$$\bar{L}_N^*(t) = \frac{L_N^*(t)}{N}, \qquad t \in \mathbb{R}_+,$$

are in the interval $[0, 1]$. At time $t = 0$, it is assumed that the condition

$$\lim_{N\to+\infty} \bar{L}_N^*(0) = x \in [0, 1]$$

holds. As long as $(L_N^*(t))$ has not reached N, the process $(L_N^*(t))$ is similar to the process $(L_N(t))$ of the $M/M/\infty$ queue. According to Theorem 6.13 the quantity $L_N(t)/N$ is approximately ρ for t sufficiently large. As it will be seen, this queue has, depending on the values of ρ, three different regimes.

An Informal Description of the Different Regimes.

1. If $\rho > 1$, this is the *super-critical* case. The law of large numbers (6.13) shows that the process $L_N(t)$ reaches the capacity N in finite time. The queue is quickly filled in this regime.
2. If $\rho < 1$, the *sub-critical* case. Since $(L_N(t) - \rho N)/\sqrt{N}$ is approximately a diffusion, the process $(L_N^*(t))$ lives in a region like $\left[\rho N - A\sqrt{N}, \rho N + A\sqrt{N}\right] \subset [0, N]$ for some non-negative A. The probability that a call is rejected is therefore small in this case.

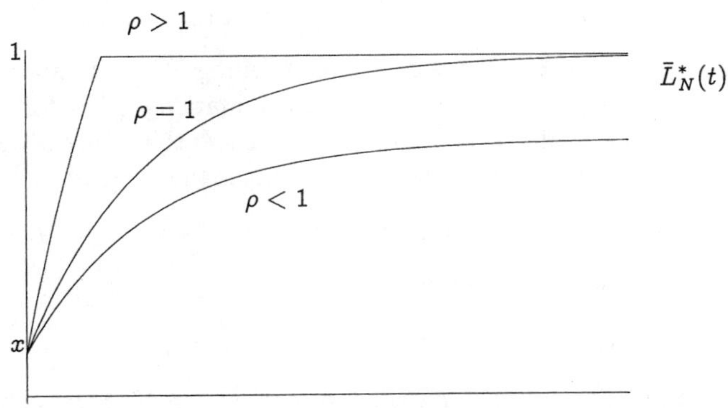

FIG. 6.1. The Three Regimes of the $M/M/N/N$ Queue

3. If $\rho = 1$, the *critical case.* The fluid limit $(x(t))$ tends to 1 as t gets large, hence the queue is asymptotically full at the fluid level. This rough picture will be improved in the following.

Each of these regimes is analyzed separately, the behavior of the hitting time T_N^* of N is analyzed and a functional law of large numbers is proved in each case.

6.7.1 The Super Critical Case $\rho > 1$

When $\rho > 1$, the queue is very quickly saturated. It can be heuristically described as follows. If $L_N^*(0)$ is approximately Nx with $x \leq 1$, as long as the number of customers is around Nx, the process $(L_N^*(t))$ looks like a transient birth and death process whose jumps of size $+1$ occur at rate λN strictly greater than the rate $\mu x N$ (since $\mu x N < \lambda x N < \lambda N$) of the jumps of size -1. Therefore, the process is converging to the boundary value N. The fluid limit $(\rho + (x - \rho)\exp(-\mu t))$ is at 1 (the fluid capacity of the queue) at time

$$\frac{1}{\mu} \log\left(\frac{\rho - x}{\rho - 1}\right).$$

The following proposition proves the convergence in distribution of the hitting time of N. It has been seen that if $L_N^*(t) \equiv Nx$, the process behaves locally as the number of customers of a transient $M/M/1$ queue with arrival rate λN and service rate $\mu x N$.

PROPOSITION 6.15. *If $\rho > 1$, as N goes to infinity, the random variable T_N^* converges in probability to*

$$\frac{1}{\mu} \log\left[(\rho - x)/(\rho - 1)\right],$$

when the initial condition $L_N^*(0)$ *is such that* $\lim_{N\to+\infty} L_N^*(0)/N = x$.

PROOF. Corollary 6.5 and Doob's optional stopping Theorem show that the process $(\exp(\mu(t\wedge T_N))(\rho N - L_N(t\wedge T_N)))$ is a martingale. Before time T_N^*, the processes $(L_N(t))$ and $(L_N^*(t))$ can be coupled so that they coincide, and therefore $T_N = T_N^*$. The martingale property gives the identity

$$\mathbb{E}\left[\exp(\mu(t\wedge T_N^*))\,(\rho N - L_N^*(t\wedge T_N^*))\right] = \rho N - L_N^*(0),$$

and since $L_N^*(t\wedge T_N^*) \le N$, the inequality

$$(\rho N - N)\mathbb{E}\left[\exp(\mu(t\wedge T_N^*))\right] \le \rho N - L_N^*(0)$$

holds. By letting t go to infinity, monotone convergence Theorem shows that

$$\mathbb{E}\left(\exp(\mu T_N^*)\right) \le (\rho - \bar{L}_N^*(0))/(\rho - 1).$$

The random variable $\exp(\mu T_N^*)$ is therefore integrable, and since it is non-negative,

$$0 \le \exp\left(\mu(t\wedge T_N^*)\right)\left[\rho N - L_N^*(t\wedge T_N^*)\right] \le \rho N \exp\left(\mu T_N^*\right).$$

The martingale $[\exp(\mu(t\wedge T_N^*))(\rho N - L_N^*(t\wedge T_N^*))]$ is therefore uniformly integrable, and the following identity

$$\mathbb{E}\left(\exp\left(\mu T_N^*\right)\right) = \frac{\rho - \bar{L}_N^*(0)}{\rho - 1} \tag{6.31}$$

holds. The function $x \to (x - N\rho)^2 - x$ having values between $N^2(\rho-1)^2 - N$ and $(N\rho)^2$ for x between 0 and N, the same kind of argument can be applied to the second martingale of Relationship (6.9) stopped at T_N^*:

$$\left(\exp(2\mu(t\wedge T_N^*))\left[(L_N^*(t\wedge T_N^*) - N\rho)^2 - L_N^*(t\wedge T_N^*)\right]\right).$$

This yields

$$\mathbb{E}\left(\exp\left(2\mu T_N^*\right)\right) = \frac{(\rho - \bar{L}_N^*(0))^2 - \bar{L}_N^*(0)/N}{(\rho-1)^2 - 1/N}.$$

Equation (6.31) and the last identity show that $\operatorname{var}\left(\exp(\mu T_N^*)\right) = O(1/N)$. By using Chebishov's Inequality, for $\varepsilon > 0$, one gets

$$\mathbb{P}\left(\left|\exp(\mu T_N^*) - \mathbb{E}(\exp(\mu T_N^*))\right| > \varepsilon\right) \le \operatorname{var}\left(\exp(\mu T_N^*)\right)/\varepsilon^2.$$

Hence, according to Relationship (6.31), the sequence (T_N^*) converges in probability to

$$\frac{1}{\mu}\log\frac{\rho - x}{\rho - 1}.$$

The proposition is proved. □

The next lemma will be used to prove that the renormalized process stays at 1 when 1 is reached.

LEMMA 6.16. *The process* $(L_N^*(t))$ *can be lower bounded by a birth and death process* $(Z_N(t))$ *starting from* $L_N^*(0)$ *whose Q-matrix* (R_{ij}) *is given by* $R_{i,i+1} = \lambda N$, *for* $i < N$ *and* $R_{i,i-1} = \mu N$ *for* $i > 0$.

PROOF. The proof uses a coupling argument. The process $(Z_N(t))$ is constructed with $(L^*_N(t))$. The instants of jumps of $(Z_N(t))$ have to be defined. The jumps of size $+1$ are the same for $(Z_N(t))$ and $(L^*_N(t))$. If at time t_0, the process $(L^*_N(t))$ has a jump of size 1 then $(Z_N(t))$ has the same jump. Moreover, independently of the other jumps, the process $(Z_N(t))$ jumps to $Z_N(t)-1$ at rate $\mu(N-L_N(t))$ when $Z_N(t)>0$. In this manner, the jumps of size -1 of $(Z_N(t))$ occur at rate μN. Since only downward jumps have been added, the inequality $L_N(t)\geq Z_N(t)$ obviously holds. The lemma is proved. □

PROPOSITION 6.17 (Law of Large Numbers). *If $\rho>1$, for $\varepsilon>0$ and $t\in\mathbb{R}_+$,*

$$\lim_{N\to+\infty}\mathbb{P}\left(\sup_{0\leq s\leq t}\left|\frac{L^*_N(s)}{N}-y(s)\right|>\varepsilon\right)=0,$$

with

$$y(t)=\left(\rho+(x-\rho)e^{-\mu t}\right)\wedge 1.$$

PROOF. If $\Delta_N(s)=|L^*_N(s)/N-y(s)|$ for $s\in\mathbb{R}_+$,

$$\text{(6.32)}\quad \mathbb{P}\left(\sup_{0\leq s\leq t}\Delta_N(s)>\varepsilon\right)\\ \leq \mathbb{P}\left(\sup_{0\leq s\leq T^*_N\wedge t}\Delta_N(s)>\varepsilon\right)+\mathbb{P}\left(\sup_{T^*_N\leq s\leq t}\Delta_N(s)>\varepsilon\right).$$

The second term of the right hand side of Inequality (6.32) can be written as

$$\mathbb{P}\left(\sup_{T^*_N\leq s\leq t}\left|\frac{L^*_N(s)}{N}-y(s)\right|>\varepsilon\right)\\ \leq \mathbb{P}\left(\sup_{T^*_N\leq s\leq t}\left|\frac{L^*_N(s)}{N}-1\right|>\frac{\varepsilon}{2}\right)+\mathbb{P}\left(\sup_{T^*_N\leq s\leq t}|y(s)-1|>\frac{\varepsilon}{2}\right).$$

By using the strong Markov property, this expression can be bounded by

$$\text{(6.33)}\qquad \mathbb{P}_N\left(\sup_{0\leq s\leq t}\left(1-\frac{L^*_N(s)}{N}\right)>\frac{\varepsilon}{2}\right)+\mathbb{P}\left(\sup_{T^*_N\leq s\leq t}|y(s)-1|>\frac{\varepsilon}{2}\right).$$

The function $s\to y(s)$ being continuous at

$$s_0=\frac{1}{\mu}\log\left(\frac{\rho-x}{\rho-1}\right),$$

with $y(s)=1$ for $s\geq s_0$, there exists $\delta>0$ such that

$$\sup_{s\geq s_0-\delta}|y(s)-1|\leq\varepsilon/2.$$

It implies the inequality

$$\mathbb{P}\left(\sup_{T_N^* \le s \le t} |y(s) - 1| > \frac{\varepsilon}{2}\right) \le \mathbb{P}\left(T_N^* \le s_0 - \delta\right).$$

Proposition 6.15 shows that the last term converges to 0 as N tends to infinity. The other term of Expression (6.33) can be bounded as follows

$$\mathbb{P}_N\left(\sup_{0 \le s \le t}\left(1 - \frac{L_N^*(s)}{N}\right) > \frac{\varepsilon}{2}\right) \le \mathbb{P}_N\left(T_{N-\lfloor \varepsilon N/2 \rfloor}^* \le t\right).$$

According to Lemma 6.16, the process $(L_N^*(t))$ can be lower-bounded by a birth and death process $(Z_N(t))$ reflected at N whose Q-matrix (R_{ij}) is given by $R_{i,i+1} = \lambda N$ and $R_{i,i-1} = \mu N$. If τ_N is the hitting time of $N - \lfloor \varepsilon N/2 \rfloor$ by $(Z_N(t))$, one gets

$$\mathbb{P}_N\left(\sup_{0 \le s \le t}\left(1 - \frac{L_N^*(s)}{N}\right) > \frac{\varepsilon}{2}\right) \le \mathbb{P}_N\left(\tau_N \le t\right).$$

The process $(N - Z_N(t/N))$ can be described as the number of customers of an ergodic $M/M/1$ queue where μ is the arrival rate and λ the service rate and $N\tau_N$ is the hitting time of $\lfloor \varepsilon N/2 \rfloor$ starting from 0. Proposition 5.11 page 119 shows that the variable $(\mu/\lambda)^{\lfloor \varepsilon N/2 \rfloor} N\tau_N$ converges in distribution to an exponentially distributed random variable. In particular, $\mathbb{P}_N\left(\tau_N \le t\right)$ converges to 0 as N gets large. Therefore, the left hand side of Inequality (6.33) converges to 0 and the same is true for the second term of the right hand side of Inequality (6.20).

With a similar method, since L_N^* and L_N coincide until T_N^*, the first term of the right hand side of Inequality (6.20) converges also to 0, hence the same is true for the left hand side of this inequality. The proposition is proved. □

The renormalized process starting from 1 remains at 1. It is quite natural to study the fluctuations of the initial process $(L_N^*(t))$ below the boundary N. The following proposition investigates the behavior of the process

$$(N - L_N^*(t/N)),$$

which is also the number of empty places of the queue. Basically it is shown that there is only a finite number of empty places, the queue is almost always full.

PROPOSITION 6.18. *If $\rho > 1$, the finite distributions of the Markov process $(Y_N(t)) = (N - L_N^*(t/N))$ with the initial condition $Y_N(0) = a \in \mathbb{N}$ converge in distribution to the corresponding finite distributions of a birth and death process $(Y(t))$ whose Q-matrix $Q = (q_{i,j})$ is given by $q_{i,i+1} = \mu$ and $q_{i+1,i} = \lambda$ for $i \in \mathbb{N}$.*

PROOF. The embedded Markov chain $(D_N(n))$ of $(Y_N(t))$ (see Section B.2 page 363) has the transition matrix $p_N = (p_N(i,j))$ given by

$$p_N(i,i+1) = \mu(N-i)/\left(\lambda N + \mu(N-i)\right) \quad \text{if } i < N,$$
$$p_N(i,i-1) = \lambda N/\left(\lambda N + \mu(N-i)\right) \quad \text{if } 0 < i \leq N,$$

and the other coefficients are zero. When N tends to infinity, the coefficients of the matrix p_N converge to the coefficient of the matrix p of the embedded Markov chain $(D(n))$ of a birth and death process with $\mathcal{Q}$-matrix Q,

$$p(i,i+1) = \frac{\mu}{\lambda+\mu}, \quad p(i,i-1) = \frac{\lambda}{\lambda+\mu}, \quad i > 0.$$

Therefore, the sequence of Markov chains $(D_N(n))$, $N \in \mathbb{N}$ converges in distribution to $(D(n))$.

With the initial condition $Y_N(0) = a$, the instant of the next jump is exponentially distributed with parameter $(\lambda + \mu(N-a))/N$ and thus converges in distribution to an exponentially distributed random variable with parameter $\lambda + \mu$. The proof of the proposition is then a consequence of the construction of a jump process with its embedded Markov chain (see Appendix B). □

The invariant measure of the process $(N - L_N^*(t))$ is also the invariant measure of the process $(N - L_N^*(t/N))$. The last proposition suggests that this invariant distribution converges to the invariant distribution of $(Y(t))$.

PROPOSITION 6.19. *If $\rho > 1$, the invariant distribution of the number of empty places $(N - L_N^*(t))$ converges in distribution to a geometric distribution with parameter $1/\rho$, in particular the blocking probability $\pi_N(N)$ satisfies the relation*

$$\lim_{N\to+\infty} \pi_N(N) = 1 - 1/\rho.$$

PROOF. The invariant distribution π_N being a truncated Poisson distribution with parameter $N\rho$, for $k \leq N$,

$$\pi_N(N-k) = \frac{(\rho N)^{N-k}}{(N-k)!} \frac{1}{\sum_{i=0}^N (\rho N)^{N-i}/(N-i)!} = \left(\frac{1}{\rho}\right)^k \frac{1}{K_N}$$

with

$$K_N = \sum_{i=0}^{k} \frac{1}{\rho^i} \frac{N^k(N-k)!}{N^i(N-i)!} + \sum_{i=k+1}^{N} \frac{1}{\rho^i} \frac{N^k(N-k)!}{N^i(N-i)!};$$

the expression $N^k(N-k)!/N^i(N-i)!$ converges to 1 for $i \leq N$ and is bounded by 1 for $i \geq k$. Lebesgue's Theorem applied to the second sum of the last identity shows the convergence

$$\lim_{N\to+\infty} K_N = \frac{1}{1-1/\rho},$$

which proves the convergence in distribution

$$\lim_{N\to+\infty} \pi_N(N-k) = \left(\frac{1}{\rho}\right)^k \left(1 - \frac{1}{\rho}\right).$$

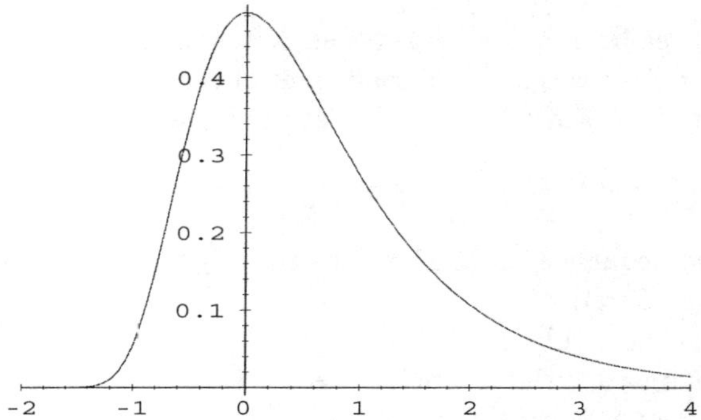

FIG. 6.2. The Asymptotic Density of $\mu T_N^* - (\log N)/2$.

□

6.7.2 The Critical Case $\rho = 1$

An informal description is given first. If $L_N^*(0)$ is just below N, the jumps of size -1 and $+1$ occur with the same rate, approximately λN. The equation for the hitting time of saturation for the fluid model is $1+(x-1)\exp(-\mu t) = 1$, and its solution is, of course, $t = +\infty$. Theorem 6.14 shows that the process $(L_N(t))$ is a Gaussian perturbation around the fluid limit, hence N is in fact hit in finite time. The central limit theorem proved in Theorem 6.14 shows the approximation

$$L_N(t) \sim N\left(1 + (x-1)e^{-\mu t}\right) + \sqrt{N} X(t),$$

if one admits that $(X(t))$ stay in a bounded domain with a high probability (this process is in fact ergodic), the process $(L_N(t))$ reaches N for t of the order $(\log \sqrt{N})/\mu$. The next proposition shows that the hitting time of N is indeed very close to $(\log N)/(2\mu)$.

PROPOSITION 6.20. *If $\rho = 1$ and $L_N^*(0) = 0$, the random variable*

$$\frac{\mu T_N^* - \log N}{2}$$

converges in distribution to a random variable whose density is given by

$$\sqrt{\frac{2}{\pi}}\, e^{-x} e^{-\exp(-2x)/2}, \quad x \geq 0.$$

PROOF. Proposition 6.9 shows that for $\alpha > 0$,

$$\mathbb{E}_0\left(e^{-\alpha\mu T_N^*}\right) = \Gamma(\alpha)/\Phi_N(\alpha),$$

with

$$\Phi_N(\alpha) = \int_0^{+\infty} (1 + c/N)^N \, c^{\alpha-1} e^{-c} \, dc.$$

The change of variables $c = \sqrt{2Nx}$ gives

$$\Phi_N(\alpha) = N^{\alpha/2} \, 2^{\alpha/2-1} \int_0^{+\infty} \left(1 + \sqrt{2x/N}\right)^N x^{\alpha/2-1} e^{-\sqrt{2Nx}} \, dx.$$

The last integral can be expressed as

$$A_N = \int_0^{+\infty} e^{N \log\left(1+\sqrt{2x/N}\right) - \sqrt{2Nx}} x^{\alpha/2-1} \, dx.$$

The function $y \to y^2 \log(1 + 1/y) - y$ being non-increasing on $\mathbb{R}_+$, for $x \geq 0$, the sequence

$$\left(N \log\left(1 + \sqrt{2x/N}\right) - \sqrt{2Nx}\right)$$

is non-increasing and converges to $-x$ as N tends to infinity. Monotone convergence Theorem yields the convergence

$$\lim_{N \to +\infty} A_N = \int_0^{+\infty} e^{-x} x^{\alpha/2-1} \, dx = \Gamma(\alpha/2).$$

Consequently,

$$\lim_{N \to +\infty} N^{\alpha/2} \, 2^{\alpha/2-1} \mathbb{E}_0 \left(\exp\left(-\alpha \mu T_N^*\right)\right) = \frac{\Gamma(\alpha)}{\Gamma(\alpha/2)}.$$

By using Legendre's duplication Formula for Gamma functions (see Whittaker and Watson [**53**] page 240), for $x \geq 0$,

$$2^{2x-1} \Gamma(x) \Gamma(x + 1/2) = \sqrt{\pi} \Gamma(2x),$$

one gets the relation

$$\lim_{N \to +\infty} \mathbb{E}_0 \left[\exp\left(-\alpha(\mu T_N^* - (\log N)/2)\right)\right] = \frac{2^{\alpha/2}}{\sqrt{\pi}} \Gamma\left(\frac{\alpha+1}{2}\right).$$

The last expression is a Laplace transform, since

$$\frac{2^{\alpha/2}}{\sqrt{\pi}} \Gamma\left(\frac{\alpha+1}{2}\right) = \frac{1}{\sqrt{\pi}} \int_0^{+\infty} \exp\left(\alpha \log(2y)/2 - \log y/2 - y\right) \, dy,$$

the change of variables $x = -\log(2y)/2$ yields

$$\frac{2^{\alpha/2}}{\sqrt{\pi}} \Gamma\left(\frac{\alpha+1}{2}\right) = \sqrt{\frac{2}{\pi}} \int_{-\infty}^{+\infty} e^{-\alpha x} \exp\left(-x - e^{-2x}/2\right) \, dx.$$

The proposition is proved. □

The law of large numbers for $(L_N^*(t))$ is similar to the result obtained for the process $(L_N(t))$.

PROPOSITION 6.21. *If $\rho = 1$, for $t \in \mathbb{R}_+$ and $\varepsilon > 0$,*

$$\lim_{N\to+\infty} \mathbb{P}\left(\sup_{0\le s\le t}\left|\frac{L_N^*(s)}{N} - (1+(x-1)e^{-\mu s})\right| > \varepsilon\right) = 0.$$

PROOF. The proof is simple: With a high probability, T_N^* is larger than t and, between 0 and T_N^*, the process $(L_N^*(t))$ is identical to $(L_N(t))$. Rigorously, the inequality

$$\mathbb{P}\left(\sup_{0\le s\le t}\left|\frac{L_N^*(s)}{N} - (1+(x-1)e^{-\mu s})\right| > \varepsilon\right)$$
$$\le \mathbb{P}(T_N^* \le t) + \mathbb{P}\left(\sup_{0\le s\le t}\left|\frac{L_N(s)}{N} - (1+(x-1)e^{-\mu s})\right| > \varepsilon\right),$$

Theorem 6.13 and the last proposition are used to get the result. □

The fluctuations of $(L_N^*(t))$ around the fluid limit $(1+(x-1)\exp(-\mu t))$ are of the order $\sqrt{N}$. If $L_N(0) = N$, it is possible to prove a central limit theorem for $(L_N^*(t))$, the limit in distribution of $((N - L_N^*(t))/\sqrt{N})$ is in fact a reflected Ornstein-Uhlenbeck process, i.e. a non-negative process satisfying the stochastic differential Equation (6.35) with $\alpha = 1$ and $\sigma = 1$. A rigorous proof of such a result can be obtained by using the results of Appendix D page 375.

This part is concluded with an asymptotic estimate of the blocking probability.

PROPOSITION 6.22. *The invariant distribution of the Markov process*

$$\left(\frac{N - L_N^*(t)}{\sqrt{N}}\right)$$

converges in distribution, when N tends to infinity, to the absolute value of a centered Gaussian random variable whose variance is 1.

Moreover, the blocking probability $\pi_N(N)$ satisfies

$$\lim_{N\to+\infty} \sqrt{N}\pi_N(N) = \sqrt{2/\pi}.$$

PROOF. Formula (6.30) for π_N shows the relation

$$\pi_N(k) = \frac{\mathbb{P}(\mathcal{N}_1([0,N]) = k)}{\mathbb{P}(\mathcal{N}_1([0,N]) \le N)},$$

for $k \in \mathbb{N}$ where, as usual, $\mathcal{N}_1$ is a Poisson process with parameter 1. The central limit Theorem for Poisson processes shows that the quantity

$$\mathbb{P}(\mathcal{N}_1([0,N]) \le N) = \mathbb{P}\left(\frac{\mathcal{N}_1([0,N]) - N}{\sqrt{N}} \le 0\right)$$

converges to $1/2$ as N tends to infinity, and for $x \in \mathbb{R}_+$

$$\lim_{N\to+\infty} \pi_N\left(k : 0 \le \frac{N-k}{\sqrt{N}} \le x\right) = 2 \lim_{N\to+\infty} \mathbb{P}\left(0 \le \frac{N - \mathcal{N}_1([0,N])}{\sqrt{N}} \le x\right)$$
$$= \sqrt{\frac{2}{\pi}} \int_0^x e^{-u^2/2}\, du.$$

The first part of the proposition is proved.

The proof of the last assertion uses Stirling's formula

$$\mathbb{P}(\mathcal{N}_1([0,N]) = N) = \frac{N^N e^{-N}}{N!} \sim 1/\sqrt{2\pi N}.$$

□

At equilibrium the loss rate (the intensity of the arrivals which find the queue full) is given by $\lambda N \pi_N(N)$, it is thus equivalent to $\lambda\sqrt{2N/\pi}$. In the super-critical case the loss rate is linear with respect to N, i.e. $\sim \lambda N(1-1/\rho)$ by Proposition 6.19.

6.7.3 The Sub-Critical Case $\rho < 1$

This case is quite simple, the process $(L_N^*(t))$ lives mainly in a region of size $\sqrt{N}$ around ρN and therefore coincides with the process $(L_N(t))$. The next proposition shows that Theorems 6.13 and 6.14 can be easily obtained. The proof is similar to the proof of Proposition 6.21 for the process $(L_N^*(t))$. A convergence result for the hitting time of N is also obtained.

PROPOSITION 6.23. *If $\rho < 1$, with the initial condition $L_N^*(0) = 0$, the random variable $(N\rho^N \mu T_N^*)$ converges in distribution to an exponentially distributed random variable with parameter $1 - \rho$.*

PROOF. With the notations used in the proof of Proposition 6.10, by replacing ρ by ρN in Identity (6.14), for $\alpha \ge 0$, one has

$$\phi_N(\alpha) = 1/\mathbb{E}_0\left[\exp\left(-\alpha\mu T_N^*\right)\right] = 1 + \alpha \sum_{k=1}^{N} \binom{N}{k} \frac{1}{N^k \rho^k} \prod_{i=1}^{k-1} (\alpha + i).$$

If $\alpha_N = \alpha N \rho^N$, $\phi_N(\alpha_N)^{-1}$ is the Laplace transform of $N\rho^N \mu T_N^*$ in α and this quantity can be expressed as follows

$$\phi_N(\alpha_N) = 1 + \alpha \sum_{k=1}^{N} \frac{N}{k} \rho^{N-k} \prod_{i=1}^{k-1} \left(1 - \frac{i}{N}\right)\left(1 + \frac{\alpha_N}{i}\right)$$

$$= 1 + \alpha \sum_{k=0}^{N-1} (k+1)\rho^k \frac{N}{(k+1)(N-k)} \prod_{i=1}^{N-k-1} \left(1 - \frac{i}{N}\right)\left(1 + \frac{\alpha_N}{i}\right). \tag{6.34}$$

The expression $N/(k+1)(N-k)$ is bounded by 1 for $0 \le k \le N-1$, hence

$$\frac{N}{(k+1)(N-k)} \prod_{i=1}^{N-k-1} \left(1 - \frac{i}{N}\right)\left(1 + \frac{\alpha_N}{i}\right) \le (1+\alpha_N)^N \le e^{\alpha N^2 \rho^N},$$

since $\rho < 1$, this last term is bounded by a constant independent of N. Lebesgue's Theorem applied to the right hand side of Equation (6.34) gives

$$\lim_{N\to+\infty} \phi_N(\alpha_N) = 1 + \alpha \sum_{k=0}^{+\infty} \rho^k = 1 + \alpha/(1-\rho),$$

hence

$$\lim_{N\to+\infty} \mathbb{E}_0 \left(\exp(-\alpha_N \mu T_N^*)\right) = (1-\rho)/(1-\rho+\alpha).$$

The proposition is proved. □

6.8 Appendix

6.8.1 Ornstein-Uhlenbeck Processes

This process is the solution of the stochastic differential equation

$$dX(t) = -\alpha X(t)\, dt + \sigma\, dB(t), \tag{6.35}$$

where α, $\sigma > 0$ and $(B(t))$ is the standard Brownian motion. This equation can be explicitly solved, the solution $(X(t))$ can be represented as

$$X(t) = X(0)e^{-\alpha t} + \sigma \int_0^t e^{-\alpha(t-s)}\, dB(s), \quad t \in \mathbb{R}_+.$$

This is an ergodic Markov process (far away from the origin, the drift $-\alpha X(t)$ pushes in direction of 0). Its invariant measure is a centered Gaussian distribution whose variance is $\sigma^2/(2\alpha)$. If the process is at equilibrium, $(X(t))$ is a stationary Gaussian process with the same distribution as the rescaled Brownian motion

$$\left[\sigma/\sqrt{2\alpha} \exp(-\alpha t) B\left(\exp(-2\alpha t)\right)\right].$$

See Revuz and Yor [**42**] or Rogers and Williams [**43**] for more details on these processes.

6.8.2 Gronwall's Lemma

LEMMA 6.24. *If h and f are non-negative Borelian functions on $\mathbb{R}_+$ and $\varepsilon > 0$ such that, for $s \le t$,*

$$f(s) \le \varepsilon + \int_0^s f(u)h(u)\, du, \tag{6.36}$$

then, for $s \le t$,

$$f(s) \le \varepsilon \exp\left(\int_0^s h(u)\, du\right).$$

PROOF. By iterating Relationship (6.36), if $s \leq t$ and $n \geq 1$,

$$
\begin{aligned}
f(s) &\leq \varepsilon + \sum_{i=1}^{n-1} \varepsilon \int_0^s h(u_1)\,du_1 \int_0^{u_1} h(u_2)\,du_2 \cdots \int_0^{u_{i-1}} h(u_i)\,du_i \\
&\quad + \int_0^s h(u_1)\,du_1 \int_0^{u_1} h(u_2)\,du_2 \cdots \int_0^{u_{n-1}} h(u_n)\,du_n \\
&= \varepsilon + \varepsilon \sum_{i=1}^{n-1} \left(\int_0^s h(u)\,du \right)^i \int_{\mathbb{R}^n} 1_{\{0 \leq u_1 \leq \cdots \leq u_i \leq s\}} \prod_{k=1}^{i} \frac{h(u_k)\,du_k}{\int_0^s h(u)\,du} \\
&\quad + \left(\int_0^s h(u)\,du \right)^n \int_{\mathbb{R}^n} 1_{\{0 \leq u_1 \leq \cdots \leq u_n \leq s\}} \prod_{k=1}^{n} \frac{h(u_k)\,du_k}{\int_0^s h(u)\,du} \\
&= \varepsilon + \varepsilon \sum_{i=1}^{n-1} \left(\int_0^s h(u)\,du \right)^i \mathbb{P}(X_1 \leq X_2 \leq \cdots \leq X_i \leq s) \\
&\quad + \left(\int_0^s h(u)\,du \right)^n \mathbb{P}(X_1 \leq X_2 \leq \cdots \leq X_n \leq s),
\end{aligned}
$$

where (X_i) is an i.i.d. sequence of random variables with density

$$h(u) / \int_0^s h(v)\,dv$$

on the interval $[0, s]$. By symmetry, one gets

$$\mathbb{P}(X_1 \leq X_2 \leq \cdots \leq X_i \leq s) = 1/i!$$

hence

$$f(s) \leq \varepsilon + \varepsilon \sum_{i=1}^{n-1} \left(\int_0^s h(u)\,du \right)^i \frac{1}{i!} + \left(\int_0^s h(u)\,du \right)^n \frac{1}{n!},$$

the result is obtained by letting n go to infinity. □

7. Queues with Poisson Arrivals

Contents

Throughout this chapter, the arrival process is assumed to be a marked Poisson point process. See Proposition 1.11 page 11 and Section 1.3.2 page 13 for the definition and the main properties of Poisson marked point processes. In this setting, four queueing models are analyzed: The queue with an infinite number of servers (the $M/G/\infty$ queue) and the single server queue with the following service disciplines: FIFO, LIFO and Processor-Sharing. The Processor-Sharing queue receives a detailed treatment because of the central role played by an interesting branching process in the derivation of the distribution of the sojourn time. It is also an important discipline in modern stochastic models of communication networks. The last section is devoted to a common, important property of queues having a Poisson input.

This chapter is also an occasion to work with Markov processes with values in more exotic state spaces, i.e. non-finite dimensional state spaces. A Markovian description of LIFO discipline involves a state space of non-negative finite sequences. For Processor-Sharing, the state space is a set of point measures on $\mathbb{R}_+$.

Notations. The marked Poisson arrival process is defined by

$$\mathcal{N}_\lambda = \sum_{n \in \mathbb{Z}} \delta_{(t_n, \sigma_n)},$$

its intensity on $\mathbb{R} \times \mathbb{R}_+$ is given by the measure $\lambda dt \times \sigma(dx)$. As usual, if A is a Borelian subset of $\mathbb{R} \times \mathbb{R}_+$, the quantity $\mathcal{N}_\lambda(A)$ denotes the number of points (t_n, σ_n) in A. If I is an interval of $\mathbb{R}$, with a slight abuse of notation the variable $\mathcal{N}_\lambda(I)$ denotes $\mathcal{N}_\lambda(I \times \mathbb{R}_+)$. The sequence $(t_n; n \in \mathbb{Z})$ of the arrival times is non-decreasing and numbered so that $t_0 \leq 0 < t_1$. Proposition 1.11 page 11 of Chapter 1 shows that the point process $\{t_n\}$ is Poisson with intensity $\lambda\, dx$ on $\mathbb{R}$ and is independent of the i.i.d. sequence (σ_n). The common distribution of the sequence of services (σ_n) is denoted by $\sigma(dx)$. The filtration associated to this marked point process is denoted by $(\mathcal{F}_t)$,

$$\mathcal{F}_t = \sigma \left\langle \mathcal{N}_\lambda(]-\infty, s] \times A) : s \leq t,\ A \in \mathcal{B}(\mathbb{R}_+) \right\rangle,$$

for $t \geq 0$, $\mathcal{F}_t$ is the σ-field of the events occurring before time t.

7.1 Single Server M/G/1 FIFO Queue

The stationary waiting time and the number of customers of this queue have already been analyzed in Chapter 2. A nice branching property due to Kendall is presented. It gives directly the Laplace transform of a busy period as the solution of a fixed point equation.

The variable B denotes the (possibly infinite) duration of a busy period starting with one initial customer with service σ_0. The variable B is the first time t when all customers arrived between 0 and t, the initial customer included, have been served. Equivalently, at time B, all services of customers arrived before B have been processed,

$$B = \inf\left\{ t > 0 : \sigma_0 + \int_{]0,t] \times \mathbb{R}_+} x\, \mathcal{N}_\lambda(ds, dx) - t \leq 0 \right\}.$$

The variable B is obviously a stopping time for the filtration $(\mathcal{F}_t)$.

If two customers are initially present with respective service times σ_1 and σ_2, the busy period generated by these customers can be decomposed in two parts. If the second customer is ignored, the variable B_1 denotes the duration of the busy period generated by the service σ_1. At time $t = B_1$, the total work load of the queue is σ_2, the service of the second initial customer. The busy period generated by σ_2 is the busy period associated with the point process $T^{B_1}\mathcal{N}_\lambda$ which is the point process $\mathcal{N}_\lambda$ translated by B_1. Since B_1 is a stopping time, the strong Markov property of Poisson point processes (Proposition 1.18 page 18) shows that $T^{B_1}\mathcal{N}_\lambda$ is a Poisson point process with the same distribution as $\mathcal{N}_\lambda$ and independent of B_1. The duration of the busy period generated by two customers can thus be expressed as the sum of two independent variables B_1 and B_2 with the same distribution.

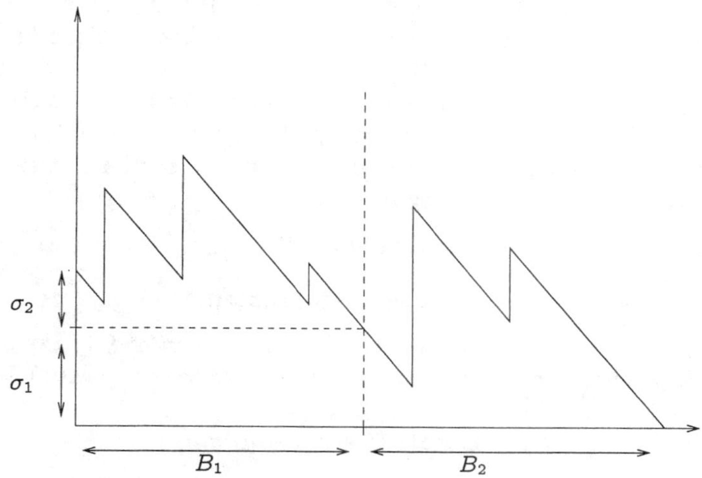

FIG. 7.1. Busy Periods of the $M/G/1$ Queue

Going back to the busy period with one initial customer, at time $t = \sigma_0$, after the departure of this customer, there are $\mathcal{N}_\lambda(]0, \sigma_0])$ customers in the queue requiring respectively the services σ_i, $i = 1, \ldots, \mathcal{N}_\lambda(]0, \sigma_0])$. An easy generalization of what has been proved for two customers shows that the duration B of the busy period can then be expressed as

$$B = \sigma_0 + \sum_{i=1}^{\mathcal{N}_\lambda(]0,\sigma_0])} B_i, \tag{7.1}$$

where the B_i's are the duration of the busy periods generated by the $\mathcal{N}_\lambda(]0, \sigma_0])$ customers present at time $t = \sigma_0$. These variables are independent with the same distribution as B and independent of σ_0. In the language of branching processes, the customer with service σ_0 generates $\mathcal{N}_\lambda(]0, \sigma_0])$ clients. The last identity gives the relation

$$\begin{aligned}
\widetilde{B}(\xi) = \mathbb{E}(\exp(-\xi B)) &= \mathbb{E}\left[\int \exp\left(-\xi\left[x + \sum_{1}^{\mathcal{N}_\lambda(]0,x])} B_i\right]\right)\sigma(dx)\right] \\
&= \mathbb{E}\left[\int e^{-\xi x}\widetilde{B}(\xi)^{\mathcal{N}_\lambda(]0,x])}\,\sigma(dx)\right] \\
&= \mathbb{E}\left[\int \exp\left[-\xi x - \lambda x\left(1 - \widetilde{B}(\xi)\right)\right]\,\sigma(dx)\right] \\
&= \widetilde{\sigma}\left[\xi + \lambda\left(1 - \widetilde{B}(\xi)\right)\right],
\end{aligned}$$

with the convention that the Laplace transform of some random variable X taken at $\xi \in \mathbb{C}$ is denoted by $\widetilde{X}(\xi)$. It is easy to check that, for $x > 0$, the function $f : y \to \widetilde{\sigma}(\xi+\lambda(1-y))$ is convex, takes the value $\mathbb{E}[\exp(-(\xi+\lambda)\sigma)]$ for $y = 0$ and $\mathbb{E}(\exp(-\xi\sigma))$ for $y = 1$. It has therefore a unique fixed point in the interval $[0, 1]$.

The stability condition $\lambda\mathbb{E}(\sigma) < 1$ plays a role at $\xi = 0$. Indeed, if the inequality $\lambda\mathbb{E}(\sigma) < 1$ holds, by convexity, the equation $y = \widetilde{\sigma}(\lambda(1-y))$ has only one solution $y = 1$ in the interval $[0, 1]$. Otherwise, there exists another fixed point $y_c < 1$. It is then easy to show that

$$\mathbb{P}(B < +\infty) = \lim_{\xi \to 0} \widetilde{B}(\xi) = y_c < 1.$$

These results are summarized in the next proposition.

PROPOSITION 7.1. *For $\xi > 0$, the Laplace transform at ξ of the duration of the busy period is the unique solution y in the interval $[0, 1]$ of the equation*

$$y = \mathbb{E}[\exp(-(\xi + \lambda(1-y))\sigma_0)].$$

The Laplace transform of the busy period is not explicit since it is given as the solution of a fixed point equation. By taking successive derivatives at 0 of this equation, it is possible to get the moments of this variable. More directly, Equation (7.1) gives, via integration,

$$\mathbb{E}(B) = \mathbb{E}(\sigma)/(1 - \lambda\mathbb{E}(\sigma)),$$

by induction all the moments of B can be obtained in this way.

For the $M/M/1$ queue, the Laplace transform of the distribution of a busy period is derived in Proposition 5.4 page 109, for $\xi > 0$, its Laplace transform is given by

$$\mathbb{E}\left[\exp(-\xi B)\right] = \frac{\lambda + \mu + \xi - \sqrt{(\lambda + \mu + x)^2 - 4\lambda\mu}}{2\lambda}, \tag{7.2}$$

where μ is the service rate of the queue.

The same branching process is studied in more detail in Section 7.3. Is is used to derive the distribution of the sojourn time of the single server queue with Processor-Sharing discipline.

7.2 Infinite Server Queue

For this queue, an arriving customer is served when it arrives. In particular, its sojourn time is its service time. If, for $t \geq 0$, $L(t)$ is the number of customers at time t, under the initial condition $L(0) = 0$, then

$$L(t) = \sum_{n \in \mathbb{N}} 1_{\{t_n \leq t,\, t_n + \sigma_n > t\}} = \mathcal{N}_\lambda(A_t),$$

where $A_t = \{(s, x) \in \mathbb{R}_+ \times \mathbb{R}_+ : s \leq t, s + x > t\}$. Since $\mathcal{N}_\lambda$ is a Poisson point process with intensity $\mu(dt, dx) = \lambda\, dt\, \sigma(dx)$ on $\mathbb{R}_+ \times \mathbb{R}_+$ (Proposition 1.11 page 11), the number of customers at times t has a Poisson distribution with parameter $\mu(A_t)$,

$$\mu(A_t) = \int_{A_t} \lambda\, ds\, \sigma(dx) = \lambda \int (t - (t-x)^+)\, \sigma(dx) = \lambda \int (x \wedge t)\, \sigma(dx).$$

Therefore, the number of customers converges in distribution to a Poisson distribution with parameter $\rho = \lambda \mathbb{E}(\sigma)$.

The Departure Process. This is the point process $D = \{t_n + \sigma_n\}$ and, according to Dobrushin's Proposition 1.13 page 12 of Chapter 1, D is a Poisson process with parameter λ. Notice that when services are exponentially distributed, this property can be shown in the same way as for the $M/M/1$ queue, since the process $(L(t))$ is also reversible in this case.

PROPOSITION 7.2. *For an infinite server queue with Poisson arrivals, the stationary distribution of the number of customers is given by* Pois(ρ), *the Poisson distribution with parameter ρ, and the departure process is Poisson with intensity λ. Moreover, for $t \geq 0$,*

$$||\mathbb{P}_0(L(t) \in \cdot) - \text{Pois}(\rho)||_{tv} \leq 1 - \exp\left[-\lambda \mathbb{E}\left((\sigma - t)^+\right)\right],$$

where $||\cdot||_{tv}$ denotes the total variation norm (see Appendix C).

PROOF. Only the inequality on the rate of convergence has to be proved. Since the distribution of the number of customers at time t is Poisson with parameter $\lambda \mathbb{E}(\sigma \wedge t)$, it is sufficient to show that for $\delta > 0$ and $\eta > 0$ such that $\eta \leq \delta$, the inequality

$$||\,\text{Pois}(\eta) - \text{Pois}(\delta)||_{tv} \leq 1 - e^{-(\delta - \eta)}$$

holds. Let $\mathcal{N}_1$ be a Poisson process with parameter 1 on $\mathbb{R}$,

$$||\,\text{Pois}(\eta) - \text{Pois}(\delta)||_{tv} = \frac{1}{2} \sum_{k \geq 0} |\mathbb{P}(\mathcal{N}_1([0,\eta]) = k) - \mathbb{P}(\mathcal{N}_1([0,\delta]) = k)|.$$

Since

$$\left|1_{\{\mathcal{N}_1([0,\eta])=k\}} - 1_{\{\mathcal{N}_1([0,\delta])=k\}}\right| \leq 1_{\{\mathcal{N}_1(]\eta,\delta])\neq 0\}} \left(1_{\{\mathcal{N}_1([0,\delta])=k\}} + 1_{\{\mathcal{N}_1([0,\eta])=k\}}\right),$$

one gets

$$||\,\text{Pois}(\eta) - \text{Pois}(\delta)||_{tv} \leq \frac{1}{2} \sum_{k \geq 0} \mathbb{P}\left(\mathcal{N}_1([0,\delta]) = k, \mathcal{N}_1(]\eta,\delta]) \neq 0\right) + \sum_{k \geq 0} \mathbb{P}\left(\mathcal{N}_1([0,\eta]) = k, \mathcal{N}_1(]\eta,\delta]) \neq 0\right).$$

This last quantity is given by $\mathbb{P}(\mathcal{N}_1(]\eta,\delta]) \neq 0) = 1 - \exp(-(\delta - \eta))$. The proposition is proved. □

7.3 Processor-Sharing M/G/1 Queue

This service discipline can be described as follows: if, between $s \geq 0$ and $s+h$, $h \geq 0$, the number of customers in the queue is N, then each of them receives the service h/N (i.e. their residual service time is decreased by this quantity). If $(L(t))$ is the process of the number of customers in the queue, the fact that a customer arriving at t_0 with service σ_0 leaves at t_0+S_0 implies

$$\int_{t_0}^{t_0+S_0} \frac{1}{L(s)}\, ds = \sigma_0.$$

Note that $(L(s))$ cannot be 0 between t_0 and t_0+S_0. As before, $\mathcal{N}_\lambda$ is a Poisson point process with parameter λ representing the arrivals of customers.

This service discipline is egalitarian, it accomplishes a fair sharing of the resource among the users. It plays a very important role in a large number of models of stochastic networks. Technically, branching processes and time change arguments are used to get the various distributions of equilibrium. This approach is mainly due to Grishechkin, Kitaev and Yashkov[Yas83], it is implicit in some earlier works on the subject.

To get a simple picture of these methods, Section 7.3.2 considers an $M/M/1$ queue with Processor-Sharing discipline. The case of a general service distribution is investigated in Section 7.3.3. It is more involved since a Markovian description requires the use of a measure-valued process, but once this difficulty is solved, the analysis is not much more difficult than the exponential case. A branching process, already encountered in Section 7.1 is presented. It plays an important role to derive sojourn time distributions.

7.3.1 A Branching Process

Definition 14. *The process $(Z(t))$ denotes the number of individuals of a branching process with the following characteristics:*

- *There is one ancestor, $Z(0) = 1$.*
- *During its lifetime, an individual gives birth at rate λ (i.e. according to a Poisson process with this parameter).*
- *The lifetime of an individual has the distribution $\sigma(dx)$.*

As usual with branching processes, 0 is an absorbing state for $(Z(t))$, the population may become extinct. This branching process is referred to as Crump-Mode-Jagers branching process in literature. See Athreya and Ney [4] for a presentation of classical results on branching processes. Note that, in general, $(Z(t))$ is not a Markov process when the distribution $\sigma(dx)$ of lifetime is not exponential.

[Yas83] S.F. Yashkov, *A derivation of response time distribution for an $M/G/1$ processor-sharing queue*, Problems of Control and Information Theory **12** (1983), no. 2, 133–148.

The next proposition is the branching argument which will be used to derive the Laplace transform of the sojourn time.

PROPOSITION 7.3 (Branching Property). *If $(Z(t))$ is the branching process with initial condition $Z(0)=1$ and, for $t \geq 0$,*

$$H(t) \stackrel{def.}{=} \int_0^t Z(s)\, ds,$$

then the equality in distribution of processes

$$(H(t)) \stackrel{dist.}{=} \left(\sigma \wedge t + \int_0^{\sigma\wedge t} H_s(t-u)\, \mathcal{N}_\lambda(du)\right) \tag{7.3}$$

holds, where σ is a random variable whose distribution is $\sigma(dx)$ and independent of an independent family of processes $(H_s(t))$, $s \geq 0$, with the same distribution as $(H(t))$.

The random variable $H(t)$ can be seen as the sum of the ages, at time t, of the individuals born of before time t. This is the sum of the lengths of the intervals between 0 and t in Figure 7.2.

PROOF. The proposition is a direct consequence of the branching property. The initial individual generates children during its lifetime σ according to a Poisson process with parameter λ. To a child born at time s corresponds an independent branching process $(Z_s(t-s) : t \geq s)$. For $s \leq u \leq t$, the variable $Z_s(u-s)$ is the contribution of this child to the population at time u. The processes $(Z_s(t))$ for $s \geq 0$ are i.i.d. with the same distribution as $(Z(t))$ therefore, the identity of distribution of processes

$$(Z(t)) \stackrel{\text{dist.}}{=} \left(1_{\{t\leq\sigma\}} + \int_0^{t\wedge\sigma} Z_s(t-s)\, \mathcal{N}_\lambda(ds)\right)$$

holds. If this relation is integrated between 0 and t, one gets Equation (7.3). The branching process associated to an individual born at time $s \in [0,t]$ contributes to $H(t)$ for a duration of time $t-s$. □

The following technical result is used several times in the following.

PROPOSITION 7.4. *If $(H_s(t))$, $s \geq 0$ are i.i.d processes with the same distribution as $(H(t))$, then the identity*

$$\mathbb{E}\left[\exp\left(-\xi \int_0^x H_u(t-u)\, \mathcal{N}_\lambda(du)\right)\right] = \exp\left(-\lambda \int_0^x \left[1 - \mathbb{E}\left(e^{-\xi H(t-u)}\right)\right] du\right) \tag{7.4}$$

holds for non-negative x, t and ξ.

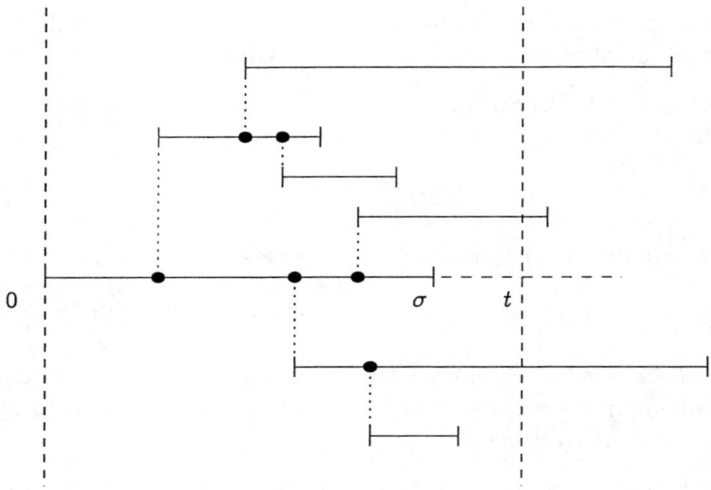

FIG. 7.2. A Branching Process on the Interval $[0, t]$.

PROOF. For positive x, t and ξ, first note that

$$\mathbb{E}\left[\exp\left(-\xi\int_0^x H_u(t-u)\,\mathcal{N}_\lambda(du)\right)\right] = \mathbb{E}\left[\prod_{t_n\le x}\exp\left(-\xi H_{t_n}(t-t_n)\right)\right]$$

where the sequence (t_n) are the points of the Poisson point process $\mathcal{N}_\lambda$. The independence of the Poisson point process $\mathcal{N}_\lambda$ and the processes $H_s(u)$, $u \ge 0$ gives

$$\begin{aligned}\mathbb{E}\left[-\xi\int_0^x H_s(t-u)\,\mathcal{N}_\lambda(du)\right] &= \mathbb{E}\left[\prod_{t_n\le x}\mathbb{E}\left[\exp(-\xi H(t-t_n))\,|\,\mathcal{N}_\lambda\right]\right]\\ &= \mathbb{E}\left[\prod_{t_n\le x} e^{-g(t_n)}\right],\end{aligned}$$

where $g(u) = -\log\mathbb{E}[\exp(-\xi H(t-u))]$ for $u \le t$. The last expression can thus be written as

$$\mathbb{E}\left[\exp\left(-\int_0^x g(u)\,\mathcal{N}_\lambda(du)\right)\right] = \exp\left(-\lambda\int_0^x\left[1-e^{-g(u)}\right]du\right),$$

by using the expression of the Laplace transform of a Poisson point process (see Proposition 1.5 page 6). Therefore, the identity

$$\mathbb{E}\left[\exp\left(-\int_0^x g(u)\,\mathcal{N}_\lambda(du)\right)\right] = \exp\left(-\lambda\int_0^x\left[1-\mathbb{E}\left(e^{-\xi H(t-u)}\right)\right]du\right)$$

holds. The lemma is proved. □

7.3.2 Exponential Service Distribution

Throughout this section, services are assumed to be exponentially distributed with parameter μ. When a customer is in the queue with $x-1$ others customers, $x \geq 1$, it is served at rate μ/x. The process of the number of customers $(L(t))$ in a Processor-Sharing queue is, in this case, a birth and death process with the same Q-matrix as the $M/M/1$ queue: For $x > 0$, a jump from x to $x-1$ occurs at rate $x \times \mu/x = \mu$. When $\rho = \lambda/\mu < 1$, the stationary distribution of $(L(t))$ is a geometric distribution with parameter ρ.

In the study of the Processor-Sharing queue, it is useful to consider a Markov process $(\overline{Z}(t))$ which is this branching process but with an external source giving birth at rate λ so that the process never dies out. It is easily seen that the branching processes $(Z(t))$ and $(\overline{Z}(t))$ are Markov processes with the following characteristics.

PROPOSITION 7.5. *The processes $(Z(t))$ and $(\overline{Z}(t))$ are Markovian birth and death processes with Q-matrices $Q = (q_{xy})$ and $\overline{Q} = (\overline{q}_{x,y})$ defined by*

$$q_{x,x+1} = \lambda x \text{ and } q_{x,x-1} = \mu x,$$

and

$$\overline{q}_{x,x+1} = \lambda(x+1) \text{ and } \overline{q}_{x,x-1} = \mu x,$$

for $x \geq 0$.

For $t \geq 0$, define the variable $A(t)$ by

$$A(t) = \int_0^t \left(1 + \overline{Z}(s)\right) ds. \tag{7.5}$$

This is an increasing function associated to $(\overline{Z}(t))$. The process $(C(t))$ denotes the inverse of $(A(t))$, i.e. for $t \geq 0$, $A(C(t)) = C(A(t)) = t$. The process $(A(t))$ is an *additive functional* of the Markov process $(\overline{Z}(t))$. See Rogers and Williams [**43**] on this subject.

PROPOSITION 7.6. *Define $(\overline{L}(t))$ by, for $t \geq 0$, $\overline{L}(t) = \overline{Z}(C(t))$. The process $(\overline{L}(t))$ is a Markov process whose Q-matrix $R = (r_{xy})$ is given by*

$$r_{x,x+1} = \lambda \quad \text{and} \quad r_{x,x-1} = \mu \frac{x}{x+1}.$$

Equivalently, $(\overline{L}(t))$ is the number of non-permanent customers in a Processor-Sharing queue with a permanent customer.

Moreover, the time necessary for a permanent customer to receive service x when there are initially $y \in \mathbb{N}$ customers has the same distribution as

$$A(x) = \int_0^x \left(1 + \overline{Z}(s)\right) ds,$$

when the Markov process $\overline{Z}(s)$ has the initial state $\overline{Z}(0) = y$.

The interesting part of the proposition is the fact that $A(x)$ is the sojourn time of a customer having a duration of service x in a Processor-Sharing queue. So the sojourn time which is, a priori, a complicated functional of the arrival process can be expressed as a simple functional of a branching process. Independence properties of the branching process $(\overline{Z}(t))$ and the representation with $(A(x))$ simplify much the analysis of the stationary distribution of sojourn time.

PROOF. Note that, for $s \geq 0$, the variable $C(s)$ is a stopping time for the natural filtration of $(\overline{Z}(t))$: For $a \geq 0$,

$$\{C(s) \leq a\} = \{A(a) \geq s\} = \left\{\int_0^a \left(1 + \overline{Z}(u)\right)\, du \geq s\right\},$$

this last event is clearly measurable with respect to the σ-field generated by the random variables $X(u)$, $u \leq a$.

The Markov Property. For $0 \leq s \leq t$,

$$\begin{aligned}\mathbb{P}\left[\overline{L}(t) \in \cdot \mid \overline{L}(u), u \leq s\right] &= \mathbb{P}\left[\overline{Z}(C(t)) \in \cdot \mid \overline{Z}(u), u \leq C(s)\right] \\ &= \mathbb{P}\left[\overline{Z}(C(t)) \in \cdot \mid \overline{Z}(C(s))\right] = \mathbb{P}\left[\overline{L}(t) \in \cdot \mid \overline{L}(s)\right],\end{aligned}$$

by the strong Markov property of $(\overline{Z}(t))$ for the stopping time $C(s)$.

The Q-Matrix. If f is some function on $\mathbb{N}$ and $\overline{Q}$ is the Q-matrix of $(\overline{Z}(t))$, by Proposition B.4 page 364, the process

$$(M_0(t)) \stackrel{\text{def.}}{=} \left(f\left(\overline{Z}(t)\right) - \int_0^t \overline{Q}(f)\left(\overline{Z}(s)\right)\, ds\right)$$

is a local martingale with respect to the natural (augmented) filtration $(\mathcal{F}_t)$ of $(\overline{Z}(t))$. If f is a bounded function then, according to Proposition A.7 page 354, $(M_0(t))$ is a martingale. Since $C(t) \leq t$ (because $A(t) \geq t$), the variable $M_0(C(t))$ is integrable. Doob's optional stopping Theorem gives, for $t \geq 0$ and $u \geq 0$,

$$\mathbb{E}[M_0(C(t)) \mid \mathcal{F}_u] = M_0(C(t) \wedge u).$$

By using again Doob's optional stopping Theorem for the uniformly integrable martingale $u \to \mathbb{E}[M(C(t)) \mid \mathcal{F}_u]$ and the stopping time $C(s)$, one gets that the identity

$$\mathbb{E}\left[M_0(C(t)) \mid \mathcal{F}_{C(s)}\right] = M_0(C(s))$$

holds for $s \leq t$. Consequently, the process

$$(M(t)) \stackrel{\text{def.}}{=} \left(f\left[\overline{Z}(C(t))\right] - \int_0^{C(t)} Q(f)\left(\overline{Z}(s)\right)\, ds\right)$$

is a martingale for the natural filtration of $(\overline{Z}(C(t))) = (\overline{L}(t))$ for any bounded function f. A change of variables shows that the process

$$(M(t)) = \left(f\left(\overline{L}(t)\right) - \int_0^t \frac{\overline{Q}(f)(\overline{L}(s))}{1 + \overline{L}(s)}\, ds\right)$$

is a local martingale for any function f on $\mathbb{N}$. From Proposition B.6 page 365, one gets that the infinitesimal generator R of $(\overline{L}(t))$ is given by

$$R(f)(x) = \frac{\overline{Q}(f)(x)}{1+x},$$

for $x \in \mathbb{N}$. This proves the first part of the proposition.

Between 0 and $A(x)$, due to the Processor-Sharing policy, the permanent customer receives the amount of service

$$\int_0^{A(x)} \frac{1}{1+\overline{L}(s)}\, ds = \int_0^x \frac{1}{1+\overline{Z}(u)} A'(u)\, du = x.$$

Thus, $A(x)$ is the time necessary to receive the amount of service x in the Processor-Sharing queue. □

COROLLARY 7.7. *When $\rho < 1$, the mean sojourn time at equilibrium of a customer requiring the service $x \geq 0$ is $x/(1-\rho)$.*

PROOF. It is easily checked that the stationary distribution of the branching process is a geometric distribution with parameter ρ. See Section 4.2.1 page 84 for example. Thus, at equilibrium, the sojourn time of a customer with a duration of service x has the same distribution as the variable

$$\int_0^x \left(1+\overline{Z}(s)\right)\, ds,$$

where the branching process $(\overline{Z}(s))$ is at equilibrium. The expected value of this quantity is thus given by

$$\mathbb{E}\left(\int_0^x \left(1+\overline{Z}(s)\right)\, ds\right) = \int_0^x \mathbb{E}\left(1+\overline{Z}(s)\right)\, ds = x\,\mathbb{E}\left(1+\overline{Z}(0)\right) = \frac{x}{1-\rho}.$$

□

The Distribution of Sojourn Time. The key quantity to derive the distribution of the sojourn time of the Processor-Sharing queue is the distribution of the random variable $H(t)$.

PROPOSITION 7.8 (The Distribution of $H(t)$). *For $\xi \geq 0$, the Laplace transform at ξ of the variable*

$$H(t) = \int_0^t Z(s)\, ds$$

is given by

$$\mathbb{E}\left[e^{-\xi H(t)}\right] = \frac{(\mu-\phi_2(\xi))e^{\phi_1(\xi)t} - (\mu-\phi_1(\xi))e^{\phi_2(\xi)t}}{(\phi_1(\xi)-\lambda)e^{\phi_1(\xi)t} - (\phi_2(\xi)-\lambda)e^{\phi_2(\xi)t}}, \tag{7.6}$$

where $\phi_i(\xi)$, $i = 1, 2$ are the solutions of $y^2 - (\lambda+\mu+\xi)y + \lambda\mu = 0$, with

$$\phi_1(\xi) = \frac{1}{2}\left(\lambda+\mu+\xi+\sqrt{(\lambda+\mu+\xi)^2 - 4\lambda\mu}\right).$$

PROOF. Denote by $\widetilde{H}(t)$ the Laplace transform of $H(t)$ at ξ. Equations (7.3) and (7.4) give the relation

$$\widetilde{H}(t) = \int_0^t \mu \exp\left(-\mu v - \xi v - \lambda \int_0^v (1 - \widetilde{H}(t-u))\,du\right) dv \\ + \exp\left(-\mu t - \xi t - \lambda \int_0^t (1 - \widetilde{H}(t-u))\,du\right),$$

hence

$$\widetilde{H}(t) \exp\left((\lambda + \mu + \xi)t - \lambda \int_0^t \widetilde{H}(u)\,du\right), \\ = 1 + \mu \int_0^t \exp\left((\lambda + \mu + \xi)v - \lambda \int_0^v \widetilde{H}(u)\,du\right) dv. \tag{7.7}$$

If the function $t \to F(t)$ is defined by

$$\log F(t) = (\lambda + \mu + \xi)t - \lambda \int_0^t \widetilde{H}(u)\,du, \tag{7.8}$$

then, by differentiating, one gets

$$\frac{F'(t)}{F(t)} = \lambda + \mu + \xi - \lambda \widetilde{H}(t), \tag{7.9}$$

Equation (7.7) can be rewritten as

$$\frac{1}{\lambda}\left(\lambda + \mu + \xi - \frac{F'(t)}{F(t)}\right) F(t) = 1 + \mu \int_0^t F(u)\,du.$$

By differentiating again, it follows that F satisfies the linear second order differential equation

$$F''(t) - (\lambda + \mu + \xi)F'(t) + \lambda\mu F(t) = 0, \tag{7.10}$$

with the initial conditions $F(0) = 1$ and $F'(0) = \mu + \xi$, since $\widetilde{H}(0)$ is 1. Therefore F can be expressed as

$$F(t) = a e^{\phi_1(\xi)t} + b e^{\phi_2(\xi)t}, \tag{7.11}$$

with $a + b = 1$ and $a\phi_1(\xi) + b\phi_2(\xi) = \mu + \xi$, hence

$$a = \frac{\phi_1(\xi) - \lambda}{\phi_1(\xi) - \phi_2(\xi)} \quad \text{and} \quad b = \frac{\phi_2(\xi) - \lambda}{\phi_2(\xi) - \phi_1(\xi)},$$

by using that $\mu + \xi = \phi_1(\xi) + \phi_2(\xi) - \lambda$. The last identity and Equation (7.9) yield

$$\lambda \widetilde{H}(t) = \lambda + \mu + \xi - \frac{a\phi_1(\xi)e^{\phi_1(\xi)t} + b\phi_2(\xi)e^{\phi_2(\xi)t}}{ae^{\phi_1(\xi)t} + be^{\phi_2(\xi)t}} \\ = \phi_1(\xi) + \phi_2(\xi) - \frac{(\phi_1(\xi) - \lambda)\phi_1(\xi)e^{\phi_1(\xi)t} - (\phi_2(\xi) - \lambda)\phi_2(\xi)e^{\phi_2(\xi)t}}{(\phi_1(\xi) - \lambda)e^{\phi_1(\xi)t} - (\phi_2(\xi) - \lambda)e^{\phi_2(\xi)t}},$$

one concludes by using that $\phi_1(\xi)\phi_2(\xi) = \lambda\mu$. The proposition is proved. □

By letting t go to infinity in Equation (7.6), one obtains that

$$\mathbb{E}\left[e^{-\xi H(\infty)}\right] = \frac{\mu - \phi_2(\xi)}{\phi_1(\xi) - \lambda} = \frac{\phi_2(\xi)}{\lambda} = \frac{\lambda + \mu + \xi - \sqrt{(\lambda+\mu+\xi)^2 - 4\lambda\mu}}{2\lambda}.$$

Hence, by Equation (7.2), the variable $H(\infty)$ has the same distribution as the busy period of an $M/M/1$ queue. This is not surprising since, in this case, the branching argument used in this section is the same as the one used in Section 7.1.

PROPOSITION 7.9. *If $\rho < 1$ and $S(x)$ is the sojourn time of a customer with service time x in an $M/M/1$ Processor-Sharing at equilibrium then, for $\xi \geq 0$, the Laplace transform of $S(x)$ at ξ is given by*

$$\mathbb{E}\left[e^{-\xi S(x)}\right] = \frac{(1-\rho)\left(\phi_1(\xi) - \phi_2(\xi)\right)\, \mu e^{\mu x}}{\left(\mu\phi_1(\xi) + \lambda\phi_2(\xi) - 2\lambda\mu\right) e^{\phi_1(\xi)x} - \left(\mu\phi_2(\xi) + \lambda\phi_1(\xi) - 2\lambda\mu\right) e^{\phi_2(\xi)x}},$$

where λ and μ are the arrival and the service rates and

$$\phi_i(\xi) = \frac{1}{2}\left(\lambda + \mu + \xi \pm \sqrt{(\lambda+\mu+\xi)^2 - 4\lambda\mu}\right),$$

for $i = 1$ and 2.

PROOF. Proposition 7.6 shows that $S(x)$ has the same distribution as

$$A(x) = \int_0^x \left(1 + \overline{Z}(u)\right)\, du,$$

when the distribution of $\overline{Z}(0)$ is a geometric with parameter $\rho = \lambda/\mu$ (the equilibrium distribution of the $M/M/1$ Processor-Sharing queue). At time 0, there are $\overline{Z}(0)$ initial individuals who will generate $\overline{Z}(0)$ independent branching processes $(Z_i^0(t))$, $i = 1, \ldots, \overline{Z}(0)$ with the same distribution as $(Z(t))$. On the time interval $[0, x]$, the external source associated with $(\overline{Z}(t))$ generates individuals according to a Poisson process with parameter λ who will also generate branching processes $(Z_s(t))$ if $s \in [0, x]$ is an arrival time. Therefore, the identity in distribution

$$\begin{aligned} A(x) - x &\stackrel{\text{dist.}}{=} \sum_{i=1}^{\overline{Z}(0)} \int_0^x Z_i^0(u)\, du + \int_0^x \int_u^x Z_u(s-u)\, ds\, \mathcal{N}_\lambda(du) \\ &\stackrel{\text{dist.}}{=} \sum_{i=1}^{\overline{Z}(0)} H_i^0(x) + \int_0^x H_u(x-u)\, \mathcal{N}_\lambda(du) \end{aligned}$$

holds, where the variables H_i^0 and H_u, $u \leq x$ correspond to the variable H of Proposition 7.3 for the processes $(Z_i^0(t))$ and $(Z_u(t))$, $u \leq x$ respectively. As before, denote by $\widetilde{H}(t)$ the Laplace transform of $(H(t))$ at $\xi \geq 0$. The various independence properties of these branching processes and Equation (7.4) give the relation

$$e^{\xi x}\mathbb{E}\left[e^{-\xi S(x)}\right] = \frac{1-\rho}{1-\rho\widetilde{H}(x)} \exp\left(-\lambda \int_0^x \left(1 - \widetilde{H}(x-u)\right) du\right)$$
$$= \frac{1-\rho}{1-\rho\widetilde{H}(x)} \frac{e^{(\mu+\xi)x}}{F(x)},$$

where $x \to F(x)$ is the function defined by Equation (7.8) in the proof of the previous proposition. Equation (7.11) yields

$$\mathbb{E}\left[e^{-\xi S(x)}\right] = \frac{1-\rho}{1-\rho\widetilde{H}(x)} \frac{e^{\mu x}(\phi_1(\xi)-\phi_2(\xi))}{(\phi_1(\xi)-\lambda)e^{\phi_1(\xi)x} - (\phi_2(\xi)-\lambda)e^{\phi_2(\xi)x}},$$

the proof is concluded by using Identity (7.6). □

An Alternative Approach. There is an alternative way to derive the Laplace transform of the sojourn time with Representation (7.5). To get the distribution of $A(x)$, the sojourn time of a customer with service time x, the process $(A(t))$ is analyzed in a Markovian setting. To do so, consider

$$(M(t)) = \left(\overline{Z}(t), x + A(t)\right) = \left(\overline{Z}(t), x + \int_0^t \left(1 + \overline{Z}(s)\right) ds\right),$$

with $M(0) = (n,x) \in \mathbb{N} \times \mathbb{R}_+$. It is easily checked that $(M(t))$ is indeed a Markov process on $\mathbb{N} \times \mathbb{R}_+$. If f is a bounded function on $\mathbb{N} \times \mathbb{R}_+$ which is differentiable with respect to the second coordinate, then the infinitesimal generator Ω of $(M(t))$ is defined for such a function f by

$$\Omega(f)(n,x) = \lim_{h\to 0} \frac{1}{h}\left(\mathbb{E}_{(n,x)}\left[f(M(h))\right] - f(n,x)\right).$$

Straightforward calculations give that

$$\Omega(f)(n,x) = \lambda(n+1)\left(f(n+1,x) - f(n,x)\right)$$
$$+ \mu n\left(f(n-1,x) - f(n,x)\right) + (n+1)\frac{\partial f}{\partial x}(n,x).$$

Kolmogorov's Equations for this Markov process,

$$\frac{\partial}{\partial t}\mathbb{E}_{(n,x)}\left[f\left(M(t)\right)\right] = \mathbb{E}_{(n,x)}\left[\Omega(f)\left(M(t)\right)\right],$$

taken at $f(n,x) = u^n \exp(-\xi x)$, with $u \in [0,1]$ and $\xi \geq 0$ show that, for any initial distribution ν on $\mathbb{N} \times \mathbb{R}_+$, the function

$$\psi(u,t) = \mathbb{E}_\nu\left[u^{\overline{Z}(t)} e^{-\xi A(t)}\right]$$

satisfies the partial differential equation

$$\frac{\partial}{\partial t}\psi(u,t) = \left(\lambda u^2 - (\lambda+\mu+\xi)u + \mu\right)\frac{\partial}{\partial u}\psi(u,t) + \left(\lambda(u-1)+\xi\right)\psi(u,t).$$

When the queue is at equilibrium, the boundary condition is given by

$$\psi(u,0) = \frac{1-\rho}{1-u\rho}.$$

This differential equation can be solved with classical methods, hence the Laplace transform $\psi(1,x)$ of sojourn time when service time is x can be obtained. This is, basically, the approach followed originally by Coffman, Muntz and Trotter (See Asmussen [**3**], Section A, III.9). The branching process point of view has the advantage of generalizing more easily to the case of a general distribution of service time.

7.3.3 General Service Distribution

The state of the queue. The state of the queue is given by residual times of customers present in the queue. Contrary to the LIFO discipline (see Section 7.4), there is no need to know the order of arrival of customers to determine the amount of service that each customer receives. Point measures on $\mathbb{R}_+^* =]0,+\infty[$ are used to describe this queue. Section 1.1 of Chapter 1 page 2 gives the basic definitions in this domain.

For $N \in \mathbb{N}$ and $x_1, \ldots, x_N \in \mathbb{R}_+^*$, the measure

$$\Gamma = \delta_{x_1} + \delta_{x_2} + \cdots + \delta_{x_N}$$

describes the state of the queue as follows: There are N customers with respective service times $x_1, \ldots, x_N \in \mathbb{R}_+$. The number of customers whose residual service time is less than $x \geq 0$ is given by $\Gamma(]0,x])$, the quantity $\Gamma(\mathbb{R}_+^*)$ is just the total number of customers and

$$\int_0^{+\infty} x\,\Gamma(dx) = x_1 + x_2 + \cdots + x_N$$

is the amount of work in the queue. Note that the point measures considered do not have mass on 0 since a customer with a null residual service leaves the queue. The set $\{x_1, \ldots, x_N\}$ gives an equivalent representation of the queue. Point measures are used mainly because topologies, σ-fields are already available for measures on $\mathbb{R}_+^*$.

If the initial state of the queue is given by some point measure $\Gamma_0 = \mu$ and Γ_t is the state of the queue at time $t \geq 0$ then, if Γ_t is non-trivial, the points of Γ_{t+dt} are the points of Γ_t translated on the left by the amount $dt/\Gamma_t(\mathbb{R}_+^*)$ (remember that $\Gamma_t(\mathbb{R}_+^*)$ is the total number of customers at time t). A Dirac measure at σ is added to the state if a new customer with service time σ arrives. Mathematically, it can be expressed as follows: For any continuous function f with compact support on $\mathbb{R}_+^*$,

$$\begin{aligned} (7.12)\quad \langle f, \Gamma_{t+dt}\rangle &\stackrel{\text{def.}}{=} \int_{\mathbb{R}_+} f(u)\,\Gamma_{t+dt}(du) \\ &= \int_{\mathbb{R}_+} f\left(u - \frac{dt}{\Gamma_t(\mathbb{R}_+^*)}\right)\Gamma_t(du) + f(\sigma)\,\mathcal{N}_\lambda(dt, d\sigma), \end{aligned}$$

with the convention that the first term of the right hand side is 0 if Γ_t is the null measure.

The above equation shows that a process (Γ_t) can be defined when the initial state $\Gamma_0 = \mu$ is an arbitrary non-negative Radon measure and not only a point measure. This is done by induction as follows: For any element f of $C_c(\mathbb{R}_+^*)$, the space of continuous functions with compact support, the functional $t \to < f, \Gamma_t >$ is assumed to be defined for $t \leq t_n$ (the nth point of the arrival process). It can then be also defined on the interval $[t_n, t_{n+1}]$ by using Equation (7.12). Hence the function $f \to <f, \Gamma_t>$ is then defined for any f of $C_c(\mathbb{R}_+^*)$ f and any $t \geq 0$. Since the operator $f \to <f, m>$ on the space of continuous functions with compact support determines completely any Radon measure m on $\mathbb{R}_+^*$, the process (Γ_t) is then uniquely determined.

In this way, a process (Γ_t) with the Markov property is defined on the locally compact, separable, metric space $\mathcal{M}(\mathbb{R}_+^*)$ of non-negative Radon measures on $\mathbb{R}_+^*$ (equipped with the weak topology). As usual, the notation $\mathbb{E}_\mu(\cdot)$ denotes the expectation with respect to the distribution of (Γ_t) when $\Gamma_0 = \mu$. General results concerning measure valued Markov processes are given in the survey by Dawson [**15**], see also Dynkin [**18**].

DEFINITION 15. *If f is a non-negative measurable function on the set $\mathbb{R}_+^* =]0, +\infty[$, the functional F_f on $\mathcal{M}(\mathbb{R}_+^*)$ is defined by*

$$F_f(m) = \exp\left(-\int f(u)\, m(du)\right) = \exp(-<f, m>),$$

for $m \in \mathcal{M}(\mathbb{R}_+^*)$.

The set of continuous functions f with compact support on $\mathbb{R}_+^*$ is denoted by $C(\mathbb{R}_+^*)$ and the subset of differentiable functions in $C(\mathbb{R}_+^*)$ is denoted by V.

PROPOSITION 7.10. *The set $\mathcal{V}$ of linear combinations of F_f where f is in V is dense in the set of continuous operators on $\mathcal{M}(\mathbb{R}_+^*)$.*

PROOF. The operator F_f is clearly continuous on $\mathcal{M}(\mathbb{R}_+^*)$ equipped with the weak topology. It is not difficult to show that the set V is dense for the uniform norm in the set of continuous functions on $]0, +\infty[$ with compact support. Lemma 3.2.5 of Dawson [**15**] gives the density property of the subset $\mathcal{V}$ in the set of continuous operators on $\mathcal{M}(\mathbb{R}_+^*)$. □

PROPOSITION 7.11. *The process (Γ_t) describing the Processor-Sharing discipline is a Feller-Dynkin process.*

The definition of a Feller-Dynkin process is given in Appendix 7.7 page 205.

PROOF. Lemma 3.5.1 and Corollary 3.5.2 of Dawson [**15**] show that the Markov process (Γ_t) has the Feller-Dynkin property if

1. the mapping $\mu \to \mathbb{E}_\mu[F_f(\Gamma_t)]$ is continuous on $\mathcal{M}(\mathbb{R}_+^*)$ for any $f \in V$;

2. if 1 denotes the function which is constant and equal to 1, the quantity $\mathbb{E}_\mu[F_1(\Gamma_t)]$ converges to 0 when the total mass of μ on $\mathbb{R}_+^*$ converges to infinity;
3. Continuity at the origin, for any $f \in V$ and $\mu \in \mathcal{M}(\mathbb{R}_+^*)$,
$$\lim_{t\to 0} \mathbb{E}_\mu[F_f(\Gamma_t)] = \exp(-\langle f, \mu\rangle).$$

Conditions 2) and 3) are direct consequences that, by Equation (7.12), the total mass of the process (Γ_t) decreases at rate 1 as long as it is positive.

The process $(\Gamma_{\mu,t})$ is defined as the solution of Equation (7.12) when $\Gamma_{\mu,0} = \mu$. The function $\mu \to F_f(\Gamma_{\mu,t})$ is clearly continuous (by induction on the number of arrivals in the interval $[0,t]$ as before). By Lebesgue's Theorem, the above mapping is thus continuous. The proposition is proved. □

The Infinitesimal Generator. Recall that the infinitesimal generator Ω is defined, for a continuous functional G on $\mathcal{M}(\mathbb{R}_+^*)$, if the limit
$$\Omega(G)(m) = \lim_{h\to 0} \frac{1}{h}\left[\mathbb{E}_m(G(\Gamma_h)) - G(m)\right],$$
exists for any $m \in \mathcal{M}(\mathbb{R}_+^*)$. See Ethier and Kurtz [**19**].

It is assumed that $\Gamma_0 = m$ is a non-trivial and non-negative Radon measure. For $h > 0$, if (t_1, σ_1) is the first point of the arrival process and f is an element of V, then
$$\begin{aligned}\mathbb{E}_m&[F_f(\Gamma_h)] - F_f(m)\\ &= \mathbb{E}\left[\exp\left(-f(\sigma_1)1_{\{t_1\le h\}} - \int_{\mathbb{R}_+^*} f\left(u - \frac{h}{m(\mathbb{R}_+^*)}\right) m(du)\right)\right.\\ &\qquad\qquad \left. - \exp\left(-\int_{\mathbb{R}_+^*} f(u)\, m(du)\right)\right] + o(h).\end{aligned}$$
By dividing by h and letting h go to 0, one gets the following proposition.

PROPOSITION 7.12. *The infinitesimal generator Ω of the Markov process (Γ_t) on $\mathcal{M}(\mathbb{R}_+^*)$ is defined by,*
$$\Omega(F_f)(m) = \left(\lambda\mathbb{E}[\exp(-f(\sigma)) - 1] - \frac{1}{m(\mathbb{R}_+^*)}\int_{\mathbb{R}_+^*} f'(u)\, m(du)\right) F_f(m),$$
for $m \in \mathcal{M}(\mathbb{R}_+^*)$ and f is a differentiable function with compact support on $\mathbb{R}_+$. By convention, the measure $m(dx)/m(\mathbb{R}_+^*)$ is identically 0 when m is the null measure.

Invariant Distribution of the Processor-Sharing Queue. The next proposition gives the existence of an equilibrium distribution for the number of customers and their respective residual service times.

PROPOSITION 7.13. *When $\rho = \lambda\mathbb{E}(\sigma) < 1$, at equilibrium, the number of customers of the Processor-Sharing queue is geometrically distributed with parameter ρ and, when the queue is not empty, residual service times are independent with the density $\mathbb{P}(\sigma \geq x)/\mathbb{E}(\sigma)$ on $\mathbb{R}_+^*$.*

More formally, if $\mathbb{Q}$ is the distribution on $\mathcal{M}(\mathbb{R}_+^*)$ whose Laplace transform is given by

$$\int_{\mathcal{M}(\mathbb{R}_+^*)} \exp\left(-\int_{R_+} g(x)\, m(dx)\right) \mathbb{Q}(dm) \\ = (1-\rho) + \frac{(1-\rho)\lambda \int_{\mathbb{R}_+^*} \exp(-g(x))\mathbb{P}(\sigma \geq x)\, dx}{1 - \lambda \int_{\mathbb{R}_+^*} \exp(-g(x))\mathbb{P}(\sigma \geq x)\, dx},$$

for a non-negative Borelian function g on $\mathbb{R}_+^*$, then $\mathbb{Q}$ is an invariant distribution for the Markov process (Γ_t) defined by Equation (7.12).

PROOF. The distribution π is an invariant measure for the process (Γ_t) if the identity

$$\mathbb{E}_\pi[G(\Gamma_t)] = \pi(G) \stackrel{\text{def.}}{=} \int_{\mathcal{M}(\mathbb{R}_+^*} G(m)\, \pi(dm),$$

holds for any $t \geq 0$ and any continuous bounded functional on G on $\mathcal{M}(\mathbb{R}_+^*)$.

Since (Γ_t) is a Feller-Dynkin Markov process, Proposition 9.17 of Ethier and Kurtz [**19**] and the density of $\mathcal{V}$ show that it is enough to prove that, for any $f \in V$, the identity

$$\int_{\mathcal{M}(\mathbb{R}_+^*)} \Omega(F_f)(m)\, \mathbb{Q}(dm) = 0, \tag{7.13}$$

holds, where Ω is the infinitesimal generator defined by Proposition 7.12. By definition of the distribution $\mathbb{Q}$,

$$\frac{1}{1-\rho}\int_{\mathcal{M}(\mathbb{R}_+^*)} \Omega(F_f)(m)\, \mathbb{Q}(dm) \\ = \lambda\left(\int_{\mathbb{R}_+^*} \exp(-f(x))\, \sigma(dx) - 1\right) + A + B, \tag{7.14}$$

with

$$A = \lambda \int_{\mathbb{R}_+^*} (\exp(-f(z_0)) - 1)\, \sigma(dz_0) \\ \times \sum_{n\geq 1} \rho^n \int_{\mathbb{R}^n} \exp\left(-\sum_{j=1}^n f(z_j)\right) \prod_{j=1}^n \frac{\mathbb{P}(\sigma \geq z_j)}{\mathbb{E}(\sigma)}\, dz_i$$

and

$$B = -\sum_{n\geq 1} \rho^n \int_{\mathbb{R}^n} \frac{1}{n} \sum_{i=1}^n f'(z_i) \exp\left(-\sum_{j=1}^n f(z_j)\right) \prod_{j=1}^n \frac{\mathbb{P}(\sigma \geq z_j)}{\mathbb{E}(\sigma)}\, dz_i.$$

For $n \geq 1$, by Fubini's Theorem the relationship

$$\int f'(z) e^{-f(z)} \mathbb{P}(\sigma \geq z)\, dz$$
$$= \mathbb{E}\left(\int_0^\sigma f'(z) e^{-f(z)}\, dz\right) = \int_{\mathbb{R}_+^*} (\exp(-f(z)) - 1)\, \sigma(dz),$$

holds, therefore, B can be rewritten as

$$\begin{aligned} B &= -\sum_{n\geq 1} \rho^n \int_{\mathbb{R}^n} f'(z_1) \exp\left(-\sum_{j=1}^n f(z_j)\right) \prod_{j=1}^n \frac{\mathbb{P}(\sigma \geq z_j)}{\mathbb{E}(\sigma)}\, dz_i \\ &= -\lambda \int_{\mathbb{R}_+^*} (\exp(-f(z)) - 1)\, \sigma(dz) \\ &\quad \times \sum_{n\geq 1} \rho^{n-1} \int_{\mathbb{R}^{n-1}} \exp\left(-\sum_{j=1}^{n-1} f(z_j)\right) \prod_{j=1}^{n-1} \frac{\mathbb{P}(\sigma \geq z_j)}{\mathbb{E}(\sigma)}\, dz_i. \end{aligned}$$

The quantity $A + B$ is thus given by

$$A + B = -\lambda \int_{\mathbb{R}_+^*} (\exp(-f(z)) - 1)\, \sigma(dz),$$

hence Equation (7.14) shows that Identity (7.13) holds. The proposition is proved. □

It can be proved, by using Theorem 3.8 page 154 of Asmussen [**3**] for example, that the Markov chain (Γ_t) is Harris ergodic when $\lambda\mathbb{E}(\sigma) < 1$, so that $\mathbb{Q}$ is the only invariant probability distribution and that the distribution of Γ_t converges to $\mathbb{Q}$ as t tends to infinity.

Time Change Arguments. The generalization of time change arguments of Section 7.3.2 is only sketched. The branching process of Section 7.3.1 can also be represented as a measure valued process. If, for $N \in \mathbb{N}$ and a N-tuple $x_1, \ldots, x_N \in \mathbb{R}_+^*$, the point measure

$$\mathcal{Z} = \delta_{x_1} + \delta_{x_2} + \cdots + \delta_{x_N}$$

can be seen as the state of a branching process: there are N individuals with respective lifetimes $x_1, \ldots, x_N \in \mathbb{R}_+^*$. In a similar way as for Processor-Sharing queue, it is possible to define a Markov process $(\overline{\mathcal{Z}}_t)$ when the initial state $\overline{\mathcal{Z}}_0$ is an arbitrary element of $\mathcal{M}(\mathbb{R}_+^*)$ such that if $\overline{\mathcal{Z}}_0$ is a positive point measure then $(\overline{\mathcal{Z}}_t)$ is the state of the branching process at time t with an external source. The Markov process $(\overline{\mathcal{Z}}_t)$ has also the Feller-Dynkin property and its infinitesimal generator $\Omega_{\overline{\mathcal{Z}}}$ is defined by

$$\Omega_{\overline{\mathcal{Z}}}(F_f)(m) = \left[\lambda[1+m(\mathbb{R}_+^*)]\mathbb{E}\left(e^{-f(\sigma)}-1\right) - \int_{\mathbb{R}_+^*} f'(u)\,m(du)\right] F_f(m),$$

for $m \in \mathcal{M}(\mathbb{R}_+^*)$ and any function f in V. In the exponential case, the quantity $1+m(\mathbb{R}_+^*)$ is simply the rate of increase of the number of customers when the state of the queue is $m \in \mathcal{M}(\mathbb{R}_+^*)$ (see Proposition 7.5).

An additive functional $(A(t))$ is defined by

$$A(t) = \int_0^t \left[1 + \overline{\mathcal{Z}}_s(\mathbb{R}_+^*)\right] ds,$$

for $t \geq 0$ and $(C(t))$ denotes its inverse. This is the equivalent of Definition 7.5 for a general service distribution. Proposition 7.6 is generalized as follows: $(\overline{\mathcal{Z}}_{C(t)})$ is a Feller-Dynkin Markov process on $\mathcal{M}(\mathbb{R}_+^*)$ whose infinitesimal generator $\widetilde{\Omega}$ is given by

$$\widetilde{\Omega}(F_f) = \left[\lambda\mathbb{E}\left(e^{-f(\sigma)}-1\right) - \frac{1}{1+m(\mathbb{R}_+^*)}\int_{\mathbb{R}_+^*} f'(u)\,m(du)\right] F_f(m),$$

for $m \in \mathcal{M}(\mathbb{R}_+^*)$ and any function f in V. This is also the infinitesimal generator of the Markov process (Γ_t) with a permanent customer (see Proposition 7.12). These facts being established, the branching property of Proposition 7.3 can then be used to derive the distribution of the sojourn of a customer. Analytically, the situation is more complicated than in the exponential case, the equivalent of linear differential Equation (7.10) being more involved. See Yashkov [Yas83].

Remark.

The identification of the distribution of $(\overline{\mathcal{Z}}_{C(t)})$ with the distribution of the number of customers of a Processor-Sharing queue with a permanent customer is quite easy to prove when their respective infinitesimal generators are identified. Section 7.3.2 offers a simple illustration of this fact since, in this case both processes are birth and death processes. The characterization with infinitesimal generators involves nevertheless some minimal requirements on the topology of the state space and on the regularity of the Markov process itself. This is the price of this appealing approach, the next section on LIFO discipline shows that it is not always easy. Note that the proof of the Markov property, and even of the strong Markov property of (Γ_t) do not require such a technical apparatus. The book by Ethier and Kurtz [**19**] is a very good reference in that domain.

7.4 LIFO M/G/1 Queue with Preemptive Service

For this discipline, a customer cannot be served if it is not the latest arrived in the queue. Customers arrived during its sojourn time must have been already

[Yas83] Op. cit. page 182.

served at its departure. The sojourn time of a customer whose service is σ_0 is thus simply the duration of the busy period generated by this customer. The duration of the busy period is the same for FIFO and LIFO disciplines. The variable B define in Section 7.1 and the sojourn time of a customer in an $M/G/1$ LIFO preemptive queue have the same distribution, its Laplace transform is given (implicitly) by Proposition 7.1.

State Space. Since customers may be interrupted during their services, a Markovian description of this queue must at least include residual service times of customers present in the queue and also their respective order of arrival. Hence, a vector of the form $(z_1, \ldots, z_n)$ can represent the state of the queue: for $1 \leq k \leq n$, the quantity z_k is the residual service time of the kth customer, the older customer having index n. As it will be seen, especially to define the infinitesimal generator, it is more convenient to see the vector $(z_1, \ldots, z_n)$ as a doubly infinite sequence.

A convenient state space $\mathcal{S}$ for LIFO queue is the set of non-negative sequences with finite sum and vanishing on the left,

$$\mathcal{S} = \left\{ (x_i) \in \mathbb{R}_+^{\mathbb{Z}} : \sum_{i \in \mathbb{Z}} x_i < +\infty, \exists N \text{ such that } x_i = 0 \text{ when } i \leq= -N \right\}.$$

Sequences x and $y \in \mathcal{S}$ differing by 0 components are identified. A formal way of saying that is to consider the quotient space for the equivalence relation $\sim$. This relation is defined as follows, for x and $y \in \mathbb{R}_+^{\mathbb{Z}}$, then $x \sim y$ if there exist subsets I and J of $\mathbb{Z}$ such that

- $x_i = 0$ for $i \in I$ and $y_j = 0$ for $j \in J$;
- $(x_i : i \notin I) = (y_j : j \notin J)$. The element $(x_i : i \notin I)$ is just the sequence x after the components with index in I have been removed.

In particular, two sequences which differ by a translation are identical for this relation. An element x of $\mathcal{S}$ is represented as $x = [x_0, \ldots, x_m]$ whenever $x_i = 0$ if $i \notin \{0, \ldots, m\}$ (of course m can be infinite). Note that with this convention $[x_0, \ldots, x_m] = [0, x_0, \ldots, x_m]$. The quantity $[0]$ denotes the element of $\mathcal{S}$ having all its components equal to 0.

Topological considerations. A topology has to be defined on $\mathcal{S}$ so that it is a locally compact topological Hausdorff space. The topology has to be defined carefully to take into account the specificity of LIFO discipline: for a small $\varepsilon > 0$ and $u > 0$, the elements $[\varepsilon, u]$ and $[u]$ have to be close since that, if they are taken as initial states, after time ε, without any new arrival, the state of the queue will be respectively $[u]$ and $[u - \varepsilon]$. In the same way for $v > 0$, the states $[u, \varepsilon, v]$ and $[u, v]$ have to be close and so on.

Since basically $\mathcal{S}$ is a space of sequences, it is naturally embedded into the space of convergent series $L_1(\mathbb{Z})$ endowed with the distance d such that

$$d_{L_1}(a, b) = \sum_{i \in \mathbb{Z}} \frac{1}{2^{|i|}} \min(1, |a_i - b_i|),$$

when $a = (a_i)$ and $b = (b_i) \in L_1(\mathbb{Z})$. Convergence for this topology is equivalent to the convergence of every coordinate. The topology on $\mathcal{S}$ is taken as a quotient topology: if $x \in \mathcal{S}$, for $r > 0$, define

$$O_{x,r} = \{y \in \mathcal{S} : \exists\, a, b \in L_1(\mathbb{Z}) \text{ such that } x \sim a, y \sim b \text{ and } d_{L_1}(a,b) < r\},$$

then, the family $\{O_{x,r} : x \in \mathcal{S},\, r > 0\}$ is taken as a fundamental system of open neighborhoods of $\mathcal{S}$. In this way, for positive u, v and ε, the element $[u, \varepsilon, v]$ is close to $[u, v]$ since $[u, v] \sim [u, 0, v]$ and $d_1([u, 0, v], [u, \varepsilon, v])$ is arbitrary small with ε.

With this topology, for $C > 0$, the set

$$K_C = \left\{ x = (x_i) \in \mathcal{S} : \sum_{i \in \mathbb{Z}} x_i < C \right\} \tag{7.15}$$

is compact. A sketch of the proof is given. Take a sequence (U_n) of K_C. For $n \in \mathbb{N}$, U_n can be written as $U_n = (x_k^n : k \geq 0)$. For any $k \geq 0$, the sequence is $(x_k^n; n \geq 0)$ is bounded by C, therefore it has a convergent subsequence. By using a diagonal argument, see Rudin [**46**] for example, there exists a subsequence (n_p) such that, for any $k \geq 0$, the sequence $(x_k^{n_p}; p \geq 0)$ is converging to some $w_k \geq 0$. Fatou's Lemma gives that

$$\sum_{k \geq 0} w_k \leq \liminf_{p \to +\infty} \sum_{k \geq 0} x_k^{n_p} \leq C,$$

hence, (w_k) is in $\mathcal{S}$. Consequently, the sequence (U_n) has a convergent subsequence in $\mathcal{S}$. The proof of the fact that, if every sequence of K_C has a convergent subsequence then K_C is compact, is left as an exercise. It has been proved that, in this way, $\mathcal{S}$ is a locally compact space. Finally, $\mathcal{S}$ is equipped with the associated Borel σ-field.

The state space is not embedded in a finite dimensional vector space. Queueing systems with an infinite dimensional state space have been already encountered, for example see Section 4.4.3 page 97 and the last example of Section 9.5 page 258.

Transitions. For $\sigma \in \mathbb{R}_+$ and $x \in \mathcal{S}$ such that $x = [x_1, \ldots, x_n]$ with $x_1 > 0$ and $0 < \varepsilon < x_1$,

- $\sigma \cdot x \in \mathcal{S}$ denotes the element of $\mathcal{S}$ which is obtained by concatenation of σ at the beginning of the vector x, i.e. $\sigma \cdot x = [\sigma, x_1, \ldots, x_n]$;
- $x - \varepsilon \in \mathcal{S}$ is the element $[x_1 - \varepsilon, x_2, \ldots, x_n]$.

If $R(t) = (R_i(t)) \in \mathcal{S}$ denotes the state of the LIFO queue at time $t \geq 0$, then it satisfies the following stochastic differential equation:
If $\mathcal{N}_\lambda([t, t+dt]) = 0$, only the customer at the head of the queue is served,

$$\begin{cases} R_1(t+dt) = R_1(t) - dt, & \text{if } R(t) \neq [0], \\ R_k(t+dt) = R_k(t), & \text{for } k \neq 1. \end{cases}$$

Otherwise, for some $\sigma \in \mathbb{R}_+$, $\mathcal{N}_\lambda(\{t\} \times \{\sigma\}) \neq 0$, a new customer with service time σ arrives at time t, then

$$\begin{cases} R_1(t) = \sigma, \\ R_k(t) = R_{k-1}(t-), & \text{for } k > 1. \end{cases}$$

Feller-Dynkin Property. This section shows that the process $(R(t))$ of residual service times can be described as a Feller-Dynkin Markov process (see Rogers and Williams [**44**]). The set of real-valued continuous functions on $\mathcal{S}$ vanishing at infinity is denoted by $C_0(\mathcal{S})$: if f is such a function, for $\varepsilon > 0$, there exists some $C > 0$ such that $|f(x)| \leq \varepsilon$ when $x \notin K_C$, where K_C is the compact set defined by Equation (7.15). On the state space $\mathcal{S}$, for $t \geq 0$, the operator P_t is defined on $C_0(\mathcal{S})$ by

$$P_t(f)(x) = \mathbb{E}_x[f(R(t))],$$

for $f \in C_0(\mathcal{S})$ and $x \in \mathcal{S}$.

PROPOSITION 7.14. *The process $(R(t))$ describing the LIFO discipline is a Feller-Dynkin process.*

PROOF. By using the definition of the Feller-Dynkin property of Section 7.7 page 205, the properties (ii) and (iv) are clear. Condition (iii) is just the strict Markov property which is a consequence of the Poisson property of the arrival process. Only Condition (i) has to be checked, among other things, it says that the process cannot return from infinity in a finite time.

In the following, f is a function of $C_0(\mathcal{S})$. If $R_x(t)$ is the process at time $t \geq 0$ with the initial condition $R(0) = x$, by induction on the number of jumps of the point process $\mathcal{N}_\lambda$ in the interval $[0, t]$, it is easily seen that $x \to R_x(t)$ is a continuous operator on $\mathcal{S}$. The function $x \to \mathbb{E}_x[f(R(t))]$ is therefore continuous on $\mathcal{S}$.

Since f is vanishing at infinity, for $\varepsilon > 0$, there exists some $C > 0$ such that $|f(x)| \leq \varepsilon$ when $x \notin K_C$, see Equation (7.15). Since the server works at rate at most 1, if $R(0) = x \notin K_{C+t}$, then $R(t) \notin K_C$ so $\mathbb{E}_x[f(R(t))] \leq \varepsilon$. The function $x \to P_t(f)(x)$ is thus vanishing at infinity. The proposition is proved. □

Infinitesimal Generator.

DEFINITION 16. *The set $\mathcal{D}$ is the set of functions $f \in C_0(\mathcal{S})$ such that for any $n \geq 1$ and $z_2 > 0, \ldots z_n > 0$, the function*

$$\begin{aligned}]0, +\infty[&\to \mathbb{R} \\ z_1 &\to f([z_1, z_2, \ldots, z_n]), \end{aligned}$$

is differentiable. The derivative is denoted by $D_1(f)$.

The next proposition shows that $\mathcal{D}$ is a set of functions on which the infinitesimal generator of $(R(t))$ can be defined.

PROPOSITION 7.15. *The infinitesimal generator Ω of the Markov process $(R(t))$ is defined on $\mathcal{D}$ by*

$$\begin{aligned}\Omega(f)(x) &= -1_{\{x\neq[0]\}}D_1(f)(x) + \lambda\int_0^{+\infty}[f(a\cdot x)-f(x)]\,\sigma(da)\\ &= -1_{\{x\neq[0]\}}D_1(f)(x) + \lambda\mathbb{E}\left[f(\sigma\cdot x)-f(x)\right]\end{aligned} \tag{7.16}$$

for $f\in\mathcal{D}$ and $x\in\mathcal{S}$.

PROOF. The proof is straightforward since, for $f\in\mathcal{D}$ and $x\in\mathcal{S}$, the limit

$$\lim_{h\to 0}\frac{1}{h}\left[\mathbb{E}_x\left(f(R(h))-f(x)\right]\right.$$

exists and is equal to the right hand side of Equation (7.16), it is therefore $\Omega(f)(x)$. The operator D_1 is simply associated to the service on the first non-zero component. □

By using the fact that differentiable functions are dense in the set of continuous functions on $\mathbb{R}_+^n$, it is not difficult to prove that the set $\mathcal{D}$ is dense in $C_0(\mathcal{S})$. See Ethier and Kurtz [**19**] for more details in this domain.

Invariant Distribution.

PROPOSITION 7.16. *Under Condition $\rho=\lambda\mathbb{E}(\sigma)<1$, at equilibrium, the number of customers of the LIFO queue is geometrically distributed with parameter ρ and the residual service times are independent random variables with density $\mathbb{P}(\sigma\geq x)/\mathbb{E}(\sigma)$ on $\mathbb{R}_+$.*
Equivalently, the distribution π on the state space

$$\mathcal{S}=\{x\in\mathcal{S}: x=[x_1,\dots,x_n], x_1>0,\dots,x_n>0\}$$

such that $\pi([0])=1-\rho$ and, for $n\geq 1$,

$$\pi([dx_1,\dots,dx_n])=\rho^n(1-\rho)\prod_{i=1}^n\frac{\mathbb{P}(\sigma\geq x_i)}{\mathbb{E}(\sigma)}\,dx_i,$$

is invariant for the Markov process $(R(t))$.

PROOF. Denote, for $g\in C_0(\mathcal{S})$,

$$E_\pi(g)\stackrel{\text{def.}}{=}\int_{\mathcal{S}}g(x)\,\pi(dx).$$

With the same arguments as in the proof of Proposition 7.13, the invariance property of π is true if, for any $f\in\mathcal{D}$, the quantity $\mathbb{E}_\pi(\Omega(f))$ is 0.

Since

$$\int_{\mathcal{S}}\Omega(f)(x)\pi(dx)=(1-\rho)\lambda\mathbb{E}\left[f([\sigma])-f([0])\right]+A+B, \tag{7.17}$$

with

$$A = \sum_{n=1}^{+\infty} \lambda\rho^n(1-\rho)$$
$$\int \mathbb{E}\left(f([\sigma, x_1, \dots, x_n]) - f\left([x_1, \dots, x_n]\right)\right) \prod_{i=1}^{n} \frac{\mathbb{P}(\sigma \geq x_i)}{\mathbb{E}(\sigma)}\, dx_i,$$

and

$$B = -\sum_{n=1}^{+\infty} \rho^n(1-\rho) \int_{\mathbb{R}_+^n} D_1(f)\left([x_1, \dots, x_n]\right) \prod_{i=1}^{n} \frac{\mathbb{P}(\sigma \geq x_i)}{\mathbb{E}(\sigma)}\, dx_i.$$

By using the definition of the operator D_1, one gets, for $x_1, x_2, \dots, x_n \in \mathbb{R}_+$,

$$\begin{aligned}\int_0^{+\infty} & D_1(f)\left([a, x_1, \dots, x_n]\right)\mathbb{P}(\sigma \geq a)\, da \\ &= \mathbb{E}\left(\int_0^{\sigma} D_1(f)\left([a, x_1, \dots, x_n]\right)\, da\right) \\ &= \mathbb{E}\left(f\left([\sigma, x_1, \dots, x_n]\right) - f\left([0, x_1, \dots, x_n]\right)\right),\end{aligned}$$

hence

$$\begin{aligned}\int_0^{+\infty} & D_1(f)\left([a, x_1, \dots, x_n]\right) \\ &= \mathbb{E}\left(f\left([\sigma, x_1, \dots, x_n]\right) - f\left([x_1, \dots, x_n]\right)\right),\end{aligned}$$

since $[0, x_1, \dots, x_n] = [x_1, \dots, x_n]$. Therefore, with a change of variables, B can be written as

$$B = -\sum_{n=1}^{+\infty} \lambda\rho^{n-1}(1-\rho)$$
$$\times \int \mathbb{E}\left(f\left([\sigma, x_1, \dots, x_{n-1}]\right) - f\left([x_1, \dots, x_{n-1}]\right)\right) \prod_{i=1}^{n-1} \frac{\mathbb{P}(\sigma \geq x_i)}{\mathbb{E}(\sigma)}\, dx_i.$$

In the sum $A+B$, all terms of the series cancel out except the first one in B which is

$$-(1-\rho)\lambda\mathbb{E}\left(f\left([\sigma]\right) - f([0])\right),$$

from Equation (7.17), one gets that

$$\int \Omega(f)(x)\pi(dx) = 0.$$

The proposition is proved. □

As for Processor-Sharing queue, it can be shown that $(R(t))$ is in fact a Harris ergodic Markov process when $\lambda\mathbb{E}(\sigma) < 1$.

7.5 The Insensitivity Property

The invariant distribution of the number of customers of LIFO $M/G/1$ queue depends on the distribution of services only through their mean $\mathbb{E}(\sigma_0)$. For the infinite server queue and $M/G/1$ queue with Processor-Sharing discipline, the situation is the same. This property is usually called *insensitivity* of the queue. A queue is insensitive if the stationary distribution of the number of customers does not change when the distribution of service times is replaced by another distribution with the same mean.

The insensitivity property does not hold for the FIFO queue. For FIFO discipline, Relationship (2.19) page 50 shows that the distribution of the number of customers depends on the distribution of the stationary waiting time which is itself a non-trivial functional of the service distribution (Equation (2.17)): It does not depend only of its mean.

For $M/G/1$ queue with LIFO or Processor-Sharing discipline, the insensitivity property has been proved by using brute force arguments, i.e. by calculating the explicit stationary distribution for an arbitrary service distribution. This property can be proved in a more general framework, (some queueing networks for example), just by checking some simple algebraic relations when all services are exponentially distributed. See Burman[Bur81] or Whittle [**54**] for example.

7.6 The Distribution Seen by Customers

This section considers a general queueing system with Poisson arrivals. The queues with a Poisson input $\mathcal{N}_\lambda$ have the following remarkable property: At equilibrium the state of the queue at an arbitrary instant and the state of the queue before the arrival of a customer have the same distribution. The term "PASTA" is sometimes used in the literature for this property, it stands for "Poisson Arrivals See Times Averages".

This property is not natural. If arrivals occur every time unit and services are deterministic and equal to 1/2, the state of the queue before an arrival is always empty but the probability that the queue is empty at an arbitrary instant is 1/2.

This property is not mysterious at all: Proposition 1.15 page 14 has established that the distribution of a Poisson process seen from one of its points (this point excluded) is identical to the distribution of the same Poisson process seen from an arbitrary point (0 say). If the equilibrium state of the queue at some moment t is a functional of the arrival process outside the point t, then clearly the stationary distribution at an arbitrary moment and stationary distribution at the arrival of a customer are the same. The following

[Bur81] D.Y. Burman, *Insensitivity in queueing systems*, Advances in Applied Probability **13** (1981), 846–859.

section gives a presentation of this simple fact. It is somewhat formal (and a little annoying) since one has to define what is the "state" of the queue and how the equilibrium is reached. It must be kept in mind that Proposition 1.15 is the real explanation of this phenomenon.

The State of the Queue. Recall that $\mathcal{M}_p(\mathbb{R}\times\mathbb{R}_+)$ is the space of point measures on $\mathbb{R}\times\mathbb{R}_+$ (see Chapter 1). For $s \le t \in \mathbb{R}$, it is assumed that if the queue is empty at time s, the state of the queue just before time t is given by $\phi(s,t,\mathcal{N}_\lambda)$, where

$$\phi : \{(s,t)\in\mathbb{R}\times\mathbb{R} : s\le t\}\times\mathcal{M}_p(\mathbb{R}\times\mathbb{R}_+)\to\mathbb{R}$$

is a non-negative measurable functional such that, for $s \le t$,

$$\phi(s,t,\mathcal{N}_\lambda) = \phi\left(0,t-s,T^s\mathcal{N}_\lambda\right), \tag{7.18}$$

$$\phi(s,t,\mathcal{N}_\lambda) = \phi(t_1(s),t,\mathcal{N}_\lambda([s,t[\cap\cdot)), \tag{7.19}$$

where $T^s\mathcal{N}_\lambda$ denotes the translation of the point process $\mathcal{N}_\lambda$ by s (see Definition 33 page 306), $t_1(s)$ is the first point of $\mathcal{N}_\lambda$ greater than s and the notation $\mathcal{N}_\lambda([s,t[\cap\cdot)$ is for the point process $\mathcal{N}_\lambda$ restricted to the set $[s,t[\times\mathbb{R}_+$.

Relationship (7.18) just states that the behavior of the queue is time homogeneous. Relationship (7.19) implies that $\phi(s,t,\mathcal{N}_\lambda)$ depends only on the restriction of $\mathcal{N}_\lambda$ to the interval $[s,t[$. To simplify notations, the initial state has been assumed to be empty. It is nevertheless easy to include it by adding this variable in the arguments of ϕ.

DEFINITION 17. *A queue has the* backward coupling property *for the functional ϕ if there exists a random variable $T_0\le 0$, $\mathbb{P}$-almost surely finite, such that for $s\le T_0$,*

$$\phi(s,0,\mathcal{N}_\lambda) = \phi(T_0,0,\mathcal{N}_\lambda).$$

PROPOSITION 7.17. *If Conditions (7.18) and (7.19) are satisfied, and functional ϕ has the backward coupling property, if $S(t)=\phi(0,t,\mathcal{N}_\lambda)$ denotes the state of the queue at time t, then the identity*

$$\lim_{t\to+\infty} S(t) \overset{dist.}{=} \lim_{n\to+\infty} S(t_n)$$

holds.

If $S(t)$ is defined as the state of the queue just before time t, the above proposition shows that the state observed at an arbitrary instant t, with t large, has the same distribution as the state observed by the nth customer, with n large.

PROOF. For $n\ge 1$, Conditions (7.18) and (7.19) give the identities

$$\begin{aligned}\phi(0,t_n,\mathcal{N}_\lambda) &= \phi(t_1,t_n,\mathcal{N}_\lambda)\\ &= \phi\left(t_1-t_n,0,T^{t_n}\mathcal{N}_\lambda\right) = \phi\left(t_1-t_n,0,T^{t_n}\mathcal{N}_\lambda([t_1-t_n,0[\cap\cdot)\right).\end{aligned}$$

The distances between the t_n's form an i.i.d. exponentially distributed sequence, therefore

$$\left(t_1 - t_n, T^{t_n}\mathcal{N}_\lambda([t_1 - t_n, 0[\cap \cdot)\right) \stackrel{\text{dist.}}{=} \left(t_{-n+1} - t_0, T^{t_0}\mathcal{N}_\lambda([t_{-n+1} - t_0, 0[\cap \cdot)\right).$$

Since the location of the first point of $\mathcal{N}_\lambda$ on the left of 0 is also exponentially distributed, the identity

$$\left(t_{-n+1} - t_0, T^{t_0}\mathcal{N}_\lambda([t_{-n+1} - t_0, 0[\cap \cdot)\right) \stackrel{\text{dist.}}{=} (t_{-n+2}, \mathcal{N}_\lambda([t_{-n+2}, 0[\cap \cdot))$$

also holds. Notice the importance of the fact that the interval $[t_{-n+1}, 0[$ does not include the point 0. In the left hand side of the last relation, the point process $T^{t_0}\mathcal{N}_\lambda$ has a point at 0 which is not the case for the right hand side since 0 is (almost surely) not a point of $\mathcal{N}_\lambda$. Hence the identity

$$\phi(0, t_n, \mathcal{N}_\lambda) \stackrel{\text{dist.}}{=} \phi(t_{-n+2}, 0, \mathcal{N}_\lambda([t_{-n+2}, 0[\cap \cdot)) = \phi(t_{-n+2}, 0, \mathcal{N}_\lambda)$$

holds. The backward coupling property shows that the variable $\phi(0, t_n, \mathcal{N}_\lambda)$ converges in distribution to $\phi(T_0, 0, \mathcal{N}_\lambda)$ as t tends to infinity.

In the same way, for $t \geq 0$, Relationship (7.18) gives the relation

$$\phi(0, t, \mathcal{N}_\lambda) = \phi(-t, 0, \mathcal{N}_\lambda),$$

hence the variable $\phi(0, t, \mathcal{N}_\lambda)$ also converges in distribution to $\phi(T_0, 0, \mathcal{N}_\lambda)$ as n tends to infinity. The proposition is proved. □

The $M/G/1$ FIFO Queue. If $V(t)$ is the virtual waiting time of the queue just before time t, Relationships (7.18) and (7.19) are satisfied.

According to the remark of Chapter 12 page 336, if the queue is empty at $s \leq 0$, the variable $V(0)$ can be represented as

$$\sup_{s \leq u \leq 0} \left(u + \int_{]u,0[\times\mathbb{R}_+} x\,\mathcal{N}_\lambda(dv, dx) \right) = \sup_{n, s \leq t_n < 0} \left(\sum_{i=n}^{0} \sigma_i + t_n \right).$$

If $\lambda\mathbb{E}(\sigma) < 1$, the law of large numbers shows that, $\mathbb{P}$-almost surely,

$$\lim_{n \to -\infty} \frac{1}{n} \sum_{i=n}^{0} \sigma_i + t_n = \mathbb{E}(\sigma) - \frac{1}{\lambda} < 0.$$

There exists N_0 such that for $n \leq N_0$,

$$\sum_{i=n}^{0} \sigma_i + t_n < 0,$$

hence, for $s \leq t_{N_0}$, one gets the relation

$$\sup_{s \leq u \leq 0} \left(u + \int_{]u,0[\times\mathbb{R}_+} x\,\mathcal{N}_\lambda(du, dx) \right) = \sup_{t_{N_0} \leq u \leq 0} \left(u + \int_{]u,0[\times\mathbb{R}_+} x\,\mathcal{N}_\lambda(du, dx) \right),$$

the supremum being non-negative, the backward coupling property is satisfied in this case. Proposition 7.17 shows that the stationary virtual waiting time and the stationary waiting time have the same distribution. This result has been proved in Corollary 2.15 page 49 via the calculations of these distributions. Similar results also holds for other quantities such as the number of customers.

The backward coupling property is also true for an $M/G/1$ queue with a conservative discipline (see the definition page 344) since the virtual waiting time process is the same for all conservative disciplines. Consequently, Proposition 7.17 can be applied to any variable satisfying Properties (7.18) and (7.19). Therefore, the above results concerning virtual waiting time are also true for the disciplines LIFO and Processor-Sharing.

The $M/G/\infty$ Queue. If the queue is empty at time s, the number of customers just before time t is given by the functional

$$L(t) = \int_{[s,t[\times\mathbb{R}_+} 1_{\{u+x>t\}}\, \mathcal{N}_\lambda(du, dx)$$

which clearly satisfies Relationships (7.18) and (7.19). In particular, the variable $L(0)$ is given by

$$\int_{[s,0[\times\mathbb{R}_+} 1_{\{u+x>0\}}\, \mathcal{N}_\lambda(du, dx) = \sum_{n\leq 0} 1_{\{t_n\geq s,\, t_n+\sigma_n>0\}},$$

The law of large numbers shows that the sequence $(t_n + \sigma_n)/|n|$ converges P-almost surely to $-1/\lambda$ as n tends to infinity. In the same way as before, there exists $N_0 \in \mathbb{N}$ such that for $s \leq t_{N_0}$,

$$\int_{[s,0[} 1_{\{u+x>0\}}\, \mathcal{N}_\lambda(du, dx) = \int_{[t_{N_0},0[} 1_{\{u+x>0\}}\, \mathcal{N}_\lambda(du, dx).$$

The backward coupling property holds for this queue. Proposition 7.17 can be applied to the number of customers of this queue. For $C \in \mathbb{N}$, the $M/G/C/C$ queue can be seen as the $M/G/\infty$ queue but with only C servers. Similar results also hold for this queue.

REMARK.

The backward coupling is not really necessary to have the equality of the distributions seen at an arbitrary instant or seen by a customer. It has been introduced mainly to reduce the notations and conditions of a more general result. Note that Relationships (7.18) and (7.19) are quite natural. The main ingredient for the equality of these distributions is the fact that a marked Poisson point process seen from 0 or seen from one of its points looks the same (the reference point excepted). All the other arguments of the proof rely on the convergence properties of the queueing system under investigation.

7.7 Appendix: Feller-Dynkin Markov Processes

The state space $\mathcal{S}$ is assumed to be a locally compact space which is covered by a non-decreasing sequence of compact subsets (K_n). The set of real-valued continuous functions on $\mathcal{S}$ vanishing at infinity is denoted by $C_0(\mathcal{S})$. If f is a function of $C_0(\mathcal{S})$, for $\varepsilon > 0$, there exists some $n \geq 1$ such that $|f(x)| \leq \varepsilon$ when $x \notin K_n$.

DEFINITION 18. *A family of linear operators (P_t) is said to have the Feller-Dynkin property if it verifies the following conditions*

(i) $P_t : C_0(\mathcal{S}) \to C_0(\mathcal{S})$;
(ii) *For any* $f \in C_0(\mathcal{S})$ *such that* $0 \leq f \leq 1$ *then,* $0 \leq P_t(f) \leq 1$;
(iii) $P_s \circ P_t = P_{s+t}$ *for any* $s \geq 0$, *and* $t \geq 0$, *and* P_0 *is the identity function on* $C_0(\mathcal{S})$;
(iv) *For any* $f \in C_0(\mathcal{S})$ *and* $x \in \mathcal{S}$, *then* $P_t(f)(x) \to f(x)$ *as* $t \searrow 0$.

If $(X(t))$ is some stochastic process with values in $\mathcal{S}$, the corresponding operators (P_t) are defined by

$$P_t(f)(x) = \mathbb{E}_x[f(X(t))],$$

for $x \in \mathcal{S}$ and $f \in C_0(\mathcal{S})$. See Rogers and Williams [**44**] for more details.

8. Recurrence and Transience of Markov Chains

Contents

The general problem of convergence in distribution of Markov chains with a countable state space is investigated in this chapter. For the $GI/GI/1$ queue in Chapter 2, the convergence in distribution of the Markov chain (W_n) has been obtained by using an explicit representation of the random variable W_n as a functional of the random walk associated with the interarrival times and the service times. Markov chains describing the behavior of most queueing systems cannot, in general, be represented in such a simple way. In this chapter, simple criteria are given to determine whether a given Markov chain is ergodic or transient. The main results are Theorem 8.6 for ergodicity and Theorem 8.10 for transience. In practice, these results can be used in many applications. These criteria can be seen as an extension, in a probabilistic setting, of a classical stability result of ordinary differential equations due to Lyapunov[Lia07] in 1892. In a stochastic context, the first results of this type are apparently due to Khasminskii for diffusions (see Khasminskii [**26**]). The stability criterion by Lyapunov is recalled at the end of the chapter (see Hirsch and Smale [**27**] for a detailed presentation of these questions). In Chapter 9, an important scaling method is introduced that will give a more precise picture of the relation between the stability of ordinary differential equations and the stability properties of Markov

The study is restricted to the case of countable state spaces, mainly to avoid the usual technical complications of Markov chains with a continuous

[Lia07] A.M. Liapunov, *Problème général de la stabilité du mouvement*, Annales de la Faculté des Sciences de l'Université de Toulouse **9** (1907), 203–475.

state space (see Nummelin [**39**] on this subject). The books by Fayolle *et al.* [**20**] and Meyn and Tweedie [**34**] are completely devoted to these questions.

Throughout this chapter, (M_n) is an homogeneous irreducible and aperiodic Markov chain defined on a probability space $(\Omega, \mathcal{F}, \mathbb{P})$ and with values in a countable state space $\mathcal{S}$, that is,

– *Markov Property.*

$$\text{(8.1)}\quad \mathbb{P}(M_{n+1} = y \mid M_n = x_n, \dots, M_0 = x_0) = \mathbb{P}(M_{n+1} = y \mid M_n = x_n),$$

for any $n \in \mathbb{N}$ and y, $x_i \in \mathcal{S}$, $i = 1, \dots, n$.

– *Homogeneity.*

$$\mathbb{P}(M_{n+1} = y \mid M_n = x) = \mathbb{P}(M_1 = y \mid M_0 = x),$$

for any $n \in \mathbb{N}$ and x and $y \in \mathcal{S}$.

– *Irreducibility.* For any x, $y \in \mathcal{S}$, there exists $n \in \mathbb{N}$ such that

$$p(x, y) = \mathbb{P}(M_n = x \mid M_0 = y) > 0.$$

The transition matrix is denoted by P, with elements

$$p(x, y) = \mathbb{P}(M_1 = y \mid M_0 = x),$$

for x and $y \in \mathcal{S}$, and $\mathcal{F} = (\mathcal{F}_n)$ is the filtration generated by the sequence (M_n), for $n \geq 0$, i.e.

$$\mathcal{F}_n = \sigma \langle M_0, \dots, M_n \rangle .$$

It is well known and easily shown that such a Markov chain has the *strong Markov property*, that is, Relationship (8.1) is still true when n is replaced by an arbitrary stopping time T :

$$\mathbb{P}(M_{T+1} = y \mid M_T = x, M_{T-1}, \dots, M_0) = \mathbb{P}(M_1 = y \mid M_0 = x),$$

for any x and $y \in \mathcal{S}$. (See Feller [**21**] for example.)

The notations $\mathbb{P}_x(\cdot)$ and $\mathbb{E}_x(\cdot)$ are respectively for the probability and the expected value for the Markov chain starting from x, i.e. conditioned on the event $\{M_0 = x\}$. More generally, if ν is a probability distribution on $\mathcal{S}$, the notations $\mathbb{P}_\nu(\cdot)$ and $\mathbb{E}_\nu(\cdot)$ are for the distribution of the Markov chain (M_n) when the distribution of M_0 is ν. Let θ: $\Omega \to \Omega$ be the time shift by one unit. For $p \in \mathbb{N}$, θ^p is the pth iterate of θ. In particular, the sequence $(M_n \circ \theta^p)$ is original sequence (M_n) shifted by p, i.e.

$$(M_n \circ \theta^p; n \geq 0) = (M_{p+n}; n \geq 0) .$$

Notice that the random variables related to the Markov chain are defined on the probability space Ω, in particular for any random variable X on Ω, the notation $X \circ \theta$ is the usual composition of mappings.

By choosing an appropriate probability space, such a function θ exists. (See Section 10.4 page 297 and a related construction in Section 11.2 page 304.)

DEFINITION 19. *If τ is a stopping time with respect to filtration $\mathcal{F}$, the* induced *sequence associated with τ is the increasing sequence (t_n) defined by $t_0 = 0$ and $t_n = t_{n-1} + \tau \circ \theta^{t_{n-1}}$, for $n \geq 1$. The* induced chain *is the sequence (M_{t_n}) and, finally, the* induced filtration *$\mathcal{F}^\tau$ is defined by $\mathcal{F}^\tau = (\mathcal{F}_n^\tau) = (\mathcal{F}_{t_n})$.*

When the Markov chain (M_n) is ergodic and stationary, (i.e. the distribution of M_0 is the invariant probability), the shift θ is an endomorphism of the associated probability space (Proposition 10.17 page 298). When τ is the hitting time of some subset by the Markov chain, the function θ^τ is the induced endomorphism defined page 282.

PROPOSITION 8.1. *The variables $(t_n; n \geq 0)$ are stopping times and the induced chain (M_{t_n}) is a homogeneous Markov chain.*

PROOF. First it is shown that t_n is a stopping time : By induction, if t_{n-1} is a stopping time for $k \geq 1$, then

$$\{t_n = k\} = \bigcup_{i=0}^{k} \{t_{n-1} = i\} \cap \{\tau \circ \theta^i = k - i\},$$

and since $\{\tau \circ \theta^i = k - i\}$ is $\mathcal{F}_k$-measurable, t_n is a stopping time.

The strong Markov property of (M_n) gives the Markov property of the induced chain, for $x \in \mathcal{S}$

$$\mathbb{P}\left(M_{t_{n+1}} = x \mid \mathcal{F}_n^\tau\right) = \mathbb{P}(M_{t_{n+1}} = x \mid \mathcal{F}_{t_n}) = \mathbb{P}_{M_{t_n}}(M_\tau = x).$$

□

8.1 Recurrence of Markov Chains

Recall that an irreducible Markov chain is *recurrent* if for any x and $y \in \mathcal{S}$,

$$\mathbb{P}(\exists n \in \mathbb{N}, M_n = y \mid M_0 = x) = 1.$$

In other words, the hitting time T_y of y by the Markov chain,

$$T_y = \inf\{k \geq 0 : M_k = y\}, \tag{8.2}$$

is $\mathbb{P}_x$-almost surely finite. The next proposition is an elementary result on the recurrence of Markov chains.

PROPOSITION 8.2. *If there exists a function $f : \mathcal{S} \to \mathbb{R}_+$ and a constant $K > 0$ such that*

a) $\{x : f(x) \leq L\}$ is finite for any $L > 0$,

b) $\mathbb{E}_x(f(M_1)) \leq f(x)$ if $f(x) > K$,

then the Markov chain (M_n) is recurrent.

PROOF. Doob's classical argument is used. If

$$T_K = \inf\{k \geq 0 : f(M_k) \leq K\},$$

then T_K is clearly a stopping time. Since $f(M_n) > K$ on the $\mathcal{F}_n$-measurable event $\{n < T_K\}$, Condition b) and the Markov property imply

$$\mathbb{E}(f(M_{n+1}) \mid \mathcal{F}_n) = \mathbb{E}_{M_n}(f(M_1)) \leq f(M_n) \text{ on } \{n < T_K\}.$$

The sequence $(f(M_{T_K \wedge n}))$ is therefore a non-negative supermartingale. Hence, by Theorem A.3 page 352, it converges $\mathbb{P}$-almost surely to a finite limit.

In particular, $\mathbb{P}$-a.s. on the event $\{T_K = +\infty\}$, the sequence $(f(M_n))$ is converging. For any $L > 0$, by Condition a), the set $\{x : f(x) \leq L\}$ is finite and the irreducibility of the Markov chain shows that the sequence (M_n) visits almost surely the set $\{x : f(x) > L\}$, this implies the convergence

$$\limsup_{n \to +\infty} f(M_n) = +\infty.$$

One concludes that $\mathbb{P}(T_K = +\infty) = 0$, the variable T_K is $\mathbb{P}$-almost surely finite. The Markov chain is recurrent. □

DEFINITION 20. *A function* $f : \mathcal{S} \to \mathbb{R}$ *is* superharmonic *for the Markov chain* (M_n) *when* $\mathbb{E}_x(f(M_1)) \leq f(x)$ *for any* $x \in \mathcal{S}$. *A function is* harmonic *if this inequality is an equality.*

PROPOSITION 8.3. *A function f on $\mathcal{S}$ is superharmonic if and only if, for* $M_0 = x \in \mathcal{S}$, *the sequence* $(f(M_n))$ *is a supermartingale.*

PROOF. If f is superharmonic, for $n \geq 1$, the Markov property and the superharmonicity of f give

$$\mathbb{E}(f(M_{n+1}) \mid \mathcal{F}_n) = E_{M_n}(f(M_1)) \leq f(M_n),$$

hence the sequence $(f(M_n))$ is a supermartingale.

Conversely, if for $M_0 = x \in \mathcal{S}$, the sequence $(f(M_n))$ is a supermartingale, then $\mathbb{E}(f(M_1) \mid \mathcal{F}_0) \leq f(M_0)$ and therefore $\mathbb{E}_x(f(M_1)) \leq f(x)$. The function f is superharmonic. □

Condition b) of Proposition 8.2 can be expressed as a superharmonicity property of function f on a subset of $\mathcal{S}$. Nevertheless, the next proposition shows that this property cannot be true on the whole space if the Markov chain is recurrent.

PROPOSITION 8.4. *A Markov chain (M_n) has a non-constant, non-negative superharmonic function if and only if it is transient.*

Recall that a Markov chain is *transient* if, for any $y \in \mathcal{S}$, the variable T_y defined by Equation (8.2) is infinite with positive probability.

PROOF. If there exists some non-constant non-negative superharmonic function f, Proposition 8.3 shows that $(f(M_n))$ is a non-negative supermartingale, in particular it converges almost surely to a finite limit.

Since f is not constant, there exist x and $y \in \mathcal{S}$ such that $f(x) \neq f(y)$. Assume that the Markov chain is recurrent. With probability 1, it visits infinitely often x *and* y; hence the sequence $(f(M_n))$ cannot converge. Contradiction, the Markov chain cannot be recurrent.

Conversely, if the Markov chain is transient, for $z \in \mathcal{S}$ and T_z the hitting time of z defined by Equation (8.2), the function $f_z(y) = \mathbb{P}_y(T_z < +\infty)$ is superharmonic; indeed if $x \neq z$,

$$\mathbb{E}_x\left(f_z(M_1)\right) = \mathbb{E}_x\left[\mathbb{P}_{M_1}(T_z < +\infty)\right] = \mathbb{E}_x\left(1_{\{T_z<+\infty\}}\right) = f_z(x).$$

Since $f_z(z) = 1$ and f_z is bounded by 1, f_z is a non-negative superharmonic function. It is non-constant since, for $y \neq z$, $f_z(y) < 1$ by transience of the Markov chain. □

8.2 Ergodicity

Recall that an irreducible and aperiodic Markov chain (M_n) is ergodic if one of the two equivalent conditions holds:

1. For any $y \in \mathcal{S}$, when $M_0 = y$, the variable T_y^+,

$$T_y^+ = \inf\{k > 0 : M_k = y\} \tag{8.3}$$

   is integrable.
2. There exists an invariant probability distribution π, i.e.

$$\mathbb{E}_\pi(f(M_1)) = \mathbb{E}_\pi(f),$$

   holds for any non-negative function f on $\mathcal{S}$.

Notice that if $M_0 = y$, then $T_y^+ > 0$ and the variable T_y defined by Equation (8.2) is 0. The basic result on the ergodicity of Markov chains is presented in the next proposition.

PROPOSITION 8.5. *If F is a finite subset of $\mathcal{S}$ and*

$$T_F = \inf\{k \geq 0 : M_k \in F\},$$

if $g(x) = \mathbb{E}_x(T_F)$ is finite for any $x \in \mathcal{S}$ and $\mathbb{E}_x(g(M_1)) < +\infty$ for all $x \in F$, the Markov chain is ergodic.

PROOF. Since the variable T_F is almost surely finite, the Markov chain visits F infinitely often with probability 1. For $p \geq 1$, X_p is the element of F visited by (M_n) at its pth passage in F. In other words the sequence (X_n) is the induced Markov chain associated with the hitting time of F. Notice that the state space of this Markov chain is finite. A well-known (and elementary) result on irreducible finite Markov chains shows that (X_n) has a unique invariant probability distribution π_F (see Feller [**21**]). The hitting time T_F^+ of F is defined as

$$T_F^+ = \inf\{k > 0 : M_k \in F\},$$

and the non-negative measure π on $\mathcal{S}$ by

$$\pi(f) = \mathbb{E}_{\pi_F}\left(\sum_{i=0}^{T_F^+-1} f(M_i)\right), \tag{8.4}$$

for any bounded function f on $\mathcal{S}$.

The measure π is invariant for the Markov chain (M_n). Indeed, if x is an element of $\mathcal{S}$, Definition (8.4) used for the indicator function of the set $\{x\}$ gives the identity

$$\pi(x) = \mathbb{E}_{\pi_F}\left(\sum_{i=0}^{T_F^+-1} 1_{\{M_i=x\}}\right).$$

Hence

$$\pi(x)\mathbb{E}_x\left(f(M_1)\right) = \mathbb{E}_{\pi_F}\left(\sum_{i=0}^{+\infty} 1_{\{M_i=x,\ 0\le i<T_F^+\}}\mathbb{E}_x\left[f(M_1)\right]\right).$$

For $i \in \mathbb{N}$, the event $\{0 \le i < T_F^+, M_i = x\}$ being $\mathcal{F}_i$-measurable, thus the Markov property of (M_n) implies that

$$\begin{aligned}
\pi(x)\mathbb{E}_x\left(f(M_1)\right) &= \mathbb{E}_{\pi_F}\left(\sum_{i=0}^{+\infty} 1_{\{M_i=x,\ 0\le i<T_F^+\}}\mathbb{E}_{\pi_F}\left[f(M_{i+1}) \mid \mathcal{F}_i\right]\right)\\
&= \mathbb{E}_{\pi_F}\left(\sum_{i=0}^{+\infty} \mathbb{E}_{\pi_F}\left[1_{\{M_i=x,\ 0\le i<T_F^+\}} f(M_{i+1}) \,\Big|\, \mathcal{F}_i\right]\right)\\
&= \mathbb{E}_{\pi_F}\left(\sum_{i=0}^{T_F^+-1} 1_{\{M_i=x\}} f(M_{i+1})\right).
\end{aligned}$$

The measure π satisfies the relationship

$$\begin{aligned}
\mathbb{E}_\pi\left(f(M_1)\right) = \sum_{x\in\mathcal{S}} \pi(x)\mathbb{E}_x\left(f(M_1)\right) &= \mathbb{E}_{\pi_F}\left(\sum_{i=0}^{T_F^+-1} f(M_{i+1})\right)\\
&= \mathbb{E}_{\pi_F}\left(\sum_{i=1}^{T_F^+-1} f(M_i)\right) + \mathbb{E}_{\pi_F}\left[f\left(M_{T_F^+}\right)\right].
\end{aligned}$$

By definition of the sequence (X_n), $X_1 = M_{T_F^+}$ and, since π_F the invariant probability distribution of (X_n), one gets

$$\mathbb{E}_{\pi_F}\left[f\left(M_{T_F^+}\right)\right] = \mathbb{E}_{\pi_F}[f(X_1)] = \mathbb{E}_{\pi_F}[f(X_0)] = \mathbb{E}_{\pi_F}[f(M_0)].$$

Therefore, the identity $\mathbb{E}_\pi(f(M_1)) = \pi(f)$ holds, π is an invariant measure for the Markov chain (M_n).

By decomposing the first step of the Markov chain, one gets

$$\begin{aligned}\pi(1) &= \mathbb{E}_{\pi_F}\left(T_F^+\right)\\ &= 1 + \mathbb{E}_{\pi_F}\left(\mathbb{E}_{M_1}\left(T_F^+\right)1_{\{M_1\notin F\}}\right) = 1 + \mathbb{E}_{\pi_F}\left(\mathbb{E}_{M_1}(T_F)1_{\{M_1\notin F\}}\right)\end{aligned} \tag{8.5}$$

hence, the inequality

$$\pi(1) \leq 1 + \mathbb{E}_{\pi_F}\left(g(M_1)\right) \leq 1 + \sup_{x\in F}\mathbb{E}_x(g(M_1)) < +\infty.$$

holds. The irreducible Markov chain (M_n) has a finite invariant measure, it is therefore ergodic. □

Representation (8.4) of the invariant distribution is generalized in Propositions 10.2 and 10.3 of Chapter 10 in the framework of ergodic theory. The main criterion for the ergodicity of a Markov chain can now be established. Under the present form, it is due to Filonov[Fil89].

THEOREM 8.6. *If there exist a function $f : S \to \mathbb{R}_+$, constants $K, \gamma > 0$ and an integrable stopping time $\tau \geq 1$ such that*

a) $\mathbb{E}_x(f(M_\tau) - f(x)) \leq -\gamma\mathbb{E}_x(\tau)$ holds when $f(x) > K$,

by denoting $F = \{x : f(x) \leq K\}$, the hitting time T_F of F,

$$T_F = \inf\{k \geq 0 : f(M_k) \leq K\},$$

is integrable and

$$\mathbb{E}_x(T_F) \leq \frac{f(x)}{\gamma}, \tag{8.6}$$

for $x \notin F$. Moreover, if the set F is finite and

b) $\mathbb{E}_x(f(M_1)) < +\infty$ when $f(x) \leq K$,

then the Markov chain (M_n) is ergodic.

A function f satisfying the above condition a) is decreasing in average when the initial state is "far away". Such a function is called a *Lyapunov function* for the Markov chain (M_n).

This type of criterion using a Lyapunov function exists under various formulations. Filonov's presentation is probably the most achieved form of these ergodicity criteria. It covers most of practical cases. The situation is much different concerning transience criteria. See next section. The following corollary, known as Foster's criterion, is obtained by taking the trivial stopping time $\tau \equiv 1$.

[Fil89] Y. Filonov, *A criterion for the ergodicity of discrete homogeneous Markov chains*, Akademiya Nauk Ukrainskoi SSR. Institut Matematiki. Ukrainskii Matematicheskii Zhurnal **41** (1989), no. 10, 1421–1422.

COROLLARY 8.7. *If there exist a function $f : \mathcal{S} \to \mathbb{R}_+$ and constants K, $\gamma > 0$ such that*

1) $\mathbb{E}_x(f(M_1) - f(x)) \leq -\gamma$ *when* $f(x) > K$ *;*

2) $\mathbb{E}_x(f(M_1)) < +\infty$ *when* $f(x) \leq K$ *;*

3) the set $\{x \in \mathcal{S} : f(x) \leq K\}$ *is finite,*

the Markov chain (M_n) *is then ergodic.*

Condition a) in Theorem 8.6 is the main property to derive the ergodicity of the Markov chain. The integrability Condition b), is also important as the following example shows. It is related to the Markov chain associated with a discrete renewal process (see the section on renewal processes page 26). The Markov chain (M_n) is defined by

$$M_{n+1} = M_n - 1 \qquad \text{if } M_n > 0,$$
$$M_{n+1} = Z_n \qquad \text{if } M_n = 0,$$

where (Z_n) is an i.i.d. sequence of random variables. The function $f(x) = x$ satisfies the equation $\mathbb{E}_x(f(M_1)) - f(x) = -1$ when $x > 0$. Condition a) is therefore fulfilled. But if $M_0 = 0$, the variable Z_0 is just the hitting time of 0 for this Markov chain, hence it is ergodic only if Z_0 is integrable, which is precisely the additional condition b).

PROOF OF THEOREM 8.6. If (t_n) is the induced sequence associated with the stopping time τ (see Definition 19) and $\mathcal{F}^\tau = (\mathcal{F}_k^\tau)$,

$$\nu = \inf\{k \geq 0 : t_k \geq T_F\},$$

the set $\{\nu > k\} = \{t_k < T_F\}$ is $\mathcal{F}_k^\tau$-measurable since T_F is a stopping time with respect to the filtration $(\mathcal{F}_n)$ (recall that $\mathcal{F}_k^\tau = \mathcal{F}_{t_k}$). Consequently, ν is a stopping time with respect to the filtration $\mathcal{F}^\tau$.

Define $X_n = f(M_{t_n}) + \gamma t_n$. The variable (X_n) is $\mathcal{F}_n^\tau$-measurable and

$$\mathbb{E}_x(X_{n+1} \mid \mathcal{F}_n^\tau) = \mathbb{E}_x\left[f(M_{t_n+\tau(\theta_{t_n})}) + \gamma\left(t_n + \tau\left(\theta^{t_n}\right)\right) \;\middle|\; \mathcal{F}_n^\tau\right].$$

The strong Markov property shows the relationship

$$\mathbb{E}_x(X_{n+1} \mid \mathcal{F}_n^\tau) = \gamma t_n + \mathbb{E}_{M_{t_n}}\left[f(M_\tau) + \gamma\tau\right];$$

the inequality $f(M_{t_n}) > K$ is true on the event $\{\nu > n\} = \{t_n < T_F\}$, therefore

$$\mathbb{E}_x(X_{n+1} \mid \mathcal{F}_n^\tau) - X_n = \mathbb{E}_{M_{t_n}}\left[f(M_\tau) - f(M_{t_n}) + \gamma\tau\right] \leq 0$$

on the event $\{\nu > n\}$ according to Condition a). The sequence $(X_{\nu\wedge n})$ is therefore a non-negative supermartingale for the filtration $(\mathcal{F}_n^\tau)$. In particular, $\mathbb{E}_x(X_{\nu\wedge n}) \leq X_0$, hence

$$\mathbb{E}_x\left[f(M_{t_{\nu\wedge n}}) + \gamma t_{\nu\wedge n}\right] \leq f(x).$$

The function f being non-negative, monotone convergence Theorem gives the inequality $\mathbb{E}_x(t_\nu) \leq f(x)/\gamma$. Since $t_\nu \geq T_F$, one obtains the desired upper bound for the average hitting time of F,

$$\mathbb{E}_x(T_F) \leq \frac{f(x)}{\gamma},$$

for any $x \in S$. For $x \in F$, the above inequality and Condition b) yield

$$\mathbb{E}_x\left[\mathbb{E}_{M_1}(T_F)\right] \leq \mathbb{E}_x\left[\frac{f(M_1)}{\gamma}\right] < +\infty.$$

Proposition 8.5 implies that the Markov chain is ergodic. □

Relationship (8.6) is not really useful for $x \in F$, since $T_F = 0$ when $M_0 = x$. By taking

$$T_F^+ = \inf\{k > 0 : M_k \in F\},$$

when $F = \{x\}$, this variable is the hitting time T_x^+ of x defined by Equation (8.3). The variables T_F^+ and T_F coincide when $M_0 = x \notin F$ and $\mathbb{E}_x(T_F^+) \leq f(x)/\gamma$ in this case. For $x \in F$, the inequality

$$\mathbb{E}_x(T_F^+) \leq 1 + \mathbb{E}_x\left[\mathbb{E}_{M_1}(T_F)\right]$$

is obtained in the same way as Relationship (8.5). Finally, Relationship (8.6) shows the upper bound

$$\mathbb{E}_x(T_F^+) \leq 1 + \mathbb{E}_x\left[f(M_1)\right]/\gamma.$$

Moments of Hitting Times of Finite Subsets

The next proposition generalizes Inequality (8.6) of Theorem 8.6. It is due to Malyshev and Men'shikov [20]. It gives a criterion on the probability distribution of the Markov chain for the existence of moments of order $d \geq 1$ and exponential moments.

PROPOSITION 8.8. *Under Assumption 1) of Corollary 8.7,*

1. *if there is an $a \geq 1$ such that,*

$$\sup_{x\in S} \mathbb{E}_x\left(|f(M_1) - f(x)|^a\right) < +\infty,$$

then for $x \notin F$, there exists a constant C_1 such that

$$\mathbb{E}_x\left(T_F^a\right) \leq C(1 + f(x)^a).$$

2. *In the same way, if there is a $\lambda > 0$,*

(8.7) $$C = \sup_{x\in S} \mathbb{E}_x\left[\exp(\lambda|f(M_1) - f(x)|)\right] < +\infty,$$

then for $x \notin F$, there exists λ_1 and $\lambda_2 > 0$ depending only of the constants γ, C and λ such that

$$\mathbb{E}_x\left(e^{\lambda_1 T_F}\right) \leq \exp(\lambda_2 f(x)).$$

PROOF. Only part 2) of the proposition is shown; the proof of 1) is similar (see the indications below). For $n \in \mathbb{N}$ and $\lambda_1 \in]0, \lambda[$, one sets

$$Z_n = e^{\lambda_1 f(M_n)},$$

by induction Z_n is an integrable random variable: This is obvious for $n = 0$ and if it is true at rank $n-1$ then

$$\begin{aligned}\mathbb{E}(Z_n) &= \mathbb{E}\left(e^{\lambda_1 f(M_n)}\right) = \mathbb{E}\left(Z_{n-1}e^{\lambda_1(f(M_n)-f(M_{n-1}))}\right)\\ &= \mathbb{E}\left[Z_{n-1}\mathbb{E}_{M_{n-1}}\left(e^{\lambda_1(f(M_n)-f(M_{n-1}))}\right)\right]\end{aligned}$$

by the Markov property, Bound (8.7) gives the inequality $\mathbb{E}(Z_n) \leq C\mathbb{E}(Z_{n-1})$, hence Z_n is integrable. By using again the Markov property, one gets

$$\mathbb{E}(Z_{n+1} - Z_n \mid \mathcal{F}_n) = Z_n \mathbb{E}_{M_n}\left(e^{\lambda_1[f(M_1)-f(M_0)]} - 1\right).$$

For $x \in \mathbb{R}$, the elementary inequality

$$|e^x - 1 - x| \leq \frac{|x|^2}{2} e^{|x|},$$

shows the relationship

$$\mathbb{E}(Z_{n+1} - Z_n \mid \mathcal{F}_n) \leq Z_n \left[\lambda_1 E_{M_n}(\Delta) + \lambda_1^2 \mathbb{E}_{M_n}\left(\frac{\Delta^2}{2} e^{\lambda_1|\Delta|}\right)\right]$$

with $\Delta = f(M_1) - f(M_0)$. For $\lambda_0 < \lambda$, Bound (8.7) shows that

$$C_1 = \sup_{x\in S} \mathbb{E}_x\left(\frac{\Delta^2}{2} e^{\lambda_0|\Delta|}\right) < +\infty.$$

There exists a non-negative $\lambda_1 \leq \lambda_0$ sufficiently small such that the inequality $-\gamma\lambda_1 + \lambda_1^2 C_1 \leq -\lambda_1\gamma/2$ holds. Such a λ_1 is fixed. Condition a) of Corollary 8.7 and Bound (8.7) show that, on the event $\{T_F > n\}$, the inequalities

$$\mathbb{E}(Z_{n+1} - Z_n \mid \mathcal{F}_n) \leq -Z_n\gamma/2, \text{ and } \mathbb{E}(Z_{n+1} \mid \mathcal{F}_n) \leq Z_n(1 - \lambda_1\gamma/2)$$

hold. If λ_2 is such that $\exp(\lambda_2) = (1 - \lambda_1\gamma/2)$, then $\lambda_2 > 0$ and the last inequality can be rewritten as

$$\mathbb{E}(e^{\lambda_2(n+1)} Z_{n+1} \mid \mathcal{F}_n) \leq e^{\lambda_2 n} Z_n,$$

on the event $\{T_F > n\}$. The sequence $(e^{\lambda_2 n\wedge T_F} Z_{n\wedge T_F})$ is therefore a supermartingale, hence if $f(x) > K$,

$$\mathbb{E}_x\left(e^{\lambda_2 n\wedge T_F}\right) \leq \mathbb{E}_x\left(e^{\lambda_2 n\wedge T_F} Z_{n\wedge T_F}\right) \leq \mathbb{E}_x(Z_0) = e^{\lambda_1 f(x)},$$

one concludes by letting n go to infinity. □

Part 1) of Proposition 8.8 can be proved as follows. For $n \in \mathbb{N}$, Define $Z_n = (1 + \varepsilon(f(M_n) + n\gamma/2))^a$ and prove that Z_n is integrable for $n \in \mathbb{N}$. From the elementary inequality

$$(1+x)^a - 1 - ax \leq a(a-1)(1+|x|)^{a-2}|x|^2/2,$$

deduce that there exists some $\varepsilon > 0$ such that $(Z_{n\wedge T_F})$ is a supermartingale when $M_0 = x$ and $f(x) \geq K$.

8.3 Transience

This section is devoted to transience criteria for Markov chains. The first result gives a non-ergodicity criterion due to Tweedie.

PROPOSITION 8.9. *If there exists a non-constant function* $f : \mathcal{S} \to \mathbb{R}_+$ *and a constant* $K > 0$ *such that* $K \leq \sup\{f(x) : x \in \mathcal{S}\}$ *and*

a) $\mathbb{E}_x(f(M_1)) \geq f(x)$ *when* $f(x) \geq K$ *;*

b) $\sup_{x\in\mathcal{S}} \mathbb{E}_x\left(|f(M_1) - f(x)|\right) = C < +\infty$,

then the Markov chain (M_n) *cannot be ergodic.*

PROOF. The function f being non-constant, up to a change of K, it can be assumed that $\inf\{f(x) : x \in \mathcal{S}\} < K$ holds. The sequence $(f(M_n))$ is integrable, it is proved by induction:

$$\mathbb{E}_x\left(f(M_1)\right) \leq \mathbb{E}_x\left(|f(M_1) - f(x)|\right) + f(x) \leq C + f(x);$$

now, if for $n \geq 1$ the variable $f(M_n)$ is integrable,

$$\mathbb{E}_x\left(f(M_{n+1})\right) \leq \mathbb{E}_x\left(|f(M_{n+1}) - f(M_n)|\right) + \mathbb{E}_x(|f(M_n)|)$$

since

$$\mathbb{E}_x\left(|f(M_{n+1}) - f(M_n)|\right) = \mathbb{E}_x\left[\mathbb{E}_{M_n}\left(|f(M_1) - f(M_0)|\right)\right] \leq C,$$

by Condition b). The integrability is proved. If $\nu = \inf\{k \geq 0 : f(M_k) < K\}$, Condition a) expresses that the sequence $(f(M_{n\wedge\nu}))$ is a non-negative sub-martingale. In particular for $x \in \mathcal{S}$,

$$\mathbb{E}_x(f(M_{\nu\wedge n})) \geq f(x). \tag{8.8}$$

If the chain is ergodic, the variable ν is integrable and the convergence

$$f(M_{\nu\wedge n}) \overset{L_1}{\to} f(M_\nu)$$

holds as n tends to infinity. Indeed, the inequality

$$\begin{aligned}\mathbb{E}_x\left[|f(M_{\nu\wedge n}) - f(M_\nu)|\right] &\leq \mathbb{E}_x\left[\sum_{\nu\wedge n}^{\nu-1} |f(M_{k+1}) - f(M_k)|\right]\\ &= \mathbb{E}_x\left[\sum_{n}^{+\infty} \mathbb{E}_x\left(|f(M_{k+1}) - f(M_k)\,|\,\mathcal{F}_k\right) 1_{\{\nu>k\}}\right],\end{aligned}$$

and the Markov property show the inequality

$$\mathbb{E}_x\left[|f(M_{\nu\wedge n}) - f(M_\nu)|\right] \leq \mathbb{E}_x\left[\sum_{n}^{+\infty} \mathbb{E}_{M_k}(|f(M_1) - f(M_0)|)1_{\{\nu>k\}}\right].$$

Since the inequality $f(M_k) \geq K$ is satisfied on the event $\{\nu > k\}$, Condition b) shows that this last term is upper bounded by

$$C\,\mathbb{E}_x\left[\sum_{k\geq n} 1_{\{\nu>k\}}\right] = C\,\mathbb{E}_x\left((\nu - n)^+\right),$$

and thus converges to 0 as n tends to infinity. The L_1-convergence is thus true if the Markov chain is ergodic. By letting n go to infinity in Equation (8.8), one gets the inequality $\mathbb{E}_x(f(M_\nu)) \geq K$ for all $x \in \mathcal{S}$. If $f(x) \geq K$, this contradicts the definition of ν since $f(M_\nu) < K$. The proposition is proved. □

A condition on the jumps similar to condition b) cannot be avoided: If (M_n) is the Markov chain on $\mathbb{N}$ such that

- $p(0,1) = 1$;
- $p(n,n+1) = 1 - p(n,n-1) = p < 1/2, \qquad n \geq 1.$

By taking the function $g(x) = x$, for $x \geq 1$,

$$\mathbb{E}_x(g(M_1) - g(x)) = 2p - 1 < 0.$$

This function g satisfies the hypotheses of Theorem 8.6, this chain is thus ergodic. If, for $x \geq 0$, $f(x) = a^x$ with $a = (1-p)/p$, then it is easily seen that $\mathbb{E}_x(f(M_1)) = f(x)$ for $x \geq 1$. Condition a) is satisfied for this function f, nevertheless the Markov chain is ergodic.

Conditions a) and b) are not sufficient to imply the transience of the Markov chain. The symmetrical simple random walk on $\mathbb{Z}$ satisfies a) and b) with $f(x) = |x|$ and $K = 1$ but is recurrent.

The following theorem gives a simple criterion for transience. It is a version of a more general result due to Lamperti[Lam60] (who attributes his ideas to Doob).

Theorem 8.10. *If there exist a function* $f : \mathcal{S} \to \mathbb{R}_+$ *and constants* K, $\gamma > 0$ *such that* $\sup\{f(x) : x \in \mathcal{S}\} \geq K$ *and*

a) $\mathbb{E}_x(f(M_1) - f(x)) \geq \gamma$ *when* $f(x) \geq K$,

b) $\sup_{x\in\mathcal{S}} \mathbb{E}_x\left(|f(M_1) - f(x)|^2\right) < +\infty$,

then the Markov chain (M_n) *is transient.*

Proof. Since $\sup\{f(x) : x \in \mathcal{S}\} \geq K$, from Condition a) there exists $x_0 \in \mathcal{S}$ such that $f(x_0) \geq K + \gamma$. The transition matrix $q(\cdot,\cdot)$ is defined as follows,

$$q(x,y) = \mathbb{P}_x(M_1 = y), \quad f(x) \geq K, \qquad (8.9)$$
$$q(x,y) = \mathbb{P}_x(M_\nu = y) = \delta_{x_0}(y), \quad f(x) < K,$$

[Lam60] J. Lamperti, *Criteria for the recurrence or transience of stochastic process I.*, Journal of Mathematical Analysis and Applications **1** (1960), 314–330.

where ν is the hitting time of x_0 for the Markov chain. If the Markov chain (M_n) is transient, the variable ν is assumed to be almost surely finite. A Markov chain (N_n) associated with q satisfies

$$\mathbb{E}_x(f(N_1) - f(x)) \geq \gamma, \tag{8.10}$$

for *all* $x \in \mathcal{S}$, moreover

$$\begin{aligned} A = \sup_{x\in\mathcal{S}} \mathbb{E}_x \left[|f(N_1) - f(x)|^2\right] \\ \leq \sup_{x\in\mathcal{S}} \mathbb{E}_x \left[|f(M_1) - f(x)|^2\right] + (f(x_0) + K)^2 < +\infty. \end{aligned} \tag{8.11}$$

For $B > 0$ and $n \in \mathbb{N}$, if

$$Z_n = \frac{1}{B + f(N_n)},$$

and $(\mathcal{G}_n)$ denotes the filtration associated with the sequence (Z_n). By the Markov property of (N_n),

$$\mathbb{E}(Z_{n+1}|\mathcal{G}_n) - Z_n = \mathbb{E}_{N_n}\left[\frac{1}{B + f(N_1)} - \frac{1}{B + f(N_0)}\right] = \Delta(N_n), \tag{8.12}$$

where, for $x \in \mathcal{S}$, $\Delta(x) = \mathbb{E}_x\left[1/(B + f(N_1)) - 1/(B + f(x))\right]$. The identity

$$(B + f(x))\Delta(x) + \frac{\mathbb{E}_x(f(N_1) - f(x))}{B + f(x)} = \mathbb{E}_x\left[\frac{(f(N_1) - f(x))^2}{(B + f(N_1))(B + f(x))}\right]$$

is easily verified, Inequalities (8.10) and (8.11) give the upper bound

$$\Delta(x) \leq \frac{1}{(B + f(x))^2}\left(-\gamma + \frac{A}{B}\right).$$

Consequently, if B is large enough, the quantity $\Delta(x)$ is negative for all $x \in \mathcal{S}$. From Relationship (8.12), one gets that the sequence (Z_n) is a non-negative supermartingale. The function $g(x) = 1/(B + f(x))$ is therefore superharmonic for the Markov chain (N_n) (Proposition 8.3) and, it is non-constant on its state space, (N_n) is therefore transient (Proposition 8.4). Consequently, if $N_0 = x_0$, the set $\Omega_0 = \{N_n \neq x_0, \forall n \geq 1\}$ has a positive probability. Definition (8.9) of the transition matrix $q(\cdot,\cdot)$ shows that, on the set Ω_0, the sequence (N_n) does not visit the set $\{x : f(x) < K\}$. The transition matrices $p(\cdot,\cdot)$ and $q(\cdot,\cdot)$ are identical outside $\{x : f(x) < K\}$, one concludes that $\mathbb{P}_{x_0}(M_n \neq x_0, \forall n \geq 1) > 0$. The Markov chain (M_n) is transient. □

Condition a) is, of course, the central condition to have the transience of the Markov chain. A condition on jumps, such as Condition b), is nevertheless necessary: For $\alpha \geq 0$, the transition matrix $p(\cdot,\cdot)$ on $\mathbb{N} - \{0, 1\}$ is defined by

- $p(n, n + 1) = 1 - 1/(n^\alpha \log n)$ for $n \geq 2$,
- $p(n, 2) = 1/(n^\alpha \log n)$ for $n > 2$.

If $f(x) = x$, a Markov chain (M_n) associated with the stochastic matrix $p(\cdot,\cdot)$ satisfies the identity

$$\mathbb{E}_x(f(M_1) - f(x)) = 1 - \frac{1}{x^\alpha \log x} - \frac{x-2}{x^\alpha \log x}$$

for $x > 2$. There are two cases:

- $\alpha < 1$. Since
$$\lim_{x \to +\infty} \mathbb{E}_x(f(M_1) - f(x)) = -\infty,$$
by using Theorem 8.6, the ergodicity of the Markov chain is proved.
- $\alpha \geq 1$ and for x sufficiently large $\mathbb{E}_x(f(M_1) - f(x)) \geq 1/2$. If T is the return time of 2 starting from 2, its distribution is given by
$$\mathbb{P}(T \geq n) = \prod_2^{n+2} \left(1 - \frac{1}{k^\alpha \log k}\right) = \exp\left[-\sum_2^{n+2} \log\left(1 - \frac{1}{k^\alpha \log k}\right)\right],$$
for $n \in \mathbb{N}$.
When $\alpha > 1$, it is easily checked that $\mathbb{P}(T = +\infty) > 0$, the Markov chain is thus transient. If $\alpha = 1$, Condition a) of the last theorem is satisfied but $\mathbb{P}(T = +\infty) = 0$, the Markov chain is not transient but null recurrent.

Convergence Rate to Infinity

Condition a) of Theorem 8.11 suggests a growth rate of $f(M_n)$ greater than $n\gamma$. This result is not true in general. For example, if the set $\{x : f(x) < K\}$ is infinite, with positive probability the Markov chain may stay forever in this set. The following theorem gives a criterion in this domain.

THEOREM 8.11. *If there exist a function* $f : S \to \mathbb{R}_+$ *and constants* K, $\delta_0, \gamma > 0$ *such that* $\sup\{f(x) : x \in S\} \geq K$ *and*

a) $\mathbb{E}_x(f(M_1) - f(x)) \geq \gamma$ *when* $f(x) \geq K$,

b) $\sup_{x \in S} \mathbb{E}_x\left[\exp\left(\delta_0 \left|f(M_1) - f(x)\right|\right)\right] < +\infty$,

c) *the set* $\{x : f(x) \leq K + \gamma\}$ *is finite,*

then the Markov chain (M_n) *is transient and for* $x \in S$, $\mathbb{P}_x$*-almost surely*

$$\liminf_{n \to +\infty} \frac{f(M_n)}{n} \geq \gamma. \tag{8.13}$$

PROOF. The proof uses the same method as the proof of Theorem 8.10 : construct a related Markov chain such that Condition a) holds on the whole state space and then exhibit a convenient supermartingale.

The Markov chain is irreducible and the set $\{x : f(x) < K + \gamma\}$ is finite, hence, the exit time

$$\nu = \inf\{k \geq 0 : f(M_k) \geq K + \gamma\},$$

is almost surely finite. The transition matrix $q(\cdot,\cdot)$ is defined as follows :

(8.14)
$$q(x,y) = \mathbb{P}_x(M_1 = y), \quad f(x) \geq K,$$
$$q(x,y) = \mathbb{P}_x(M_\nu = y), \quad f(x) < K.$$

If (N_n) is a Markov chain associated with the matrix q, for $x \in \mathcal{S}$,

(8.15)
$$E_x(f(N_1) - f(x)) \geq \gamma.$$

If $M_0 = N_0$, the jump $|f(N_1) - f(N_0)|$ is $|f(M_1) - f(M_0)|$ if $f(M_0) \geq K$, otherwise it is bounded by the variable

$$\max_{x, f(x) \leq K} |f(M_1) - f(x)|,$$

which has an exponential moment of order δ_0 since the set $\{x : f(x) \leq K\}$ is finite by Condition c). Consequently,

(8.16)
$$A = \sup_{x \in \mathcal{S}} \mathbb{E}_x \left[\exp\left(\delta_0 \,|\, f(N_1) - f(N_0)|\right)\right] < +\infty.$$

For $n \geq 0$, $\delta > 0$ and if γ_1 is fixed such that $0 < \gamma_1 < \gamma$, the random variable Z_n defined by

$$Z_n = \exp\left[\delta(n\gamma_1 - f(N_n))\right]$$

verifies

$$\mathbb{E}_x(Z_{n+1} \,|\, \mathcal{G}_n) - Z_n = Z_n \mathbb{E}_x\left(\exp\left[\delta(\gamma_1 - f(N_{n+1}) + f(N_n))\right] - 1 \,\middle|\, \mathcal{G}_n\right),$$

if $(\mathcal{G}_n)$ is the filtration associated with the sequence (Z_n). The last equation and the Markov property of (N_n) imply that

(8.17)
$$\mathbb{E}_x(Z_{n+1} \,|\, \mathcal{G}_n) - Z_n = Z_n \mathbb{E}_{N_n}(\exp(\delta X) - 1),$$

with $X = \gamma_1 - f(N_1) + f(N_0)$. For $t \in \mathbb{R}$, the elementary inequalities

$$e^{\delta t} - 1 + \delta t \leq \frac{\delta^2 |t|^2}{2} e^{\delta |t|} \leq \delta^2 \left(C + e^{\delta_0 |t|}\right)$$

are easily checked for some constant C independent of $\delta < \delta_0/2$ and $t \in \mathbb{R}$. For $x \in \mathcal{S}$, using the last inequality and Relationships (8.15) and (8.16), one gets the upper bound,

$$\mathbb{E}_x(\exp(\delta X) - 1) \leq \delta(\gamma_1 - \gamma + \delta(C + A)).$$

If δ is chosen sufficiently small, the left hand side of this inequality is therefore negative for all $x \in \mathcal{S}$.

Relationship (8.17) shows that the sequence (Z_n) is a non-negative supermartingale, and consequently, it converges almost surely to a finite limit. (Theorem A.3 page 352). In particular, the sequence $(n\gamma_1 - f(N_n))$ is almost surely bounded, hence almost surely,

$$\liminf_{n \to +\infty} \frac{f(N_n)}{n} \geq \gamma_1.$$

By letting γ_1 go to γ, γ_1 can be replaced by γ in this inequality.

To prove the same result for the Markov chain (M_n), the subsequence (Y_n) of (M_n) is defined by induction as follows,

- $Y_0 = M_0$;
- for $n \geq 1$, $Y_n = M_p$ for some $p \in \mathbb{N}$,
  - $Y_{n+1} = M_{p+1}$ if $f(Y_n) \geq K$;
  - otherwise, $Y_{n+1} = M_{\nu_p}$ where $\nu_p = \inf\{k > p : f(Y_n) \geq K + \gamma\}$.

According to Definition (8.14) the sequence (Y_n) is a Markov chain whose transition matrix is q therefore, almost surely

$$\liminf_{n\to+\infty} \frac{f(Y_n)}{n} \geq \gamma.$$

In particular, almost surely, $f(Y_n) \geq K$ after some rank, since $p(x,\cdot)$ and $q(x,\cdot)$ coincide when $f(x) \geq K$, there exist two (random) integers p_0 and n_0 such that $Y_n = M_{p_0+n}$ for all $n \geq n_0$. Thus, the sequence (M_n) satisfies Relationship (8.13). □

8.4 Ergodicity of Markov Processes

In this section, $(X(t))$ is an irreducible Markov jump process on the countable state space $\mathcal{S}$ whose $\mathcal{Q}$-matrix is given by $Q = (q_{ij})$ and $(\mathcal{F}_t)$ denotes the associated filtration. If f is a function on $\mathcal{S}$, for $x \in \mathcal{S}$, $Q(f)(x)$ is given by

$$Q(f)(x) = \sum_{i\in\mathcal{S}} q_{xi}(f(i) - f(x)),$$

provided that the series converge. Appendix B gives the main definitions and results concerning these Markov processes. An ergodicity criteria is proved for these Markov jump processes. If, formally, by considering the embedded Markov chain, the results of Section 8.2 could be used, it is in practice much more convenient to work directly with the continuous time processes (think of the case of Jackson networks for example). In the discrete time case, results consider quantities such that $\mathbb{E}_x(f(M_1)) - f(x)$ for a convenient function f. In the continuous time case, it is quite natural to study the quantity

$$\left.\frac{d}{dt}\mathbb{E}_x\left[f(X(t))\right]\right|_{t=0}$$

which is given by $Q(f)(x)$ according to Relationship (B.1). By analogy with discrete time, one can guess that if $Q(f)(x)$ is negative for x outside some finite subset, then the Markov process is recurrent. Proposition 8.14 is the rigorous formulation of such a result. The following proposition is the continuous time version of Proposition 8.5.

PROPOSITION 8.12. *If T_F is the hitting time of a finite subset F of $\mathcal{S}$,*

$$T_F = \inf\{s \geq 0 : X(s) \in F\},$$

and if $g(x) = \mathbb{E}_x(T_F)$ is finite on $\mathcal{S}$ and $\mathbb{E}_x[g(X(1))] < +\infty$ for any $x \in F$, then the Markov process $(X(t))$ is ergodic.

PROOF. The proof uses the same arguments as the proof of Proposition 8.5. The increasing sequence (t_n) of stopping times is defined by induction: $t_0 = 0$ and

$$t_n = \inf\{s > t_{n-1} + 1 : X(s) \in F\}$$

for $n \geq 1$. Notice the translation by 1 in the definition of these stopping times, in particular $t_{n+1} - t_n \geq 1$ for $n \in \mathbb{N}$. The Markov property of $(X(t))$ shows that the sequence $(X(t_n))$ is a Markov chain on F. It is easily seen that this Markov chain is also irreducible, π_F denotes its invariant probability distribution. If f is a non-negative function on $\mathcal{S}$, π is the measure on $\mathcal{S}$ defined by

$$\begin{aligned}\mathbb{E}_\pi(f) &= \int_{\mathcal{S}} f(x)\,\pi(dx)\\ &= \mathbb{E}_{\pi_F}\left(\int_0^{t_1} f(X(s))\,ds\right) = \sum_{x\in F} \pi_F(x)\mathbb{E}_x\left(\int_0^{t_1} f(X(s))\,ds\right).\end{aligned}$$

It is sufficient to prove that this measure is finite and invariant with respect to the Markov process $(X(t))$, i.e. $\mathbb{E}_\pi[f(X(t))] = \mathbb{E}_\pi(f)$ for all $t \geq 0$.

The Markov property of $(X(t))$ gives

$$\begin{aligned}\pi(\mathcal{S}) = \mathbb{E}_{\pi_F}(t_1) &= 1 + \mathbb{E}_{\pi_F}\left[\mathbb{E}_{X(1)}(T_F)\right]\\ &= 1 + \mathbb{E}_{\pi_F}\left[g(X(1))\right] \leq 1 + \sup_{x\in F} \mathbb{E}_x\left[g(X(1))\right] < +\infty.\end{aligned}$$

The measure π has therefore a finite mass.

If f is a non-negative function on $\mathcal{S}$ and $t \geq 0$,

$$\begin{aligned}\mathbb{E}_\pi[f(X(t))] &= \sum_{x\in\mathcal{S}} \pi(x) E_x[f(X(t))]\\ &= \sum_{x\in\mathcal{S}} \mathbb{E}_{\pi_F}\left(\int_0^{t_1} 1_{\{X(s)=x\}}\,ds\right) E_x[f(X(t))]\\ &= \sum_{x\in\mathcal{S}} \int_0^{+\infty} \mathbb{E}_{\pi_F}\left(1_{\{X(s)=x,\, t_1>s\}}\right) E_x[f(X(t))]\,ds.\end{aligned}$$

The event $\{X(s) = x,\, t_1 > s\}$ being $\mathcal{F}_s$-measurable, using again the Markov property, one gets

$$\begin{aligned}\mathbb{E}_\pi[f(X(t))] &= \sum_{x\in S}\mathbb{E}_{\pi_F}\left(\int_0^{+\infty} 1_{\{X(s)=x,\, t_1>s\}} f(X(t+s))\,ds\right)\\ &= \mathbb{E}_{\pi_F}\left(\int_0^{t_1} f(X(t+s))\,ds\right) = \mathbb{E}_{\pi_F}\left(\int_t^{t+t_1} f(X(s))\,ds\right).\end{aligned}$$

Hence, the identity

$$\mathbb{E}_\pi[f(X(t))] = \mathbb{E}_{\pi_F}\left(\int_t^{t_1} f(X(s))\,ds\right) + \mathbb{E}_{\pi_F}\left(\int_{t_1}^{t+t_1} f(X(s))\,ds\right) \tag{8.18}$$

holds. By the strong Markov property of $(X(t))$ (Proposition B.1),

$$\mathbb{E}_{\pi_F}\left(\int_{t_1}^{t+t_1} f(X(s))\,ds\right) = \mathbb{E}_{\pi_F}\left[\mathbb{E}_{X(t_1)}\left(\int_0^{t} f(X(s))\,ds\right)\right],$$

since π_F is the invariant probability distribution of the Markov chain $(X(t_n))$. If the distribution of the variable $X(0)$ is π_F, the distribution of $X(t_1)$ is also π_F, therefore

$$\mathbb{E}_{\pi_F}\left(\int_{t_1}^{t+t_1} f(X(s))\,ds\right) = \mathbb{E}_{\pi_F}\left(\int_0^{t} f(X(s))\,ds\right).$$

Identity (8.18) gives finally

$$\mathbb{E}_\pi[f(X(t))] = \mathbb{E}_{\pi_F}\left(\int_0^{t_1} f(X(s))\,ds\right) = \mathbb{E}_\pi(f),$$

and consequently the invariance of the measure π for the Markov process $(X(t))$. The proposition is proved. □

THEOREM 8.13. *If there exist a function $f : S \to \mathbb{R}_+$, constants K, $\gamma > 0$ and an integrable stopping time τ such that for $f(x) > K$,*

$$\mathbb{E}_x\left[f(X(\tau))\right] - f(x) \le -\gamma\mathbb{E}_x(\tau),$$

if $F = \{x : f(x) \le K\}$, the hitting time T_F of F is integrable and

$$\mathbb{E}_x(T_F) \le f(x)/\gamma \tag{8.19}$$

for $x \notin F$. Moreover, if F is a finite subset and $\mathbb{E}_x[f(X(1))] < +\infty$ for $x \in S$, then the Markov process $(X(t))$ is ergodic.

PROOF. The proof is completely identical to the proof of Theorem 8.6 (by using the ergodicity criterion of Proposition 8.12). □

The constant 1 which appears in the integrability condition of $g(X(1))$ in Proposition 8.12 or of $f(X(1))$ in Theorem 8.13 can, of course, be replaced by any arbitrary constant. The next proposition presents an ergodicity criterion using explicitly the Q-matrix of the Markov process.

PROPOSITION 8.14. *If there exist a function $f : S \to \mathbb{R}_+$ and constants K, $\gamma > 0$ such that*

a) if $f(x) > K$, then

$$Q(f)(x) \leq -\gamma;$$

b) the random variables $\sup\{f(X(s)) : s \leq 1\}$ and $\int_0^1 |Q(f)(X(s))|\, ds$ are integrable;

c) the set $F = \{x : f(x) \leq K\}$ is finite,

then the Markov process $(X(t))$ is ergodic.

PROOF. If T_F is the hitting time of F for the process $(X(t))$, the variable $\tau = T_F \wedge 1$ is clearly a stopping time and the process

$$\left(f\left(X(\tau \wedge t)\right) - f(X(0)) - \int_0^{\tau\wedge t} Q(f)(X(s))\, ds \right)$$

a local martingale (Proposition B.4 page 364). The integrability assumptions b) and Proposition A.7 show that this is in fact a martingale. If $x \notin F$ and $t \geq 0$, one thus gets

$$\mathbb{E}_x \left[f\left(X(\tau \wedge t)\right)\right] - f(x) = \mathbb{E}_x \left(\int_0^{\tau\wedge t} Q(f)(X(s))\, ds \right).$$

By letting t go to infinity, Condition b) implies that Lebesgue's Theorem can be used to obtain the identity

$$\begin{aligned} \mathbb{E}_x \left[f\left(X(\tau)\right)\right] - f(x) &= \mathbb{E}_x \left(\int_0^{\tau} Q(f)(X(s))\, ds \right) \\ &\leq -\gamma \mathbb{E}_x(\tau), \end{aligned}$$

by Condition a). It is sufficient to apply the last theorem to conclude. □

8.5 Some Applications

The $M/G/1$ Queue. If L_n is the number of customers in this queue at time T_n of the nth departure, then (L_n) is an irreducible Markov chain such that

$$L_n = L_{n-1} + \mathcal{N}_\lambda([T_n, T_n + \sigma_n[) - 1,$$

when $L_{n-1} > 0$, where σ_n is the service of the nth customer and $\mathcal{N}_\lambda$ a Poisson point process with intensity λ. On the event $\{L_0 > 0\}$, the relation

$$\mathbb{E}(L_1 - L_0 \mid L_0) = \mathbb{E}(\mathcal{N}_\lambda([0, \sigma[)) - 1 = \lambda\mathbb{E}(\sigma) - 1,$$

holds. In particular the identity function is a Lyapunov function if $\lambda\mathbb{E}(\sigma) < 1$, the Markov chain is thus ergodic in this case. Conversely, if $\lambda\mathbb{E}(\sigma) > 1$, there exists K such that $\lambda\mathbb{E}(\sigma \wedge K) > 1$. If the services (σ_n) are replaced by the services $(\sigma_n \wedge K)$, it is easy to check that the corresponding Markov chain $(\widetilde{L}_n)$ is a lower bound for the sequence (L_n), i.e. $\widetilde{L}_n \leq L_n$ for all $n \geq 1$. Since the jumps of $(\widetilde{L}_n)$ have a bounded second moment, Theorem 8.11 applied with Lyapunov function $f(x) = x$ shows that the Markov chain is transient. Consequently, (L_n) is also transient in this case.

Transience of the ALOHA Protocol. This model is a classical example of a communication protocol in a distributed environment. Transmitters share a communication channel, they receive messages that must be sent on this channel. If more than two messages are sent in the same time unit, there is a collision of messages on the channel, they are lost and the corresponding transmitters know that transmission failed (they can listen the collision). Transmitters cannot cooperate to share the channel (imagine they are spread throughout islands), therefore they cannot queue *globally* messages to be transmitted. At the beginning of every time unit, if they have a message to transmit, they have to decide if they try to send the message or not. (Clearly enough, to avoid repetitive collisions, they cannot try to send a message every time unit).

ALOHA is a simple algorithm that gives a (partial) solution to this decentralized problem. At the beginning of every time unit, a transmitter having a message tosses a coin. If the outcome is head (with probability p) it tries to sent the message, otherwise it does nothing. The number of new messages arrived during the nth time unit is denoted by a_n and the sequence (a_n) is assumed to be i.i.d. with $\mathbb{E}(a_0) > 0$. The total number of messages waiting for transmission at time n is denoted by L_n. The sequence (L_n) is clearly a Markov chain whose transitions are given by

$$L_1 = L_0 + a_0 - 1_{\left\{B_1^1 + B_2^1 + \cdots + B_{L_0}^1 = 1\right\}},$$

where (B_i^1) are i.i.d Bernoulli variables with parameter $p < 1$ (the case $p = 1$ is clearly trivial). If $n_0 \in \mathbb{N}$ is such that for $n \geq n_0$, $np(1-p)^{n-1} \leq \mathbb{E}(a_0)/2$, then

$$\mathbb{E}(L_1 - L_0 \mid L_0 \geq n_0) = \mathbb{E}(a_0) - np(1-p)^{n-1} \geq \mathbb{E}(a_0)/2.$$

As in the last example, by replacing (a_i) by $(a_i \wedge K)$, it is then easily seen that the Markov chain (L_n) is transient. This communication protocol is unstable independently of the value of p. Moreover, Theorem 8.11 shows that

$$\liminf_{n \to +\infty} L_n/n \geq \mathbb{E}(a_0).$$

Since $L_n \leq \sum_0^n a_i$, the law of large numbers shows that $\mathbb{P}$-almost surely,

$$\limsup_{n \to +\infty} L_n/n \leq \mathbb{E}(a_0).$$

The sequence (L_n/n) converges almost surely to $\mathbb{E}(a_0)$.

The $G/M/k$ FIFO Queue. For $n \geq 0$, L_n is the total number of customers (i.e. in the queue or being served) at the arrival of the nth customer. Clearly (L_n) is an irreducible Markov chain. The variables $\mathcal{N}_\mu^i$, $1 \leq i \leq k$ denote independent Poisson point process with intensity μ. Distances between the points of $\mathcal{N}_\mu^i$ represent the values of services delivered by the ith server. The sequence of arrivals of customers is denoted by (t_n) with $t_0 = 0$, it is assumed that the variables $(t_{n+1} - t_n)$ is i.i.d. and that t_1 has a moment of order 2.

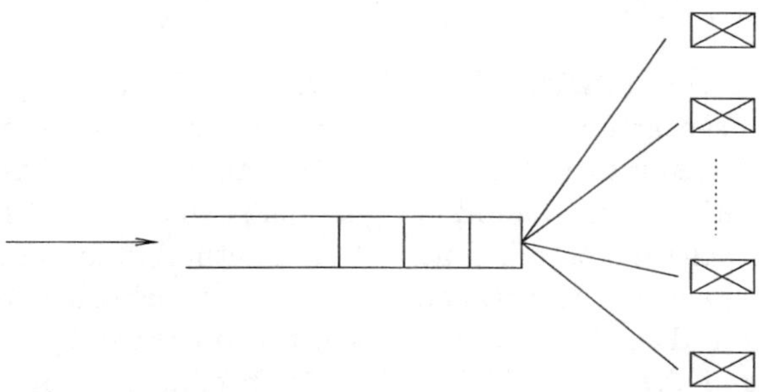

FIG. 8.1. The $G/M/k$ Queue

The transitions of this Markov chain can be described as follows,

$$L_{n+1} \leq k \vee \left(L_n + 1 - \sum_{i=1}^{k} N_\mu^i \left(]t_n, t_{n+1}]\right) \right), \tag{8.20}$$

for $n \in \mathbb{N}$. Indeed, if the queue is never empty between t_n and t_{n+1}, this implies that none of the servers has been idle during that time, therefore

$$L_{n+1} = L_n + 1 - \sum_{i=1}^{k} \mathcal{N}_\mu^i \left(]t_n, t_{n+1}]\right).$$

Otherwise, if at least one of the servers is idle at some moment between t_n and t_{n+1}, the queue is then necessarily empty at that time and remains so until next arrival, at time t_{n+1}, consequently $L_{n+1} \leq k$. Relationship (8.20) is proved. The inequality

$$\mathbb{E}(L_1 - L_0 \mid L_0 = x) \leq \mathbb{E}\left[\left(1 - \sum_{i=1}^{k} \mathcal{N}_\mu^i \left(]0, t_1]\right) \right) \vee (k - x) \right]$$

and Lebesgue's Theorem show the relation

$$\lim_{x \to +\infty} \mathbb{E}(L_1 - L_0 \mid L_0 = x) \leq 1 - k\mu\mathbb{E}(t_1),$$

thus, by Theorem 8.6, the Markov chain is ergodic if $1 - k\mu\mathbb{E}(t_1) < 0$.

Similarly, it is easy to check that, for $n \in \mathbb{N}$,

$$L_{n+1} \geq L_n + 1 - \sum_{1}^{k} \mathcal{N}_\mu^i \left(]t_n, t_{n+1}]\right),$$

hence

$$L_{n+1} - L_n \geq 1 - \sum_{i=1}^{k} \mathcal{N}_\mu^i\left(]t_n, t_{n+1}]\right).$$

When $1/\mu\mathbb{E}(t_1) < k$, the Markov chain (L_n) is ergodic and it is transient when $1/\mu\mathbb{E}(t_1) > k$.

A Single Server Queue with Variable Arrivals. Up to now, the stability of systems has been analyzed with a convenient Lyapunov function f for which the instantaneous drift $f(M_{n+1}) - f(M_n)$ has a negative average when $f(M_n)$ is sufficiently large. The following model is an example of a situation where the Markov chain is ergodic but, but for a natural function $f = \|\cdot\|$, the average of the instantaneous drift can be positive for arbitrary large states.

The model is a discrete time queue. Customers arrive by i.i.d. batches at the beginning of the time units. Each of them requires a time unit to be served. The queue has two basic states: The "normal" state where the size of the batches are distributed as some random variable a_0 and a "perturbated" state where this size is distributed as a random variable a_1. When the state is normal, it becomes perturbated with probability $1 - \alpha_0$ at the next time unit. In a perturbated state, the state queue becomes normal with probability $1 - \alpha_1$.

A Markov chain describing this queue is given by $M_n = (I_n, L_n)$ where I_n is the state of the queue at time $t = n$ (0, normal and 1, perturbated) and L_n is the number of customers at that time. The transitions of the second coordinate are given by $L_1 = l + a_i - 1_{\{l>0\}}$, when $(I_0, L_0) = (i, l)$ and for $i = 0, 1$, the variable a_i denotes a random variable with the same distribution as the number of arrivals in a time unit when the state of the queue is i.

If $(I_0, L_0) = (i, l)$, τ denotes the first time n that (I_n) reaches i and $I_{n-1} = 1 - i$ and f is the function defined by $f(i, l) = l$. The variable τ is clearly a stopping time and can be expressed as $1 + G_0 + 1 + G_1$ where G_0 and G_1 are two independent random variables with a geometric distribution with parameter α_0 and α_1. A simple upper bound gives the inequality

$$\begin{aligned} f(M_\tau) &\leq \sum_{k=1}^{1+G_0} a_{0,k} + \sum_{k=1}^{1+G_1} a_{1,k} + (L_0 - \tau)^+ \\ &= f(M_0) + \sum_{k=1}^{1+G_0} (a_{0,k} - 1) + \sum_{k=1}^{1+G_1} (a_{1,k} - 1) + (\tau - L_0)^+, \end{aligned}$$

where, for $i = 0, 1$ and $k \geq 1$, $a_{i,k}$ denotes the number of arrivals during the kth time unit when the state of the queue is i. For $\varepsilon > 0$, if K is chosen such that $\mathbb{E}((\tau - l)^+) < \varepsilon$ for $l \geq K$, by integrating the last inequality, one gets,

$$\mathbb{E}_{(i,l)}(f(M_\tau) - f(M_0)) \leq \frac{\mathbb{E}(a_0) - 1}{1 - \alpha_0} + \frac{\mathbb{E}(a_1) - 1}{1 - \alpha_1} + \varepsilon,$$

for $l \geq K$. Consequently, if

$$\frac{\mathbb{E}(a_0) - 1}{1 - \alpha_0} + \frac{\mathbb{E}(a_1) - 1}{1 - \alpha_1} < 0,$$

the Markov chain (I_n, L_n) is ergodic.

For this queue, the expected value of the instantaneous drift is given by

$$\mathbb{E}(L_1 - L_0 \mid I_0 = i, L_0 > 0) = \mathbb{E}(a_i) - 1, \tag{8.21}$$

for $i = 0, 1$. The ergodicity condition which has been obtained is much weaker that the conditions $\mathbb{E}(a_0) < 1$ and $\mathbb{E}(a_1) < 1$ which could be suggested by Relationships (8.21). When the queue is ergodic, the average drift on a time unit is not necessarily negative, even if the initial state is large. In any case, the average drift on a sufficiently long period is always negative when the initial state is large.

8.6 The Classical Version of Lyapunov's Theorem

If W is an open set of $\mathbb{R}^n$ containing 0 and $H : W \to \mathbb{R}^n$ a continuous function such that

$$H(0) = 0,$$

the maximal solution of the ordinary differential equation (i.e. a solution defined on the largest possible interval of $\mathbb{R}_+$)

$$X'(t) = H(X(t))$$

such that $X(0) = x \in W$ is denoted by $(X(x,t))$. The assumption on H implies that 0 is an equilibrium point of this differential equation, i.e. $X(0,t) = 0$ for all $t \geq 0$. The following theorem gives a criterion for the stability of 0 : If the initial value x of $(X(x,t))$ is not too far from 0, then the function $t \to X(x,t)$ converges to 0 as t tends to infinity.

THEOREM 8.15. *If there exist a neighborhood U of 0 and a continuous function $f : U \to \mathbb{R}$ differentiable on $U - \{0\}$ such that*

$$f(0) = 0 \text{ and } f(x) > 0, \text{ if } x \in U - \{0\},$$

$$S(H(x))f'(x) < 0 \text{ for } x \in U - \{0\}, \tag{8.22}$$

where $S(y)$ is the sum of the coordinates of $y \in \mathbb{R}^n$, the point 0 is a stable equilibrium point, i.e. there exists a neighborhood U_1 of 0 such that for any $x \in U_1$,

$$\lim_{t \to +\infty} X(x,t) = 0.$$

Condition (8.22) shows that, if the initial point is in a neighborhood of 0, the function $t \to f(X(t))$ has a negative derivative. It forces $X(t)$ to converge to 0. If $f(M_1) - f(x)$ is interpreted as the derivative of $n \to f(M_n)$ at $n = 0$, Condition b) of Corollary 8.7 simply states that this derivative is *in average* less than $-\gamma$ outside a finite set.

9. Rescaled Markov Processes and Fluid Limits

Contents

9.1 Introduction

It is in general quite difficult to have a satisfactory description of an ergodic Markov process describing a stochastic network. When the dimension of the state space d is greater than 1, the geometry complicates a lot any investigation : Analytical tools of Chapter 2 for dimension 1 cannot be easily generalized to higher dimensions. Note that the natural order on the real line plays an important role for Wiener-Hopf methods. The example of queueing networks seen in Chapter 4 for which the stationary distribution has a product form should be seen as an interesting exception, but an exception. In the same way as in Chapter 3, it is possible nevertheless to get some insight on the behavior of these processes through some limit theorems. In this chapter, limit results consist in speeding up time and scaling appropriately the process itself with some parameter. The behavior of such rescaled stochastic processes is analyzed when the scaling parameter goes to infinity. In the limit one gets a sort of caricature of the initial stochastic process which is defined as a *fluid limit* (see the rigorous definition below). As it will be seen, a fluid limit keeps the main characteristics of the initial stochastic process while some stochastic fluctuations of second order vanish with this procedure. In "good cases", a fluid limit is a deterministic function, solution of some ordinary differential equation. As it can be expected, the general situation is somewhat more complicated. These ideas of rescaling stochastic processes have emerged recently in the analysis of stochastic networks, to study their ergodicity properties in particular. See Rybko and Stolyar[RS92] for example. In statistical physics, these methods are quite classical, see Comets[Com91].

The chapter is organized as follows: Section 9.2 gives the basic definitions concerning the scaling of the stochastic process, fluid limits in particular are introduced. Several examples are presented and discussed. Section 9.3 introduces a large class of Markov processes for which general results on the existence of fluid limits hold. Section 9.4 investigates the relation between fluid limits and solutions of some Skorohod problem (see Appendix D). Section 9.5 establishes a relation between ergodicity of the initial stochastic process and stability of fluid limits at 0. An interesting example of a null-recurrent Markov process having fluid limits converging to 0 concludes the section. Section 9.6 studies fluid limits of Markov processes having a subset of their coordinates at equilibrium.

In the following, $(X(x,t))$ denotes an irreducible càdlàg Markov jump process on a countable state space $\mathcal{S}$ starting from $x \in \mathcal{S}$, i.e. such that $X(x,0) = x \in \mathcal{S}$. The notation $(X(t))$ can be used when the dependence on the initial point is not ambiguous. As before, $\mathcal{N}_\xi(\omega, dx)$, $\omega \in \Omega$, denotes a Poisson point process on $\mathbb{R}$ with parameter $\xi \in \mathbb{R}_+$, all Poisson processes used are assumed to be a priori independent. The topology on the space of probability distributions induced by the Skorohod topology on the space of càdlàg functions $D([0,T], \mathbb{R}^d)$ is used. The reading of Appendix C.2 is recommended, the definition and the main results on this topology are briefly recalled there.

Note that the assumption of a countable state space forbids to consider queueing networks with general service distributions or general interarrival distributions. This is not a real restriction since, in general, fluid limits do not really depend on distributions but on their averages. Hence the assumption that all distributions considered are exponential is only formally restrictive. As usual, it simplifies much the description of the underlying Markov process: Forward recurrence times vectors for interarrival intervals and services can be safely ignored. Note that among classical models there are some exceptions

[RS92] A.N. Rybko and A.L. Stolyar, *On the ergodicity of random processes that describe the functioning of open queueing networks*, Problems on Information Transmission **28** (1992), no. 3, 3–26.

[Com91] F. Comets, *Limites hydrodynamiques*, Astérisque (1991), no. 201-203, Exp. No. 735, 167–192 (1992), Séminaire Bourbaki, Vol. 1990/91.

to this fact: The $G/G/1$ Processor-Sharing queue has a fluid limit which depends on the distribution of services; see Jean-Marie and Robert[JMR94].

Since this topic is quite recent and still evolving (some important questions are still open, see Section 9.7 for a quick survey), this chapter is only an introduction to fluid limits. Generally speaking, the main goal of this chapter is to advertise the use of scaling methods in the study of complex Markov processes. For this reason, simple examples are scattered in the text to illustrate some of the problems encountered.

A different scaling is presented in Section 6.5 of Chapter 6 for the simple case of the $M/M/\infty$ queue, see Hunt and Kurtz[HK94] for a presentation of the general case. For this scaling, some of the methods of this chapter can be used (martingale parts of the associated Markov processes also vanish asymptotically for example).

9.2 Rescaled Markov Processes

Throughout this chapter, it is assumed that the state space $\mathcal{S}$ can be embedded in a subset of some normed space ($\mathbb{R}^d$ for example), $\|\cdot\|$ denotes the associated norm.

Definition 21. *For $x \in \mathcal{S}$, $(\overline{X}(x,t))$ denotes the process $(X(x,t))$ renormalized so that for $t \geq 0$,*

$$\overline{X}(x,t) = \frac{1}{\|x\|} X\left(x, \|x\| t\right).$$

As usual, if there is no ambiguity on x, the notation $(\overline{X}(t))$ will be used.

The scaling consists in speeding up time by the norm of the initial state and the space variable by its inverse. The rescaled process starts from $x/\|x\|$, a state of norm 1.

Only continuous time Markov processes are considered in this chapter. In discrete time, if (X_n) is a Markov chain, the corresponding rescaled process can be also defined by

$$\overline{X}(x,t) = \frac{X_{\lfloor \|x\| t \rfloor}}{\|x\|},$$

if $X_0 = x \in \mathcal{S}$ and $t \in \mathbb{R}_+$, where $\lfloor y \rfloor$ is the integer part of $y \in \mathbb{R}$. Most of the results of this chapter are also valid in this case.

[JMR94] A. Jean-Marie and Ph. Robert, *On the transient behavior of some single server queues*, Queueing Systems, Theory and Applications **17** (1994), 129–136.

[HK94] P.J. Hunt and T.G Kurtz, *Large loss networks*, Stochastic Processes and their Applications **53** (1994), 363–378.

9.2.1 Fluid Limits

Definition 22. *A* fluid limit *associated with the Markov process* $(X(t))$ *is a stochastic process which is one of the limits of the process*

$$(\overline{X}(x,t)) = \left(\frac{X(x, \|x\|t)}{\|x\|}\right)$$

when $\|x\|$ *goes to infinity.*

Strictly speaking, if $\mathbb{Q}_x$ is the distribution of $(\overline{X}(x,t))$ on the space of càdlàg functions $D(\mathbb{R}_+, \mathcal{S})$, a fluid limit is a probability distribution $\widetilde{\mathbb{Q}}$ on $D(\mathbb{R}_+, \mathbb{R}^d)$ such that

$$\widetilde{\mathbb{Q}} = \lim_n \mathbb{Q}_{x_n}$$

for some sequence (x_n) of $\mathcal{S}$ whose norm converges to infinity. By choosing an appropriate probability space, it can be represented as a càdlàg stochastic process $(W(t))$ whose distribution is $\widetilde{\mathbb{Q}}$. A fluid limit is thus an asymptotic description of sample paths of a Markov process with a large initial state.

9.2.2 Examples

A Random Walk on $\mathbb{Z}$. For $\lambda > 0$, $\mathcal{N}_\lambda$ is a Poisson process on $\mathbb{R}$ with intensity λ (as usual $\mathcal{N}_\lambda(A)$ denotes the number of points of Poisson process $\mathcal{N}_\lambda$ in a Borelian subset A of $\mathbb{R}$). Let (Y_i) be an i.i.d. sequence of integer valued random variables having a finite second moment. For $x \in \mathbb{Z}$, the process $(X(x,t))$ defined by

$$X(x,t) = x + \sum_{i=1}^{\mathcal{N}_\lambda(]0,t])} Y_i$$

is clearly a Markov process. The associated rescaled process $(\overline{X}(x,t))$ is given by

$$\overline{X}(x,t) = \frac{1}{|x|}\left(x + \sum_{i=1}^{\mathcal{N}_\lambda(]0,|x|t])} Y_i\right),$$

if $t \geq 0$ and $x \in \mathbb{Z}$. The independence properties of the Poisson process and the sequence (Y_i) give that

$$\left(\overline{X}(x,t) - \frac{x}{|x|} - \lambda \mathbb{E}(Y_0)t\right)$$
$$= \left(\frac{1}{|x|}\sum_{i=1}^{\mathcal{N}_\lambda(]0,|x|t])} (Y_i - \mathbb{E}(Y_0)) + \frac{\mathbb{E}(Y_0)}{|x|}\left[\mathcal{N}_\lambda(]0,|x|t]) - \lambda|x|t\right]\right)$$

is a martingale whose increasing process is

$$\left[\frac{1}{|x|}\lambda t\left[\operatorname{var}(Y_0) + \mathbb{E}(Y_0)^2\right]\right].$$

Doob's Inequality (Theorem A.4 page 352) shows that for $\varepsilon > 0$ and $t \geq 0$,

$$\mathbb{P}\left(\sup_{0\leq s\leq t} \left|\overline{X}(x,s) - \operatorname{sgn} x - \lambda \mathbb{E}(Y_0)\, s\right| \geq \varepsilon \right) \leq \frac{1}{|x|\varepsilon^2} \lambda t \left[\operatorname{var}(Y_0) + \mathbb{E}(Y_0)^2\right].$$

where $\operatorname{sgn} z = 1$ if $z \geq 0$ and -1 otherwise. Therefore, Proposition C.5 shows that the process $(\overline{X}(x,t) - \operatorname{sgn} x)$ converges to $(-\lambda\mathbb{E}(Y_0)\,t)$ as $|x|$ goes to infinity. This process has therefore two fluid limits given by $(-1 + \lambda\mathbb{E}(Y_0)\,t)$ and $(1 + \lambda\mathbb{E}(Y_0)\,t)$.

Using Donsker's Theorem, it is easily seen that for a large non-negative x then

$$\overline{X}(x,t) = \frac{1}{|x|} X(x, t|x|) \sim (1 + \lambda\mathbb{E}(Y_0)\,t) + \frac{1}{\sqrt{|x|}} \sigma B(t),$$

where $(B(t))$ is a standard Brownian motion and $\sigma = \sqrt{\lambda \operatorname{var}(Y_0)}$. Scaling the Markov process has the effect of erasing stochastic (Brownian) fluctuations around a linear deterministic sample path.

When $\mathbb{E}(Y_0) = 0$, the scaling does not give much information since the corresponding fluid limits (1) and (-1) are degenerate as stochastic processes. In this case the natural scaling is the central limit scaling,

$$\left(\frac{1}{|x|} \left(X\left(x, |x|^2 t\right) - x\right)\right) \tag{9.1}$$

which converges in distribution to $(\sqrt{\lambda \operatorname{var}(Y_0)} B(t))$. To catch interesting events, time scale has to be much faster than space scale. In fact, the scalings considered in this chapter are of first order (functional law of large numbers also called Euler's scaling). On the contrary, Scaling (9.1) is of second order (functional central limit theorem).

The $M/M/1$ Queue. The arrival rate is λ and the service rate μ, $L(t)$ is the number of customers of the queue at time $t \geq 0$. The rescaled process is given by

$$\overline{L}(N,t) = \frac{1}{N} L(Nt),$$

when $L(0) = N \in \mathbb{N}$. Proposition 5.17 page 128 of Chapter 5 on the $M/M/1$ queue shows the convergence

$$(\overline{L}(N,t)) \longrightarrow \left((1 + (\lambda - \mu)\, t)^+ \right),$$

when N goes to infinity, again by using Proposition C.5 page 371. The function $(1 + (\lambda - \mu)\, t)^+$ is therefore the unique fluid limit of this Markov process.

Two Queues in Tandem. The arrival process is Poisson with parameter λ_1, customers are served at rate μ_1 in the first queue and rate μ_2 in the second queue. For $i = 1$ and 2, the quantity $l_i(t)$ denotes the number of customers in queue i at time $t \geq 0$. The associated Markov process is $(L(t)) = (l_1(t), l_2(t))$. It is assumed that $\lambda_1 < \mu_2 \leq \mu_1$, these conditions imply in particular the Markov process $(L(t))$ is ergodic (Proposition 4.5 page 87).

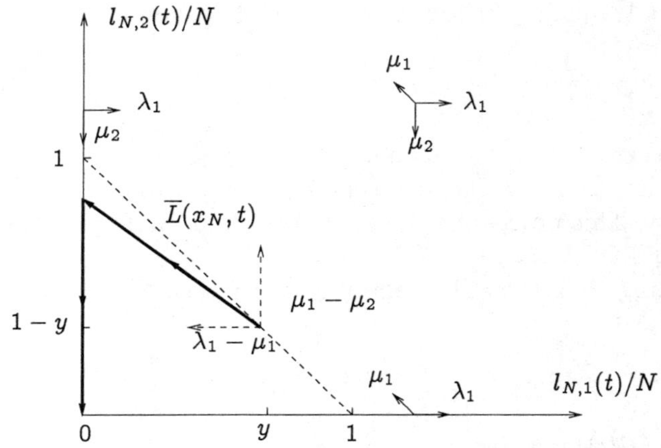

FIG. 9.1. Two Queues in Tandem : A Limiting Trajectory

The discussion that follows on the corresponding fluid limits is not entirely rigorous, but it gives a practical point of view on how fluid limits can be identified. By formulating the problem as a Skorohod problem, it will be rigorously established in Section 9.4. Section 9.6 presents another approach to characterize fluid limits.

The state space is $\mathbb{N}^2$ and the norm is defined by $\|z\| = |z_1| + |z_2|$, for $z = (z_1, z_2) \in \mathbb{R}^2$. Let (x_N) be a sequence of initial states whose norm converges to infinity. By taking a subsequence, it can be assumed that the sequence $(x_N/\|x_N\|)$ converges to some point in the unit sphere of $\mathbb{R}_+^2$, hence for $N \in \mathbb{N}$,

- $x_N = (l_{N,1}(0), l_{N,2}(0))$;
- $l_{N,1}(0) + l_{N,2}(0) = N$;
- the sequence $(l_{N,1}(0)/N)$ converges to $y \in [0,1]$, in particular the sequence of the rescaled initial states $(x_N/\|x_N\|)$ converges to $(y, 1-y)$.

The associated rescaled process is defined by

$$\overline{L}(x_N, t) = \left(\frac{l_{N,1}(Nt)}{N}, \frac{l_{N,2}(Nt)}{N}\right).$$

The first component of $\overline{L}(x_N, t)$ is simply a rescaled $M/M/1$ queue, consequently the convergence

$$\left(\frac{l_{N,1}(Nt)}{N}\right) \to \left((y + (\lambda_1 - \mu_1)t)^+\right),$$

holds when N goes to infinity. If τ_N is the first time when queue number 1 is empty, then almost surely the variable τ_N/N converges to $t_1 = y/(\mu_1 - \lambda_1)$ as N tends to infinity. As long as the first queue is non-empty, the second queue receives a Poissonian input with rate μ_1 therefore, the convergence

$$\left(\frac{l_{N,2}(Nt)}{N}\right) \longrightarrow (1 - y + (\mu_1 - \mu_2)t)\,,$$

holds for $t < t_1$ as N goes to infinity.

After time τ_N, the first queue starts from 0 and reaches equilibrium rapidly (Proposition 5.8 page 114), its output process is therefore approximately Poisson. Because of the fast time scale of the rescaled process, the second queue can therefore be thought as an $M/M/1$ queue receiving a Poisson process with rate λ_1 and starting from $l_{N,2}(\tau_N)$ moreover, almost surely

$$y_1 \stackrel{\text{def.}}{=} \lim_{N\to+\infty} \frac{l_{N,2}(\tau_N)}{N} = 1 - y + \frac{\mu_1 - \mu_2}{\mu_1 - \lambda_1} y = 1 + (\lambda_1 - \mu_2)t_1.$$

Consequently, for $t > t_1$,

$$\begin{aligned}\left(\frac{l_{N,2}(Nt)}{N}\right) &\to (y_1 + (\lambda_1 - \mu_2)(t - t_1))^+ \\ &= (1 + (\lambda_1 - \mu_2)t)^+.\end{aligned}$$

To summarize, when N tends to infinity the convergence

$$(9.2)\qquad (\overline{L}(x_N,t)) \to \begin{cases} [y + (\lambda_1 - \mu_1)t, 1 - y + (\mu_1 - \mu_2)t] & \text{for } t \le t_1, \\ \\ [0, (1 + (\lambda_1 - \mu_2)t)^+] & \text{for } t \ge t_1, \end{cases}$$

holds, with $t_1 = y/(\mu_1 - \lambda_1)$.

It is easily checked that the only possible limit for the rescaled process $\{\overline{L}(x,t) : x \in \mathbb{N}^2\}$ are the piecewise linear functions defined by Equation (9.2) for some y in $[0, 1]$.

Non-Uniqueness of Fluid Limits. For the examples investigated up to now, if the rescaled initial states $(x_N/\|x_N\|)$ converges to some x as $\|x_N\|$ goes to infinity, the sequence $(\overline{X}_N(x_N,t))$ of rescaled processes converges. The corresponding fluid limits depend only on x and not on the particular sequence $(x_N/\|x_N\|)$ of initial states converging to x.

An example shows that this is not always the case. One considers a Markov jump process on $\mathcal{S} = \mathbb{Z} \times \mathbb{N}$ whose Q-matrix $Q = (q_{xy})$ is given by (see Figure 9.2),

$$\begin{aligned}
&\text{for } i \in \mathbb{Z}, && q_{(i,j)(i,j-1)} = \alpha \quad \text{if} \quad j \ge 1, && q_{(i,j)(i,j+1)} = \beta \quad \text{if} \quad j \ge 0,\\
&\text{for } j \in \mathbb{N}, && q_{(i,j)(i+1,j)} = \lambda \quad \text{if} \quad i > 0, && q_{(i,j)(i-1,j)} = \mu \quad \text{if} \quad i > 0,\\
& && q_{(i,j)(i-1,j)} = \lambda \quad \text{if} \quad i < 0, && q_{(i,j)(i+1,j)} = \mu \quad \text{if} \quad i < 0,\\
& && q_{(0,j)(1,j)} = \lambda/2, && q_{(0,j)(-1,j)} = \lambda/2.
\end{aligned}$$

It is assumed that $\alpha > \beta$ and $\lambda > \mu$. If $(X(t)) = (X_1(t), X_2(t))$ is the associated Markov process, $(X_2(t))$ can be thought as the number of customers of a stable $M/M/1$ queue whose arrival rate is β and service rate α. For $t \ge 0$, define $(\overline{X}_N(t)) = (\overline{X}_{N,1}(t), \overline{X}_{N,2}(t))$ the associated rescaled process. Its second component clearly converges to $((1 + (\alpha - \beta)t)^+)$.

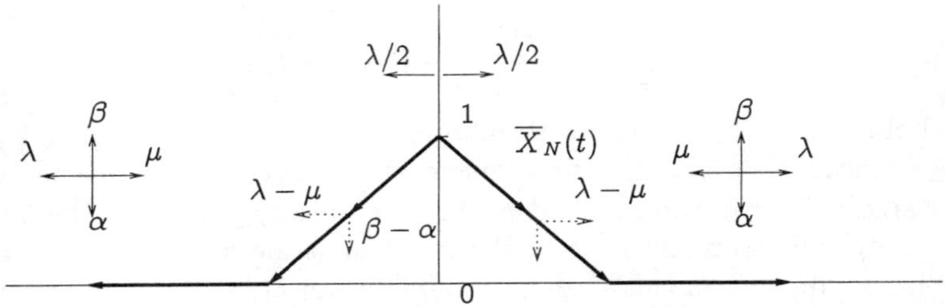

FIG. 9.2. Limiting Trajectories for a Markov Process on $\mathbb{Z} \times \mathbb{N}$

The process $(|X_1(t)|)$ is clearly the number of customers of an $M/M/1$ queue whose arrival rate is λ and service rate μ. According to the assumption $\lambda > \mu$, it is transient. Therefore, almost surely, $(|X_1(t)|)$ visits 0 only a finite number of times, thus $(X_1(t))$ has a constant sign after some time.

If $X(0) = (\lfloor\sqrt{N}\rfloor, N)$ then $X_{N,1}(t)$ is non-negative for any $t \geq 0$ with a probability converging to 1 as N goes to infinity. Indeed, a transient $M/M/1$ starting from $\sqrt{N}$ does not visit 0 with a probability converging to 1 as N tends to infinity. Consequently, the sign of the first component of the rescaled process is positive,

$$\lim_{N\to+\infty} \overline{X}_N(0) = (0,1) \quad \text{and} \quad (\overline{X}_N(t)) \to \left[\lambda-\mu)t, (1+(\alpha-\beta)t)^+\right].$$

In the same way if $X(0) = (-\lfloor\sqrt{N}\rfloor, N)$ then

$$\lim_{N\to+\infty} \overline{X}_N(0) = (0,1) \quad \text{and} \quad (\overline{X}_N(t)) \to \left[-(\lambda-\mu)t, (1+(\alpha-\beta)t)^+\right].$$

The two sequences $(\lfloor\sqrt{N}\rfloor, N)$ and $(-\lfloor\sqrt{N}\rfloor, N)$ of initial states have the same limit $(0,1)$ for the initial point of the rescaled process but they give rise to two different fluid limits.

A Non-Deterministic Fluid Limit. The scaling considered in this chapter removes some of the stochastic fluctuations (it has been seen for the case of random walks, see Section 9.3 for a more general picture) Nevertheless, this procedure does not remove *all* the randomness of the initial stochastic process as it will be shown.

In the last example, one takes the sequence of initial states $(x_N) = (0, N)$. By symmetry, the first coordinate converges to $+\infty$ or $-\infty$ with probability 1/2. Therefore, as N goes to infinity, the first component of the rescaled process converges to $((\lambda-\mu)t)$ with probability 1/2 and to $(-(\lambda-\mu)t)$ with probability 1/2. The limit of the rescaled process $(\overline{X}_N(t))$ is not deterministic in this case since the scaling does not affect the final sign of the first component.

A Process Without a Fluid Limit: The $M/M/\infty$ Queue. The arrival and service rates are given respectively by λ and μ, and $(L(t))$ is the process of the number of customers of this queue. If the initial state is N, the rescaled process is given by

$$\overline{L}_N(t) = \frac{1}{N} L(Nt).$$

For $a \in \mathbb{N}$, T_a denotes the hitting time of a for $(L(t))$, Propositions 6.8 and 6.10 page 154 show that for any $\varepsilon > 0$ and $K > 0$,

$$\lim_{N \to +\infty} \mathbb{P}_N \left(\frac{T_0}{N} \geq \varepsilon \right) = 0, \quad \text{and} \quad \lim_{N \to +\infty} \mathbb{P}_0 \left(\frac{T_{\varepsilon N}}{N} \leq K \right) = 0.$$

Thus, for $0 < s \leq t$,

$$\mathbb{P}_N \left(\sup_{s \leq u \leq t} \overline{L}_N(u) > \varepsilon \right) \leq \mathbb{P}_N \left(T_0/N \geq s \right) + \mathbb{P}_N \left(T_0/N \leq s, \sup_{s \leq u \leq t} > \varepsilon \right),$$

the strong Markov property shows that the second term of the right hand side of this inequality is bounded by

$$\mathbb{P}_0 \left(T_{\varepsilon N}/N \leq t \right).$$

Therefore, the convergence

$$\lim_{N \to +\infty} \mathbb{P}_N \left(\sup_{s \leq u \leq t} \overline{L}_N(u) > \varepsilon \right) = 0$$

holds for $\varepsilon > 0$ and $0 < s \leq t$.

The process $(1_{\{0\}}(t))$ is therefore a candidate for a possible limit of the rescaled process $(\overline{L}_N(t))$. Indeed, using the notation of Section C.2 on Skorohod topology, the inequality

$$d_T(f, g) \geq |f(0) - g(0)|$$

holds for any function f and g of $D([0, T], \mathbb{R})$ where $d_T(f, g)$ is the distance of Skorohod topology. The mapping $(z(t)) \to z(0)$ is therefore continuous on $D(\mathbb{R}_+, \mathbb{R})$. A fluid limit of this queue is necessarily 1 at $t = 0$. One concludes that the $M/M/\infty$ queue does not have any fluid limit since the only possible limit is not right continuous at 0.

If these (quite formal) topological considerations are ignored, the possible "fluid limit" for this scaling is hardly of any interest to understand the qualitative behavior of this queue. The scaling considered in Chapter 6 page 158 is much more interesting from this point of view.

9.2.3 Fluid Limits as Dynamical Systems

The state space $\mathcal{S}$ is supposed to be included in a cone H of a normed space (if $\alpha \in H$ and $\lambda > 0$, then $\lambda\alpha \in H$). It is assumed that $\mathcal{S}$ "approximates" H, i.e. for any $\alpha \in H - \{0\}$, there exists some sequence (x_N) of $\mathcal{S}$ such that (x_N/N) converges to α.

For any $\alpha \in H$ such that $\|\alpha\| = 1$, it is assumed that there exists some continuous deterministic process $(T_t(\alpha))$ such that the convergence of processes

$$\lim_{\substack{\|x\|\to+\infty \\ x/\|x\|\to\alpha}} \left(\frac{X(x,\|x\|t)}{\|x\|}\right) = (T_t(\alpha)). \tag{9.3}$$

holds. This amounts to say that fluid limits of the Markov process $(X(x,t))$ are indexed by the unit ball restricted to H. It will be seen that this is the case for Jackson networks for example.

The operator T_t is defined, a priori, on the unit ball. It can be extended to $H-\{0\}$ in the following way: For $\alpha \in H-\{0\}$ and $t \geq 0$, set

$$T_t(\alpha) = \|\alpha\|\, T_{t/\|\alpha\|}\left(\frac{\alpha}{\|\alpha\|}\right).$$

By writing, for $N \geq 1$,

$$\frac{X(x,Nt)}{N} = \|\alpha\| \frac{X(x, N\|\alpha\|\, t/\|\alpha\|)}{N\|\alpha\|},$$

it is easily seen that the extension of T_t is a natural generalization of Equation (9.3) since the convergence

$$\lim_{\substack{\|x\|,\,N\to+\infty \\ x/N\to\alpha}} (X(x,Nt)/N) = (T_t(\alpha)) \tag{9.4}$$

holds for $\alpha \in H-\{0\}$.

PROPOSITION 9.1 (Semi-Group Property). *If Convergence* (9.3) *holds, the operators* (T_t) *have the following properties, for* s, $t \geq 0$,

$$T_{s+t} = T_t \circ T_s.$$

For $u > 0$, $t \geq 0$ *and* $\alpha \in H-\{0\}$

$$T_{ut}(\alpha) = uT_t(\alpha/u).$$

PROOF. A sequence (x_N) of $\mathcal{S}$ such that x_N/N converges to $\alpha \in H-\{0\}$ is fixed. By using Skorohod's representation Theorem C.8 page 373, it can be assumed that $(X(x_N,Nt)/N)$ converges almost surely to $(T_t(\alpha))$ for the Skorohod topology when N goes to infinity. The identity in distribution, for s, $t \geq 0$,

$$X(x_N, N(s+t)) \stackrel{\text{dist.}}{=} X(X(x_N,Ns),Nt) \tag{9.5}$$

is a consequence of the Markov property. By Proposition 5.2 page 118 of Ethier and Kurtz [**19**] and the fact that the function $u \to T_u(\alpha)$ is continuous, the sequence $(X(x_N,Ns)/N)$ converges almost surely to $T_s(\alpha)$. By dividing Equation (9.5) by N and by taking the limit, one gets that $T_{s+t}(\alpha) = T_t(T_s(\alpha))$. The semi-group property is proved. The last identity of the proposition directly follows from Equation (9.4). □

Under the assumptions of the above theorem, the Markov process is transformed, via the scaling, into a dynamical system (T_t) on H. The discrete structure of $\mathcal{S}$ is thus replaced by the "continuous" manifold H of a normed space. This is an important aspect of fluid limits since the continuous structure of H may lead to some classical calculus. A typical situation is when (T_t) is the dynamical system generated by a deterministic differential equation which is a classical situation in ergodic theory. See Arnol'd [**2**].

9.3 Fluid Limits of a Class of Markov Processes

In this section, general properties of fluid limits of a large class of Markov processes with values in $\mathbb{N}^d$ are investigated. If some other Markov processes do not fit exactly in this class, the technical arguments used here can be applied in many situations.

From now on, dimension $d \in \mathbb{N}$ is fixed, the L_1-norm on $\mathbb{Z}^d$ is denoted by $\|\cdot\|$, i.e. $\|m\| = |m_1| + \cdots + |m_d|$ if $m = (m_k) \in \mathbb{R}^d$.

Definition 23. *A Markov jump process $(X(t))$ belongs to the class $(\mathcal{C})$ if its Q-matrix $Q = (q_{xy}\,;\ x, y \in \mathcal{S})$ given by*

$$q_{x,x+m} = \sum_{i\in I} 1_{\{x\in C_i\}}\lambda_i \nu_i(\{m\}),$$

for $x \in \mathbb{N}^d$, $m \in \mathbb{Z}^d$ and $q_{xy} = 0$, where I is at most countable and, for $i \in I$,

- *C_i is a subset of $\mathbb{R}^d$;*
- *ν_i is a probability distribution on $\mathbb{N}^d$;*
- *$\lambda_i \geq 0$;*
- *$\sum_{i\in I} \lambda_i \int_{\mathbb{Z}^d} \|m\|^2 \nu_i(dm) < +\infty$.*

For $i \in I$, x and $m \in \mathbb{N}^d$, when the process is at x is in the set C_i, it jumps from x to $x + m$ according to a Poisson process with intensity $\lambda_i \nu_i(\{m\})$. The dynamic of these Markov processes is locally constant: The transition rates do not change as long as it is inside or outside C_i.

Such a Markov process can also be seen as the solution of the stochastic differential equation

$$dX(t) = \sum_{i\in I} 1_{\{X(t-)\in C_i\}} \int_{\mathbb{Z}^d} m\, V_i(dt, dm), \tag{9.6}$$

where V_i is a Poisson process on $\mathbb{R}_+ \times \mathbb{Z}^d$ with intensity $\lambda_i\, dt \otimes \nu_i(dm)$. The point processes V_i, $i \in I$ are assumed to be independent. The integral on $\mathbb{Z}^d$ in Equation (9.6) is with respect to the variable m. Note that for a fixed t, the marked point process $V_i(dt, dm)$ is either the null measure or a Dirac mass (see Proposition 1.11 page 11).

Differential Equation (9.6) is vector-valued, if $j = 1, \dots, d$, for the jth coordinate, it can be written

$$dX_j(t) = \sum_{i\in I} 1_{\{X(t-)\in C_i\}} \int_{\mathbb{Z}^d} m_j\, V_i(dt, dm). \tag{9.7}$$

Proposition A.11 page 356 in Appendix A establishes existence and uniqueness of the solutions of such differential equations.

Examples.

The $M/M/1$ queue. The number of customers $(L(t))$ can be represented as the solution of the equation

$$dL(t) = \mathcal{N}_\lambda(dt) - 1_{\{L(t-)\neq 0\}}\,\mathcal{N}_\mu(dt),$$

if the arrival and service rates are respectively given by λ and μ (see Equation (5.1) page 104). Set the point processes $V_1(dt, dm) = \mathcal{N}_\lambda(dt) \otimes \delta_1$ and $V_2(dt, dm) = \mathcal{N}_\mu(dt) \otimes \delta_{-1}$, with $C_1 = \mathbb{N}$ and $C_2 = \mathbb{N} - \{0\}$ (δ_x is the Dirac mass in x). This equation is clearly of type (9.6). More generally, the vector of the number of customers in a Jackson network has also this property.

Jackson Networks With d Nodes. By using the notations of Section 4.4.1 page 92, $P = (p_{ij})$ is the routing matrix, (λ_i) and (μ_i) the vectors of the arrival and service rates at the nodes of the network. A column and a row is added to the routing matrix so that $p_{i0} = 1 - p_{i1} - \cdots - p_{id}$ for $1 \leq i \leq d$, $p_{0i} = 0$ and $p_{00} = 1$, R denotes this $(d+1) \times (d+1)$ matrix. The vectors e_i, $i = 1, \ldots, d$ are the unit vectors in $\mathbb{R}^d$ and $e_0 = 0$.

For i, $j \leq d$ with $i > 0$, $C_{ij} = \{n = (n_k) : n_i > 0\}$ and V_{ij} is a Poisson process with intensity $\mu_i p_{ij}\, dt \otimes \delta_{e_j - e_i}$; $C_{0j} = \mathbb{N}^d$ and V_{0j} is Poisson with rate $\lambda_j\, dt \otimes \delta_{e_j}$. The process V_{ij} represents the arrival process at queue j coming from queue i, V_{0i} is the external arrivals at node i and V_{j0} is associated with the departures of the network from node j. With the same method used to prove Equations (5.1) page 104 and Equation (6.1) page 143, it is easily checked that the solution $(X(t))$ of the stochastic differential equation

$$dX(t) = \sum_{\substack{0\leq i,j\leq d,\\ (i,j)\neq(0,0)}} 1_{\{X(t-)\in C_{ij}\}} \int_{\mathbb{Z}^d} m\, V_{ij}(dt, dm)$$

has the same distribution as a Markov process describing a Jackson network with the parameters defined above.

A Priority Queue. Two classes of customers, class 1 and class 2, arrive at this queue. For $i = 1, 2$, customers of class i arrive according to a Poisson process with parameter λ_i and are served at rate μ_i. Class 1 customers have absolute priority on the class 2 customers, i.e. a customer of class 2 can only be served when no customer of class 1 is present in the queue.

By taking $C_1^- = \{n = (n_1, n_2) : n_1 > 0\}$, $C_2^- = \{n = (0, n_2) : n_2 > 0\}$ and for $i = 1$, 2, $C_i^+ = \mathbb{N}^2$, V_i^+ a Poisson process with rate $\lambda_i\, dt \otimes \delta_{e_i}$ and V_i^- a Poisson process with rate $\mu_i\, dt \otimes \delta_{-e_i}$, the solution $L(t) = (L_1(t), L_2(t))$ of the differential Equation (9.6) has the same distribution as a Markov process describing the priority queue. In the same way, Markov processes describing networks of priority queues belong also to class $(\mathcal{C})$.

If X is a Markov process of class $(\mathcal{C})$, for $x \in \mathbb{N}^d$, $\left(\overline{X}(x,t)\right)$ denotes the rescaled process, for $t \geq 0$

$$\overline{X}(x,t) = \frac{1}{\|x\|} X(x, \|x\|t).$$

An elementary, but useful, property of these Markov processes is proved in the next proposition. (See Section 9.5 page 258.)

PROPOSITION 9.2. *If $(X(t))$ is a Markov process of class $(\mathcal{C})$, for $T > 0$, the random variables*

$$\{\|X\|(x,T)\,;\ x \in \mathbb{N}^d\} = \left\{ \frac{\|X(x, \|x\|T)\|}{\|x\|}\,;\ x \in \mathbb{N}^d \right\}$$

are uniformly integrable.

Uniform integrability is defined in Appendix A page 351.

PROOF. It is enough to prove that the second moments of these variables are bounded, i.e.

$$\sup_{x\in\mathbb{N}^d} \mathbb{E}\left((\|X\|(x,T))^2\right) < +\infty.$$

For $i \in I$, Proposition 1.22 page 21 shows that the process

$$\left(\int_{]0,t]\times\mathbb{Z}^d} \|m\| \; [V_i(ds,dm) - \lambda_i\, ds\, \nu_i(dm)] \right) \tag{9.8}$$

is a martingale with the increasing process

$$\left(\lambda_i \int_{\mathbb{Z}^d} \|m\|^2\, \nu_i(dm)\, t \right).$$

By independence of the Poisson point processes V_i, $i \in I$), the martingales defined by Equation (9.8) are independent, with zero mean and therefore orthogonal.

For $t \geq 0$, Relationship (9.6) gives the upper bound

$$\|X(x,t)\| \leq \|x\| + \sum_{i\in I} \int_{]0,t]\times\mathbb{Z}^d} \|m\|\, V_i(ds,dm).$$

The elementary inequality $(x_1 + x_2 + x_3)^2 \leq 9(x_1^2 + x_2^2 + x_3^2)$ yields

$$\frac{1}{3}\mathbb{E}\left((\|X(x,t)\|)^2\right) \leq \|x\|^2 + \left(\sum_{i\in I} \lambda_i \int_{\mathbb{Z}^d} \|m\|\, \nu_i(dm)\, t \right)^2$$
$$+\,\mathbb{E}\left(\sum_{i\in I} \int_{]0,t]\times\mathbb{Z}^d} \|m\| \; [V_i(ds,dm) - \lambda_i\, ds\, \nu_i(dm)] \right)^2.$$

By independence of the martingales defined by Equation (9.8), the right hand side of this inequality is

$$\|x\|^2 + \left(\sum_{i\in I} \lambda_i \int_{\mathbb{Z}^d} \|m\|\, \nu_i(dm)\, t \right)^2 + \sum_{i\in I} \lambda_i \int_{\mathbb{Z}^d} \|m\|^2\, \nu_i(dm)\, t,$$

consequently for $T > 0$,

$$\mathbb{E}\left(\left(\frac{\|X(x,\|x\|T)\|}{\|x\|}\right)^2\right)$$
$$\leq 3\left(1+\frac{1}{\|x\|}\sum_{i\in I}\lambda_i\int_{\mathbb{Z}^d}\|m\|^2\,\nu_i(dm)\,T+\left(\lambda_i\int_{\mathbb{Z}^d}\|m\|\,\nu_i(dm)\,T\right)^2\right),$$

the second moments of the rescaled processes at time T are thus bounded as $\|x\|$ tends to infinity. The proposition is proved. □

For $1 \leq j \leq d$, stochastic differential Equation (9.7) shows that the process

$$(M_j(x,t)) = \left(X_j(t)-X_j(0)-\sum_{i\in I}\lambda_i\int_{\mathbb{Z}^d}m_j\,\nu_i(dm)\int_0^t 1_{\{X(s)\in C_i\}}\,ds\right)$$
$$= \left(\sum_{i\in I}\int_{]0,t]\times\mathbb{Z}^d}1_{\{X(s-)\in C_i\}}m_j\,\left[V_i(ds,dm)-\lambda_i\,ds\,\nu_i(dm)\right]\right) \tag{9.9}$$

is a martingale. In the same way as in the above proof, Proposition A.9 page 355 shows that its increasing process is given by

$$(\langle M_j\rangle(x,t)) = \left(\sum_{i\in I}\int_{]0,t]\times\mathbb{Z}^d}1_{\{X(s)\in C_i\}}\lambda_i\,ds\;m_j^2\,\nu_i(dm)\right).$$

The process $\left(\overline{M}_j(x,t)\right)$ is the martingale (9.9) rescaled, i.e.

$$\overline{M_j}(x,t)=\frac{1}{\|x\|}M_j(x,\|x\|t).$$

The increasing process of this martingale is

$$\left\langle\overline{M}_j\right\rangle(x,t)=\frac{1}{\|x\|^2}\sum_{i\in I}\int_{]0,\,\|x\|t]\times\mathbb{Z}^d}1_{\{X(x,s)\in C_i\}}\lambda_i\,ds\;m_j^2\,\nu_i(dm)$$
$$=\frac{1}{\|x\|}\sum_{i\in I}\int_{]0,t]\times\mathbb{Z}^d}1_{\{X(x,\|x\|s)\in C_i\}}\lambda_i\,ds\;m_j^2\,\nu_i(dm).$$

From Doob's Inequality one gets that for $\varepsilon>0$ and $t\geq 0$,

$$\mathbb{P}\left(\sup_{0\leq s\leq t}|\overline{M}_j(x,t)|\geq\varepsilon\right)\leq\frac{1}{\varepsilon^2}\,E\left(\overline{M}_j(x,t)^2\right)=\frac{1}{\varepsilon^2}\,E\big(\langle\overline{M}_j(x,t)\rangle\big)$$
$$\leq\frac{t}{\|x\|\varepsilon^2}\sum_{i\in I}\lambda_i\int_{\mathbb{Z}^d}m_j^2\,\nu_i(dm).$$

Thus, the martingale $(\overline{M}_j(x,t))$ converges in probability to 0 uniformly on compact sets when $\|x\|$ tends to infinity. By using Representation (9.9) of $(M(t))$, this yields

$$(9.10)\quad \overline{X}_j(x,t) = \frac{x_j}{\|x\|} + \overline{M}_j(x,t) + \sum_{i\in I} \lambda_i \int_{\mathbb{Z}^d} m_j\, \nu_i(dm) \int_0^t 1_{\{X(x,\|x\|s)\in C_i\}}\, ds.$$

For T, δ, $\eta > 0$, if w_h is the modulus of continuity of a function h on $[0,T]$,

$$w_h(\delta) = \sup(|h(t)-h(s)|\,;\ s,t \le T, |t-s| \le \delta),$$

Equation (9.10) shows that the inequality

$$(9.11)\quad \mathbb{P}\left(w_{\overline{X}_j(x,\cdot)}(\delta) \ge \eta\right) \le \mathbb{P}\left(w_{\overline{M}_j(x,\cdot)}(\delta) \ge \eta/2\right) + \mathbb{P}\left(\sup_{s,t\le T, |t-s|\le\delta} \sum_{i\in I} \lambda_i \int_{\mathbb{Z}^d} |m_j|\, \nu_i(dm) \int_s^t 1_{\{X(x,\|x\|u)\in C_i\}}\, du \ge \eta/2\right)$$

holds. The second term of the right hand side of the last inequality is zero as soon as

$$2\delta \sum_{i\in I} \int_{\mathbb{Z}^d} |m_j|\, \nu_i(dm) < \eta.$$

The inequality

$$\mathbb{P}\left(w_{\overline{M}_j(x,\cdot)}(\delta) \ge \eta/2\right) \le \mathbb{P}\left(\sup_{0\le s\le t} |\overline{M}_j(x,t)| \ge \eta/2\right)$$

and the convergence of $(\overline{M}_j(x,t))$ to 0 show that the first term of the right hand side of Inequality (9.11) tends to 0 when $\|x\|$ goes to infinity. Therefore, there exists some $K > 0$ such that if $\|x\| \ge K$, then

$$\mathbb{P}\left(w_{\overline{M}_j(x,\cdot)}(\delta) \ge \eta/2\right) \le \varepsilon.$$

The hypotheses of Theorem C.9 page 373 on the relative compactness of a set of càdlàg functions are thus verified. Hence, if (x_n) is a sequence of initial state whose norm converges to infinity, the set of rescaled processes

$$\{(\overline{X}_j(x_n,t)),\quad n\in\mathbb{N}\}$$

is therefore relatively compact and all its limits are continuous functions. For $\varepsilon > 0$, there exists a compact subset K_j of $D(\mathbb{R}_+,\mathbb{R})$ such that

$$\inf_{n\in\mathbb{N}} \mathbb{P}\left(\overline{X}_j(x_n,\cdot) \in K_j\right) \ge 1-\varepsilon/d,$$

hence,

$$\inf_{n\in\mathbb{N}} \mathbb{P}\left(\left(\overline{X}_j(x_n,\cdot)\ ;\ j=1,\dots,d\right) \in \prod_1^d K_j\right) \ge 1-\varepsilon.$$

Since the product of compact sets is also compact, one concludes that the set of the rescaled processes $(\overline{X}_j(x_n,t)\,;\, j=1,\dots,d)$, $n\in\mathbb{N}$ is relatively compact and all its limits are continuous processes. This result is summarized in the proposition.

PROPOSITION 9.3. *If $(X(x,t))$ is a Markov process of class $(\mathcal{C})$ and (x_n) a sequence of elements of S whose norm converges to infinity, the set of rescaled processes*

$$\left\{\left(\overline{X}(x_n,t)\right)\;;\; n\in\mathbb{N}\right\} = \left\{\frac{1}{\|x_n\|}(X(x_n,\|x_n\|t))\,;\; n\in\mathbb{N}^d\right\}$$

is relatively compact and its limiting points, the fluid limits of $(X(x,t))$, are continuous stochastic processes. The process

$$(9.12)\quad \left(\overline{M}(x,t)\right) = \left(\overline{X}(x_n,t) - \frac{x_n}{\|x\|} - \sum_{i\in I}\lambda_i\int_{\mathbb{Z}^d} m\,\nu_i(dm)\int_0^t 1_{\{X(x_n,\|x_n\|s)\in C_i\}}\,ds\right)$$

is a martingale converging in probability to 0 uniformly on compact sets when n tends to infinity, i.e. for any $T\geq 0$ and $\varepsilon>0$,

$$\lim_{n\to+\infty}\mathbb{P}(\sup_{0\leq s\leq T}\|\overline{M}(x_n,s)\|\geq\varepsilon)=0.$$

The next step is to characterize fluid limits of these processes. Since the class $(\mathcal{C})$ of Markov processes is fairly large, a precise answer in this domain is quite unlikely. The original process can be, in general, expressed as a solution of a stochastic differential equation. It may also be true that any fluid limit is a solution of stochastic differential equation and perhaps, in some cases, a deterministic differential equation if all the stochastic fluctuations disappear after the scaling.

To get interesting results one has to assume that the subsets (C_i) are cones of $\mathbb{R}^d$, i.e. for $i\in I$, if $x\in C_i$ then $\alpha x\in C_i$ when $\alpha>0$, Identity (9.10) can then be rewritten as

$$(9.13)\quad \overline{X}(x,t) = \frac{x}{\|x\|} + \overline{M}(x,t) + \sum_{i\in I}\lambda_i\int_{\mathbb{Z}^d} m\,\nu_i(dm)\int_0^t 1_{\{\overline{X}(x,s)\in C_i\}}\,ds.$$

By assuming that limits can be taken in the indicator functions and that $x/\|x\|\to\alpha$ as $\|x\|$ tends to infinity, a fluid limit $(\overline{z}(t))$ should satisfy the equation

$$\overline{z}(t) = \alpha + \sum_{i\in I}\lambda_i\int_{\mathbb{Z}^d} m\,\nu_i(dm)\int_0^t 1_{\{\overline{z}(s)\in C_i\}}\,ds,$$

or, as a differential equation,

$$d\overline{z}(t) = \sum_{i\in I}\lambda_i\int_{\mathbb{Z}^d} m\,\nu_i(dm)\;1_{\{\overline{z}(t)\in C_i\}}\,dt.$$

The next proposition shows that such an equation is indeed satisfied when the fluid limit is not on the set of discontinuities of the differential equation, i.e. when $\overline{z}(t)$ is not on the boundary of one of the C_i, $i\in I$.

PROPOSITION 9.4. *If $(\overline{z}(t))$ is a fluid limit of a Markov process of class $(\mathcal{C})$, when the sets $(C_i\,;\, i \in I)$ are cones of $\mathbb{R}^d$ and there exists some time interval $[0, t_0]$ such that with probability 1, for any $t \in [0, t_0]$, the process $(\overline{z}(t))$ is not on the boundary of any set (C_i), then almost surely,*

$$\overline{z}(t) = \overline{z}(0) + \sum_{i \in I} \lambda_i \int_{\mathbb{Z}^d} m\, \nu_i(dm) \int_0^t 1_{\{\overline{z}(u) \in C_i\}}\, du, \tag{9.14}$$

holds for all $t \in [0, t_0]$.

The above proposition can be rephrased as follows: If z_0 is not on the boundary of the cones C_i and $\overline{z}(0) = z_0$ then the differential equation

$$d\overline{z}(t) = \sum_{i \in I} \left(\lambda_i \int_{\mathbb{Z}^d} m\, \nu_i(dm)\right) 1_{\{\overline{z}(t) \in C_i\}}\, dt$$

holds for t in a neighborhood of 0.

PROOF. There exists a sequence (x_n) of $\mathbb{N}^d$ such that $(\|x_n\|)$ converges to infinity and the sequence of rescaled processes $(\overline{X}(x_n, t))$ converges in distribution to $(\overline{z}(t))$. As usual if h is function on $\mathbb{R}_+$, and $t \geq 0$,

$$\|h\|_{\infty, t_0} = \sup\{|h(u)| : u \leq t_0\}.$$

By taking an appropriate probability space, Skorohod's representation Theorem (Theorem C.8 page 373) shows that one can assume that $(\overline{X}(x_n, t))$ converges almost surely to $(\overline{z}(t))$ for Skorohod Topology. Almost surely, there exist continuous increasing functions (α_n) from $[0, t_0]$ into $[0, t_0]$ such that $\alpha_n(0) = 0$ and $\alpha_n(t_0) = t_0$ and the inequalities

$$\lim_{n \to +\infty} \sup_{s,\, t \in [0, t_0]\,;\, s \neq t} \left| \log \frac{\alpha_n(s) - \alpha_n(t)}{s - t} \right| = 0,$$

$$\lim_{n \to +\infty} \sup_{s \in [0, t_0]} \left\| \overline{X}(x_n, \alpha_n(s)) - \overline{z}(s) \right\| = 0, \tag{9.15}$$

hold, in particular,

$$\lim_{n \to +\infty} \|\dot{\alpha}_n - 1\|_{\infty, t_0} = 0,$$

by Proposition C.2 page 369, where $\dot{\alpha}_n$ is some version of the Radon-Nikodym derivative of the increasing function α_n.

For $i \in I$, f_i denotes the indicator function of the set C_i. If $0 \leq t \leq t_0$, and $n \in \mathbb{N}$, with a change of variables, one gets that

$$\int_0^{\alpha_n(t)} f_i\left(\overline{X}(x_n, u)\right)\, du = \int_0^t f_i\left[\overline{X}(x_n, \alpha_n(u))\right] \dot{\alpha}_n(u)\, du,$$

therefore,

$$\Lambda_n(t) \stackrel{\text{def.}}{=} \left| \int_0^{\alpha_n(t)} f_i\left(\overline{X}(x_n,u)\right)\,du - \int_0^t f_i(\overline{z}(u))\,du \right|$$
$$\leq \left| \int_0^t \left(f_i\left[\overline{X}(x_n,\alpha_n(u))\right] - f_i\left(\overline{z}(u)\right)\dot{\alpha}_n(u)\,du \right|$$
$$+ \left| \int_0^t f_i\left(\overline{z}(u)\right)\left(\dot{\alpha}_n(u) - 1\right)du \right|,$$

hence,

$$\Lambda_n(t) \leq \|\dot{\alpha}_n\|_{\infty,t_0} \int_0^{t_0} \left| f_i\left[\overline{X}(x_n,\alpha_n(u))\right] - f_i(\overline{z}(u)) \right| du + t_0\|\dot{\alpha}_n - 1\|_{\infty,t_0}.$$

The quantity $|f_i(\overline{X}[x_n,\alpha_n(u)]) - f_i(\overline{z}(u))|$ being bounded by 2, from Fatou's Lemma one gets

$$\limsup_{n\to+\infty} \int_0^{t_0} \left| f_i(\overline{X}[x_n,\alpha_n(u)]) - f_i(\overline{z}(u)) \right| du$$
$$\leq \int_0^{t_0} \limsup_{n\to+\infty} \left| f_i\left[\overline{X}(x_n,\alpha_n(u))\right] - f_i\left(\overline{z}(u)\right) \right| du,$$

the assumption on $\overline{z}$ and Inequality (9.15) imply that the expression under the integral is zero. Therefore, the almost sure convergence

$$\lim_{n\to+\infty} \sup_{t\leq t_0} \Lambda_n(t) = 0,$$

holds, and thus the convergence

$$\left(\int_0^t f_i\left(\overline{X}(x_n,u)\right)\,du\,;\ i \in I \right) \longrightarrow \left(\int_0^t f_i\left(\overline{z}(u)\right)\,du\,; i \in I \right)$$

holds for the Skorohod topology on $D([0,t_0],\mathbb{R}^d)$. One deduces the convergence of the corresponding probability distributions on $D([0,t_0],\mathbb{R}^d)$.

Back on the initial probability space, Identity (9.13) gives for $t \leq t_0$,

$$\overline{X}(x_n,t) = \frac{x_n}{\|x_n\|} + \overline{M}(x_n,t) + \sum_{i\in I} \lambda_i \int_{\mathbb{Z}^d} m\,\nu_i(dm) \int_0^t f_i\left(\overline{X}(x_n,u)\right)\,du.$$

The left hand side of this identity converges in distribution to $(\overline{z}(t))$. Since the martingale $(\overline{M}(x_n,t))$ converges to 0 uniformly on the interval $[0,t_0]$ (Proposition 9.12), Proposition C.5 page 371 shows that on this time interval the process on the right hand side converges in distribution to

$$\left(\overline{z}(0) + \sum_{i\in I} \lambda_i \int_{\mathbb{Z}^d} m\,\nu_i(dm) \int_0^t 1_{\{\overline{z}(u)\in C_i\}}\,du \right).$$

Note that it has been used that if $(\overline{X}(x_n,t))$ converges to $(z(t))$, then the sequence of random variables $(\overline{X}(x_n,0))$ converges in distribution to $z(0)$, (see the discussion of the example of the $M/M/\infty$ queue page 239). The proposition is proved. □

The above proposition shows that the real difficulty of the analysis of fluid limits of a given Markov process concerns mainly the discontinuities of the associated dynamical system, i.e. when a given fluid limit reaches or leaves the boundaries of the sets (C_i). Outside these points the rescaled process is only the solution of some classical ordinary deterministic differential equation (which of course does not imply that it is easy to handle, but the framework is somewhat more familiar).

This is one of the most important properties of the scaling used in this chapter. It has the merit of reducing the study of a Markov process to its analysis around its discontinuities, where the real problems are. These points of discontinuity can keep a part of the complexity of the original Markov process, like some random component which has not vanished for example. In general, they can be studied by analyzing the macroscopic behavior, i.e. before scaling. Up to now, there is no generic method in this domain. In the case of some queueing networks, Dai[Dai95] gives a useful set of equations satisfied by the fluid limits for *all* $t \geq 0$. The next section presents an important tool to study fluid limits of a sub-class of Markov processes.

9.4 Relations with Skorohod Problems

If $(Y(t))$ is a càdlàg functions with values in $\mathbb{R}^d$ such that $Y(0) \geq 0$ and $P = (p_{ij})$ is a $d \times d$-matrix, a solution to Skorohod Problem associated with $(Y(t))$ and P is a couple of functions $(X(t)) = (X_i(t)\,;\, 1 \leq i \leq d)$ and $(R(t)) = (R_i(t); 1 \leq i \leq d)$ in $D(\mathbb{R}_+, \mathbb{R}^d)$ such that, for any $t \geq 0$,

a) $X(t) = Y(t) + (I - {}^tP)R(t)$;
b) for $1 \leq i \leq d$, $X_i(t) \geq 0$, the function $t \to R_i(t)$ is non-decreasing and $R_i(0) = 0$;
c) the reflection condition:

$$\int_0^{+\infty} X_i(s)\, dR_i(s) = 0.$$

The function $(Y(t))$ is one of the data of the problem, it is the "free" process. The function $(X(t))$ is the first component of the solution, it is constrained to live in the positive orthant of $\mathbb{R}^d$. By Condition a), the function $(R(t))$ is the *pushing process*, it keeps $(X(t))$ in the positive orthant whenever $(Y(t))$ wanders too much outside this domain. Condition c) suggests that this must be done in a minimal way: The support of the positive measure dR_i, the points of increase of $t \to R_i(t)$ is the set of t where $X_i(t)$ is 0.

[Dai95] J.G. Dai, *On positive Harris recurrence of multiclass queueing networks: a unified approach via fluid limit models*, Annals of Applied Probability **5** (1995), no. 1, 49–77.

The process $(R(t))$ can also be interpreted as a local time of $(X(t))$. In the one-dimensional case, this problem has been already been considered for the $M/M/1$ queue (See Definition 11 page 106). Skorohod Problem is sometimes also called oblique reflection mapping Problem.

Appendix D gives some conditions of existence and uniqueness and regularity properties of the solutions to Skorohod Problem (see Proposition D.4 page 381). When the results of Appendix D can be used, the method is as follows: The Markov process $(X(x,t))$ under study is shown to be the first component of the solution of Skorohod Problem associated with some free process $(Y(t)$ and matrix P, i.e. for $x \in \mathcal{S}$ and $t \geq 0$,

$$X(x,t) = Y(t) + \left(I - {}^tP\right) R(t).$$

The next step is to show that the rescaled process $(\overline{X}(x,t))$ is also the first component of the solution of Skorohod Problem associated with the rescaled process $(\overline{Y}(x,t))$

$$\overline{X}(x,t) = \overline{Y}(x,t) + \left(I - {}^tP\right) \overline{R}(x,t). \tag{9.16}$$

If the process $(\overline{Y}(x,t))$ converges to some process $(y(t))$ when $\|x\|$ converges to infinity then Proposition D.4 implies that $(\overline{X}(x,t), \overline{R}(x,t))$ converges in distribution to the solution of the Skorohod Problem associated with $(y(t))$. The convergence of the rescaled Markov process is thus proved in this way, furthermore it is shown that its fluid limit is *the* solution of some Skorohod Problem. Looking back to Equation (9.16), one may think that it should have been simpler to prove directly the convergence of $(R(x,t))$ as $\|x\|$ tends to infinity. This is not true at all, the process $(R(x,t))$ (the pushing process) is connected to the instants when the Markov reaches the boundary of the domain of the Markov process. In practice, it is difficult to prove directly convergence results for $(R(x,t))$ (see the examples below). The process $(\overline{Y}(x,t))$ contains martingales which become negligible as $\|x\|$ gets large, so that the convergence of $(\overline{Y}(x,t))$ is not a problem in general. The example of the $M/M/1$ queue illustrates this method in a simple setting.

9.4.1 The M/M/1 Queue

Let $(L(t))$ be the process for the number of customers of the $M/M/1$ queue whose arrival rate is λ and service rate μ. Equation (5.1 page 104) for this process is, for $t \geq 0$,

$$\begin{aligned}
L(t) &= L(0) + \mathcal{N}_\lambda\left(]0,t]\right) - \int_0^t 1_{\{L(s-)>0\}}\, \mathcal{N}_\mu(ds) \\
&= L(0) + \mathcal{N}_\lambda(]0,t]) - \mathcal{N}_\mu(]0,t]) + \int_0^t 1_{\{L(s-)=0\}}\, \mathcal{N}_\mu(ds),
\end{aligned}$$

$$L(t) = Y(t) + \mu \int_0^t 1_{\{L(s-)=0\}}\, ds = Y(t) + \mu \int_0^t 1_{\{L(s)=0\}}\, ds, \tag{9.17}$$

with

$$Y(t) = L(0) + (\lambda - \mu)t + M(t),$$

and $(M(t))$ is the martingale defined by

$$M(t) = [\mathcal{N}_\lambda(]0,t]) - \lambda t] - [\mathcal{N}_\mu(]0,t]) - \mu t] + \int_0^t 1_{\{L(s-)=0\}} [\mathcal{N}_\mu(ds) - \mu\, ds].$$

Relationship (9.17) shows that the couple

$$\left[(L(t)), \left(\mu \int_0^t 1_{\{L(s)=0\}}\, ds\right)\right]$$

solves Skorohod Problem associated with the free process $(Y(t))$.

If $L(0) = N$ and W is one of the processes L, Y and M, the quantity $\overline{W}_N(t) = W(Nt)/N$ denotes the corresponding rescaled process. By Relationship (9.17), the processes

$$(S_N(t)) = \left[\left(\overline{L}_N(t)\right), \left(\mu \int_0^t 1_{\{\overline{L}_N(s)=0\}}\, ds\right)\right]$$

are solutions of Skorohod Problem associated with the free process $(\overline{Y}_N(t))$. Proposition D.4 shows that if the sequence $(\overline{Y}_N(t))$ converges appropriately to some $(y(t))$, so does the sequence $(S_N(t))$ to the solution of Skorohod Problem associated with $(y(t))$. With the same argument as in the proof of Proposition 9.3, it is easy to get the convergence, for $\varepsilon > 0$,

$$\lim_{N\to+\infty} \mathbb{P}\left(\sup_{0\le s\le T} |\overline{M}_N(s)| \ge \varepsilon\right) = 0.$$

Consequently, if $(y(t)) \stackrel{\text{def.}}{=} 1 + (\lambda - \mu)t$, one gets the convergence

$$\lim_{N\to+\infty} \mathbb{P}\left(\sup_{0\le s\le T} |\overline{Y}_N(s) - y(s)| \ge \varepsilon\right) = 0.$$

Proposition D.4 page 381 shows that if (X_y, R_y) is the solution of Skorohod Problem associated with y, there exists $K_T > 0$ such that

$$\|\overline{L}_N - \overline{X}_y\|_{\infty,T} \le K_T \|\overline{Y}_N - y\|_{\infty,T},$$

therefore

$$\limsup_{N\to+\infty} \mathbb{P}\left(\sup_{0\le s\le T} |\overline{L}_N(s) - X_y(s)| \ge \varepsilon\right) \le \limsup_{N\to+\infty} \mathbb{P}\left(\sup_{0\le s\le T} |\overline{Y}_N(s) - y(s)| \ge \varepsilon/K_T\right) = 0.$$

It is easily shown that $X_y(t) = (1 + (\lambda - \mu)t)^+$ (see the example analyzed page 377). Therefore, the function $((1 + (\lambda - \mu)t)^+)$ is *the* fluid limit of $(L(t))$. This method can be generalized to the case of Jackson networks.

9.4.2 Jackson Networks

The use of Theorem D.3 page 378 to study fluid limits of Jackson networks is due to Chen and Mandelbaum[CM91]. The term oblique reflection mapping is used in this reference instead of Skorohod Problem. The previous notations for Jackson networks with d nodes are used: The vectors (λ_i) and (μ_i) for arrival and service rates and a Markovian matrix associated to the routing in the network $(p_{ij}\,;\, i,j = 0,\dots,d)$ on $\{0,\dots,d\}$ such that 0 is the only absorbing state. The vector of the number of customers at time t is denoted by $X(t) = (X_i(t)\,;\, i = 1,\dots,d)$. In this case, Equation (9.6) for the process $(X(t))$ is the solution of the integral equation, for $i = 1,\dots,d$,

$$(9.18)\quad X_i(t) = X_i(0) + \mathcal{N}_{\lambda_i}(]0,t]) \\ + \sum_{j=1}^{d}\int_{]0,t]} 1_{\{X_j(s-)\neq 0\}}\,\mathcal{N}_{\mu_j p_{ji}}(ds) - \sum_{j=0}^{d}\int_{]0,t]} 1_{\{X_i(s-)\neq 0\}}\,\mathcal{N}_{\mu_i p_{ij}}(ds).$$

By denoting $(M_i(t))$ the martingale

$$(9.19)\quad M_i(t) = [\mathcal{N}_{\lambda_i}(]0,t]) - \lambda_i t] \\ + \sum_{j=1}^{d}\int_{]0,t]} 1_{\{X_j(s-)\neq 0\}}\left[\mathcal{N}_{\mu_j p_{ji}}(ds) - \mu_j p_{ji}\,ds\right] \\ - \sum_{j=0}^{d}\int_{]0,t]} 1_{\{X_i(s-)\neq 0\}}\left[\mathcal{N}_{\mu_i p_{ij}}(ds) - \mu_i p_{ij}\,ds\right],$$

Equation (9.18) can be rewritten as

$$X_i(t) = X_i(0) + M_i(t) + \left(\lambda_i - \mu_i + \sum_{j=1}^{d}\mu_j p_{ji}\right)t \\ + \sum_{j=0}^{d}\int_0^t 1_{\{X_i(s)=0\}}\mu_i p_{ij}\,ds - \sum_{j=1}^{d}\int_0^t 1_{\{X_j(s)=0\}}\mu_j p_{ji}\,ds.$$

If $1 \le i \le d$, if

$$(9.20)\qquad R_i(t) = \mu_i \int_0^t 1_{\{X_i(s)=0\}}\,ds,$$

$$(9.21)\qquad Y_i(t) = X_i(0) + M_i(t) + \left(\lambda_i - \mu_i + \sum_{j=1}^{d}\mu_j p_{ji}\right)t,$$

one gets the identity

[CM91] H. Chen and A. Mandelbaum, *Discrete flow networks: bottleneck analysis and fluid approximations*, Mathematics of Operation Research **16** (1991), no. 2, 408–446.

$$X_i(t) = Y_i(t) + R_i(t) - \sum_{j=1}^{d} R_j(t) p_{ji},$$

which can be rewritten more compactly as

$$X(t) = Y(t) + \left(I - {}^tP\right) R(t),$$

where I is the identity matrix of $\mathbb{R}^d$. By assumption, the routing matrix $P = (p_{ij}\,;\, 1 \le i,j \le d)$ is sub-Markovian without any recurrent state. For $i = 1, \dots, d$, one has $Y_i(0) \ge 0$ and the functions $(X_i(t))$ and $(R_i(t))$ are non-negative and càdlàg. The function $t \to R_i(t)$ is continuous, non-decreasing, zero at 0 and increases only when X_i is zero. The couple (X, R) is thus the solution to the Skorohod Problem associated with Y.

Let (x_N) be a sequence of $\mathbb{N}^d$ such that $\|x_N\| = N$ and

$$\lim_{N \to +\infty} \frac{x_N}{N} = \alpha = (\alpha_i) \in \mathbb{R}_+^d,$$

in particular $\|\alpha\| = 1$. As before, if for $x \in \mathbb{N}^d$, $(W(x,t))$ is a process on $\mathbb{N}^d$, the notation $(\overline{W}_N(t)) = (W(x_N, Nt)/N$ is used.

For $N \ge 1$, $(Y_N(t))$, $(R_N(t))$ and $(M_N(t))$ are the processes defined by Equations (9.19), (9.20) and (9.21) with the initial condition $X(0) = x_N$. The associated rescaled processes are $(\overline{Y}_N(t))$, $(\overline{R}_N(t))$ and $(\overline{M}_N(t))$. In this way $(\overline{X}_N, \overline{R}_N)$ is the solution of Skorohod Problem for function $\overline{Y}_N$, in particular,

$$\overline{X}_N(t) = \overline{Y}_N(t) + \left(I - {}^tP\right) \overline{R}_N(t).$$

It is easily checked that the martingale $(\overline{M}_N(t))$ is the martingale $(\overline{M}(x_N, t))$ of Proposition 9.3, thus it converges in probability to 0 uniformly on compact sets. Therefore, the process $(\overline{Y}_N(t))$ converges to the deterministic function $(\overline{Y}_\infty(t)) = (y_i(t))$ defined by, for $1 \le i \le d$ and $t \ge 0$,

$$y_i(t) = \alpha_i + \left(\lambda_i - \mu_i + \sum_{j=1}^{d} \mu_j p_{ji}\right) t,$$

i.e., if $\varepsilon > 0$,

$$\lim_{N \to +\infty} \mathbb{P}\left(\sup_{0 \le s \le T} \left\|\overline{Y}_N(s) - \overline{Y}_\infty(s)\right\| \ge \varepsilon\right) = 0.$$

If $(\overline{X}_\infty, \overline{R}_\infty)$ is the solution of Skorohod Problem associated with the matrix P and the function $\overline{Y}_\infty$. Proposition D.4 page 381 shows that there exists some $K_T > 0$ such that

$$\left\|\overline{X}_N - \overline{X}_\infty\right\|_{\infty,T} \le K_T \left\|\overline{Y}_N - \overline{Y}_\infty\right\|_{\infty,T},$$
$$\left\|\overline{R}_N - \overline{R}_\infty\right\|_{\infty,T} \le K_T \left\|\overline{Y}_N - \overline{Y}_\infty\right\|_{\infty,T},$$

therefore

$$\mathbb{P}\left(\left\|\overline{X}_N - \overline{X}_\infty\right\|_{\infty,T} \geq \varepsilon\right) \leq \mathbb{P}\left(\left\|\overline{Y}_N - \overline{Y}_\infty\right\|_{\infty,T} \geq \varepsilon/K_T\right),$$
$$\mathbb{P}\left(\left\|\overline{R}_N - \overline{R}_\infty\right\|_{\infty,T} \geq \varepsilon\right) \leq \mathbb{P}\left(\left\|\overline{Y}_N - \overline{Y}_\infty\right\|_{\infty,T} \geq \varepsilon/K_T\right).$$

Thus, the sequence $(\overline{X}_N(t), \overline{R}_N(t))$ converges in probability uniformly on compact sets to the function $(\overline{X}_\infty(t), \overline{R}_\infty(t))$, which is deterministic since $(\overline{Y}_\infty(t))$ is. Proposition C.5 page 371 shows that if the sequence of the rescaled initial states converges, a Jackson network has only one fluid limit and it is deterministic. The following proposition has been established.

PROPOSITION 9.5. *If $(X(x,t)) = ((X_i(x,t)\,;\, i=1,\ldots,d))$ is the Markov process associated with a Jackson network whose routing matrix is P and initial state $x \in \mathbb{N}^d$, and (x_N) is a sequence of $\mathbb{N}^d$ such that*

$$\lim_{N\to+\infty} \|x_N\| = +\infty \quad \text{and} \quad \lim_{N\to+\infty} \frac{x_N}{\|x_N\|} = \alpha = (\alpha_i),$$

if $\overline{X}(x_N,t) = X(x_n, \|x_N\|t)/\|x_N\|$, the process

$$\left[\left(\overline{X}(x_N,t)\right), \left(\mu_i \int_0^t 1_{\{\overline{X}_i(x_N,u)=0\}}\,du\right)\right]$$

converges in distribution uniformly on compact sets to the deterministic function $(X_{y_\alpha}, R_{y_\alpha})$, solution of Skorohod Problem for the matrix P and the function $(y_\alpha(t))$ defined by

$$y_\alpha(t) = \alpha + \left(\lambda - \left(I - {}^tP\right)\mu\right)t,$$

$\lambda = (\lambda_i)$ and $\mu = (\mu_i)$ being respectively the vectors of arrival and service rate at the nodes of the network. The set of functions $(X_{y_\alpha}(t))$ with $\alpha \in \mathbb{R}_+^d$ of norm 1 is the set of all fluid limits of $(X(x,t))$.

REMARKS.

1. The description of the Markov process associated with a Jackson network as one component of the solution of Skorohod Problem gives readily the convergence of the rescaled processes and also the convergence of
$$\left(\mu_i \int_0^t 1_{\{\overline{X}_i(x_N,u)=0\}}\,du\right)$$
to the second component of Skorohod Problem. Note that a "bare-hand" proof of this result is not completely trivial. This is the real benefit of Skorohod Problem formulation that it is not necessary to prove the convergence of pushing processes.
2. The Skorohod Problem formulation can also be used not only to get functional law of large numbers (fluid limits) but also to derive functional central limit theorems, much in the same way as it is done in this section: If a functional central limit theorem holds for the free process $(Y(t))$, it is

also true for its reflected version. See Williams[Wil95] for a survey of these questions.

Two $M/M/1$ Queues in Tandem. The fluid limits of the network of Example 9.2.2 page 235 is analyzed as Skorohod Problem. The parameters of this Jackson network are given by $\lambda = (\lambda_1, 0)$, $\mu = (\mu_1, \mu_2)$ and the routing matrix $P = ((0,1),(0,0))$. If $\alpha = (\alpha_1, 1-\alpha_1) \in [0,1]^2$, the function $(y_\alpha(t)) = (y_{\alpha,1}(t), y_{\alpha,2}(t))$ is defined by

$$
\begin{aligned}
y_{\alpha,1}(t) &= \alpha_1 + (\lambda_1 - \mu_1)t,\\
y_{\alpha,2}(t) &= 1 - \alpha_1 + (\mu_1 - \mu_2)t.
\end{aligned}
$$

The fixed point Equation (D.4) page 378 satisfied by $(R(t))$ gives the identities

$$
\begin{aligned}
X_{y_\alpha,1}(t) &= (\alpha_1 + (\lambda_1 - \mu_1)t)^+,\\
R_{y_\alpha,1}(t) &= (\alpha_1 + (\lambda_1 - \mu_1)t)^-,\\
R_{y_\alpha,2}(t) &= 0 \vee \sup_{0\le s\le t} \left[R_{y_\alpha,1}(s) - y_{\alpha,2}(s)\right]\\
&= 0 \vee \sup_{0\le s\le t} \left[(\alpha_1 + (\lambda_1 - \mu_1)s)^- - (1-\alpha_1) - (\mu_1 - \mu_2)s\right],
\end{aligned}
$$

and the function $(X_{y_\alpha,2}(t))$ is given by

$$
X_{y_\alpha,2}(t) = y_{\alpha,2}(t) + R_{y_\alpha,2}(t) - R_{y_\alpha,1}(t).
$$

The fluid limit is therefore completely determined. When $\lambda_1 < \mu_2 \le \mu_1$, if $t_1 = \alpha_1/(\mu_1 - \lambda_1)$ and $t_2 = 1/(\mu_2 - \lambda_1)$, the above equations yield, for $t \le t_1$,

$$
R_{y_\alpha}(t) = (0,0), \quad X_{y_\alpha}(t) = [\alpha_1 + (\lambda_1 - \mu_1)t, 1 - \alpha_1 + (\mu_1 - \mu_2)t],
$$

for $t_1 \le t \le t_2$,

$$
R_{y_\alpha}(t) = [-\alpha_1 - (\lambda_1 - \mu_1)t, 0], \quad X_{y_\alpha}(t) = [0, 1 + (\lambda_1 - \mu_2)t],
$$

and for $t \ge t_2$,

$$
R_{y_\alpha}(t) = [-\alpha_1 - (\lambda_1 - \mu_1)t, -1 - (\lambda_1 - \mu_2)t], \quad X_{y_\alpha}(t) = (0,0),
$$

Convergence (9.2) page 237 is therefore established rigorously.

Fluid limits presented in Proposition 9.5 are an asymptotic description of the Markov process associated with a Jackson network. In a second step, it is natural to investigate how some of the properties of the original Markov process are inherited by fluid limits and conversely. The next proposition is connected to the stability properties of a Jackson network. It is shown that if the parameters of the network are such that the associated Markov process is ergodic (Theorem 4.11 page 95), then all its fluid limits are identically 0

[Wil95] R.J. Williams, *Semimartingale reflecting Brownian motions in the orthant*, Stochastic networks (New York) (F.P. Kelly and R.J. Williams, eds.), IMA Volumes in Mathematics and Its Applications, no. 71, Springer Verlag, 1995.

after some finite time. It is due to Meyn and Down. This property and its converse are analyzed in Section 9.5.

PROPOSITION 9.6. *If* $(\bar{\lambda}_i\,;\, i = 1,\dots,d)$ *is the unique solution of traffic equations*

$$\bar{\lambda}_i = \lambda_i + \sum_{j=1}^{N} \bar{\lambda}_j p_{ji}, \quad i = 1,\dots,d, \tag{9.22}$$

and if $\bar{\lambda}_i < \mu_i$ *for all* $i = 1,\dots,d,$, *there exists some constant* T *such that all fluid limits are* 0 *after time* T, *i.e.* $X_{y_\alpha}(t) = 0$ *for all* $t \geq T$ *and any* $\alpha \in \mathbb{R}_+^d$ *such that* $\|\alpha\| = 1$.

PROOF. By Lemma 4.10 page 93, for any vector $\lambda = (\lambda_i)$ of $\mathbb{R}_+^d$, there exists a unique $\bar{\lambda} = (\bar{\lambda}_i)$ with non-negative coordinates such that

$$\left(I - {}^tP\right)\bar{\lambda} = \lambda \quad \text{or} \quad \bar{\lambda} = A\lambda, \tag{9.23}$$

with $A = (I - {}^tP)^{-1}$, this is the vectorial expression of Equations (9.22). The matrix A is a non-negative operator, i.e. the coordinates of Ax are non-negative if those of x are also non-negative. The inequalities $\bar{\lambda}_i < \mu_i$, for all $i = 1,\dots,d$ imply the existence of an invariant probability distribution (Theorem 4.11 page 95).

Let $\alpha \in \mathbb{R}_+^d$ such that $\|\alpha\| = 1$. Coordinates of the function

$$(y_\alpha(t)) = (y_{\alpha,i}(t)\,;\, i = 1,\dots,d)$$

are clearly Lipschitz, Proposition D.4 page 381 shows that the same property is true for coordinates of the functions

$$(X_{y_\alpha}(t)) = (X_{y_\alpha,i}(t)\,;\, i = 1,\dots,d), \quad (R_{y_\alpha}(t)) = (R_{y_\alpha,i}(t)\,;\, i = 1,\dots,d).$$

In particular if h is one of these functions, h is absolutely continuous with respect to Lebesgue's measure on $\mathbb{R}$ (see the proof of Proposition C.2 page 369), then

$$h(t) = h(0) + \int_0^t \dot{h}(s)\,ds$$

holds Lebesgue almost everywhere on $\mathbb{R}_+$, where $\dot{h}$ is the Radon-Nikodym derivative of h. If

$$V_i(t) \stackrel{\text{def.}}{=} \sup_{0\leq s<t} \frac{R_{y_\alpha,i}(t) - R_{y_\alpha,i}(s)}{t-s}$$

and $V(t) = (V_i(t))$, in particular for $i = 1,\dots,d$, the inequality

$$\dot{R}_{y_\alpha,i}(t) \leq V_i(t)$$

holds Lebesgue-almost everywhere.

An Inequality for the Pushing Process. It is proved that, for $i = 1, \dots, d$, the relationship

$$\dot{R}_{y_\alpha,i}(t) + \bar{\lambda}_i - \mu_i \leq 0, \tag{9.24}$$

holds Lebesgue-almost everywhere on $\mathbb{R}_+$. This can be interpreted as follows: On each node i the arrival rate is at most $\bar{\lambda}_i$, thus the local pushing process does not need to compensate at a rate greater than $\mu_i - \bar{\lambda}_i$.

By Proposition D.3, the function R_{y_α} satisfies Equations (D.4) page 378,

$$R_{y_\alpha,i}(t) = 0 \vee \sup_{0 \leq u \leq t} \left(\sum_{j=1}^{d} p_{ji} R_{y_\alpha,j}(u) - y_{\alpha,i}(u) \right),$$

for $i = 1, \dots, d$. By looking at the various cases, it is easily checked that

$$\begin{aligned} R_{y_\alpha,i}(t) - &R_{y_\alpha,i}(s) \\ &\leq \sup_{s \leq u \leq t} \left(\sum_{j=1}^{d} p_{ji}(R_{y_\alpha,j}(u) - R_{y_\alpha,j}(s)) - (y_{\alpha,i}(u) - y_{\alpha,i}(s)) \right) \\ &\leq \sum_{j=1}^{d} p_{ji}(R_{y_\alpha,j}(t) - R_{y_\alpha,j}(s)) + \sup_{s \leq u \leq t} \left(-(y_{\alpha,i}(u) - y_{\alpha,i}(s))\right), \end{aligned}$$

holds for $0 \leq s \leq t$. The definition of y_α and Relationship (9.23) yield for u and $s \geq 0$,

$$y_\alpha(u) - y_\alpha(s) = (u - s) \left(I - {}^tP\right) (\bar{\lambda} - \mu).$$

The last inequality can thus be rewritten as

$$\begin{aligned} R_{y_\alpha}(t) - R_{y_\alpha}(s) &\leq (t - s)\, {}^tPV(t) + \sup_{s \leq u \leq t} \left[-(u - s) \left(I - {}^tP\right) (\bar{\lambda} - \mu)\right] \\ &\leq (t - s) \left[{}^tPV(t) - \left(I - {}^tP\right) (\bar{\lambda} - \mu)\right], \end{aligned}$$

with the convention that inequalities hold coordinate by coordinate. Since $\bar{\lambda} \leq \mu$, one gets

$$V \leq {}^tPV(t) - \left(I - {}^tP\right) \left(\bar{\lambda} - \mu\right), \text{ hence } \left(I - {}^tP\right) \left(V(t) + \bar{\lambda} - \mu\right) \leq 0.$$

Since the matrix A, inverse of $(I - {}^tP)$, is a non-negative operator, the inequality $V(t) + \bar{\lambda} - \mu \leq 0$ holds, and therefore Inequality (9.24) since $\dot{R}_{y_\alpha}(t) \leq V(t)$.

The function $(X_{y_\alpha}(t))$ satisfies the reflection condition

$$X_{y_\alpha}(t) = y_\alpha(t) + \left(I - {}^tP\right) R_{y_\alpha}(t)$$

for $t \geq 0$. By using $A = (I - {}^tP)^{-1}$, it follows that

$$AX_{y_\alpha}(t) = A\alpha + A\lambda t - A \left(I - {}^tP\right) \mu t + A \left(I - {}^tP\right) R_{y_\alpha}(t),$$

$$AX_{y_\alpha}(t) = A\alpha + \left(\bar{\lambda} - \mu\right) t + R_{y_\alpha}(t), \tag{9.25}$$

by Relationship (9.23). For $x \in \mathbb{R}_+^d$, define

$$f(x) = \sum_i (Ax)_i,$$

where $(Ax)_i$ is the ith coordinate of Ax. The function $g : t \to f(X_{y_\alpha}(t))$ is non-negative (using again that A is a non-negative operator), absolutely continuous since it is a linear combination of functions having this property. Identity (9.25) shows that the identity

$$\dot{g}(t) = \sum_{i=1}^{d} \left(\bar{\lambda}_i - \mu_i + \dot{R}_{y_\alpha,i}(t)\right)$$

holds Lebesgue-almost everywhere. Inequality (9.24) implies that this last quantity is non-positive for any $t \geq 0$. The function g is thus non-negative and non-increasing. If t_0 is such that $g(t_0)$ is non-zero and the functions g and R_{y_α} are differentiable at t_0, there exists some coordinate $i_0 \in \{1, \ldots, d\}$ such that $X_{y_\alpha,i_0}(t_0) \neq 0$. From Condition c) of Skorohod Problem formulation, one gets $\dot{R}_{y_\alpha,i_0}(t_0) = 0$, and with Inequality (9.24), this gives the relationship

$$\dot{g}(t_0) \leq \bar{\lambda}_{i_0} - \mu_{i_0}.$$

Thus, the inequality

$$\dot{g}(t) \leq -\eta \stackrel{\text{def.}}{=} \sup_{1 \leq i \leq d} \left(\bar{\lambda}_i - \mu_i\right) < 0$$

holds Lebesgue-almost everywhere on the set where g is non-zero. If

$$T_0 = \inf\{t \geq 0 : g(t) = 0\},$$

then for $t \leq T_0$,

$$g(t) = g(0) + \int_0^t \dot{g}(s)\, ds \leq g(0) - \eta t,$$

hence T_0 is certainly less or equal to $g(0)/\eta$. Since g is non-negative and non-increasing, it must be 0 for $t \geq T_0$. Coordinates of α being bounded by 1, $g(0)$ is also bounded with respect to α in the unit ball of $\mathbb{R}_+^d$. The constant T_0 can thus be chosen independently of α. The proposition is proved. □

9.5 Criteria for Ergodicity Properties

A general Markov process $(X(t))$ in a countable state space $\mathcal{S}$ is considered in this chapter. The main theme of this section is the relation between the ergodicity of a Markov chain and the stability of its fluid limits.

9.5.1 A Criterion for Ergodicity

The following proposition is a combination of two results, one due to Filonov (Theorem 8.6 page 213) and the other to Rybko and Stolyar[RS92].

THEOREM 9.7. *If there exists a stopping time τ and positive constants A and ε such that*

$$\limsup_{\|x\|\to+\infty} \frac{\mathbb{E}_x(\tau)}{\|x\|} \le A,$$

$$\limsup_{\|x\|\to+\infty} \mathbb{E}_x\left(\frac{\|X(\tau)\|}{\|x\|}\right) \le 1-\varepsilon, \tag{9.26}$$

and if

- *the random variable $\|X(a)\|$ is integrable for some $a>0$,*
- *for any $K\ge 0$ the set $F_K=\{x\in\mathcal{S}:\|x\|\le K\}$ is finite,*

then the Markov process $(X(t))$ is ergodic.

For K sufficiently large, if T_{F_K} is the hitting time of F_K by $(X(t))$, the inequality

$$\frac{\mathbb{E}_x(T_{F_K})}{\|x\|} \le \frac{4A}{\varepsilon}, \tag{9.27}$$

holds for any $x\in\mathcal{S}$ such that $\|x\|>K$.

PROOF. There exists $K\ge 0$ such that, the relationships

$$\mathbb{E}_x(\|X(\tau)\|)\le(1-\varepsilon/2)\|x\| \quad\text{and}\quad \mathbb{E}_x(\tau)\le 2A\|x\|$$

hold for $\|x\|\ge K$.

Therefore, if $\|x\|>K$, one gets the inequality

$$\mathbb{E}_x(\|X(\tau)\|)-\|x\|\le -\frac{\varepsilon}{2}\|x\|\le -\frac{\varepsilon}{4A}\mathbb{E}_x(\tau).$$

Since the set $\{x\in\mathcal{S}:\|x\|\le K\}$ is finite, Theorem 8.13 page 224 applied with the function $f(x)=\|x\|$ shows that $(X(t))$ is an ergodic Markov process. The bound on the expected value of T_{F_K} is a consequence of Inequality (8.19) of this theorem. □

Condition (9.26) requires that after some time τ, of the order of $\|x\|$, the function $t\to\|X(x,t)\|/\|x\|$ is strictly less than 1, the norm of the initial state.

COROLLARY 9.8. *If, for any $K\ge 0$, the set*

$$F_K=\{x\in\mathcal{S}:\|x\|\le K\}$$

is finite and there exists some constant T such that

[RS92] Op. cit. page 232.

$$\limsup_{\|x\|\to+\infty} \mathbb{E}_x\left(\left\|\overline{X}(T)\right\|\right) = \limsup_{\|x\|\to+\infty} \frac{\mathbb{E}_x\left(\left\|X(\|x\|T)\right\|\right)}{\|x\|} = 0,$$

then the Markov process $(X(t))$ *is ergodic.*

PROOF. It is sufficient to remark that $\tau = \|X(0)\|T$ is a stopping time such that $\mathbb{E}_x(\tau)/\|x\| = T$ and to apply the above theorem. □

The criterion of Chapter 8, Theorem 8.13 page 224, for the ergodicity of Markov processes requires an appropriate Lyapunov function. In practice it is not always easy to exhibit such a function. Despite that Theorem 9.7 is a straight application of Theorem 8.13, it turns out that it is easier, in general, to show that the fluid limits converge to 0 rather than trying to find some good energy function for the Markov process. Indeed, quite often, fluid limits can be described (partially) through some set of ordinary differential equations allowing then such a study.

A typical situation to apply this corollary is as follows: The Markov process $(X(t))$ is ergodic if

1. the state space $\mathcal{S}$ is included in $\mathbb{R}^d$ for $d \in \mathbb{N}$ and the assumptions on the F_K's are defined in the corollary.
2. There exists some $T \in \mathbb{R}_+$ such that $z(T) = 0$ for any fluid limit $(z(t))$.

It should be kept in mind that this is *not* a general situation. There are some cases where the Markov process is ergodic but there does not exist a deterministic time at which all fluid limits are at 0. (See Section 9.5.3.)

Example : Jackson Networks. Proposition 9.6 shows that if the solution $\bar{\lambda} = (\bar{\lambda}_i)$ of traffic Equations (9.22) page 256 is such that $\bar{\lambda}_i < \mu_i$ holds for $i = 1, \ldots, d$, then all the fluid limits are deterministic and 0 after some fixed instant. By using the uniform integrability result of Proposition 9.2 (for the L_1-convergence of the rescaled processes to the fluid limits), one gets the ergodicity of the Markov process describing the Jackson network. This result has been already obtained in Chapter 4 because this Markov process has a product invariant measure (which has a finite mass under the above conditions). If the distributions of services, or inter-arrivals are not anymore exponential but general with the same expected values, invariant distributions are not known in this case. It is not difficult to show that, nevertheless, the corresponding fluid limits do not change for this network and therefore the (Harris) ergodicity can also be proved in that case too (by adding forward recurrence times and using some general technical results in this domain). Other examples are presented in Section 9.6.

More Complex Phenomena: Multi-class FIFO Networks. The state space of the Markov processes encountered up to now were embedded into some finite dimensional vector space. The following infinite dimensional example (already considered in Section 4.4.1 page 92) shows that the geometry of the state space is a very important factor. For $J \in \mathbb{N}$, $J \geq 2$ and $j = 1, \ldots J$, λ_j and μ_j are positive real numbers. It is assumed that J Poissonian arrivals flows arrive at a single server queue with rate λ_j for $j = 1, \ldots, J$ and customers from the jth flow require an exponentially distributed service with parameter μ_j. All the arrival flows are assumed to be independent. The service discipline is FIFO.

A natural way to describe this process is to take the state space of the finite strings with values in the set $\{1, \ldots, J\}$, i.e. $\mathcal{S} = \cup_{n \geq 0}\{1, \ldots, J\}^n$, with the convention that $\{1, \ldots, J\}^0$ is the set of the null string. See Section 7.4 page 196 for a similar representation of the $M/G/1$ queue with the LIFO discipline. See also Serfozo [**47**]. If $n \geq 1$ and $x = (x_1, \ldots, x_n) \in \mathcal{S}$ is the state of the queue at some moment, the customer at the kth position of the queue comes from the flow with index x_k, for $k = 1, \ldots, n$.

The length of a string $x \in \mathcal{S}$ is defined by $\|x\|$. Note that $\|\cdot\|$ is not, strictly speaking, a norm. For $n \geq 1$ there are J^n vectors of length n; the state space has therefore an exponential growth with respect to that function. Hence, if the string valued Markov process $(X(t))$ describing the queue is transient then certainly the length $\|X(t)\|$ converges to infinity as t gets large. Because of the large number of strings with a fixed length, the process $(X(t))$ itself has, a priori, infinitely many ways to go to infinity.

In the finite-dimensional state space, the growth of the volume of the ball of radius n is only polynomial with respect to n. When the dynamic is locally homogeneous as for Markov processes of class $(\mathcal{C})$, there is a finite number of categories of large states: If a state is large then some of its coordinates are large. In practice, the fluid limits are obtained by letting the norm of the initial state go to infinity, hence one has to consider all the subsets of the coordinates that may converge to infinity and study the corresponding fluid limits. This approach is not possible when strings are considered. Note that this does not mean that the finite dimensional case is easy since dynamical systems in $\mathbb{R}^d$ can be, in general, quite complicated.

To conclude, fluid limits are a very important ingredient to prove the ergodicity of Markov processes but other methods may be required to describe correctly the dynamic of Markov process.

In the case of a single server queue, all these questions are quite formal since the workload of the queue at time $t \geq 0$ (the sum of services of customers in the queue) is a simple one dimensional Markov process. The situation is entirely different when a network of two such queues is considered. Bramson[Bra94] has shown that complicated phenomena could indeed occur.

[Bra94] M. Bramson, *Instability of FIFO queueing networks*, Annals of Applied Probability 4 (1994), no. 2, 414–431.

9.5.2 Absorption at 0 of Fluid Limits

The examples of ergodic Markov processes seen before have the same property: Their fluid limits reach and stay at 0 after some finite time. (See also the discussion at the end of the proof of Proposition 5.16 page 125 for the $M/M/1$ queue.)

There is an intuitive explanation to this phenomenon: The Markov process $(X(x,t))$ reaches a fixed finite set in a duration of time asymptotically linear with respect to $\|x\|$. Then, after some finite time (negligible for the fluid time scale), the process is at equilibrium and therefore essentially stays in the neighborhood of this finite set, consequently $\|\overline{X}\|(x,t) = \|X(x,\|x\|\,t))\|/\|x\|$ tends to 0 as $\|x\|$ gets large.

The next proposition shows that under the assumptions of Proposition 9.8 and a mild hypothesis of relative compactness this general explanation is true.

PROPOSITION 9.9. *If the conditions*
a) for $K \geq 0$, the set $\{x \in \mathcal{S} : \|x\| \leq K\}$ is finite,
b) there exist some $T > 0$ such that

$$\limsup_{\|x\|\to+\infty} \frac{\mathbb{E}_x\left(\|X(\|x\|T)\|\right)}{\|x\|} = 0,$$

c) the set of càdlàg random processes

$$\left\{\left(\frac{\|X(y,\|x\|\,t)\|}{\|x\|}\right) ; x, y \in \mathcal{S}, \|y\| \leq \|x\|\right\}$$

is relatively compact,

are satisfied, then for any fluid limit $(W(t))$, there exists some random variable τ, $\mathbb{P}$-almost surely finite such that

$$\mathbb{P}\left(\sup_{t>\tau} W(t) = 0\right) = 1. \tag{9.28}$$

Conditions a) and b) are the conditions for ergodicity of the process of Corollary 9.8. Condition c) is apparently stronger. Nevertheless, in practice, the existence of fluid limits is usually obtained by showing that the set

$$\left\{\left(\overline{X}(x,t)\right) ; x \in \mathcal{S}\right\} = \left\{\left(\frac{X(x,\|x\|\,t)}{\|x\|}\right) ; x \in \mathcal{S}\right\}$$

is relatively compact. In general this property *implies* Condition c) of the proposition. See Section 9.3.

PROOF. Corollary 9.8 shows that $(X(t))$ is an ergodic Markov process. Let $K > 0$ be such that the set $F = \{x \in \mathcal{S} : \|x\| \leq K\}$ is non-empty. If T_F is the hitting time of F by the process $(X(t))$, the variable T_F is almost surely finite and Inequality (9.27) of Proposition 9.8 gives the existence of a constant C such that

$$\mathbb{E}_x(T_F) \le C\|x\|$$

when $\|x\| > K$. Consequently, if τ_F is the rescaled hitting time of F

$$\tau_F(x) = \inf\{t > 0 : \|X(x, \|x\|\, t)\| \le K\},$$

or equivalently, $\tau_F(x) = T_F/\|x\|$. One gets the inequality

$$\sup_{x:\|x\|>K} \mathbb{E}_x(\tau_F(x)) \le C,$$

the set of random variables $\tau_F(x)$, $x \in \mathcal{S}$ is relatively compact. By definition of a fluid limit, there exists a sequence (x_N) of $\mathcal{S}$ such that $(\overline{X}(x_N,t))$ converges (as a process) in distribution to $(W(t))$ when N goes to infinity. This sequence of processes is in particular relatively compact, the sequence

$$\left\{\left((\overline{X}(x_N,t)), \tau_F(x_N)\right) \; ; \; N \ge 1\right\}$$

is therefore also relatively compact in the space of probability distributions on $D(\mathbb{R}_+, \mathbb{R}) \times \mathbb{R}_+$ (the product of two compacts is compact for the product topology). By taking a sub-sequence, it can be assumed that the above sequence converges in distribution to $((W(t)), \tau)$. In particular, for $\varepsilon > 0$ and $0 < a \le b$,

$$\text{(9.29)} \quad \liminf_{N\to+\infty} \mathbb{P}\left(\sup_{a\le t\le b} \left\|\overline{X}(x_N,t)\right\| > \varepsilon,\ \tau_F(x_N) < a\right) \\ \ge \mathbb{P}\left(\sup_{a\le t\le b} W(t) > \varepsilon,\ \tau < a\right),$$

indeed, if O is an open set of $D(\mathbb{R}_+, \mathbb{R}) \times \mathbb{R}_+$, then

$$\liminf_{N\to+\infty} \mathbb{P}\left(\left[(\|\overline{X}(x_N,t)\|), \tau_F(x_N)\right] \in O\right) \ge \mathbb{P}\left(\left[(W(t)), \tau\right] \in O\right),$$

by Theorem 1.4 of Billingsley [7]. The strong Markov property of $(X(t))$ for the stopping time

$$T_F = \|X(0)\|\tau_F(X(0))$$

gives the relationship

$$\mathbb{P}\left(\sup_{a\le t\le b} \left\|\overline{X}(x_N,t)\right\| > \varepsilon,\ \tau_F(x_N) < a\right) = \\ \mathbb{E}_{x_N}\left[1_{\{\tau_F(x_N)<a\}} \mathbb{P}_{X(T_F)}\left(\sup_{a\le t\le b} \frac{\|X(t-\tau_F(x_N))\|x_N\|)\|}{\|x_N\|} > \varepsilon\right)\right] \\ \le \sum_{y\in F} \mathbb{P}_y\left(\sup_{0<t\le b} \frac{\|X(\|x_N\|\,t)\|}{\|x_N\|} > \varepsilon\right).$$

If each of the terms of the last sum tends to 0 as N goes to infinity, the set F being finite, Relationship (9.29) will give the identity, for $0 \le a \le b$,

$$\mathbb{P}\left(\sup_{a\leq t\leq b} W(t) > \varepsilon,\ \tau < a\right) = 0, \quad \text{hence } \mathbb{P}\left(\sup_{t\geq a} W(t) > \varepsilon,\ \tau < a\right) = 0,$$

by letting b go to infinity. The almost-sure right-continuity property of $(W(t))$ at any point yields the relationship

$$\mathbb{P}\left(\sup_{t>\tau} W(t) > \varepsilon\right) = \mathbb{P}\left(\bigcup_{a\in\mathbb{Q}}\left\{\sup_{t\geq a} W(t) > \varepsilon,\ \tau < a\right\}\right)$$
$$\leq \sum_{a\in\mathbb{Q}} \mathbb{P}\left(\sup_{t\geq a} W(t) > \varepsilon,\ \tau < a\right).$$

Relationship (9.28) is proved.

Thus, it remains to prove that for $y \in \mathcal{S}$ the quantity

$$\mathbb{P}_y\left(\sup_{0<t\leq b}\frac{\|X(\|x_N\|\,t)\|}{\|x_N\|} > \varepsilon\right) = \mathbb{P}\left(\sup_{0<t\leq b}\frac{\|X(y,\|x_N\|\,t)\|}{\|x_N\|} > \varepsilon\right)$$

tends to 0 when N goes to infinity.

Since the Markov process $(X(t))$ is ergodic, the random variable $X(y,t)$ converges in distribution as t tends to infinity. The same is true for $\|X(y,t)\|$ thus, for $\varepsilon > 0$, there exist K and $t_0 > 0$ such that if $t \geq t_0$, then $\mathbb{P}(\|X(y,t)\| \geq K) \leq \varepsilon$. The variable $\sup_{0\leq s\leq t_0}\|X(y,s)\|$ being almost everywhere finite (remember that X is càdlàg, hence $\mathbb{P}$-a.s. bounded on any finite interval), K can thus be chosen so that the inequality

$$\mathbb{P}\left(\sup_{0\leq s\leq t_0}\|X(y,s)\| \geq K\right) \leq \varepsilon, \quad \text{hence } \sup_{t\geq 0}\mathbb{P}(\|X(y,t)\| \geq K) \leq \varepsilon$$

holds. If $p \geq 1$, $0 \leq t_1 \leq \cdots \leq t_p$ and $\delta > 0$, the inequality

$$\mathbb{P}\left(\sup_{1\leq i\leq p}\frac{\|X(y,\|x_N\|t_i)\|}{\|x_N\|} \geq \delta\right) \leq p\varepsilon$$

is thus true as soon as $\|x_N\| \geq K/\delta$. One concludes that the finite marginal distributions of the process $(\|X(y,\|x_N\|\,t)\|/\|x_N\|)$ converge in distribution to the corresponding marginal of the process identically equal to 0. By Condition d), the sequence of processes

$$\left(\frac{\|X(y,\|x_N\|\,t)\|}{\|x_N\|}\right)$$

is relatively compact. Proposition C.7 page 372 shows that any limiting point of this sequence is necessarily the trivial process equal to 0. The sequence of processes is therefore converging to 0. For $T > 0$, the very definition of Skorohod Topology on $D([0,T],\mathbb{R})$ implies the continuity of the mapping

$$(z(t)) \to \sup_{0<t\leq T} z(t)$$

on this space, one concludes that the identity

$$\lim_{N\to+\infty} \mathbb{P}\left(\sup_{0<t\leq b} \frac{\|X(y,\|x_N\|\,t)\|}{\|x_N\|} > \varepsilon\right) = 0,$$

holds for any $b > 0$. The proposition is proved. □

Under some conditions, Proposition 9.9 shows that for any fluid limit of the Markov process, there exists some random time after which this fluid limit is identically 0. In the case of an ergodic Jackson network, a stronger property is true (Proposition 9.5): There exists some deterministic time after which *all* fluid limits are zero. The next section gives an example where the variable τ *cannot* be chosen to be deterministic. Proposition 9.9 cannot thus be improved in that direction, the variable τ is indeed random in general.

9.5.3 An Interesting Example

In this section, an example of a non-ergodic Markov process with fluid limits converging to 0 is given. At the same time, justifying this detailed development, an interesting multiplicative phenomenon is presented.

A Markov process $(X(t))$ is defined on the state space $\mathcal{S} = \mathbb{Z} \times \mathbb{N}$. The $\mathcal{Q}$-matrix $Q = (q_{ab} : a, b \in \mathcal{S})$ is as follows: In the interior of $\mathcal{S}$, for $x \geq 1$ and $y \geq 1$,

$$q_{(x,y),(x,y-1)} = 1, \qquad q_{(-x,y),(-x,y-1)} = 1,$$
$$q_{(x,y),(x+1,y)} = \beta, \qquad q_{(-x,y),(-x-1,y)} = \alpha,$$

on the vertical axis denoted by Δ,

$$q_{(0,y),(1,y)} = q_{(0,y),(-1,y)} = 1,$$

and on the horizontal axis, for $x \geq 0$,

$$q_{(x,0),(0,x+1)} = q_{(-x,0),(0,x+1)} = 1.$$

When the Markov process is on the vertical axis Δ, it jumps into the left or right orthant with equal probability. Inside an orthant its vertical component decreases at rate one and its horizontal coordinate grows at rate α or β until the horizontal axis is reached. From the horizontal axis, the Markov process goes to the vertical axis at rate 1 by exchanging the coordinates (and adding 1). The last step is formal: The two half-axes $\{(0,0),(-1,0),(-2,0),\ldots\}$ and $\{(1,0),(2,0),\ldots\}$ could be identified with the vertical axis $\{(0,1),(0,2),\ldots\}$, and the state space $\mathcal{S}$ should be seen as the union of two cones sticked along a vertical axis.

For $\xi > 0$, $\mathcal{N}_\xi$ denotes a Poisson process on $\mathbb{R}_+$ with rate ξ, when different Poisson process are used, they are assumed implicitly to be independent (an upper index is added to have Poisson processes with the same rate). The quantity $\mathcal{N}_\xi(I)$ is the number of points of $\mathcal{N}_\xi$ in the interval I.

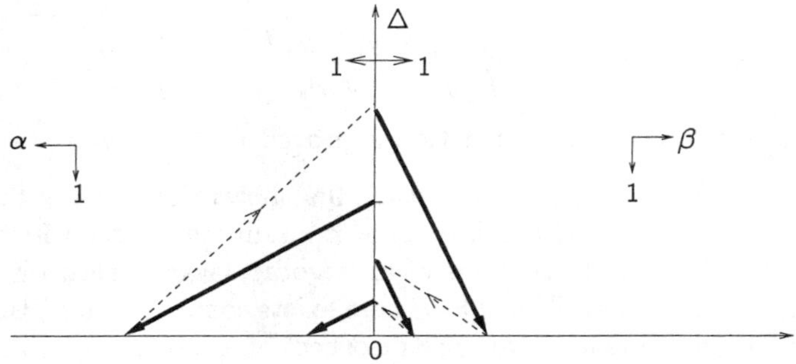

FIG. 9.3. Fluid Limits: A Multiplicative Example

The Embedded Markov Chain on Δ. Starting from the vertical axis Δ at level $M_0 = x$, the Markov process jumps into an orthant. Since the vertical component decreases at rate 1, the time to reach the horizontal axis as the same distribution as

$$t_1 = E_1 + \cdots + E_{M_0}, \tag{9.30}$$

where (E_i) is an i.i.d. sequence of exponentially distributed random variables with parameter 1. At that time, the absolute value of the first coordinate has the same distribution as $\mathcal{N}_{A_1}([0, t_1])$, where A_1 is some random variable such that $\mathbb{P}(A_1 = \alpha) = \mathbb{P}(A_1 = \beta) = 1/2$, A_1 is determined by the orthant where the Markov process has jumped. Then, the Markov process returns to the vertical axis at level

$$M_1 = \mathcal{N}_{A_1}([0, t_1]) + 1.$$

A cycle of the process $(X(t))$ has thus been described. For $n \geq 1$, M_n is defined as the second component of the process at the nth passage of $(X(t))$ on the vertical axis Δ.

LEMMA 9.10. *The sequence (M_n) is an irreducible Markov chain on the state space $\mathbb{N} - \{0\}$. If $\mathbb{E}(\log A_1) = (\log \alpha + \log \beta)/2 < 0$, it is ergodic. If, in addition, $\mathbb{E}(A_1) = (\alpha + \beta)/2 > 1$, then the first moment of the invariant probability distribution is infinite.*

PROOF. Markov property as well as irreducibility property are clear. Ergodic Theorem for i.i.d. sequences and Poisson processes shows that the convergence

$$\frac{M_1}{x} = \frac{1}{x}\mathcal{N}_{A_1}\left([0, E_1 + \cdots + E_x[\right) \to A_1, \tag{9.31}$$

holds almost surely and in L_1 as x tends to infinity. For $\eta > 1/e$, $y > 0$ and $a > 0$, Chebishov's Inequality gives the relationship

$$\mathbb{P}\left[ay - \mathcal{N}_a([0, y[) \geq \eta a y\right] \\ \leq e^{-\eta y}\mathbb{E}\left[\exp\left(ay - \mathcal{N}_a([0, y[)\right)\right] = \exp(-ay(\eta - 1/e)).$$

This shows that the probability of the event $\{\mathcal{N}_a([0, y[)/y \leq a(1 - \eta)\}$ is exponentially small with y (the ratio $\mathcal{N}_a([0, y[)/y$ is mostly away from 0). This implies that the convergence

$$\frac{1}{y}\log \mathcal{N}_a([0, y[) \to \log a$$

holds in L_1 as y goes to infinity. A similar argument with partial sums of the sequence (E_i) gives finally that $\log(M_1/x)$ converges to $\log A_1$ in $L_1(\mathbb{P}_x)$ as x goes to infinity, thus

$$\lim_{x\to+\infty} \mathbb{E}(\log(M_1/x)) = \mathbb{E}(\log A_1).$$

Therefore, if the inequality $\mathbb{E}(\log A_1) < 0$ is true, for $\varepsilon = -\mathbb{E}(\log A_1)/2$, there exists some x_0 such that if $x \geq x_0$, then

$$\mathbb{E}_x(\log M_1) - \log x \leq \mathbb{E}(\log A_1) + \varepsilon = \mathbb{E}(\log A_1)/2 < 0.$$

Corollary 8.7 can then be used with the function $f(x) = \log x$ to conclude that the Markov chain (M_n) is ergodic.

Assume now that $\mathbb{E}(A_1) > 1$ and $\mathbb{E}(\log A_1) < 0$ and that the distribution of M_0 is the invariant distribution of the Markov chain. If $\mathbb{E}(M_0)$ is finite, by invariance

$$\mathbb{E}(M_0) = \mathbb{E}(M_1) = \mathbb{E}\left[\mathcal{N}_{A_1}([0, E_1 + \cdots + E_{M_0}[)\right] + 1 \\ = \mathbb{E}(A_1)\mathbb{E}(M_0) + 1 > \mathbb{E}(M_0),$$

contradiction. Therefore $\mathbb{E}(M_0) = +\infty$. The lemma is proved. □

Equation (9.31) suggests that when the quantity M_0 is large, the next step is at $M_1 \sim A_1 M_0$. The transitions of this embedded Markov chain of $(X(t))$ are basically multiplicative. If the condition $\mathbb{E}(\log A_1) < 0$ implies ergodicity, the tail distribution of the invariant distribution of (M_n) does not have necessarily an exponential decay since its first moment is infinite when $\mathbb{E}(A_1) > 1$. This situation is quite unconventional compared to the case of Markov chains with additive transitions (like random walks) where the property of exponential decay of the invariant distribution holds in general. See Section 3.3 of Chapter 3 for example. Kesten [Kes73] analyzes the tail distributions of some related multi-dimensional Markov chains, see also Bougerol and Lacroix [8].

This property of the invariant distribution of (M_n) is crucial to prove another unconventional feature: The set of parameters α and β for which the Markov process $(X(t))$ is null-recurrent has a non-empty interior. See Proposition 9.13.

[Kes73] H. Kesten, *Random difference equations and renewal theory for products of random matrices*, Acta Mathematica **131** (1973), 207–248.

Fluid Limits. The fluid limits between two visits to the vertical axis are first considered. Let $(m_N) = (0, b_N)$ be a sequence of initial states for $(X(t))$ such that the sequence (b_N/N) converges to some $b > 0$. In particular

$$\lim_{N\to+\infty} X(m_N, 0)/N = (0, b).$$

Between time 0 and time t_1 to return to the axis Δ (t_1 is the sum of b_N i.i.d. exponential random variables with parameter 1), the norm of the process $(X(m_N, t))$ is given by

$$\|X(m_N, t)\| = \mathcal{N}_{A_1}([0, t[) + \left|b_N - \mathcal{N}_1([0, t[)\right|.$$

It is easily seen that, that, $\mathbb{P}_{m_N}$-almost surely,

$$\lim_{N\to+\infty} t_1/N \stackrel{\text{def.}}{=} \tau_1 = b, \text{ and } \lim_{N\to+\infty} X(m_N, t_1)/N = A_1 b.$$

As N tends to infinity, with the same arguments as for the example of a random walk on $\mathbb{Z}$ (page 234), the convergence in distribution of processes

$$\left[\|X(m_N, Nt)\|/N; 0 \leq t < \tau_1\right] \to \left[b + (A_1 - 1)t\,; 0 \leq t < \tau_1\right]$$

holds. The fluid limit between two visits to the vertical axis has been described. Fluid limits of the process can now be expressed explicitly.

A sequence of arbitrary initial states $(x_N) = (u_N, v_N)$ whose norm is N is now considered. By taking a subsequence, it can be assumed that there exists some $v \in [0, 1[$ such that the sequence (v_N/N) converges to v (the case $v = 1$ is similar to the case which has just been investigated). One defines γ_N as β if u_N is non-negative and α otherwise. Note that, since $v < 1$, γ_N is constant after some finite rank, γ denotes the limit of the sequence (γ_N). The first visit to the axis Δ occurs at time

$$t_1 = E_1 + \cdots + E_{v_N} \sim vN,$$

and, at that time, its second coordinate renormalized by N is such that

$$\left[|u_N| + \mathcal{N}_{\gamma_N}([0, E_1 + \cdots + E_{v_N}]) + 1\right]/n \sim b \stackrel{\text{def.}}{=} \gamma v + 1 - v.$$

If the three half-axes are identified, with the same methods as in Proposition 9.3, it is not difficult to show that there is only one fluid limit $(\overline{x}(t))$ corresponding to the sequence of initial states (x_N) and that this fluid limit $(\overline{x}(t))$ is almost surely continuous.

The fluid limit $(\overline{x}(t))$ can be defined as follows. Take (A_n) an i.i.d sequence of random variables with the same distribution as A_1, and (τ_n) is defined by $\tau_0 = v$ and

$$\tau_{n+1} = \tau_n + b\prod_{i=1}^{n} A_i = \sum_{k=1}^{n} b\prod_{i=1}^{k} A_i. \tag{9.32}$$

If t_n is the instant of the nth visit to Δ, the variable τ_n is just the limit of t_n/N as N gets large. Furthermore, for $t \in [\tau_n, \tau_{n+1}[$,

$$\frac{\|\overline{x}(t)\|}{b} = \prod_{1}^{n} A_i + (A_{n+1} - 1)(t - \tau_n),$$

in particular

$$\frac{\|\overline{x}(t)\|}{b} \leq \prod_{1}^{n} A_i + \prod_{1}^{n+1} A_i, \ t \in [\tau_n, \tau_{n+1}[. \tag{9.33}$$

Assume that $\mathbb{E}(\log A_1) < 0$. By the Law of large numbers, $\mathbb{P}$-almost surely,

$$\lim_{k \to +\infty} \frac{1}{k} \sum_{i=1}^{k} \log A_i = \mathbb{E}(\log A_1) < 0,$$

hence for $\varepsilon = -\mathbb{E}(\log A_1)/2$, there exists some (random) $k_0 \in \mathbb{N}$ such that the general term of the sum (9.32) is bounded by $\exp(-\varepsilon k)$ when $k \geq k_0$. Therefore, the sequence (τ_n) is almost surely converging to a finite limit T given by

$$T(\overline{x}) \stackrel{\text{def.}}{=} \tau_0 + b \sum_{k=0}^{+\infty} \prod_{i=1}^{k} A_i.$$

Inequality (9.33) and the continuity property of $(\overline{x}(t))$ give that $\overline{x}(T) = 0$. By using the same method as in the proof of Proposition 5.16 page 125) for the fluid limits of the $M/M/1$ queue, it is not difficult to show that $\overline{x}(t) = 0$ for $t \geq T(\overline{x})$. In fact, the norm of the original process $(X(t))$, outside the three half-axes, can be seen as an $M/M/1$ queue whose arrival rate is either α or β. Note that if $T(\overline{x})$ is almost surely finite, it is not necessarily integrable since $\mathbb{E}(A_1)$ may be greater than 1. The variable $T(\overline{x})$ depends on the sequence of initial states (x_N) only through the variables $\tau_0 = v$ and $b = \gamma v + (1 - v)$, it is therefore stochastically dominated by the variable

$$S \stackrel{\text{def.}}{=} 1 + [1 + \max(\alpha, \beta)] \sum_{k=0}^{+\infty} \prod_{i=1}^{k} A_i$$

which does not depend on the sequence of initial states. The following proposition has thus been proved.

PROPOSITION 9.11. *Under the condition* $\log \alpha + \log \beta < 0$, *for any fluid limit* $(\overline{x}(t))$ *of* $(X(t))$, *there exists some random variable* $T(\overline{x})$ *such that,* $\overline{x}(t) = 0$ *for* $t \geq T(\overline{x})$. *The variables* $T(\overline{x})$, *with* $\overline{x}$ *varying in the set of fluid limits, are stochastically dominated by a variable* S *which is integrable if* $\alpha + \beta < 2$.

Note that, as in Section 9.2.2 page 238, the fluid limits are *not* deterministic processes. For this model, this property turns to be out quite important. As it will be seen, the above proposition does not imply that the Markov process $(X(t))$ is ergodic as soon as $\mathbb{E}(\log A_1) < 0$.

Recurrence Properties. Since, outside the half-axes, the Markov process grows at rate $\lambda \mathbb{E}(A_1) - 1$, then clearly condition $\alpha + \beta < 2$ should imply ergodicity of the Markov process $(X(t))$. This is indeed the case as next proposition shows.

PROPOSITION 9.12. *Under the condition $\alpha + \beta < 2$, the Markov process $(X(t))$ is ergodic.*

PROOF. Theorem 9.7 is used. For $n \geq 1$, with the same notations as in the proof of Proposition 9.11, if t_n is the instant of the nth visit to Δ, then

$$\limsup_{\|x\| \to +\infty} \frac{\mathbb{E}_x(t_n)}{\|x\|} = \sup_{\overline{x}} \mathbb{E}(\tau_n) \leq \sup_{\overline{x}} \mathbb{E}(T(\overline{x})) \leq \mathbb{E}(S) < +\infty,$$

where the supremum is taken over all fluid limits.

Inequality (9.33) shows that

$$\begin{aligned}
\limsup_{\|x\| \to +\infty} X(x, t_n)/\|x\| &\leq \sup_{\overline{x}} \mathbb{E}(\|\overline{x}(\tau_n)\|) \\
&\leq (1 + \alpha + \beta)\, \mathbb{E}\left[\prod_1^n A_i + \prod_1^{n+1} A_i\right] \\
&= (1 + \alpha + \beta)\, [1 + \mathbb{E}(A_1)]\, \mathbb{E}(A_1)^n.
\end{aligned}$$

Hence, there exists some n_0 such that

$$\limsup_{\|x\| \to +\infty} \frac{X(x, t_{n_0})}{\|x\|} \leq 1/2.$$

Since t_{n_0} is a stopping time, the assumptions of Theorem 9.7 are satisfied, one concludes that $(X(t))$ is ergodic. □

The next result is the main motivation of this section. It shows that if all fluid limits are constant and equal to 0 after some random time, this does not imply the ergodicity of the Markov process, even if the random time can be stochastically bounded independently of the fluid limit.

PROPOSITION 9.13. *If the conditions $\log \alpha + \log \beta < 0$ and $\alpha + \beta > 2$ are satisfied, the Markov process $(X(t))$ is null-recurrent.*

The fact that all fluid limits converge to 0 is not enough to ensure ergodicity of a Markov process. (See the remark below Corollary 9.8.)

PROOF. The invariant distribution of the Markov chain (M_n) is denoted by π. The sequence $((0, M_n))$ is the set of points of the vertical axis Δ successively visited by $(X(t))$ and

$$T = \inf\{t > 0 : X(t-) \notin \Delta, X(t) \in \Delta\},$$

is the hitting time of Δ. If the process is initially on Δ, then T has the same distribution as the sum of the variable t_1 defined by Equation (9.30) and an independent exponential random variable F_1 with parameter 1 (F_1 is the time necessary to jump from the horizontal axis to Δ).

The measure μ on $\mathbb{Z}\times\mathbb{N}$ is defined by

$$\int_{\mathbb{Z}\times\mathbb{N}} f(z)\,\mu(dz) = \mathbb{E}_{\tilde{\pi}}\left(\int_0^T f(X(t))\,dt\right),$$

for a non-negative function f on $\mathbb{Z}\times\mathbb{N}$. As usual, $E_{\tilde{\pi}}(\cdot)$ denotes the expected valued with the initial distribution $\tilde{\pi} = \delta_0 \otimes \pi$ on $\mathbb{N}\times\mathbb{N}$ for $(X(t))$.

With the same arguments as in the proof of Proposition 8.12 page 223, it is easily seen that μ is an invariant measure of the Markov process $(X(t))$. To prove null-recurrence, it is enough to show that there is invariant measure μ with an infinite mass, i.e. that $\mu(\mathbb{Z}\times\mathbb{N}) = +\infty$ holds. This is a direct consequence of Lemma 9.10 since

$$\mu(\mathbb{Z}\times\mathbb{N}) = \mathbb{E}_{\tilde{\pi}}(T) = 1 + \mathbb{E}_{\tilde{\pi}}(t_1) = 1 + \mathbb{E}_{\pi}(M_0) = +\infty.$$

The proposition is proved. □

To conclude, this example shows that if a Markov is ergodic, there does not exist necessarily a deterministic time at which all fluid limits are zero. And conversely if all fluid limits are zero after some time, this does not imply ergodicity of the Markov process

9.6 Local Equilibrium of a Rescaled Markov Process

The purpose of this section is to present a quite frequent phenomenon occurring in the analysis of fluid limits. It is related to the fact that, quite often, invariant distributions of some Markov chains have to be derived to get closed form expression of fluid limits. Despite this phenomenon is quite well known, it does not seem that there is some satisfactory result of practical use in a large set of examples. For this reason, an informal discussion introduces the phenomenon to give some insight on this problem. An example is then presented and completely analyzed.

Fluid limits of a Markov process $(X(t))$ are limiting points of the rescaled process. When the state space is of dimension greater than 1, some coordinates of the initial state x, indexed by a subset J, may be negligible with respect to $\|x\|$, for example $x_j = 0$ for $j \in J$. In some cases, the vector consisting only of these coordinates can reach quickly some equilibrium which is defined as a *local equilibrium*. At the "fluid" time scale, from the point of view of the "large" coordinates $(X_j(t); j \notin J)$, the vector $(X_j(t); j \in J)$ is at equilibrium (in particular the coordinates in J of the fluid limit are 0).

- For example, take a network of two $M/M/1$ queues in tandem with arrival rate λ and service rates μ_1 and μ_2 with $\lambda < \mu_1$. Starting from the initial state $(0, N)$, at the fluid scale, the first queue is at equilibrium hence the arrival process at the second queue is Poisson with rate λ, hence the fluid limit is in this case $(0, (1 + (\lambda - \mu_2)t)^+)$.

This local equilibrium remains as long as all the components of the vector $(X_j(t); j \not\in J)$ are "large", i.e. far away from the sets where the dynamic changes. As soon as one of the coordinates indexed by J of the fluid limit is 0, the set J will be enlarged with one coordinate i_0. Note that this does not imply that the set J grows since the corresponding vector $(X_j(t); j \in J \cup \{i_0\})$ may not reach some equilibrium, and perhaps some of its component may grow without bounds.

To be more specific, the simple case of an $M/M/1$ queue governed by an ergodic Markov process is investigated.

An $M/M/1$ Queue in a Markovian Environment

In this section, $(Z(t))$ is an ergodic Markov process with values in a countable state space $\mathcal{S}$, its $\mathcal{Q}$-matrix is $Q = (q_{xy})$ and π is its invariant probability distribution. If (λ_x) and (μ_x) are two sequences of positive numbers, the Poisson processes $(\mathcal{N}_{\lambda_x}, \mathcal{N}_{\mu_x}; x \in \mathcal{S})$ are defined on the same probability space, are independent, and independent of $(Z(t))$.

A Markov process with values in $\mathcal{S} \times \mathbb{N}$ is analyzed, its $\mathcal{Q}$-matrix is given by $R = (r_{(x,i),(y,j)}\, ; x, y \in \mathcal{S},\ i, j \in \mathbb{N})$ where, for x, $y \in \mathcal{S}$ and $i \in \mathbb{N}$,

$$r_{(x,i),(y,i)} = q_{xy},$$
$$r_{(x,i),(x,i+1)} = \lambda_x \quad \text{and} \quad r_{(x,i),(x,i-1)} = \mu_x \quad \text{if} \quad i \neq 0$$

and $r_{(x,i),(y,j)}$ is 0 for all the other coordinates. The terms of the sequences (λ_x) and (μ_x) are assumed to be positive. The first component $(Z(t))$ of this Markov process has also the Markov property and its $\mathcal{Q}$-matrix is given by Q. Conditionally on the event $\{Z(t) = x\}$, the second component $(X(t))$ jumps as a birth and death process with parameters λ_x and μ_x.

Clearly, $(X(t))$ has the same distribution as the solution of the stochastic differential equation $dX(t) = \mathcal{N}_{\lambda_{Z(t-)}}(dt) - 1_{\{X(t-)\neq 0\}}\, \mathcal{N}_{\mu_{Z(t-)}}(dt)$. The couple $(Z(t), X(t))$ is a Markov process whose $\mathcal{Q}$-matrix is R.

PROPOSITION 9.14. *If $(Z(t))$ is a Markov process on $\mathcal{S}$ with an invariant probability distribution π such that $Z(0) = x \in \mathcal{S}$, for $N \in \mathbb{N}$, the process $(X_N(t))$ is defined as the solution of the stochastic differential equation*

$$dX(t) = \mathcal{N}_{\lambda_{Z(t-)}}(dt) - 1_{\{X(t-)\neq 0\}}\, \mathcal{N}_{\mu_{Z(t-)}}(dt), \tag{9.34}$$

with $Z(0) = N$. If

$$v_\pi = \sum_{x \in \mathcal{S}} (\lambda_x - \mu_x)\pi(x),$$

then, under the condition $\sup_{x\in\mathcal{S}}(\lambda_x + \mu_x) < +\infty$, the process $(X_N(Nt)/N)$ converges in distribution to $((1 + v_\pi t)^+)$ as N tends to infinity.

The quantity v_π is the mean drift of the birth and death process when the first component $(Z(t))$ is at equilibrium. The proposition shows that this quantity determines the behavior of the fluid limit of the second component $(X(t))$ when the initial condition of $(Z(t))$ is small compared to $X_N(0) = N$.

PROOF. By Equation (9.34), X_N satisfies the relationship

$$X_N(Nt) = N + M_N(Nt) + \int_0^{Nt} \lambda_{Z(s)}\, ds - \int_0^{Nt} 1_{\{X_N(s)\neq 0\}} \mu_{Z(s)}\, ds, \tag{9.35}$$

for $t \geq 0$, with

$$\begin{aligned} M_N(t) = &\int_{]0,t]} \left(\mathcal{N}_{\lambda_{Z(s-)}}(ds) - \lambda_{Z(s)}\, ds\right) \\ &- \int_{]0,t]} 1_{\{X_N(s-)\neq 0\}} \left(\mathcal{N}_{\mu_{Z(s-)}}(ds) - \mu_{Z(s)}\, ds\right). \end{aligned} \tag{9.36}$$

For $t \geq 0$, define $\mathcal{F}_t$ the σ-field generated by the variables $(Z(s); s \leq t)$ and the Poisson processes $(\mathcal{N}_{\lambda_x}, \mathcal{N}_{\mu_x}\,;\, x \in \mathcal{S})$ restricted to the interval $[0, t]$, then $(M_N(t))$ is a local martingale with respect to the filtration $(\mathcal{F}_t)$ and

$$\begin{aligned} M_N(t) = \sum_{x\in\mathcal{S}} \Bigg(&\int_{]0,t]} 1_{\{Z(s-)=x\}} \left(\mathcal{N}_{\lambda_x}(ds) - \lambda_x\, ds\right) \\ &+ \int_{]0,t]} 1_{\{X_N(s-)\neq 0, Z(s-)=x\}} \left(\mathcal{N}_{\mu_x}(ds) - \mu_x\, ds\right) \Bigg). \end{aligned}$$

Propositions A.9 and A.7 page 354 show that each term of the previous sum is a martingale. With the same method as in the proof of Proposition 6.2 (via Proposition A.10 page 356), it can be shown that the increasing process $(\langle M_N\rangle(t))$ of $(M_N(t))$ at t is

$$\left(\int_0^t \lambda_{Z(s)}\, ds + \int_0^t 1_{\{X_N(s)\neq 0\}} \mu_{Z(s)}\, ds\right).$$

In particular, $\mathbb{E}((M_N(t))^2) \leq \sup_{x\in\mathcal{S}}(\lambda_x + \mu_x)\, t < +\infty$, $(M_N(t))$ is a square integrable martingale. Doob's Inequality gives the relationship

$$\begin{aligned} \mathbb{P}\left(\sup_{0\leq s\leq t} \frac{|M_N(Ns)|}{N} \geq \varepsilon\right) &\leq \frac{1}{N^2\varepsilon^2} \mathbb{E}\left(M_N(Nt)^2\right) \\ &\leq \frac{1}{N\varepsilon^2} \mathbb{E}\left(\frac{1}{N}\int_0^{Nt} \left(\lambda_{Z(s)} + \mu_{Z(s)}\right) ds\right) \leq \frac{1}{N\varepsilon^2} \sup_{x\in\mathcal{S}}(\lambda_x + \mu_x)\, t, \end{aligned}$$

consequently,

$$\lim_{N\to+\infty} \mathbb{P}\left(\sup_{0\leq s\leq t} \frac{|M_N(Ns)|}{N} \geq \varepsilon\right) = 0.$$

Equation (9.35) can be rewritten as

$$\frac{X_N(Nt)}{N} = Y_N(t) + \frac{1}{N}\int_0^{Nt} 1_{\{X_N(s)=0\}} \mu_{Z(s)}\, ds$$

with

$$Y_N(t) = 1 + \frac{M_N(Nt)}{N} + \frac{1}{N}\int_0^{Nt} \left(\lambda_{Z(s)} - \mu_{Z(s)}\right) ds.$$

The couple

$$\left(\frac{X_N(Nt)}{N}, \frac{1}{N}\int_0^{Nt} 1_{\{X_N(s)=0\}}\mu_{Z(s)}\,ds\right) \tag{9.37}$$

is the solution of the Skorohod Problem associated with the function $(Y_N(t))$.

By using the tightness criterion of Theorem C.9 page 373 and the boundedness of the sequences (λ_x) and (μ_x), it is easily seen that the sequence of the distributions of the continuous processes

$$(U_N(t)) = \left(\frac{1}{N}\int_0^{Nt} \left(\lambda_{Z(s)} - \mu_{Z(s)}\right)\,ds\right)$$

is tight. ith If $p \leq 1$ and $0 \leq t_1 \leq t_2 \leq \cdots \leq t_p$, the ergodic Theorem for the Markov process $(Z(t))$ shows that $\mathbb{P}$-almost surely

$$\lim_{N\to+\infty} \frac{1}{N}\int_0^{Nt_i} \left(\lambda_{Z(s)} - \mu_{Z(s)}\right)\,ds = v_\pi t_i,$$

for $i = 1, \ldots, p$. The finite marginal distributions of any limiting point of the sequence are the same as those of the process $(v_\pi t)$. Therefore, the sequence $(U_N(t))$ converges in distribution to $(v_\pi t)$ (the space $C(\mathbb{R}_+, \mathbb{R})$ being endowed with the topology of uniform convergence on compact sets), the sequence $(Y_N(t))$ converges in the same way to $(1 + v_\pi t)$.

Inequalities (D.9) and (D.10) of Proposition D.4 page 381 imply that the couple (9.37) converges in distribution to the solution of Skorohod associated with the function $(1 + v_\pi t)$. In particular $(X_N(Nt)/N)$ converges in distribution to $((1 + v_\pi t)^+)$. The proposition is proved. □

Examples.

Tandem Queues. The example of Section 9.2.2 page 235 with the sequence of initial states $x_N = (0, N)$ is considered. The first component $(l_1(t))$, the number of customers of an $M/M/1$ queue with parameters λ_1 and μ_1 such that $\lambda_1 < \mu_1$, is ergodic and its invariant distribution π is a geometric distribution with parameter λ_1/μ_1. The component $(l_2(t))$ is a birth and death process whose death rate is μ_2 outside 0 and the birth rate is μ_1 when $l_1(t) > 0$ and 0 otherwise. The last proposition shows that $(l_2(Nt)/N)$ converges to $((1 + v_\pi t)^+)$ with

$$v_\pi = (1 - \pi(0))\mu_1 - \mu_2 = \frac{\lambda_1}{\mu_1}\mu_1 - \mu_2 = \lambda_1 - \mu_2,$$

hence Convergence (9.2) page 237 is rigorously established when $y = 0$.

Fluid Limits of a Jump Process in $\mathbb{N}^2$. The Markov $(X(t), Y(t))$ on $\mathbb{N}^2$ considered here has a Q-matrix $Q = (q_{mn}\,;\, m, n \in \mathbb{N}^2)$ given by (see Figure 9.4):
if $i > 0$ and $j \geq 0$,

$$q_{(i,j)(i+1,j)} = \lambda_1, \quad q_{(i,j)(i,j+1)} = \lambda_2, \quad q_{(0,j)(0,j+1)} = \lambda_v, \quad q_{(j,0)(j+1,0)} = \lambda_h;$$

if $i \geq 0$ and $j > 0$,

$$q_{(i,j),(i-1,j)} = \mu_1, \quad q_{(i,j),(i,j-1)} = \mu_2, \quad q_{(0,j)(0,j-1)} = \mu_v, \quad q_{(j,0)(j-1,0)} = \mu_h,$$

with the additional boundary conditions $q_{(0,0),(0,1)} = \lambda_0$, $q_{(0,0),(1,0)} = \lambda_0$, and the other components outside the diagonal are 0. It is easily checked the process $(X(t), Y(t))$ belongs to the class $(\mathcal{C})$ of Markov processes defined in Section 9.3.

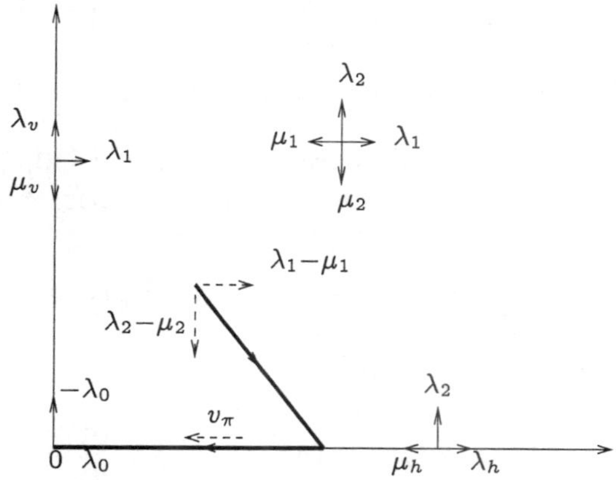

FIG. 9.4. Fluid Limit of a Jump Process in $\mathbb{N}^2$

For $N \geq 1$, if $(X_N(t)) = (X_{N,1}(t), X_{N,2}(t))$ the Markov process with Q-matrix Q and whose initial state is given by $x_N = (\lfloor \alpha N \rfloor, \lfloor (1-\alpha)N \rfloor)$ for $\alpha \in [0,1]$; in particular,

$$\lim_{N \to +\infty} \frac{\|x_N\|}{N} = 1 \quad \text{and} \quad \lim_{N \to +\infty} \frac{x_N}{N} = (\alpha, 1-\alpha).$$

In the interior of $\mathbb{N}^2$ the coordinates of the Markov process are independent and behave like two $M/M/1$ queues. The scaling properties of the $M/M/1$ queue show the process $(X_N(Nt)/N\,;\, t < t_0)$ converges in distribution to the function $(\alpha + (\lambda_1 - \mu_1)t, 1 - \alpha + (\lambda_2 - \mu_2)t\,;\, t < t_0)$ with

$$t_0 = \frac{\alpha}{(\mu_1 - \lambda_1)^+} \wedge \frac{1-\alpha}{(\mu_2 - \lambda_2)^+}.$$

If $t_0 = +\infty$, or equivalently when $\lambda_1 > \mu_1$ and $\lambda_2 > \mu_2$, the analysis of the rescaled process is finished.

If $t_0 < +\infty$, it can be assumed that, for example, $t_0 = (1-\alpha)/(\mu_2 - \lambda_2)$, in particular $\lambda_2 < \mu_2$. The time when the second coordinate of the rescaled process is 0 converges almost surely to t_0 and, by the strong Markov property, the study is reduced to the case where the initial states $(x_N) = (x_{N,1}, 0)$ satisfy

$$\lim_{N\to+\infty} \frac{x_{N,1}}{N} = 1 > 0.$$

The equilibrium distribution π of the second component is a geometric distribution with parameter λ_2/μ_2. The drift of the first component is either $\lambda_1 - \mu_1$ or $\lambda_h - \mu_h$ depending on the value of second component. Proposition 9.14 shows that $(X_{N,2}(Nt)/N)$ converges in distribution to the deterministic function $(0, (1 + v_\pi t)^+)$ with

$$v_\pi = (\lambda_1 - \mu_1)\frac{\lambda_2}{\mu_2} + (\lambda_h - \mu_h)\left(1 - \frac{\lambda_2}{\mu_2}\right).$$

In particular, if $v_\pi < 0$ (as it is the case in Figure 9.4), the fluid limit is 0 after some finite time.

The next result has been shown by Malyshev [**20**] by pasting together linear Lyapunov functions. The fluid limit approach has the advantage of giving immediately the ergodicity conditions.

PROPOSITION 9.15. *If one of the conditions above is satisfied, the Markov process with Q-matrix Q is ergodic :*

a) $\lambda_1 < \mu_1$, $\lambda_2 > \mu_2$ *and*

$$(\lambda_2 - \mu_2)\frac{\lambda_1}{\mu_1} + (\lambda_v - \mu_v)\left(1 - \frac{\lambda_1}{\mu_1}\right) < 0; \tag{9.38}$$

b) $\lambda_2 < \mu_2$, $\lambda_1 > \mu_1$ *and*

$$(\lambda_1 - \mu_1)\frac{\lambda_2}{\mu_2} + (\lambda_h - \mu_h)\left(1 - \frac{\lambda_2}{\mu_2}\right) < 0; \tag{9.39}$$

c) $\lambda_1 < \mu_1$, $\lambda_2 < \mu_2$ *and Relationships (9.38) and (9.39).*

PROOF. Conditions a), b) and c) imply that the fluid limits reach 0. Under condition c), depending on the initial state, the fluid limit may reach the horizontal axis or the vertical axis which explains the two conditions (9.38) and (9.39).

If the initial point of the fluid limit is $(\alpha, 1-\alpha)$ with $\alpha \in [0,1]$, under each of the conditions of the above proposition, it is easily checked that the associated fluid limits are 0 after some time T independent of α. By Proposition 9.2 on the uniform integrability property, the proposition is then a consequence of Corollary 9.8. □

The fluid limit of the process has been built, via strong Markov property, by pasting together pieces of trajectories. This method is quite natural and it gives an explicit expression of the fluid limit. This intuitive aspect is less obvious if one considers that the fluid limit as the solution of some Skorohod Problem. In practice, when the Skorohod Problem formulation holds, the combination of these two points of view is useful: A possible fluid limit can be identified by decomposing the trajectory into pieces, it can then be shown that this is indeed the fluid limit by showing that it solves the corresponding Skorohod problem.

9.7 Bibliographical Notes

The purpose of this section is to give some pointers to the literature. By no means, it is intended to be exhaustive.

Some puzzling examples of unstable queueing networks by Rybko and Stolyar[RS92], Bramson[Bra94] and others have given a renewed interest in scaling methods in the analysis of stochastic networks. See also Chen and Yao [**10**] and Whitt [**52**].

Scaling ideas have also been used to study ergodicity properties of some random walks, see Malyshev[Mal93] or of some reflected diffusion processes, see Dupuis and Williams[DW94]. It is quite difficult to trace these ideas back since they have been already used repeatedly in various domains. A typical example is the analysis of a deterministic process perturbated by some small stochastic noise. (See Freidlin and Wentzell [**23**].)

Quite often in the literature, the term "fluid limits" is attached to a set of deterministic equations (See Equations (9.14)). If these equations are important and useful in the analysis of some stochastic networks, they cover only some aspects of the scaling ideas.

- They do not capture the fact that fluid limits may be *random* functions instead of deterministic solutions of some ordinary differential equation. See the examples of Section 9.2.2 and Section 9.5.3. This is an important point since this asymptotic randomness occur precisely at the discontinuities of the dynamic of the Markov process.
- They may have multiple solutions among which, some can be irrelevant to the original problem. Note that this also happens in statistical mathematics, see Comets[Com91].

Example of Section 9.5.3 page 265. In Ignatyuk and Malyshev [IM93], the multiplicative phenomena have been investigated in the case of random walks in $\mathbb{Z}_+^4$. It is also sketched in Malyshev [Mal93]. A discrete time version of a

[RS92] Op. cit. page 232.

[Bra94] Op. cit. page 261.

[Mal93] V.A. Malyshev, *Networks and dynamical systems*, Advances in Applied Probability **25** (1993), no. 1, 140–175.

[DW94] P. Dupuis and R.J. Williams, *Lyapounov functions for semimartingale reflecting Brownian motions*, Annals of Applied Probability **22** (1994), no. 2, 680–702.

[Com91] Op. cit. page 232.

[IM93] I.A. Ignatyuk and V.A. Malyshev, *Classification of random walks in* $\mathbb{Z}_+^4$, Selecta Mathematica **12** (1993), no. 2, 129–194.

related multi-dimensional Markov process has been analyzed in Fayolle *et al.* [FIMM91]. The methods of these papers use Lyapunov functions.

[FIMM91] G. Fayolle, I.A. Ignatyuk, V.A. Malyshev, and M.V. Men'shikov, *Random walks in two-dimensional complexes*, Queueing Systems, Theory and Applications **9** (1991), no. 3, 269–300.

10. Ergodic Theory: Basic Results

Contents

In this chapter definitions and basic results of ergodic theory are presented in a probabilistic setting. It must be stressed that this is a *fundamental* topic in probability theory. Results proved in this chapter are classical in a Markovian framework (ergodic theorems, representations of the invariant probability,...). It is nevertheless very helpful to realize that the Markov property does not really play a role to get these results: They also hold in a much more general (and natural) setting. As it will be seen in Chapter 11, the study of stationary point processes is quite elementary if a basic construction of ergodic theory is used (the "special flow" defined page 295). Since this subject is not standard in graduate courses on stochastic processes, most of the results are proved. The reference book Cornfeld *et al.* [**13**] gives a broader point of view of this domain. In the following $(\Omega, \mathcal{F}, \mathbb{P})$ is the probability space of reference.

10.1 Discrete Dynamical Systems

DEFINITION 24. *An* endomorphism T *of* $(\Omega, \mathcal{F}, \mathbb{P})$ *is a mapping from* Ω *to* Ω *such that*

1. T *is measurable ;*
2. *the probability measure* $\mathbb{P}$ *is invariant by* T *: For any non-negative measurable function* f *on* Ω*, the equality*

$$\int_\Omega f(T(\omega))\, d\mathbb{P}(\omega) = \int_\Omega f(\omega)\, d\mathbb{P}(\omega)$$

holds.

An automorphism T *is a bijective endomorphism such that* T^{-1} *is also measurable.*

In this chapter, the sequence of iterations of a given endomorphism T is analyzed: For $\omega \in \Omega$, if T^n is the nth iterate of the endomorphism T, the way sample paths $n \to T^n(\omega)$ are visiting Ω is investigated.

If $A \in \mathcal{F}$ and $n \in \mathbb{N}$, the notation $T^{-n}(A)$ used in this chapter does not assume that T is bijective, this stands for the set $\{\omega \in \Omega : T^n(\omega) \in A\}$. When T is bijective, $T^{-n}(A)$ is also the image of A by the nth iterate T^{-n} of the endomorphism T^{-1}.

EXAMPLES.
1. If $\Omega = \{0,1\}$, $\mathcal{F}$ is the set of subsets of Ω, $\mathbb{P}(\{0\}) = \mathbb{P}(\{1\}) = 1/2$, $T(0) = 1$ and $T(1) = 0$ then T is an automorphism.
2. For $\alpha \in [0,1]$, the translation $T_\alpha : x \to x+\alpha \mod 1$ on the torus $\Omega = [0,1[$ (0 and 1 are identified) endowed with the Borel σ-field and Lebesgue measure is an automorphism.
3. With the same probability space, the operator $T : x \to (2x \mod 1)$ is also an endomorphism.
4. The transformation $(x,y) \to (x, y+\phi(x))$ where ϕ is a measurable function from $[0,1[$ to $[0,1[$, is an endomorphism of the two dimensional torus $[0,1[\times[0,1[$ endowed with Lebesgue measure.

An important example, Markovian endomorphisms, is presented in Section 10.4 page 297.

Poincaré's Recurrence Theorem

DEFINITION 25. *If* $A \in \mathcal{F}$*, the variable*

$$\nu_A(\omega) = \inf\{n > 0 : T^n(\omega) \in A\},$$

is the hitting time of the set A *by iterates of* T*. For* $n \geq 0$*, the set* $V_n(A)$ *is the set of sample paths starting from* A^c *and such that the* n *first iterates are outside* A*,*

$$V_n(A) = A^c \cap T^{-1}(A^c) \cap T^{-2}(A^c) \cap \cdots \cap T^{-n}(A^c),$$

$V_n(A)$ can also be defined as $A^c \cap \{\nu_A(\omega) > n\}$, with $A^c = \Omega - A$. For $n < 0$, $V_n(A)$ is defined as Ω.

The theorem below is the first basic result in the study of sample paths of iterates of T.

THEOREM 10.1 (Poincaré's Recurrence Theorem). *If T is an endomorphism of Ω and $A \in \mathcal{F}$ a non negligible event then, the random variable*

$$\nu_A(\omega) = \inf\{n > 0 : T^n(\omega) \in A\},$$

is $\mathbb{P}$-almost surely finite on A.

PROOF. For $n \geq 1$, the definition of ν_A gives the relationship

$$\begin{aligned} A \cap \{\nu_A > n\} &= \{\omega \in A, T(\omega) \notin A, T^2(\omega) \notin A, \dots, T^n(\omega) \notin A\} \\ &= A \cap T^{-1}(A^c) \cap T^{-2}(A^c) \cap \cdots \cap T^{-n}(A^c), \end{aligned} \tag{10.1}$$

thus

$$\begin{aligned} A \cap \{\nu_A > n\} = {} & T^{-1}(A^c) \cap T^{-2}(A^c) \cap \cdots \cap T^{-n}(A^c) \\ & - A^c \cap T^{-1}(A^c) \cap T^{-2}(A^c) \cap \cdots \cap T^{-n}(A^c), \end{aligned}$$

or, by using the sequence $(V_n(A))$ defined above,

$$A \cap \{\nu_A > n\} = T^{-1}(V_{n-1}(A)) - V_n(A). \tag{10.2}$$

The invariance of $\mathbb{P}$ by T yields the relationship

$$\mathbb{P}\left[T^{-1}(V_{n-1}(A))\right] = \mathbb{E}\left[1_{V_{n-1}(A)} \circ T\right] = \mathbb{E}\left[1_{V_{n-1}(A)}\right] = \mathbb{P}\left[V_{n-1}(A)\right],$$

and, by using Identity (10.2), one gets

$$\mathbb{P}(A \cap \{\nu_A > n\}) = \mathbb{P}(V_{n-1}(A)) - \mathbb{P}(V_n(A)).$$

The sets $V_n(A), n \in \mathbb{N}$ being non-increasing, the sequence $(\mathbb{P}[V_n(A)])$ is non-increasing, hence converging. By letting n go to infinity in the above identity, one gets the identity

$$\mathbb{P}(A \cap \{\nu_A = +\infty\}) = \lim_{n \to +\infty} \mathbb{P}(A \cap \{\nu_A > n\}) = 0,$$

the theorem is proved. □

Induced Endomorphisms

If $A \in \mathcal{F}$ is such that $\mathbb{P}(A) > 0$, the operator T_A from A to A is defined as an *induced operator*,

$$T_A : \omega \to T^{\nu_A(\omega)}(\omega).$$

The above theorem shows that T_A is $\mathbb{P}$-almost surely defined on A; by convention, one sets $T_A(\omega) = \omega$ on the negligible set where ν_A is infinite. The σ-field $\mathcal{F}_A$ is the σ-field $\mathcal{F}$ restricted to A and $\mathbb{P}_A$ is defined as the probability measure $\mathbb{P}$ conditioned on the event A, $\mathbb{P}_A = \mathbb{P}(\cdot \cap A)/\mathbb{P}(A)$. The expected value with respect to $\mathbb{P}_A$ is denoted $\mathbb{E}_A(\cdot)$.

PROPOSITION 10.2. *If T is an endomorphism $(\Omega, \mathcal{F}, \mathbb{P})$, the induced operator T_A is an endomorphism of the probability space $(A, \mathcal{F}_A, \mathbb{P}_A)$.*

PROOF. Since the mapping ν_A is measurable, hence for $B \in \mathcal{F}$,

$$\{T_A \in B\} = \bigcup_{k=1}^{+\infty} \{T^k \in B\} \cap \{\nu_A = k\},$$

the variable T_A is thus also measurable. It has to be shown that the probability distribution $\mathbb{P}_A$ is invariant with respect to T_A. For $n \in \mathbb{N}$, the indicator function of the event $\{T_A = n\}$ can be written as

$$1_{\{T_A=n\}} = 1_{A\cap T^{-1}(A^c)\cap\cdots\cap T^{-n+1}(A^c)\cap T^{-n}(A)} = 1_A \prod_{i=1}^{n-1} 1_{T^{-i}(A^c)} 1_{T^{-n}(A)}$$

$$= (1 - 1_{A^c}) \left(1 - 1_{T^{-n}(A^c)}\right) \prod_{i=1}^{n-1} 1_{T^{-i}(A^c)},$$

by using the sequence $(V_n(A))$, one gets

$$1_{\{T_A=n\}} = 1_{V_{n-2}(A)} \circ T - 1_{V_{n-1}(A)} - 1_{V_{n-1}(A)} \circ T + 1_{V_n(A)}. \tag{10.3}$$

If f is a bounded non-negative function on A,

$$\mathbb{P}(A)\mathbb{E}_A(f \circ T_A) = \sum_{n=1}^{+\infty} \mathbb{E}\left(f \circ T^n 1_{\{T_A=n\}}\right),$$

Decomposition (10.3) of the event $\{T_A = n\}$ yields

$$\begin{aligned}\mathbb{P}(A)\mathbb{E}_A(f \circ T_A) = \sum_{n=1}^{+\infty} \mathbb{E}\left[\left(1_{V_{n-2}(A)} - 1_{V_{n-1}(A)}\right) \circ T \times f \circ T^n\right] \\ + \mathbb{E}\left[\left(1_{V_n(A)} - 1_{V_{n-1}(A)}\right) f \circ T^n\right],\end{aligned}$$

and, by invariance of $\mathbb{P}$ by the endomorphism T,

$$\begin{aligned}\mathbb{P}(A)\mathbb{E}_A(f \circ T_A) = \sum_{n=1}^{+\infty} \mathbb{E}\left[\left(1_{V_{n-2}(A)} - 1_{V_{n-1}(A)}\right) f \circ T^{n-1}\right] \\ + \mathbb{E}\left[\left(1_{V_n(A)} - 1_{V_{n-1}(A)}\right) f \circ T^n\right].\end{aligned} \tag{10.4}$$

The inequality

$$\left|\left(1_{V_{n-1}(A)} - 1_{V_n(A)}\right) f \circ T^n\right| \le \|f\|_\infty 1_{\{\nu_A=n\}}$$

shows that the series

$$\sum_{n=0}^{+\infty} \mathbb{E}\left[\left(1_{V_{n-2}(A)} - 1_{V_{n-1}(A)}\right) f \circ T^{n-1}\right]$$

and

$$\sum_{n=0}^{+\infty} \mathbb{E}\left[\left(1_{V_n(A)} - 1_{V_{n-1}(A)}\right) f \circ T^n\right]$$

are converging. Identity (10.4) can thus be rewritten as

$$
\begin{aligned}
&\mathbb{P}(A)\mathbb{E}_A(f\circ T_A)\\
&\quad=\sum_{n=0}^{+\infty}\mathbb{E}\left[\left(1_{V_{n-1}(A)}-1_{V_n(A)}\right)f\circ T^n\right]\\
&\qquad\qquad+\sum_{n=1}^{+\infty}\mathbb{E}\left[\left(1_{V_n(A)}-1_{V_{n-1}(A)}\right)f\circ T^n\right]\\
&\quad=\mathbb{E}\left[\left(1_{V_{-1}(A)}-1_{V_0(A)}\right)f\right]=\mathbb{E}[1_A f]=\mathbb{P}(A)\mathbb{E}_A(f).
\end{aligned}
$$

Finally, one obtains the relationship $\mathbb{E}_A(f\circ T_A)=\mathbb{E}_A(f)$ for any bounded measurable functional. The operator T_A leaves the probability distribution $\mathbb{P}_A$ invariant, T_A is an endomorphism. □

PROPOSITION 10.3. *If T is an endomorphism and f a bounded non-negative measurable function on Ω,*

$$
\mathbb{E}\left(f1_{\{\nu_A<+\infty\}}\right)=\mathbb{P}(A)\mathbb{E}_A\left(\sum_{n=0}^{\nu_A-1}f\circ T^n\right), \tag{10.5}
$$

in particular,

$$
\mathbb{E}_A(\nu_A)=\frac{\mathbb{P}(\nu_A<+\infty)}{\mathbb{P}(A)}.
$$

Identity (10.5) expresses the probability distribution $\mathbb{P}$ on the set $\{\nu_A< +\infty\}$ as a functional of the probability distribution $\mathbb{P}_A$. The last identity is Kac's Formula.

PROOF. By Fubini's Theorem

$$
\mathbb{P}(A)\mathbb{E}_A\left(\sum_{n=0}^{\nu_A-1}f\circ T^n\right)=\sum_{n=0}^{+\infty}\mathbb{E}\left(1_{A\cap\{\nu_A>n\}}f\circ T^n\right),
$$

for $n\in\mathbb{N}$, Identity (10.4) gives the relationship

$$
A\cap\{\nu_A>n\}=T^{-1}(V_{n-1}(A))-V_n(A).
$$

The set $V_\infty(A)=\{\nu_A=+\infty\}$ verifies $V_\infty(A)\subset T^{-1}(V_\infty(A))$. By invariance of $\mathbb{P}$ by T, up to a negligible set, the sets $V_\infty(A)$ and $T^{-1}(V_\infty(A))$ are identical. Poincaré's Theorem shows that the set $V_\infty(A)$ is, modulo a negligible set, is included in the subset $\Omega-A$. Up to a negligible set, the identity

$$
A\cap\{\nu_A>n\}=T^{-1}(V_{n-1}(A)-V_\infty(A))-(V_n(A)-V_\infty(A)),
$$

is thus satisfied, hence

$$(10.6)\quad \mathbb{P}(A)\mathbb{E}_A\left(\sum_{n=0}^{\nu_A-1} f\circ T^n\right) = \mathbb{E}\left[\left(1_{V_{-1}(A)-V_\infty(A)}\circ T - 1_{V_0(A)-V_\infty(A)}\right) f\right] + \sum_{n=1}^{+\infty}\mathbb{E}\left[\left(1_{V_{n-1}(A)-V_\infty(A)}\circ T - 1_{V_n(A)-V_\infty(A)}\right) f\circ T^n\right].$$

For $N\geq 1$, the invariance of $\mathbb{P}$ by T yields the identity

$$\sum_{n=1}^{N}\mathbb{E}\left[\left(1_{V_{n-1}(A)-V_\infty(A)}\circ T - 1_{V_n(A)-V_\infty(A)}\right) f\circ T^n\right]$$
$$= \sum_{n=1}^{N}\mathbb{E}\left[1_{V_{n-1}(A)-V_\infty(A)}\, f\circ T^{n-1}\right] - \mathbb{E}\left[1_{V_n(A)-V_\infty(A)}\, f\circ T^n\right]$$
$$= \mathbb{E}\left[1_{V_0(A)-V_\infty(A)}\, f\right] - \mathbb{E}\left[1_{V_N(A)-V_\infty(A)}\, f\circ T^N\right],$$

Since $V_N(A)-V_\infty(A) = A^c\cap\{N+1\leq \nu_A<\infty\}$, one gets

$$\mathbb{E}\left[1_{V_N(A)-V_\infty(A)} f\circ T^N\right] \leq \|f\|_\infty \mathbb{P}(N+1\leq\nu_A<\infty),$$

by letting N go to infinity,

$$\sum_{n=1}^{+\infty}\mathbb{E}\left[\left(1_{V_{n-1}(A)-V_\infty(A)}\circ T - 1_{V_n(A)-V_\infty(A)}\right) f\circ T^n\right] = \mathbb{E}\left[1_{V_0(A)-V_\infty(A)}\, f\right].$$

Identity (10.6) becomes

$$\mathbb{P}(A)\,\mathbb{E}_A\left[\sum_{n=0}^{\nu_A-1} f\circ T^n\right] = \mathbb{E}\left[1_{V_{-1}(A)-V_\infty(A)}\circ T\cdot f\right] = \mathbb{E}\left[1_{\Omega-\{\nu_A=+\infty\}}\cdot f\right],$$

the first identity of the proposition is proved. Kac's Formula is obtained by taking $f=1$ in Relationship (10.5). □

10.2 Ergodic Theorems

The main convergence results are presented in this section, they show that time averages of iterates of an endomorphism converge almost surely. They generalize the law of large numbers and classical results for Markov chains.

The σ-Field of Invariant Events

DEFINITION 26. *A measurable function f is* invariant *by the endomorphism T if $\mathbb{P}$-almost surely, the identity $f\circ T=f$ holds. A measurable event A is invariant if its indicator function is invariant, or equivalently A and $T^{-1}(A)$ are identical up to a set of probability 0 that is,*

$$\mathbb{P}\left[A \cap \left(T^{-1}(A)\right)^{c} \cup T^{-1}(A) \cap A^{c}\right] = 0,$$

where $B^c = \Omega - B$, *for* $B \subset \Omega$.

Notice that the notion of invariant event is closely related, via negligible sets, to the probability distribution.

PROPOSITION 10.4. *The set* $\mathcal{I}$ *of all invariant sets by an endomorphism* T *is a* σ*-field.*

PROOF. Since $T^{-1}(\Omega) = \Omega$ clearly holds, the set Ω is in $\mathcal{I}$. If A is invariant, $T^{-1}(A)$ and A are identical up to a negligible set, then the sets A^c and $T^{-1}(A^c) = A^c$ are also identical up to a negligible set. Finally, if (A_i) is a sequence of invariant sets, for $i \in \mathbb{N}$ the subset N_i denotes the negligible set $T^{-1}(A_i) - A_i \cup A_i - T^{-1}(A_i)$. Hence the sets $T^{-1}(\cup_i A_i)$ and $\cup_i A_i$ are identical up to the negligible set $\cup_i N_i$, the set $\cup_i A_i$ is therefore invariant. □

An element A of $\mathcal{F}$ such that $A \subset T^{-1}(A)$ is clearly invariant since

$$\mathbb{P}(A) \leq \mathbb{P}\left(T^{-1}(A)\right) = \mathbb{P}(A),$$

by invariance of $\mathbb{P}$ by T therefore, the set $T^{-1}A - A$ is negligible. The sample path of an element $\omega \in \Omega$ is called the *orbit* of ω,

$$\mathcal{O}(\omega) = \{T^n(\omega) : n \geq 0\},$$

it satisfies $\mathcal{O}(\omega) \subset T^{-1}(\mathcal{O}(\omega))$, it is an invariant event.

Ergodic Endomorphisms

DEFINITION 27. *An endomorphism is* ergodic *if any invariant set by* T *has probability* 0 *or* 1.

PROPOSITION 10.5. *An endomorphism* T *is ergodic if and only if any invariant measurable function is* $\mathbb{P}$*-almost surely constant.*

PROOF. If T is ergodic and f is an invariant measurable function, since the measurable sets $\{\omega : f(\omega) = +\infty\}$ and $\{\omega : f(\omega) = -\infty\}$ are invariant, it can be assumed that they are negligible (otherwise it is finished). For $x \in \mathbb{R}$, the set $F_x = \{\omega : x \leq f(\omega)\}$ is invariant, hence $\mathbb{P}(F_x) = 0$ or 1. Since $\lim_{x\to-\infty} \mathbb{P}(F_x) = 1$ and $\lim_{x\to+\infty} \mathbb{P}(F_x) = 0$, the quantity

$$x_0 = \sup\{x : \mathbb{P}(F_x) = 1\}$$

is finite. For $z < x_0$, $\mathbb{P}$-a.s. $f(\omega) \geq z$ and if $y > x_0$, $\mathbb{P}$-a.s. $f(\omega) \leq y$. Two sequences (y_n), (z_n) can be chosen so that, for $n \in \mathbb{N}$, $z_n < x_0 < y_n$ and $\lim_{n\to+\infty} z_n = \lim_{n\to+\infty} y_n = x_0$ and the inequalities $z_n \leq f(\omega) \leq y_n$ holds $\mathbb{P}$-almost surely, thus $f(\omega) = x_0$ $\mathbb{P}$-a.s.

Conversely, if A is an invariant set, the indicator function 1_A is invariant, therefore almost surely constant, consequently $\mathbb{P}(A) = 0$ or 1, T is ergodic. □

PROPOSITION 10.6. *If T is an ergodic endomorphism, and A is a non-negligible measurable set, the hitting time*

$$\nu_A(\omega) = \inf\left\{n > 0 : T^n(\omega) \in A\right\},$$

is $\mathbb{P}$-a.s. finite on Ω. In particular the probability distribution $\mathbb{P}$ can be expressed with its restriction on A, that is

$$\mathbb{E}(f) = \mathbb{E}\left(1_A \sum_{n=0}^{\nu_A-1} f \circ T^n\right), \tag{10.7}$$

for any bounded measurable function.

The ergodicity property extends Poincaré's Theorem page 280 to the whole space. The hitting time ν_A is almost surely finite not only on A but on Ω too. In particular, almost surely the orbit $\mathcal{O}(\omega)$, $\omega \in \Omega$ is "dense" for $\mathbb{P}$, i.e. any non-negligible set A is visited.

PROOF. Indeed, the set $B = \{\nu_A(\omega) = +\infty\}$ is invariant because of the relationship $B \subset T^{-1}(B)$, consequently $\mathbb{P}(B) = 0$ or 1. Since $A \subset B^c$ by Poincaré's Theorem, necessarily $\mathbb{P}(B) = 0$, the variable ν_A is therefore $\mathbb{P}$-a.s. finite. Identity (10.7) is a straightforward consequence of Relationship (10.5) of Proposition 10.3. □

Examples

Translation on the Torus.

PROPOSITION 10.7. *The translation $T_\alpha : x \to x + \alpha \mod 1$ on the torus $[0, 1[$ is ergodic if and only if α is not a rational number.*

PROOF. If α is not a rational number, it is sufficient to prove that a function f of $L_2([0, 1[)$ invariant by T_α is necessarily constant. It is well known, see for example Rudin [**45**], that the functions $x \to \exp(2i\pi nx)$, $n \in \mathbb{Z}$ form an Hilbertian basis of $L_2([0, 1[)$. In particular, there exists a sequence (f_n) of real numbers such that

$$f(x) = \sum_{n \in \mathbb{Z}} f_n e^{2i\pi nx},$$

holds Lebesgue-almost everywhere. Since f is invariant by T_α, the identity

$$f(T_\alpha(x)) = f(x + \alpha) = \sum_{n \in \mathbb{Z}} e^{2i\pi n\alpha} f_n e^{2i\pi nx} = f(x) = \sum_{n \in \mathbb{Z}} f_n e^{2i\pi nx},$$

holds almost everywhere. The uniqueness of coordinates in the Hilbertian basis implies the identities $\exp(2i\pi n\alpha) f_n = f_n$ for all $n \in \mathbb{Z}$. For $n \neq 0$, $\exp(2i\pi n\alpha) \neq 1$ (since α is not a rational number). This implies that $f_n = 0$ for any $n \neq 0$, consequently $f = f_0$ almost everywhere. The endomorphism T_α is ergodic.

Conversely if $\alpha = p/q \in [0, 1]$ where p and q are integers without a common factor, the function $x \to \exp(2i\pi qx)$ is invariant by T_α. The endomorphism is not ergodic in this situation. □

When $\alpha = p/q \in [0,1]$ with $q \neq 0$ and p and q are relatively prime integers, the invariant sets are easy to describe. If $E = \{i/q : 0 \leq i < q\}$, $x + E$ is the smallest set F containing x such that $T^{-1}(F) = F$. Therefore, the invariant sets of T_α are, up to a negligible subset, of the form $A + E$ where A is a Borelian subset of $[0, 1/q[$.

The Transformation $x \to 2x$ on the Torus. Along the same arguments, the following proposition holds.

PROPOSITION 10.8. *The transformation $x \to 2x$ on the torus $[0,1[$ endowed with Lebesgue measure is ergodic.*

The Ergodic Theorem

The next theorem concerns time averages of orbits of an endomorphism.

THEOREM 10.9 (Birkhoff's Ergodic Theorem). *If $T : \Omega \to \Omega$ is an endomorphism and f an integrable function, $\mathbb{P}$-a.s.*

$$\lim_{n\to+\infty} \frac{1}{n}\sum_{i=1}^{n} f\left(T^i(\omega)\right) = \mathbb{E}\left(f \,|\, \mathcal{I}\right)(\omega),$$

where $\mathcal{I}$ is the σ-field of the invariant measurable sets.

PROOF. The proof below is a variation, due to Neveu[Nev83], of Garsia[Gar65]'s Proof (see also Cornfeld *et al.* [**13**] and the proof of Proposition 12.1 page 332).

For $\varepsilon > 0$, by setting $g = f - \mathbb{E}(f \,|\, \mathcal{I}) - \varepsilon$, the sequence (W_n) defined by

$$W_0 = 0, \qquad W_{n+1} = (W_n + g)^+ \circ T, \quad n \geq 0. \tag{10.8}$$

Since $W_1 \geq W_0 = 0$, then $W_2 = (W_1 + g)^+ \circ T \geq (W_0 + g)^+ \circ T = W_1$, by induction $W_{n+1} \geq W_n$ for any $n \in \mathbb{N}$, the sequence (W_n) is non-decreasing, let W be its limit. Again by induction, for $n \in \mathbb{N}$, it is easily verified that

$$W_n \geq \sum_{j=1}^{n} g \circ T^j, \tag{10.9}$$

and W_n is an integrable variable. By letting n go to infinity in Identity (10.8), one gets the relationship $W = (W + g)^+ \circ T$, the event $A = \{W = +\infty\}$ is therefore invariant by T.

Definition (10.8) of (W_n) shows that for $n \in \mathbb{N}$,

$$W_{n+1} - W_n \circ T = (g \vee -W_n) \circ T, \tag{10.10}$$

where, as usual, $a \vee b = \max(a, b)$, since A is invariant by T,

[Nev83] J. Neveu, *Construction de files d'attente stationnaires*, Lecture notes in Control and Information Sciences, 60, Springer Verlag, 1983, pp. 31–41.

[Gar65] A.M. Garsia, *A simple proof of E. Hopf's maximal ergodic theorem*, Journal of Mathematics and Mechanics **14** (1965), 381–382.

$$\mathbb{E}[(W_{n+1} - W_n \circ T)1_A] = \mathbb{E}(W_{n+1}1_A) - \mathbb{E}[(W_n 1_A) \circ T],$$

this last quantity is

$$\mathbb{E}(W_{n+1}1_A) - \mathbb{E}(W_n 1_A) = \mathbb{E}[(W_{n+1} - W_n)1_A] \geq 0.$$

By integrating Relationship (10.10) on the set A, one gets the inequality

$$\mathbb{E}[(g \vee -W_n) \circ T\, 1_A] \geq 0.$$

Lebesgue's Theorem and the invariance of A by T show that $\mathbb{E}(1_A g) \geq 0$, consequently,

$$0 \leq \mathbb{E}[1_A g] = \mathbb{E}[1_A \mathbb{E}(g \,|\, \mathcal{I})].$$

The very definition of g shows that $\mathbb{E}(g \,|\, \mathcal{I}) = -\varepsilon$, hence $0 \leq -\varepsilon \mathbb{P}(A)$, necessarily $\mathbb{P}(A) = 0$, hence $W < +\infty$, $\mathbb{P}$-a.s.

Inequality (10.9) and the fact that the sequence (W_n) is non-decreasing imply that, for $n \in \mathbb{N}$, the partial sum $g \circ T + g \circ T^2 + \cdots + g \circ T^n$ is upper bounded by W. The expression of g and the invariance of the conditional expectation $\mathbb{E}(f \,|\, \mathcal{I})$ by T, one gets

$$\sum_{j=1}^{n} [f \circ T^j - \mathbb{E}(f \,|\, \mathcal{I}) - \varepsilon] \leq W, \quad \text{thus} \quad \sum_{j=1}^{n} f \circ T^j \leq W + n[\mathbb{E}(f \,|\, \mathcal{I}) + \varepsilon],$$

consequently, $\mathbb{P}$-a.s.

$$\limsup_{n \to +\infty} \frac{1}{n} \sum_{j=1}^{n} f \circ T^j \leq \mathbb{E}(f \,|\, \mathcal{I}) + \varepsilon.$$

By letting ε go to 0, for any integrable function f, $\mathbb{P}$-almost surely

$$\limsup_{n \to +\infty} \frac{1}{n} \sum_{j=1}^{n} f \circ T^j \leq \mathbb{E}(f \,|\, \mathcal{I}),$$

by replacing f by $-f$, this yields the relationship

$$\liminf_{n \to +\infty} \frac{1}{n} \sum_{j=1}^{n} f \circ T^j \geq \mathbb{E}(f \,|\, \mathcal{I}).$$

The theorem is therefore proved. □

When the endomorphism is ergodic, the σ-field $\mathcal{I}$ of invariant sets is the trivial σ-field, in particular the conditional expectation with respect to $\mathcal{I}$ is the expectation. In this situation, the ergodic theorem gives the corollary.

COROLLARY 10.10. *If T is an ergodic endomorphism and f an integrable function, $\mathbb{P}$-a.s.,*

$$\lim_{n \to +\infty} \frac{1}{n} \sum_{i=1}^{n} f\left(T^i(\omega)\right) = \mathbb{E}(f).$$

Moreover, if $f \in L_p(\Omega)$ for $p \geq 1$, the convergence also holds with respect to the $L_p(\Omega)$ norm.

This corollary can be rephrased as follows: if the endomorphism is ergodic, then $\mathbb{P}$-a.s. the sequence of time averages of the orbit $[f(T^n(\omega))]$,

$$\left[\frac{1}{n}\sum_{i=0}^{n} f(T^i(\omega))\right],$$

converges to the expected value of $f(\omega)$, $\mathbb{E}(f)$. In particular, when $f = 1_A$, one gets that the orbit $(T^n(\omega))$ visits a region A of Ω with the frequency $\mathbb{P}(A)$. This result completes Proposition 10.6 on the density of sample paths of T. The convergence in L_2 of Cesàro means is a particular case of von Neumann's ergodic Theorem (see Neveu [**35**]).

PROOF OF COROLLARY 10.10. If f is a bounded random variable, for $n \in \mathbb{N}$, the variable $(f + f \circ T^2 + \cdots + f \circ T^n)/n$ is bounded by $\|f\|_\infty$ and converges almost surely to $\mathbb{E}(f)$ by the ergodic Theorem. Therefore, by Lebesgue's convergence Theorem, the convergence also holds in L_p.

If f is a non-negative element of L_p, for $\varepsilon > 0$, by Lebesgue's Theorem there exists $K \geq 0$ such that $\|f \wedge K - f\|_p \leq \varepsilon$, where $\|\cdot\|_p$ is the L_p norm of $L_p(\Omega)$. There exists some integer n_0 such that if $n \geq n_0$, then

$$\left\|\frac{1}{n}\sum_{i=0}^{n}(f \wedge K) \circ T^i - \mathbb{E}(f \wedge K)\right\|_p \leq \varepsilon,$$

the invariance of $\mathbb{P}$ by T gives the identity

$$\left\|\frac{1}{n}\sum_{i=0}^{n}(f \wedge K) \circ T^i - \frac{1}{n}\sum_{i=0}^{n} f \circ T^i\right\|_p$$

$$\leq \frac{1}{n}\sum_{i=0}^{n}\left\|(f \wedge K - f) \circ T^i\right\|_p = \|f \wedge K - f\|_p,$$

consequently, if $n \geq n_0$, one gets

$$\left\|\frac{1}{n}\sum_{i=0}^{n} f \circ T^i - \mathbb{E}(f)\right\|_p \leq 3\varepsilon.$$

The convergence in L_p for non-negative functions is established. Since an arbitrary function f of L_p can be decomposed as the difference of two non-negative functions in L_p, the convergence also holds for an arbitrary function. □

The Ergodic Decomposition

In this part, for a fixed mapping $T : \Omega \to \Omega$, the set $\mathcal{P}$ of probability distributions on Ω such that T is an endomorphism is investigated; $\mathcal{P}$ is the set of probability distributions on Ω left invariant by T.

A *support* of a probability distribution $\mathbb{P}$ on Ω is a measurable set S such that $\mathbb{P}(S) = 1$. Two non-negative probability distributions μ, ν on Ω

are *mutually singular* if there exists a measurable subset $S \in \mathcal{F}$ such that $\mu(S) = 1$ and $\nu(S) = 0$. (Any support of μ is ν-negligible and vice-versa). Equivalently, μ is said to be singular with respect to ν.

LEMMA 10.11. *If T is an endomorphism for two probability distributions $\mathbb{P}$ and $\mathbb{Q}$ on Ω, then*

$$d\mathbb{Q} = \alpha h \, d\mathbb{P} + (1-\alpha)\mu, \tag{10.11}$$

where $\alpha \in [0,1]$, $h \in L_1(\mathbb{P})$ is non-negative measurable function such that $\int h \, d\mathbb{P} = 1$ and μ is a probability distribution on Ω singular with respect to $\mathbb{P}$. The probability distributions $h \, d\mathbb{P}$ and μ are invariant by T (i.e. μ and $h \, d\mathbb{P} \in \mathcal{P}$) and have supports which are invariant subsets for the probability distribution $\mathbb{Q}$.

PROOF. Radon-Nikodym's Theorem (See Rudin [**45**]) gives the decomposition of the probability distribution $\mathbb{Q}$

$$d\mathbb{Q} = \alpha h \, d\mathbb{P} + (1-\alpha)\mu, \tag{10.12}$$

where $\alpha \in [0,1]$, $h \in L_1(\Omega, \mathbb{P})$ is non-negative, μ is a probability distribution on Ω singular with respect to $\mathbb{P}$: There exists a set $S \in \mathcal{F}$ such that $\mu(S) = 1$, and $\mathbb{P}(S) = 0$. By invariance of $\mathbb{P}$ by T,

$$\mathbb{P}(T^{-1}(S)) = \mathbb{P}(S) = 0,$$

if f is a bounded non-negative measurable function vanishing outside S,

$$\left| \int f \circ T \, h \, d\mathbb{P} \right| \leq \|f\|_\infty \int_{T^{-1}(S)} |h| \, d\mathbb{P} = 0,$$

by invariance of $\mathbb{Q}$ by T and since $T^{-1}(S)$ is $\mathbb{P}$-negligible,

$$\int f \circ T \, d\mu = \int f \circ T \, d\mathbb{Q} = \int f \, d\mathbb{Q} = \int f \, d\mu,$$

for any measurable function f which is 0 outside a support of μ, therefore μ is invariant by T. Relationship (10.12) shows that, since $\mathbb{Q}$ is invariant, the probability distribution $h \, d\mathbb{P}$ is also invariant by T.

A support S of μ satisfies

$$\begin{aligned} \mathbb{Q}\left(S \cap T^{-1}(S)^c\right) &= \int_{S \cap T^{-1}(S)^c} h \, d\mathbb{P} + \mu\left(S \cap T^{-1}(S)^c\right) \\ &= \mu\left(T^{-1}(S^c)\right) = \mu(S^c) = 0, \end{aligned}$$

similarly $\mathbb{Q}(S^c \cap T^{-1}(S)) = 0$. Therefore, the two sets S and $T^{-1}(S)$ are identical up to a $\mathbb{Q}$-negligible set. The set S is invariant for the probability distribution $\mathbb{Q}$. □

PROPOSITION 10.12 (Extremality Property). *If T is a measurable function from Ω to Ω, the set $\mathcal{P}$ of probability distributions on Ω invariant by T is a convex set. When $\mathcal{P}$ is non-empty, the extreme points of $\mathcal{P}$ are the probability distributions for which T is an ergodic endomorphism.*

If $\mathbb{P}$ and $\mathbb{Q}$ are two distinct probability distributions on Ω for which T is an ergodic endomorphism, they are mutually singular: There exists a subset $S \in \mathcal{F}$ such that $\mathbb{P}(S) = 1$ and $\mathbb{Q}(S^c) = 1$.

PROOF. Clearly enough, the set $\mathcal{P}$ is convex. Let $\mathbb{P}$ be a probability distribution for which T is an ergodic endomorphism. To prove that $\mathbb{P}$ is an extreme point of $\mathcal{P}$, one has to show that if there exist two probability distributions $\mathbb{P}_1$ and $\mathbb{P}_2$ invariant by T and $\alpha \in [0,1]$ such that

$$\mathbb{P} = \alpha\mathbb{P}_1 + (1-\alpha)\mathbb{P}_2, \tag{10.13}$$

then either $\alpha = 0$ or $\alpha = 1$.

According to the above lemma, by decomposing $\mathbb{P}_2$ with respect to $\mathbb{P}_1$, the probability distributions $\mathbb{P}_1$ and $\mathbb{P}_2$ can be assumed to be mutually singular and that a support S of $\mathbb{P}_1$ invariant for $\mathbb{P}$.

By ergodicity of $\mathbb{P}$, either $\mathbb{P}(S) = 1$ and then $\alpha = 1$, or $\mathbb{P}(S) = 0$ and in this case $\alpha = 0$. Decomposition (10.13) is necessarily trivial. The probability distribution $\mathbb{P}$ is an extreme point of the convex set $\mathcal{P}$.

Conversely, it is assumed that $\mathbb{P}$ is an extreme point of the convex set $\mathcal{P}$. If there exists a measurable set A invariant by T such that $0 < \mathbb{P}(A) < 1$, the probability distribution $\mathbb{P}$ can be decomposed as

$$\mathbb{P} = \mathbb{P}(A)\mathbb{Q}_A + \mathbb{P}(A^c)\mathbb{Q}_{A^c},$$

where $\mathbb{Q}_B(\cdot) = \mathbb{P}(\cdot \cap B)/\mathbb{P}(B)$ is $\mathbb{P}$ conditioned on B. Since A is invariant, the probability distributions $\mathbb{Q}_A$ and $\mathbb{Q}_{A^c}$ are invariant by T, therefore $\mathbb{P}$ cannot be an extreme point of $\mathcal{P}$. Contradiction. There does not exist a non-trivial invariant set, the endomorphism T is ergodic for $\mathbb{P}$.

To conclude, it is assumed that T is an ergodic endomorphism for two distinct probability distributions $\mathbb{P}$ and $\mathbb{Q}$, the above lemma gives the decomposition

$$\mathbb{Q} = \alpha h\, d\mathbb{P} + (1-\alpha)\mu, \tag{10.14}$$

where $\alpha \in [0,1]$, $h \in L_1(\Omega, \mathbb{P})$ and μ is singular with respect to $\mathbb{P}$: There exists $S \in \mathcal{F}$ such that $\mu(S) = 1$ and $\mathbb{P}(S) = 0$. The probability distributions μ and $h\, d\mathbb{P}$ are left invariant by T and a support S of μ is invariant for $\mathbb{Q}$. The ergodicity of T for $\mathbb{Q}$ implies that either $\mathbb{Q}(S) = 0$ or $\mathbb{Q}(S) = 1$.

If $\mathbb{Q}(S)$ is 0, Decomposition (10.14) shows that $\alpha = 1$, the probability distribution $\mathbb{Q}$ is absolutely continuous with respect to $\mathbb{P}$, $\mathbb{Q} = h\, d\mathbb{P}$. By invariance of $\mathbb{Q}$ by T,

$$\int f \circ T\, h\, d\mathbb{P} = \int f h\, d\mathbb{P},$$

for any bounded measurable function f, by iterating and summing up these equations, one gets

$$\int \frac{1}{n} \sum_{i=1}^{n} f \circ T^i\, h\, d\mathbb{P} = \int f h\, d\mathbb{P}.$$

for $n \in \mathbb{N}$. Corollary 10.10, the ergodicity of $\mathbb{P}$ and Lebesgue's Theorem show that the identity

$$\mathbb{E}(f) \int h \, d\mathbb{P} = \int f h \, d\mathbb{P},$$

holds for any bounded measurable function f, consequently h is $\mathbb{P}$-almost surely constant, equal to $\mathbb{E}(h) = 1$. The probability distributions $\mathbb{P}$ and $\mathbb{Q}$ are therefore identical, contradicting our initial assumption. Consequently $\mathbb{Q}(S)$ is equal to 1 and $\mathbb{Q} = \mu$, the probability distributions $\mathbb{P}$ and $\mathbb{Q}$ are mutually singular. The proposition is proved. □

The set of probability distributions for which T is an endomorphism is clearly a closed set for the weak topology on measures. By analogy to the finite dimensional case where every point of a closed convex subset can be written as a linear combination of extreme points, it is natural to conjecture that any probability distribution $\mathbb{P}$ invariant by T can be represented as

$$\int f \, d\mathbb{P} = \int_{\mathcal{E}} m_{\mathbb{P}}(d\mathbb{Q}) \int_{\Omega} f \, d\mathbb{Q}, \tag{10.15}$$

for any non-negative measurable function f, $\mathcal{E}$ denotes the set of probability distributions for which T is an ergodic endomorphism and $m_{\mathbb{P}}$ a measure on $\mathcal{E}$. Representation (10.15) holds under some topological assumptions, see Phelps [**41**]. Basically, this integral representation decompose $\mathbb{P}$ into ergodic components.

Example : The Translation on the Torus. The translation

$$T : x \to x + p/q,$$

on $[0, 1[$ when $q \neq 0$ and p, q are relatively prime integers, provides an elementary example of this decomposition. For $x \in [0, 1/q[$, the translation T is an ergodic automorphism on the set $x + E = \{x + i/q : 0 \leq i < q\}$, with the discrete probability distribution

$$\mu_x = \frac{1}{q} \sum_{i=0}^{q-1} \delta_{x+i/q},$$

where δ_x is the Dirac measure at x. The probability distributions μ_x indexed by $x \in [0, 1/q[$ are the only ones for which T is ergodic. Indeed, if μ is ergodic and there exist two disjoint measurable sets A_1 and A_2 of $[0, 1/q[$ non-negligible for μ, then for $i = 1, 2$, the set $B_i = \cup_{x \in A_i}(x + E)$ is invariant and non negligible. Consequently, the probability distribution μ restricted to $[0, 1/q[$ can only be a Dirac measure, the μ_x, $x \in [0, 1/q[$ are the only probability distributions for which T is ergodic. Lebesgue's measure on $[0, 1[$ can be decomposed with respect to the μ_x as follows, if f is a bounded measurable function on $[0, 1]$,

$$\int_0^1 f(x) \, dx = q \int_0^{1/q} \mu_t(f) \, dt,$$

with $\mu_t(f) = \int_0^1 f(u)\mu_t(du)$.

10.3 Continuous Time Dynamical Systems

DEFINITION 28. *A* flow (T^t) *is a set of endomorphisms of* Ω *indexed by* $\mathbb{R}$*; such that*

1. $T^t \circ T^s = T^{t+s}$ *for* $s, t \in \mathbb{R}$ *;*
2. *the mapping* $(\omega, t) \to T^t(\omega)$ *is measurable.*

A measurable function f *is* invariant *by the flow* (T^t) *if* $\mathbb{P}$*-almost surely the identity* $f \circ T^t = f$ *holds for all* $t \in \mathbb{R}$*. A set is invariant if its indicator function is invariant. A flow* (T^t) *is ergodic if the invariant subsets have probability* 0 *or* 1.

EXAMPLES.

1. If $T^t(x) = x + t \mod 1$, for $x \in [0, 1[$ and $t \in \mathbb{R}$, (T^t) is a flow on the torus $[0, 1[$ endowed with Lebesgue's measure.
2. If $\mathbb{P}$ is the distribution on $\mathcal{M}_p(\mathbb{R})$, the space of point measures on $\mathbb{R}$ of a stationary point process, the translations on the measures give a flow on $\mathcal{M}_p(\mathbb{R})$ (see Chapter 11).

THEOREM 10.13 (Ergodic Theorem for a flow). *If* (T^t) *is an ergodic flow and* $(Z(t))$ *a process such that*

1. *the variable* $S_1 = \sup\{Z(s) : s \in [0, 1]\}$ *is integrable,*
2. *the function* $t \to Z(t)$ *is additive for the flow, i.e. for* $s, t \in \mathbb{R}$,

$$Z(t + s) = Z(t) + Z(s) \circ T^t, \tag{10.16}$$

then, $\mathbb{P}$*-almost surely,*

$$\lim_{t \to +\infty} \frac{Z(t)}{t} = \mathbb{E}(Z(1)).$$

For $p \geq 1$ *the convergence is also valid for the* L_p*-norm if the variable* S_1 *is in* L_p.

The interval $[0, 1]$ in the integrability condition does not play a special role. Indeed, for $n \in \mathbb{N}$, the additivity property of the mapping $t \to Z(t)$ gives the relationship

$$\sup_{n \leq s \leq n+1} (Z(s)) = \sup_{0 \leq s \leq 1} (Z(s) \circ T^n) + Z(n)$$

and $Z(n) = \sum_{i=0}^{n-1} Z(1) \circ T^i$, consequently the variable $\sup_{0 \leq s \leq n}(Z(s))$ is also integrable for any $n \in \mathbb{N}$.

PROOF. The equalities $Z(0)=0$ and

$$Z(n)=\sum_{i=0}^{n-1} Z(1)\circ T^i$$

are direct consequences of Relationship (10.16). The ergodic Theorem applied to the endomorphism T^1 shows that, $\mathbb{P}$-a.s.

$$\lim_{n\to+\infty}\frac{Z(n)}{n}=\mathbb{E}(Z(1)\,|\,\mathcal{I}_1), \tag{10.17}$$

where $\mathcal{I}_1$ is the σ-field of invariant sets associated to T^1.

For $t\in\mathbb{R}$, the additivity property (10.16) gives

$$Z(t)=Z(\lfloor t\rfloor)+Z(t-\lfloor t\rfloor)\circ T^{\lfloor t\rfloor},$$

($\lfloor\cdot\rfloor$ is the integer part), consequently the inequality

$$\left|\frac{Z(t)-Z(\lfloor t\rfloor)}{\lfloor t\rfloor}\right|\le\frac{S_1\circ T^{\lfloor t\rfloor}}{\lfloor t\rfloor}$$

holds. It is assumed for a moment that $\mathbb{P}$-almost surely

$$\lim_{n\to+\infty}\frac{S_1\circ T^n}{n}=0. \tag{10.18}$$

The quantity $(Z(t)-Z(\lfloor t\rfloor))/\lfloor t\rfloor$ converges $\mathbb{P}$-a.s. to 0 as t tends to infinity and, according to Relationship (10.17),

$$\lim_{t\to+\infty}\frac{Z(t)}{t}=\mathbb{E}(Z(1)\,|\,\mathcal{I}_1)$$

$\mathbb{P}$-almost surely. If the random variable H is defined by

$$H=\liminf_{t\to+\infty}\frac{Z(t)}{t},$$

then $\mathbb{P}$-a.s. $H=\mathbb{E}(Z(1)\,|\,\mathcal{I}_1)$. By letting t go to infinity in the identity

$$\frac{Z(t)}{t}\circ T^s=\frac{Z(t+s)}{t}-\frac{Z(s)}{t},$$

one gets that for any $s\in\mathbb{R}$, $H\circ T^s=H$, $\mathbb{P}$-a.s. The ergodicity of the flow implies that H is a.s. constant, and that the identity $H=\mathbb{E}(H)=\mathbb{E}(Z(1))$ holds almost surely.

To conclude, Convergence (10.18) has to be proved. Since the variable S_1 is integrable, the same is true for $S_1\circ T^1-S_1$. By applying the ergodic Theorem to the endomorphism T^1 and the variable $S_1\circ T^1-S_1$, one gets that $\mathbb{P}$-almost surely

$$\lim_{n\to+\infty}\frac{S_1\circ T^n}{n}=\lim_{n\to+\infty}\frac{1}{n}\sum_{i=0}^{n-1}\left(S_1\circ T^1-S_1\right)\circ\left(T^1\right)^i=\mathbb{E}\left(S_1\circ T^1-S_1\,|\,\mathcal{I}_1\right).$$

Calculation of $\mathbb{E}(S_1\circ T^1-S_1\,|\,\mathcal{I}_1)$.

If A is an element of $\mathcal{I}_1$, the sets A and $T^{-1}(A)$ are identical up to a negligible set. The operator T^1 being also an endomorphism, the sets A and $T^1(A)$ are also identical up to a negligible set. By definition of conditional expectation

$$\mathbb{E}\left[\mathbb{E}\left(S_1 \circ T^1 \,|\, \mathcal{I}_1\right) 1_A\right] = \mathbb{E}\left(S_1 \circ T^1 1_A\right) = \mathbb{E}\left(S_1 1_{T^1(A)}\right),$$

since $1_{T^1(A)} = 1_A$ $\mathbb{P}$-a.s.

$$\mathbb{E}\left[\mathbb{E}\left(S_1 \circ T^1 \,|\, \mathcal{I}_1\right) 1_A\right] = \mathbb{E}(S_1 1_A) = \mathbb{E}[\mathbb{E}(S_1 \,|\, \mathcal{I}_1) 1_A],$$

for any A in $\mathcal{I}_1$. One gets that, almost surely, the relationship

$$\mathbb{E}\left(S_1 \circ T^1 - S_1 \,|\, \mathcal{I}_1\right) = 0,$$

holds, and thus

$$\lim_{n \to +\infty} \frac{S_1 \circ T^n}{n} = 0,$$

$\mathbb{P}$-almost surely. The convergence in L_p can be proved similarly. □

Corollary 10.14. *If $f \in L_1$ and (T^t) is an ergodic flow, then $\mathbb{P}$-a.s.*

$$\lim_{|t| \to +\infty} \frac{1}{t} \int_0^t f \circ T^s \, ds = \mathbb{E}(f).$$

Proof. If $Z(t) = \int_0^t f \circ T^s \, ds$,

$$\sup_{0 \leq s \leq 1} |Z(s)| \leq \int_0^1 |f \circ T^s| \, ds,$$

the expected value of the last term is given by $t\mathbb{E}(|f|) < +\infty$. On the other hand, since

$$Z(t+s) = Z(t) + \int_t^{t+s} f \circ T^u \, du,$$

and

$$\int_t^{t+s} f \circ T^u \, du = \int_0^s f \circ T^{u+t} \, du = Z(s) \circ T^t,$$

the ergodic Theorem 10.13 can be applied to conclude. □

Special Flows

This section presents a very important construction of ergodic theory: A continuous time dynamical system can be constructed from a discrete time dynamical system (i.e. an endomorphism).

Definition 29. *If S is an automorphism of Ω and τ a positive measurable function on the probability space $(\Omega_0, \mathcal{F}_0, \mathbb{P}_0)$, the* special flow *$(T^t)$ of height τ and base Ω_0 on the probability space $(\Omega, \mathcal{F}, \mathbb{P})$ is defined by*

1. $\Omega = \Omega_0 \times [0, \tau[= \{(\omega, x) : \omega \in \Omega_0, 0 \leq x < \tau(\omega)\}$,
2. *$\mathcal{F}$ is the σ-field restriction on Ω of the σ-field $\mathcal{F}_0 \otimes \mathcal{B}(\mathbb{R}_+)$,*

3. *The probability distribution* $\mathbb{P}$ *is the restriction to* Ω *of the product of* $\mathbb{P}_0$ *and Lebesgue's measure on* $\mathbb{R}$, *i.e.*
$$\mathbb{E}(f) = \int_\Omega f(\omega,x)\,d\mathbb{P} \stackrel{def.}{=} \frac{1}{\mathbb{E}_0(\tau)}\int_{\Omega_0} d\mathbb{P}_0 \int_0^{\tau(\omega)} f(\omega,x)\,dx,$$
for any non-negative measurable function f *defined on* Ω_0,
4. *The flow* (T^t) *can be expressed as*
$$T^t(\omega,x) = \left(S^k(\omega), x+t-t_k(\omega)\right),$$
if $t_k(\omega) - x \le t < t_{k+1}(\omega) - x$, *where*
$$t_k = \sum_{i=0}^{k-1} \tau\circ S^i \text{ if } k>0,\ t_0 = 0 \text{ and } t_k = \sum_{i=k}^{-1}\tau\circ S^i, \text{ if } k<0.$$

The fact that (T^t) is indeed a flow on the probability space $(\Omega, \mathcal{F}, \mathbb{P})$ is proved in Section 11.3 of Chapter 11 page 307.

Proposition 10.15. *If* S *is an automorphism on* Ω_0 *and* τ *a positive measurable function, the special flow* (T^t) *of height* τ *and base* Ω_0 *is ergodic if and only if* S *is ergodic.*

Proof. If S is ergodic and A is an invariant measurable set for (T^t), there exists a measurable subset A_1 of A such that $\mathbb{P}(A_1) = \mathbb{P}(A)$ and $T^t(A_1) = A_1$ for any $t \in \mathbb{R}$. If $B = \{\omega \in \Omega_0 : (\omega, 0) \in A_1\}$ is the projection of A_1 on Ω_0, for $s \in [0, \tau(\omega)[$ and $\omega \in B$ the definition of the special flow implies that $(\omega, s) = T^s(\omega, 0)$ is an element of A_1. Consequently, A_1 is the set $B \times [0, \tau[$ and the definition of $\mathbb{P}$ gives the identity
$$\mathbb{P}(A) = \mathbb{P}(A_1) = \frac{1}{\mathbb{E}_0(\tau)}\int d\mathbb{P}_0(\omega)\int_0^{\tau(\omega)} 1_{\{(\omega,x)\in B\times[0,\tau(\omega)[\}}\,dx$$
$$= \frac{1}{\mathbb{E}_0(\tau)}\int_B \tau\, d\mathbb{P}_0. \tag{10.19}$$
The set B is invariant by S : If $(\omega, 0)$ is in A_1, by invariance of A_1, $T^{\tau(\omega)}(\omega,0) = (S(\omega), 0)$ is also an element of A_1. Consequently, $S^{-1}(B) \subset B$, the set B is thus invariant by S. The ergodicity property of $\mathbb{P}_0$ implies that either $\mathbb{P}_0(B) = 0$ or 1, which also implies that either $\mathbb{P}(A) = 0$ or 1 according to Relationship (10.19). The special flow (T^t) is ergodic.

Conversely, if the flow (T^t) is ergodic and B is an invariant set for S, by setting $A = \{(\omega, x) : \omega \in B, 0 \le x < \tau(\omega)\}$, it is easy to check that A is invariant for the flow (T^t) and therefore either $\mathbb{P}(A) = 0$ or 1. Using the expression of $\mathbb{P}$, one gets that either $\mathbb{P}_0(B) = 0$ or 1. The endomorphism S is ergodic. □

Ambrose-Kakutani's Theorem[AK42], see Neveu [**37**], shows that, under some assumptions, any flow can be represented as a special flow. This result

[AK42] W. Ambrose and S. Kakutani, *Structure and continuity of measurable flows*, Duke Mathematical Journal **9** (1942), 25–42.

reduces the study of flows to the analysis of discrete dynamical systems. Totoki's paper[Tot66] investigates time changes of flows.

10.4 Markovian Endomorphisms

In this section, Markov processes are presented in the context of ergodic theory. The state space $\mathcal{S}$ is a locally compact metric space endowed with Borelian σ-field $\mathcal{B}$, p is a *transition function*, i.e. a non-negative mapping defined on $\mathcal{S} \times \mathcal{B}$ such that

1. for any $A \in \mathcal{B}$, the function $x \to p(x, A)$ is measurable,
2. for any $x \in \mathcal{S}$, $p(x, dy)$ is a probability distribution on $\mathcal{S}$.

It is assumed that the transition function has an invariant probability distribution π, that is, the identity

$$\int_{\mathcal{S}} f(x)\, \pi(dx) = \int_{\mathcal{S}^2} f(y)\, \pi(dx)\, p(x, dy), \tag{10.20}$$

holds for any bounded measurable function f on $\mathcal{S}$.

The lemma below is elementary: If the distribution of the initial point X_1 of the Markov chain (X_n) associated with p is π, for $p \geq 1$ the vector $(X_1, \dots, X_{p-1})$ has the same distribution as $(X_2, \dots, X_p)$, in other words the sequence (X_n) is stationary

LEMMA 10.16. *For $p \geq 2$, for any non-negative measurable function f on the product space $\mathcal{S}^{p-1}$,*

$$\int_{\mathcal{S}^p} f(x_2, \dots, x_p)\, \pi(dx_1) \prod_{j=1}^{p-1} p(x_j, dx_{j+1}) \tag{10.21}$$

$$= \int_{\mathcal{S}^{p-1}} f(x_2, \dots, x_p)\, \pi(dx_2) \prod_{j=2}^{p-1} p(x_j, dx_{j+1}).$$

PROOF. If $f(x_2, \dots, x_p) = g(x_2)h(x_3, \dots, x_p)$ where g and h are bounded measurable functions on their respective spaces,

[Tot66] H. Totoki, *Time changes of flows*, Memoirs of the Faculty of Sciences **20** (1966), no. 1, 27–55.

$$\int_{S^p} f(x_2,\ldots,x_p)\,\pi(dx_1)\prod_{j=1}^{p-1} p(x_j,dx_{j+1})$$

$$= \int_{S^2} g(x_2)\,\pi(dx_1)\,p(x_1,dx_2)\int_{S^{p-2}}\prod_{j=2}^{p-1} p(x_j,dx_{j+1})h(x_3,\ldots,x_p)$$

$$= \int_{S} g(x_2)\,\pi(dx_2)\int_{S^{p-2}}\prod_{j=2}^{p-1} p(x_j,dx_{j+1})h(x_3,\ldots,x_p)$$

$$= \int_{S^{p-1}} f(x_2,x_3,\ldots,x_p)\,\pi(dx_2)\prod_{j=2}^{p-1} p(x_j,dx_{j+1}),$$

according to Relationship (10.20) applied to the function

$$x_2 \to g(x_2)\int_{S^{p-1}}\prod_{j=2}^{p-2} p(x_j,dx_{j+1})h(x_3,\ldots,x_p).$$

Identity (10.21) is therefore satisfied for all the product functions f, and consequently for all the bounded measurable functions on S^p. □

DEFINITION 30. *The space $\Omega = S^{\mathbb{Z}}$ is endowed with the σ-field $\mathcal{F}$ associated with the product topology. The translation T is defined by $T(\omega) = (x_{i+1})$, for $\omega = (x_i) \in \Omega$. The element (x_{i+1}) is ω shifted on the left, that is, for $i \geq 0$, its ith component is x_{i+1}.*

A probability distribution on Ω is defined by the next proposition.

PROPOSITION 10.17. *There exists a unique probability distribution $\mathbb{P}$ on Ω such that*

$$\int_\Omega f\,d\mathbb{P} = \int_{S^{p-q+1}} f(x_p,\ldots,x_q)\,\pi(dx_p)\prod_{j=p}^{q-1} p(x_j,dx_{j+1}) \tag{10.22}$$

if f is a non-negative measurable function on Ω depending only on coordinates between p and $q \in \mathbb{Z}$, $f(\omega) = f(x_p,\ldots,x_q)$ for an element $\omega = (x_i)$ of Ω. On this probability space, the shift T is an endomorphism.

PROOF. If $\mathbb{P}_{p,q}$ is the probability distribution on S^{q-p+1} defined by the right hand side of Relationship (10.22) and if f is a non-negative measurable function depending only on coordinates between $p+1$ and $q-1$, then

$$\int f\,d\mathbb{P}_{p,q} = \int f\,d\mathbb{P}_{p+1,q} = \int f\,d\mathbb{P}_{p+1,q-1},$$

according to the above lemma. The probability distributions $\mathbb{P}_{p,q}$ for p and $q \in \mathbb{Z}$ are therefore compatible : If f depends only on coordinates between p' and q', with $p \leq p' \leq q' \leq q$, then

$$\int f\,d\mathbb{P}_{p,q} = \int f\,d\mathbb{P}_{p',q'}.$$

Daniell-Kolmogorov's Theorem, see Rogers and Williams [**44**], shows the existence and uniqueness of a probability distribution $\mathbb{P}$ on Ω satisfying Relationship (10.22).

If $f(\omega) = f(x_p, \ldots, x_q)$ for $\omega = (x_i)$, $f \circ T(\omega) = f(x_{p+1}, \ldots, x_{q+1})$, consequently

$$\int f \, d\mathbb{P} = \int f \circ T \, d\mathbb{P},$$

T is an endomorphism of the probability space $(\Omega, \mathcal{F}, \mathbb{P})$. □

The probability space $(\Omega, \mathcal{F}, \mathbb{P})$ is the *canonical* probability space associated with the Markov chain at equilibrium.

Definition 31. *For $n \in \mathbb{Z}$, the function X_n is defined as the projection on the nth coordinate, $X_n : \omega \to x_n$, if $\omega = (x_i)$.*

The sequence $(X_n; n \geq 0)$ is a Markov chain with transition function p and initial distribution π, it satisfies the identity

$$[X_{n+1}(\omega)] = [X_n(T(\omega))] \stackrel{\text{dist.}}{=} [X_n(\omega)],$$

and, by induction, for $n \in \mathbb{N}$, $X_n(\omega) = X_0 \circ T^n(\omega)$.

Examples.

1. If $\mathcal{S} = \{0, 1\}$ and $p(0,1) = p(1,0) = 1$, (X_n) is a periodic sequence with period 2. For the probability distribution $\mathbb{P}$ the two sequences : $(n \mod 2)$ and $(n+1 \mod 2)$ have probability 1/2.
2. If $p(x, dy) = \mu(dy)$ where μ is a probability distribution on $\mathcal{S}$, $(\Omega, \mathcal{F}, \mathbb{P})$ is the canonical probability space associated with an i.i.d sequence of random variables with distribution μ. In this case $\pi = \mu$.
3. If $\mathcal{S} = [0,1]^d$ is the d-dimensional torus $d \geq 1$ endowed with Lebesgue measure and $p(x, x + dy) = \mu(dy)$ for $\alpha \in \mathcal{S}$ where μ is a probability distribution on $\mathcal{S}$. The sequence (X_n) is a random walk on the torus.

Proposition 10.18. *If $p(x, dy) = \mu(dy)$, T is an ergodic endomorphism which has the* mixing property*:*

$$\lim_{n \to +\infty} \mathbb{E}(f \circ T^n g) = \mathbb{E}(f)\mathbb{E}(g),$$

for f and $g \in L_2(\Omega, \mathbb{P})$.

The mixing property implies that, for n sufficiently large the variables $f \circ T^n$ and g are almost independent. If T^n is seen as the translation at time n, it is equivalent to say that the endomorphism forgets the initial condition.

Proof. The ergodicity property is a consequence of the mixing property. Indeed, if $A \in \mathcal{F}$ is an event invariant by T and $f = 1_A$, for $g \in L_2$,

$$\mathbb{E}(f \circ T^n g) = \mathbb{E}(fg),$$

$n \in \mathbb{N}$. If the mixing property is true, the left hand side of the last identity converges to $\mathbb{E}(f)\mathbb{E}(g)$ as n tends to infinity, consequently $\mathbb{E}(fg) = \mathbb{E}(f)\mathbb{E}(g)$

for any $g \in L_2$. This implies that f is $\mathbb{P}$-a.s. constant, thus $\mathbb{P}(A) = 0$ or 1. The endomorphism T is thus ergodic.

If f_1 and $g_1 \in L_2$ depend only on a finite number of coordinates, for n sufficiently large $f_1 \circ T^n$ and g_1 depends on different coordinates. Since coordinates are independent functions in this situation (Example 2), thus $\mathbb{E}(f_1 \circ T^n g_1) = \mathbb{E}(f_1)\mathbb{E}(g_1)$ for n sufficiently large. For f, $g \in L_2$ and $\varepsilon > 0$, by definition of the product σ-field, there exist f_1 and $g_1 \in L_2$ depending only on a finite number of coordinates such that $\|f - f_1\|_2 < \varepsilon$ and $\|g - g_1\|_2 < \varepsilon$. For N_0 sufficiently large and $n \geq N_0$, $f_1 \circ T^n$ and g_1 do not depend on the same coordinates, therefore

$$\mathbb{E}(f_1 \circ T^n g_1) = \mathbb{E}(f_1)\mathbb{E}(g_1).$$

Since

$$|\mathbb{E}(f \circ T^n g) - \mathbb{E}(f)\mathbb{E}(g)| \leq |\mathbb{E}[(f - f_1) \circ T^n g]| + \\ |\mathbb{E}[f_1 \circ T^n (g - g_1)]| + |\mathbb{E}(f_1)\mathbb{E}(g_1) - \mathbb{E}(f)\mathbb{E}(g)|,$$

by Cauchy-Schwartz's Inequality and the invariance of $\mathbb{P}$ by T, one gets

$$\begin{aligned} |\mathbb{E}(f \circ T^n g) - &\mathbb{E}(f)\mathbb{E}(g)| \\ &\leq 2(\|g\|_2 + \|g_1\|_2)\|f - f_1\|_2 + 2(\|f\|_2 + \|f_1\|_2)\|g - g_1\|_2 \\ &\leq 2\varepsilon(\|g\|_2 + \|f\|_2 + 2\varepsilon). \end{aligned}$$

The mixing property is thus established. □

The law of Large numbers is a direct application of the last proposition and ergodic Theorem.

COROLLARY 10.19 (Law of Large Numbers). *If (Z_n) is an i.i.d. sequence of random variables in S and f a measurable function from S to $\mathbb{R}$ such that the variable $f(Z_0)$ is integrable, then $\mathbb{P}$-a.s.*

$$\lim_{n \to +\infty} \frac{1}{n} \sum_{i=1}^{n} f(Z_i) = \mathbb{E}(f(Z_1)).$$

The convergence also holds in L_p whenever $f(Z_0) \in L_p$ for $p \geq 1$.

PROOF. Let μ be the probability distribution of Z_0, by taking the probability space of Proposition 10.17 for which $p(x, dy) = \mu(dy)$, Definition (10.22) of $\mathbb{P}$ shows that the sequence $(X_n) = (X_0 \circ T^n)$ is i.i.d. with common distribution μ, hence it has the same distribution as (Z_n). The endomorphism T being ergodic in this case (Proposition 10.18), ergodic Theorem applied to the function $f \circ X_0$ concludes the proof. □

The next corollary is very useful to get limit results concerning point processes. A stationary renewal process is a renewal process which is invariant by non-negative translations (see the construction page 23 and Proposition 1.24).

COROLLARY 10.20. *If $\mathbb{P}$ is the probability distribution on $\mathcal{M}_p(\mathbb{R})$ stationary renewal process with common distribution μ, the flow T^t of the translations on the point measures is ergodic for $\mathbb{P}$.*

PROOF. By taking Ω_0 being the probability space associated with the transition function $p(x, dy) = \mu(dy)$, according to Proposition 11.4, the flow (T^t) is isomorphic to the special flow with base Ω_0 hand height the values of the 0th coordinate. Proposition 10.18 shows that the shift T on Ω_0 is ergodic, by Proposition 10.15 the same is true for (T^t). □

If the state space $\mathcal{S}$ of the Markov chain is finite and the Markov chain is not irreducible, there exists a non-empty subset $A \subset \mathcal{S}$ with $A \neq \mathcal{S}$ such that $p(x, A) = 1$ for any $x \in A$. The set $A^{\mathbb{Z}}$ is thus invariant by T and $T^{-1}(A^{\mathbb{Z}}) = A^{\mathbb{Z}}$. If π is an invariant probability distribution of this Markov chain such that $\pi(A) \in]0, 1[$, the associated probability distribution $\mathbb{P}$ on Ω satisfies

$$\mathbb{P}\left(A^{\mathbb{Z}}\right) = \pi(A) \in]0, 1[.$$

The endomorphism T is thus not ergodic. It is easily shown that the ergodic decomposition seen in Section 10.3 is directly related to the invariant distributions of the Markov chain on the different irreducible subsets. The following result shows that T is indeed ergodic if the Markov chain is irreducible.

PROPOSITION 10.21. *If $\mathcal{S}$ is countable and p is an irreducible transition function with an invariant probability distribution, then the automorphism T is ergodic.*

PROOF. It will be shown that for f, $g \in L_2(\Omega)$, then

$$\lim_{n\to+\infty} \frac{1}{n} \sum_{i=1}^{n} \mathbb{E}\left(f \circ T^i g\right) = \mathbb{E}(f)\mathbb{E}(g).$$

This condition, called the *weak mixing property* implies clearly that T is ergodic (same argument as in the beginning of the proof of Proposition 10.18).

The weak mixing property is first shown for functions depending only on a finite number of coordinates, for $\omega = (x_i)$, $x \in \mathcal{S}$, $f(\omega) = f(x_{p_1}, \dots, x_{q_1})$ and $g(\omega) = g(x_{p_2}, \dots, x_{q_2-1})1_{\{x_{q_2}=x\}}$, with $p_1 \le q_1$ and $p_2 \le q_2$. For more general functions the rest of the proof is similar to the corresponding part of proof of Proposition 10.18. If N_0 is sufficiently large so that $N_0 + p_1 \ge q_2 + 1$, for $i \ge N_0$,

$$\mathbb{E}\left[f \circ T^i g\right] = \mathbb{E}\left[f(X_{p_1+i}, \dots, X_{q_1+i})g(X_{p_2}, \dots, X_{q_2-1})1_{\{X_{q_2}=x\}}\right],$$

using the Markov property, the last term can be written as,

$$\mathbb{E}\left[g(X_{p_2}, \dots, X_{q_2-1})1_{\{X_{q_2}=x\}}\right] \mathbb{E}_x\left[f(X_{p_1+i-q_2}, \dots, X_{q_1-q_2+i})\right].$$

A classical result, see Feller [**21**], shows that for an irreducible Markov chain with an invariant probability distribution, Cesàro means converge to the invariant distribution, i.e.

$$\lim_{n\to+\infty} \frac{1}{n}\sum_{i=1}^{n} \mathbb{E}_x\left[f(X_{p_1+i-q_2},\ldots,X_{q_1-q_2+i})\right]$$
$$= \mathbb{E}_\pi\left[f(X_1,\ldots,X_{q_1-p_1+1})\right] = \mathbb{E}(f).$$

The weak mixing property is therefore proved. □

The next corollary can be interpreted as a law of large numbers for Markov chains.

COROLLARY 10.22. *If (X_n) is an irreducible Markov chain on a countable state space $\mathcal{S}$ having an invariant probability distribution π, then for any integrable function f on $\mathcal{S}$, $\mathbb{P}$-a.s.*

$$\lim_{n\to+\infty} \frac{1}{n}\sum_{i=1}^{n} \mathbb{E}_x\left(f(X_i)\right) = \mathbb{E}_\pi\left(f(X_0)\right).$$

PROOF. If the distribution of X_0 is π, the ergodic Theorem for the Markov chain shows that $\mathbb{P}$-a.s.

$$\mathbb{E}_\pi\left(f(X_0)\right) = \lim_{n\to+\infty} \frac{1}{n}\sum_{i=1}^{n} \mathbb{E}(f(X_i))$$
$$= \lim_{n\to+\infty} \frac{1}{n}\sum_{i=\tau_x}^{n} \mathbb{E}(f(X_i)).$$

Define τ_x the hitting time of $x \in \mathcal{S}$ (which $\mathbb{P}$-a.s. finite). The strong Markov property implies that the sequence $(X_{n+\tau_x})$ has the same distribution as the Markov chain (X_n) with $X_0 = x$. One can then easily conclude. □

11. Stationary Point Processes

Contents

11.1 Introduction

A queueing system can be seen as an operator on arrival processes. If the sequence of the arrival times of customers is (t_n) and (S_n) is the sequence of their respective sojourn times in the queue (the nth customer arrives at time t_n and leaves at $t_n + S_n$). The queue transforms a point process $\{t_n\}$ (the arrival process) in another point process $\{t_n + S_n\}$ (the departure process). In this setting, it is quite natural to investigate the properties of point processes that are preserved by such a transformation. In fact, very few properties remain unchanged. Most of the independence properties are lost for the departure process (the examples of the $M/M/1$ queue or some product form networks seen in Chapter 4 are remarkable exceptions to this general rule). For example, if the arrival process is a renewal process, the departure process is not, in general, a renewal process.

A weaker property is nevertheless preserved by queues. A point process is said to be *stationary* if the arrivals seen from an arbitrary instant $t \in \mathbb{R}$ have the same distribution as the arrival times seen from instant 0. The arrival process is homogeneous in time. (See a rigorous definition below.) It will be seen in this chapter that if the arrival process is stationary and the queue is at equilibrium, the departure process is also a stationary point process. This invariance relation is one of the rare properties which is valid for a large number of queueing systems. Moreover, these point processes are natural objects to describe the equilibrium of the $G/G/1$ queue. In this setting, there is no independence assumed for the random variables involved, only an ergodicity assumption is used. To get more explicit results, on the asymptotic distributions of these queues for example, some independence assumptions have to be introduced. (See Chapter 2.)

Stationary point processes are presented in the framework of ergodic theory of Chapter 10. See the nice paper by Ryll-Narzdewski[RN61] at the origin of this point of view. Taking advantage of the natural order on $\mathbb{R}$, stationary point processes are directly constructed from their sequence of interarrival intervals which are ordinary identically distributed sequences of random variables (stationary sequences). This direct approach has several advantages: First, it gives a simple intuitive picture of the so-called "Palm space" so that relations between the distributions at arrival instants and distributions at arbitrary instants are easier to understand. Secondly, this approach reduces to the minimum the somewhat cumbersome formalism of point processes usually required at this stage. Stationary point processes on more general spaces are (nicely) presented in Chapter 2 of Neveu [**37**].

11.2 The Palm Space of Arrival Process

The arrivals of customers are described as a sequence of non-negative random variables (τ_n, σ_n) such that

- the sequence (τ_n) is the sequence of inter-arrivals of customers, for $n \in \mathbb{Z}$, τ_n is the duration of time between the nth customer and the $(n+1)$th customer.
- For $n \in \mathbb{Z}$, σ_n is the service time required by the nth customer.
- The sequence (τ_n, σ_n) is stationary, that is, the sequence (τ_n, σ_n) has the same distribution as the sequence $(\tau_{n+1}, \sigma_{n+1})$.

Note that the indices of customers vary between $-\infty$ and $+\infty$, in Chapter 12 it is used to construct the equilibrium of the $G/G/1$ queue at time 0. The natural probability space associated with this sequence is $(\widehat{\Omega}, \widehat{\mathcal{F}}, \widehat{\mathbb{P}}, \widehat{\theta})$ where

1. $\widehat{\Omega} = (\mathbb{R}_+ \times \mathbb{R}_+)^{\mathbb{Z}}$;
2. $\widehat{\mathcal{F}}$ is the Borelian σ-field on the set $\widehat{\Omega}$;
3. The probability distribution $\widehat{\mathbb{P}}$ on $\widehat{\Omega}$ is the distribution of the sequence (τ_n, σ_n) ;

[RN61] C. Ryll-Nardzewski, *Remarks on processes of calls*, Proc. 4th Berkeley Sympos. Math. Statist. and Prob., Vol. II, Univ. California Press, Berkeley, Calif., 1961, pp. 455–465.

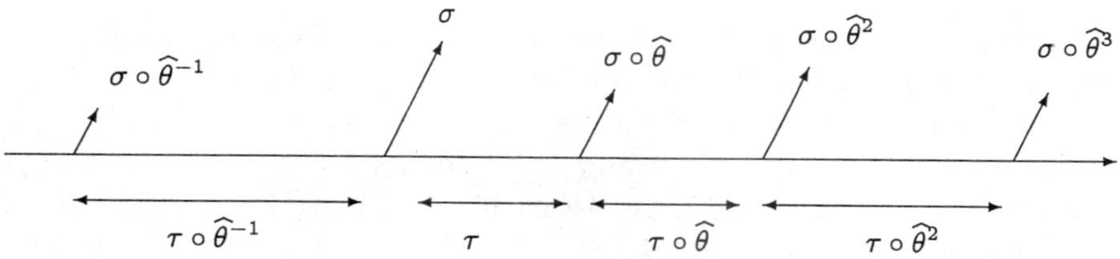

FIG. 11.1. A Point Process of Arrivals of Customers.

4. The shift $\widehat{\theta}$ on this space is given by

$$\widehat{\theta}((x_n, y_n)) = (x_{n+1}, y_{n+1}).$$

PROPOSITION 11.1. *The probability distribution $\widehat{\mathbb{P}}$ is invariant by the shift $\widehat{\theta} : \widehat{\Omega} \to \widehat{\Omega}$, i.e. for any non-negative measurable function f on $\widehat{\Omega}$,*

$$\widehat{\mathbb{E}}(f) = \int f \, d\widehat{\mathbb{P}} = \int f \circ \widehat{\theta} \, d\widehat{\mathbb{P}}. \tag{11.1}$$

PROOF. For $\omega = (x_i, y_i) \in \widehat{\Omega}$, if f is such that

$$f(\omega) = 1_{\{(x_{i_1}, y_{i_1}) \in A_1, \ldots, (x_{i_n}, y_{i_n}) \in A_n\}},$$

with $n \in \mathbb{N}$, $i_k \in \mathbb{Z}, 1 \leq k \leq n$ and $A_1, \ldots, A_n$ are Borelian subsets of $\mathbb{R}_+ \times \mathbb{R}_+$, then

$$\begin{aligned}\int f \circ \widehat{\theta} \, d\widehat{\mathbb{P}} &= \widehat{\mathbb{P}}\left[(x_{i_1+1}, y_{i_1+1}) \in A_1, \ldots, (x_{i_n+1}, y_{i_n+1}) \in A_n\right] \\ &= \widehat{\mathbb{P}}\left[(\tau_{i_1+1}, \sigma_{i_1+1}) \in A_1, \ldots, (\tau_{i_n+1}, \sigma_{i_n+1}) \in A_n\right] \\ &= \widehat{\mathbb{P}}\left[(\tau_{i_1}, \sigma_{i_1}) \in A_1, \ldots, (\tau_{i_n}, \sigma_{i_n}) \in A_n\right],\end{aligned}$$

according to the stationary property of the sequence (τ_n, σ_n), and this last quantity is the integral of f with respect to $\widehat{\mathbb{P}}$. Identity (11.1) is established for such an f. Since these functions generate the σ-field $\widehat{\mathcal{F}}$, the proposition is proved. □

The functions

$$\begin{aligned} \tau, \sigma : \widehat{\Omega} &\longrightarrow \mathbb{R}_+ \\ \omega &\longmapsto \tau(\omega) = x_0 \\ \omega &\longmapsto \sigma(\omega) = y_0 \end{aligned} \quad \text{when } \omega = (x_n, y_n)$$

are measurable, and for $n \in \mathbb{Z}$, the vector

$$\left[(\tau, \sigma), (\tau \circ \widehat{\theta}, \sigma \circ \widehat{\theta}), \ldots, (\tau \circ \widehat{\theta}^n, \sigma \circ \widehat{\theta}^n)\right]$$

has the same distribution as the vector $((\tau_0, \sigma_0), (\tau_1, \sigma_1), \ldots, (\tau_n, \sigma_n))$. In particular the random variable τ (resp. σ) has the same distribution as the variable τ_n (resp. σ_n), for $n \in \mathbb{Z}$.

The probability space $\widehat{\Omega}$ is the space of inter-arrivals and services of customers. The mapping σ is the service required by customer with index 0 and τ is the inter-arrival time with the next customer (with index 1). The function $\widehat{\theta}$ shifts indices : $\sigma = \sigma_0$ (resp. $\tau = \tau_0$) and $\sigma \circ \widehat{\theta} = \sigma_1$ (resp. $\tau \circ \widehat{\theta} = \tau_1$). The probability space $(\widehat{\Omega}, \widehat{\mathcal{F}}, \widehat{\mathbb{P}}, \widehat{\theta})$ is usually called *the Palm space of the arrivals.* The Palm space is named after a Swedish engineer in teletraffic, Conny Palm, who derived identities for arrival processes of calls[Pal43]. See also the translation [**40**]. Section 11.4 gives a formal presentation of these relationships.

Assumption. Throughout this chapter, it is assumed that $\widehat{\theta}$ is an ergodic endomorphism (see Chapter 10), the random variables σ and τ are $\widehat{\mathbb{P}}$-integrable and $\widehat{\mathbb{P}}(\tau = 0) = 0$.

The condition $\widehat{\mathbb{P}}(\tau = 0) = 0$ is not critical, it forbids the possibility of simultaneous arrivals. This is not really restrictive and it simplifies some of the statements below.

It remains to have an origin for the arrival process since only interarrival times are known. It is sufficient to define the arrival time of the last customer arrived before time 0. (This customer has index 0). The choice of the origin must be such that the point process obtained is *stationary*, i.e. the distribution of arrivals seen from time t does not depend on t. Formally, this property is defined as follows.

DEFINITION 32. *If $m \in \mathcal{M}_p(\mathbb{R} \times \mathbb{R}_+)$ is a point measure on $\mathbb{R} \times \mathbb{R}_+$ and $t \in \mathbb{R}$, $T^t m$ denotes the point measure m translated by t, i.e.*

$$T^t m(f) = \int f(x-t)\, m(dx),$$

for any non-negative measurable function f. If $x = (s, y) \in \mathbb{R} \times \mathbb{R}_+$, the quantity $x - t$ is the element $(s - t, y)$. Equivalently, for $m = \{(u_n, y_n)\}$, the translation of m by t is the point process $T^t m = \{(u_n - t, y_n)\}$.

Note that the translation only concerns the first temporal component.

DEFINITION 33. *A point process $N = \{t_n, y_n\}$ of $\mathcal{M}_p(\mathbb{R}\times\mathbb{R}_+)$ is stationary if its distribution is invariant by the operators (T^t), or equivalently*

$$\mathbb{E}[F(\{t_n, y_n\})] = \mathbb{E}[F(\{t_n - t, y_n\})],$$

for any $t \in \mathbb{R}$ and any non-negative measurable functional F on the space $\mathcal{M}_p(\mathbb{R} \times \mathbb{R}_+)$.

Corollary 1.14 page 13 shows that a Poisson point process with intensity λ on $\mathbb{R}$ is a stationary point process. The sequence of points of a point process on $\mathbb{R}$ can be represented as a non-decreasing sequence (t_n), such that

[Pal43] C. Palm, *Intensitätsschwankungen im fernsprechverkehr*, Ericsson Technics **44** (1943), 1–189.

$t_0 \leq 0 < t_1$. By convention, 0 is the index of the last customer arrived before time 0.

11.3 Construction of a Stationary Point Process

The non-decreasing sequence $(t_n(\xi))$, $\xi \in \widehat{\Omega}$ associated with partial sums of the random variables $\tau \circ \widehat{\theta}^k(\xi)$, $k \in \mathbb{Z}$ is defined by

$$\begin{aligned} t_n(\xi) &= \sum_{k=0}^{n-1} \tau \circ \widehat{\theta}^k(\xi), \qquad n \geq 1; \\ t_0(\xi) &= 0; \\ t_n(\xi) &= -\sum_{k=n}^{-1} \tau \circ \widehat{\theta}^k(\xi), \qquad n \leq -1, \end{aligned}$$

in particular $t_1 = \tau$. The ergodicity property of the endomorphism $\widehat{\theta}$ and ergodic Theorem (Theorem 10.9) show that $\widehat{\mathbb{P}}$-almost surely

$$\lim_{n\to+\infty} \frac{t_n(\xi)}{n} = \widehat{\mathbb{E}}(\tau) \text{ and } \lim_{n\to-\infty} \frac{t_n(\xi)}{n} = \widehat{\mathbb{E}}(\tau),$$

hence

$$\lim_{n\to+\infty} t_n(\xi) = +\infty \text{ and } \lim_{n\to-\infty} t_n(\xi) = -\infty,$$

$\mathbb{P}$-almost surely, the sequence (t_n) does not have a finite accumulation point. The measure $N = \{t_n\}$ is therefore almost surely a point process. It has a point at the origin $t = 0$ since $N(\{0\}) = 1$.

The problem is to fix an origin for the marked point process on $\mathbb{R} \times \mathbb{R}_+$ defined by

$$\{\bar{t}_n(\xi)\} \overset{\text{def.}}{=} \left\{ \left(t_n(\xi), \sigma \circ \widehat{\theta}^n(\xi) \right) \right\}$$

so that its distribution is invariant with respect to any deterministic translation in time. The existence of a random variable U such that for all $t \in \mathbb{R}$

$$\{\bar{t}_n(\xi) - U + t\} \overset{\text{dist.}}{=} \{\bar{t}_n(\xi) - U\}$$

should solve this problem.

This can be done by throwing the origin at random on $\mathbb{R}$, the invariance by translation of Lebesgue's measure on $\mathbb{R}$ giving then the stationarity property of the corresponding point process. This is of course not possible since Lebesgue measure has an infinite mass on $\mathbb{R}$, nevertheless this scheme can developed rigorously in the following manner. For $K \in \mathbb{N}$, if U_K is a uniformly distributed random variable on $[t_{-K}(\xi), t_K(\xi)]$, the next proposition shows that the point process $\{\bar{t}_n(\xi) - U_K\}$ converges in distribution to a stationary point process. (Recall that U_K is the origin of the point process.)

PROPOSITION 11.2. *As K tends to infinity, the marked point process*

$$\{\bar{t}_n(\xi) - U_K\} = \left\{\left(t_n(\xi) - U_K,\, \sigma \circ \widehat{\theta}^n(\xi)\right)\right\}$$

converges in distribution to a stationary point process whose distribution $\mathbb{Q}$ is given by

$$\mathbb{Q}(F) = \frac{1}{\widehat{\mathbb{E}}(t_1)} \int_{\widehat{\Omega}} d\widehat{\mathbb{P}}(\xi) \int_0^{t_1(\xi)} F\left(\{\bar{t}_n(\xi) - x\}\right) dx, \tag{11.2}$$

for any bounded measurable functional F defined on the point measures on $\mathbb{R} \times \mathbb{R}_+$.

PROOF. The functional F is assumed to depend on the restriction of point measures to some compact I interval of $\mathbb{R}$, i.e. if ν is a point measure on $\mathbb{R} \times \mathbb{R}_+$., then $F(\nu(\cdot)) = F(\nu(I \times \mathbb{R}_+ \cap \cdot))$. The proof for a general bounded functional F follows easily from this case. The quantity $m_K(F)$ denotes the expected value of the variable $F(\{\bar{t}_n - U_K\})$ conditionally on the value of the point process $\{\bar{t}_n\}$. The point U_K being uniformly distributed on $[t_{-K}, t_K]$, one gets

$$\begin{aligned}
m_K(F) &= \frac{1}{t_K - t_{-K}} \int_{t_{-K}}^{t_K} F\left(\{\bar{t}_n - x\}\right) dx \\
&= \frac{1}{t_K - t_{-K}} \sum_{k=-K}^{K-1} \int_{t_k}^{t_{k+1}} F\left(\{\bar{t}_n - x\}\right) dx \\
&= \frac{1}{t_K - t_{-K}} \sum_{k=-K}^{K-1} \int_0^{t_{k+1}-t_k} F\left(\{\bar{t}_n - t_k - x\}\right) dx. \qquad (11.3)
\end{aligned}$$

According to the definition of the sequence $(\bar{t}_n(\xi))$, for k, $n \geq 0$,

$$\bar{t}_n - t_k = \left(\sum_k^{n-1} \tau \circ \widehat{\theta}^i, \sigma \circ \widehat{\theta}^n\right) = \left(\sum_0^{n-k-1} \tau \circ \widehat{\theta}^{i+k}, \sigma \circ \widehat{\theta}^n\right),$$

and since $\widehat{\theta}^{i+k} = \widehat{\theta}^i \circ \widehat{\theta}^k$,

$$\bar{t}_n - t_k = \bar{t}_{n-k} \circ \widehat{\theta}^k.$$

This identity can be proved in the same way for a negative n. By using Relationship (11.3), one obtains

$$m_K(F) = \frac{1}{t_K - t_{-K}} \sum_{k=-K}^{K-1} \int_0^{t_1 \circ \widehat{\theta}^k} F\left(\{\bar{t}_{n-k} - x\} \circ \widehat{\theta}^k\right) dx.$$

If

$$g = \int_0^{t_1} F(\{\bar{t}_n - x\})\, dx,$$

since the point process $\{\bar{t}_{n-k} - x\}$ is identical to the point process $\{\bar{t}_n - x\}$ (the numbering does not play a role if K is sufficiently large), the quantity $m_K(F)$ can be expressed as

$$m_K(F) = \sum_{k=-K}^{K-1} g \circ \widehat{\theta}^k \Bigg/ \sum_{k=-K}^{K-1} t_1 \circ \widehat{\theta}^k.$$

The function g being integrable, the ergodicity property of $\widehat{\theta}$ and ergodic Theorem show that $m_K(F)$ converges $\widehat{\mathbb{P}}$-almost surely to $\widehat{\mathbb{E}}(g)/\widehat{\mathbb{E}}(t_1)$ as K tends to infinity. The random variable $m_K(F)$ being $\widehat{\mathbb{P}}$-almost surely bounded by $||F||_\infty$, Lebesgue's convergence Theorem gives the convergence

$$\lim_{K\to+\infty} \widehat{\mathbb{E}}(m_K(F)) = \frac{1}{\widehat{\mathbb{E}}(t_1)} \widehat{\mathbb{E}} \left(\int_0^{t_1} F(\{\bar{t}_n - x\})\, dx \right).$$

This establishes the convergence of the sequence of point processes. It remains to prove that the distribution $\mathbb{Q}$ is indeed invariant with respect to the translations (T^t). For $t \in \mathbb{R}$, the translation

$$T^t : \mathcal{M}_p(\mathbb{R} \times \mathbb{R}_+) \to \mathcal{M}_p(\mathbb{R} \times \mathbb{R}_+)$$

is continuous, therefore the distribution of the point process

$$\{\bar{t}_n - U_K - t\} = T^t \left(\{\bar{t}_n - U_K\}\right)$$

converges to $T^t\,\mathbb{Q}$, as $K \to +\infty$.

Conditioned to be in $I_K = [t_{-K}(\xi) + t, t_K(\xi)]$, the random variables U_K and $U_K + t$ have the same distribution as a uniformly distributed random variable V_K on I_K. Therefore the relationships

$$\begin{aligned}
&\widehat{\mathbb{E}}\left[F\left(\{\bar{t}_n - U_K - t\}\right)\right] \\
&\quad = \widehat{\mathbb{E}}\left[F\left(\{\bar{t}_n - (U_K + t)\}1_{\{U_K+t\in I_K\}}\right)\right] + O\left(\widehat{\mathbb{P}}(U_K + t \notin I_K)\right) \\
&\quad = \widehat{\mathbb{P}}(U_K + t \in I_K)\widehat{\mathbb{E}}\left[F\left(\{\bar{t}_n - V_K\}\right)\right] + O\left(\widehat{\mathbb{P}}(U_K + t \notin I_K)\right)
\end{aligned}$$

hold. In the same way

$$\widehat{\mathbb{E}}\left[F\left(\{\bar{t}_n - U_K\}\right)\right] = \widehat{\mathbb{P}}(U_K \in I_K)\widehat{\mathbb{E}}\left[F\left(\{\bar{t}_n - V_K\}\right)\right] + O\left(\widehat{\mathbb{P}}(U_K \notin I_K)\right),$$

by using that

$$\lim_{K\to+\infty} \widehat{\mathbb{P}}(U_K + t \in I_K) = \lim_{K\to+\infty} \widehat{\mathbb{P}}(U_K \in I_K) = 1,$$

one gets

$$\begin{aligned}
\mathbb{Q}(F) &= \lim_{K\to+\infty} \widehat{\mathbb{E}}\left[F\left(\{\bar{t}_n - U_K\}\right)\right] \\
&= \lim_{K\to+\infty} \widehat{\mathbb{E}}\left[F\left(\{\bar{t}_n - U_K - t\}\right)\right] = T^t\,\mathbb{Q}(F).
\end{aligned}$$

The proposition is proved. □

COROLLARY 11.3. *The family of endomorphisms (T^t) is a flow on the probability space $(\mathcal{M}_p(\mathbb{R}\times\mathbb{R}_+),\mathbb{Q})$, i.e. (see also Chapter 10),*

- *for $t\in\mathbb{R}$, $T^t:\mathcal{M}_p(\mathbb{R}\times\mathbb{R}_+)\to\mathcal{M}_p(\mathbb{R}\times\mathbb{R}_+)$ is bijective,*
- *the mapping $(t,m)\to T^t(m)$ is measurable from $\mathbb{R}\times\mathcal{M}_p(\mathbb{R}\times\mathbb{R}_+)$ to the set of point measures $\mathcal{M}_p(\mathbb{R}\times\mathbb{R}_+)$, both sets being endowed with Borelian σ-field;*
- *if $s,t\in\mathbb{R}$, $T^t\circ T^s=T^{t+s}$,*
- *the probability $\mathbb{Q}$ is invariant by (T^t), $T^t\mathbb{Q}=\mathbb{Q}$ for any $t\in\mathbb{R}$.*

The Probability Space of the Stationary Point Process

To obtain a stationary point process, the above proposition suggests the addition of a random variable X to the initial probability space $\widehat{\Omega}$ such that

$$0\le X<t_1(\xi),$$

and that the point process $\{\bar{t}_n(\xi)-X\}$ has distribution $\mathbb{Q}$. This can be done by considering the extension Ω of $\widehat{\Omega}$ defined by,

a) the extended probability space Ω is defined as follows

$$\Omega=\widehat{\Omega}\times[0,t_1[=\left\{(\xi,x):\xi\in\widehat{\Omega}\text{ and }0\le x<t_1(\xi)\right\},\tag{11.4}$$

b) A σ-field on the space Ω is given by the restriction $\mathcal{F}$ of the product σ-field $\widehat{\mathcal{F}}\otimes\mathcal{B}(\mathbb{R}_+)$ to the space Ω.

c) by using Relationship (11.2), it is quite natural to define $\mathbb{P}$ on Ω by

$$\mathbb{E}(f)=\int_\Omega f(\xi,x)\,d\mathbb{P}(\xi,x)\stackrel{\text{def.}}{=}\frac{1}{\widehat{\mathbb{E}}(t_1)}\int_{\widehat{\Omega}}d\widehat{\mathbb{P}}(\xi)\int_0^{t_1(\xi)}f(\xi,x)\,dx,\tag{11.5}$$

for any non-negative measurable function f on $\Omega\to\mathbb{R}_+$.

If $\omega=(\xi,x)$, the point process N is defined by

$$N(\omega,du)=\{\bar{t}_n(\omega)\}\stackrel{\text{def.}}{=}\{\bar{t}_n(\xi)-x\},$$

equivalently, for $n\in\mathbb{Z}$ and $\omega=(\xi,x)\in\Omega$,

$$\bar{t}_n(\omega)=\left(t_n(\xi)-x,\sigma\circ\widehat{\theta}^n(\xi)\right).$$

For this point process, the first point on the left of 0 is located at $\bar{t}_0(\omega)=-x$, and $(\tau[\widehat{\theta}^n(\xi)])$ is the sequence of interarrival intervals seen from this point. In this manner, for the probability distribution $\mathbb{P}$ on Ω, the distribution of the point process N is $\mathbb{Q}$: If f is a non-negative measurable function on $\mathcal{M}_p(\mathbb{R}\times\mathbb{R}_+)$, then

$$\begin{aligned}\mathbb{E}(N(\omega,f))&=\mathbb{E}\left(f\left(\{\bar{t}_n(\xi)-x\}\right)\right)\\&=\frac{1}{\widehat{\mathbb{E}}(t_1)}\int_{\widehat{\Omega}}d\widehat{\mathbb{P}}(\xi)\int_0^{t_1(\xi)}f\left(\{\bar{t}_n(\xi)-x\}\right)dx=\mathbb{Q}(f).\end{aligned}$$

the point process N is stationary.

The translation T^t of this point process induces a transformation θ^t of Ω such that the point process associated with $\theta^t(\omega)$ is given by T^tN, i.e.

$$(11.6)\qquad N(\theta^t(\omega), du) = T^t(N(\omega, du)) = \{\bar{t}_n(\xi) - x - t\} = \{\bar{t}_n(\omega) - t\}.$$

This is done as follows: Let $k \in \mathbb{Z}$ be the unique integer k such that the inequality $t_k(\omega) \leq t < t_{k+1}(\omega)$ holds. The translated point process $T^t(N(\omega, du))$ has its origin in t, the first point of the point process on its left is located at $t_k(\omega) - t$, and the distance between the two points around t is

$$t_{k+1}(\omega) - t - (t_k(\omega) - t) = t_{k+1}(\xi) - t_k(\xi) = \tau\left(\widehat{\theta}^k(\xi)\right).$$

The other points of the point process are obtained with the sequence of interarrival intervals

$$\left[\tau \circ \widehat{\theta}^{n+k}(\xi)\right] = \left[\tau \circ \widehat{\theta}^n\left(\widehat{\theta}^k(\xi)\right)\right],$$

the nth point of the translated point process is therefore at $t_{n+k}(\omega) - t$.

d) By setting

$$(11.7)\qquad \theta^t(\omega) = \left(\widehat{\theta}^k(\xi), x + t - t_k(\xi)\right), \text{ if } t_k(\xi) - x \leq t < t_{k+1}(\xi) - x,$$

then $\theta^t(\omega) \in \Omega$ and the associated point process is $T^tN(\omega, du)$. Identity (11.6) is therefore satisfied. Notice that this is the unique element of Ω such that this property holds. The sequence of interarrival intervals and the first point on the left of 0 determine the corresponding element of Ω. It is easily seen that $(\omega, t) \to \theta^t(\omega)$ is a measurable mapping from $\Omega \times \mathbb{R}$ onto Ω. For $s, t \in \mathbb{R}$, Relationship (11.6) gives

$$N\left(\theta^{t+s}(\omega), dx\right) = T^{t+s}(N(\omega, dx)) = T^t \circ T^s(N(\omega, dx)),$$

by using Corollary 11.3, hence

$$N\left(\theta^{t+s}(\omega), dx\right) = T^t(N(\theta^s(\omega), dx)) = N\left(\theta^t(\theta^s(\omega)), dx\right).$$

Consequently, Relationship (11.6) shows that (θ^t) satisfies the identity

$$(11.8)\qquad \theta^{t+s}(\omega) = \theta^t \circ \theta^s(\omega).$$

The measure $\{t_n(\omega)\}$ is $\mathbb{P}$-almost surely a point process since, $\widehat{\mathbb{P}}$-almost surely, the sequence $(|t_n(\xi)|)$ converges to infinity as $|n|$ gets large. Therefore the definition of $\mathbb{P}$ shows that, $\mathbb{P}$-almost surely, $|t_n(\omega)|$ converges to infinity as $|n|$ tends to infinity.

The basic properties of the extended probability space Ω are summarized in the proposition.

Proposition 11.4. *If $(\Omega, \mathbb{P}, (\theta^t))$ is the probability space defined by Relationships (11.4), (11.5) and (11.7), the stationary point process*

$$N : (\Omega, P, (\theta^t)) \to (\mathcal{M}_p(\mathbb{R} \times \mathbb{R}_+), \mathbb{Q}, (T^t))$$
$$\omega = (\xi, x) \longmapsto N(\omega, du) = \{\bar{t}_n(\omega)\} = \{\bar{t}_n(\xi) - x\}$$

is an isomorphism of flows, i.e.

a) N is one to one;
b) the probability distribution on $\mathcal{M}_p(\mathbb{R} \times \mathbb{R}_+)$ induced by $\mathbb{P}$ via the mapping N is $\mathbb{Q}$;
c) the point process N commutes with (θ^t),

$$N\left(\theta^t(\omega), dx\right) = T^t N(\omega, dx), \qquad \omega \in \Omega, \tag{11.9}$$

or, for any $n \in \mathbb{Z}$,

$$t_n\left(\theta^t(\omega)\right) = t_{n+k}(\omega) - t, \tag{11.10}$$

where k is the unique integer such that $t_k(\omega) \leq t < t_{k+1}(\omega)$.

The probability space defined by Relationships (11.4), (11.5) and (11.7) is a classical construction of ergodic theory, see Chapter 10. This is the continuous time dynamical system associated with the discrete time dynamical system $(\widehat{\Omega}, \widehat{\mathbb{P}}, \widehat{\theta})$ and the function τ. The construction of a point process with a fixed Palm space is equivalent to the construction of the flow associated with the function τ.

11.4 Relations Between the Palm Space and Its Extension

Embedding of $\widehat{\Omega}$ into Ω. The probability space $(\widehat{\Omega}, , \widehat{\mathbb{P}}, \widehat{\theta})$ can be naturally embedded into $(\Omega, \mathcal{F}, \mathbb{P})$, by the mapping ϕ

$$\begin{aligned} \phi : \widehat{\Omega} &\to \Omega \\ \xi &\to (\xi, 0). \end{aligned}$$

From now on, $\widehat{\Omega}$, $\widehat{\theta}$, $\widehat{\mathbb{P}}$, and the random variables σ, t_1 are identified to their image by ϕ : If f is non-negative and measurable on Ω, one sets

$$\int_{\widehat{\Omega}} f \, d\widehat{\mathbb{P}} = \int_{\widehat{\Omega}} f(\xi, 0) \, d\widehat{\mathbb{P}}(\xi),$$

and $\widehat{\theta}(\xi, 0) = (\widehat{\theta}(\xi), 0)$.

The point process associated with the element $\omega = (\xi, 0)$ is nothing but the point process $\{\overline{t}_n(\xi)\}$ which has a mass at 0. The set $\widehat{\Omega}$ is therefore the subset of Ω on which the point process N has a mass at 0. By using the definitions of θ^t and t_n, it is easily seen that for $n \in \mathbb{Z}$ and $\omega = (\xi, x) \in \Omega$,

$$\theta^{t_n}(\omega) = \left(\widehat{\theta}^n(\xi), 0\right) \in \widehat{\Omega}. \tag{11.11}$$

The mapping $\theta^{t_n} : \Omega \to \Omega$ is therefore $\widehat{\Omega}$-valued. If $0 \leq x < t_1(\xi)$, Relationship (11.7) shows that, for $\omega = (\xi, 0) \in \widehat{\Omega}$, then $\theta^x(\omega) = (\xi, x)$. Since $\widehat{\Omega}$ is a subset of Ω, Relationship (11.5) connecting the probability distributions $\mathbb{P}$ and $\widehat{\mathbb{P}}$ can be rewritten as follows.

PROPOSITION 11.5. *If f is a non-negative measurable function on Ω, then*

$$\int f \, d\mathbb{P} = \frac{1}{\widehat{\mathbb{E}}(t_1)} \int_{\widehat{\Omega}} \int_0^{t_1(\omega)} f(\theta^s(\omega)) \, ds \, d\widehat{\mathbb{P}}. \tag{11.12}$$

The next formula, due to Mecke[Mec67], gives another relation between the probability distributions $\mathbb{P}$ and $\widehat{\mathbb{P}}$, the probability distribution $\widehat{\mathbb{P}}$ can be expressed with $\mathbb{P}$.

PROPOSITION 11.6 (Mecke's Formula). *The stationary marked point process N verifies*

$$\mathbb{E}\left(\int_{\mathbb{R}\times\mathbb{R}_+} f(\theta^t(\omega), t, y) \, N(\omega, dt, dy)\right) = \lambda \int \int_{\mathbb{R}} f(\omega, t, \sigma(\omega)) \, dt \, d\widehat{\mathbb{P}}, \tag{11.13}$$

where $\lambda = 1/\widehat{\mathbb{E}}(t_1)$ and f is a non-negative measurable function on $\Omega \times \mathbb{R} \times \mathbb{R}_+$.

The next corollary is a direct consequence of the above proposition.

COROLLARY 11.7. *If f is a non-negative measurable function on $\Omega \to \mathbb{R}_+$ and $t \in \mathbb{R}$, then*

$$\mathbb{E}\left(\int_0^t f(\theta^u(\omega)) \, N(\omega, du)\right) = \lambda \widehat{\mathbb{E}}(f) t, \tag{11.14}$$

and if $g : \mathbb{R} \to \mathbb{R}_+$ is a Borelian function,

$$\mathbb{E}\left(\int g(t) \, N(\omega, dt)\right) = \lambda \int g(t) \, dt;$$

in particular, for $a \leq b \in \mathbb{R}$, then $\mathbb{E}(N[a, b]) = \lambda(b - a)$, where

$$N[a, b] = \int_{[a,b]} N(\omega, dt)$$

is the number of points of N in the interval $[a, b]$.

According to Definition 2 of Chapter 1, the point process N has intensity $\lambda \, dx$. Lebesgue measure being implicit for the intensity of stationary point processes, the point process N is said to have *intensity* λ.

PROOF OF PROPOSITION 11.6. If $\omega = (\xi, x) \in \Omega$,

$$\int_{\mathbb{R}\times\mathbb{R}_+} f\left(\theta^t(\omega), t, y\right) \, N(\omega, dt, dy) = \sum_n f\left(\theta^{t_n}(\omega), t_n(\omega), \sigma \circ \widehat{\theta}^n(\xi)\right),$$

Identity (11.11) gives

[Mec67] J. Mecke, *Stationäre zufällige Masse auf lokalkompakten Abelschen Gruppen*, Zeitschrift für Wahrscheinlichkeitstheorie und verw. Geb. **9** (1967), 36–58.

$$\int_{\mathbb{R}\times\mathbb{R}_+} f\left(\theta^t(\omega), t, y\right) N(\omega, dt, dy) = \sum_n f\left(\left(\widehat{\theta}^n(\xi), 0\right), -x + t_n(\xi), \sigma \circ \widehat{\theta}^n(\xi)\right).$$

According to Relationship (11.5), the expected value, with respect to the probability distribution $\mathbb{P}$, of this random variable is given by

$$\begin{aligned}\mathbb{E}\left(\int_{\mathbb{R}\times\mathbb{R}_+} f\left[\theta^t(\omega), t, y\right] N(\omega, dt, dy)\right)& \\ &= \lambda\widehat{\mathbb{E}}\left(\int_0^{t_1(\xi)} dx \int f\left[\theta^t(\omega), t, y\right] N(\omega, dt, dy)\right) \\ &= \lambda\widehat{\mathbb{E}}\left(\sum_n \int_{t_n(\xi)-t_1(\xi)}^{t_n(\xi)} f\left[\left(\widehat{\theta}^n(\xi), 0\right), x, \sigma \circ \widehat{\theta}^n(\xi)\right] dx\right).\end{aligned} \tag{11.15}$$

Since $t_n(\xi) - t_1(\xi) = t_{n-1} \circ \widehat{\theta}(\xi)$, for $n \in \mathbb{Z}$

$$\begin{aligned}\widehat{\mathbb{E}}\left(\int_{t_n(\xi)-t_1(\xi)}^{t_n} f\left[\left(\widehat{\theta}^n(\xi), 0\right), x, \sigma \circ \widehat{\theta}^n(\xi)\right] dx\right)& \\ &= \int_{\mathbb{R}} \widehat{\mathbb{E}}\left(1_{\{t_{n-1}\circ\widehat{\theta}(\xi)\leq x\leq t_n\}} f\left[\left(\widehat{\theta}^n(\xi), 0\right), x, \sigma \circ \widehat{\theta}^n(\xi)\right]\right) dx,\end{aligned}$$

the invariance of $\widehat{\theta}$ by $\widehat{\mathbb{P}}$ shows that the last term is equal to

$$\int_{\mathbb{R}} \widehat{\mathbb{E}}\left(1_{\{t_{n-1}\circ\widehat{\theta}^{-(n-1)}(\xi)\leq x\leq t_n\circ\widehat{\theta}^{-(n)}(\xi)\}} f[(\xi, 0), x, \sigma(\xi)]\right) dx.$$

Therefore, Relationship (11.15) becomes

$$\begin{aligned}\mathbb{E}\left(\int_{\mathbb{R}\times\mathbb{R}_+} f[\theta^t(\omega), t, y]\, N(\omega, dt, dy)\right)& \\ &= \lambda\widehat{\mathbb{E}}\left(\sum_{n\in\mathbb{Z}} \int_{t_{n-1}\circ\widehat{\theta}^{-(n-1)}(\xi)}^{t_n\circ\widehat{\theta}^{-n}(\xi)} f[\xi, x, \sigma(\xi)]\, dx\right).\end{aligned}$$

By summing up integrals inside the braces, Identity (11.13) is thus proved. □

Usually the presentation of stationary point processes is somewhat different. A stationary point process N is given and Identity (11.13), Mecke's Formula, is used to introduce the Palm measure. The existence of the Palm measure being established, the main properties of the Palm space are then proved (such as the invariance of $\widehat{\mathbb{P}}$ for the shift $\widehat{\theta}$ for example). See Neveu [**37**]. Constructing directly a stationary point process from its sequence of interarrival intervals presents the advantage of giving a simple intuitive picture of the Palm space. See also Sigman [**49**].

Nevertheless, it may happen that the stationary point process is given without any reference to the sequence of interarrival intervals, in the case of the superposition of stationary point processes for example. The appendix considers this question and the following definition solves this (small) difficulty.

DEFINITION 34. *If N is a point process on a probability space $(\Omega, \mathbb{P}, (\theta^t))$ such that Relationship (11.13) holds for some probability distribution $\widehat{\mathbb{P}}$ on Ω, this probability distribution is called Palm probability of the point process.*

Conditioning in the Neighborhood of 0

For the moment, a stationary point process has been constructed from a stationary ergodic sequence of random variables. It is not excluded that other constructions might be possible, leading to different stationary point processes. The appendix shows that there is only one possible construction. Intuitively it seems that, if the stationary point process is conditioned to have a point in a neighborhood of 0, then the original point process $(t_n(\xi))$ should be obtained. Indeed, Proposition 11.2 shows that the distribution of N is the limit in distribution (with respect to the probability distribution $\widehat{\mathbb{P}}$) of the point processes $\{t_n(\xi) - U_K\}$ when K tends to infinity. The random variable U_K, uniformly distributed on the interval $[t_{-K}(\xi), t_K(\xi)]$, is the origin of the point process. For a fixed K, if the origin is conditioned to be in a small neighborhood of one of the points of the sequence $(t_n(\xi))$, the distribution of the point process $\{\bar{t}_n(\xi) - U_K\}$ is close to $\widehat{\mathbb{P}}$, the distribution of the point process $\{\bar{t}_n(\xi)\}$ with a mass at 0. By assuming that this approximation holds when K tends to infinity, the probability distribution $\mathbb{P}$ (the distribution of the stationary point process N) conditioned to have a point in a neighborhood of 0 is asymptotically $\widehat{\mathbb{P}}$ when the size of the neighborhood tends to 0. This heuristic reasoning is established rigorously in the proposition.

PROPOSITION 11.8. *If f is a bounded measurable function defined on Ω such that $t \to f(\theta^t(\omega))$ is $\widehat{\mathbb{P}}$-almost surely right continuous at $t = 0$, then*

$$\lim_{x \to 0+} \mathbb{E}\left(f \mid N[-x, 0] \neq 0\right) = \widehat{\mathbb{E}}(f).$$

PROOF. If $x > 0$, Relationship (11.12) gives

$$\begin{aligned}\mathbb{E}\left(f 1_{\{N[-x,0]\neq 0\}}\right) &= \lambda\widehat{\mathbb{E}}\left(\int_0^{t_1} \left[f\, 1_{\{N[-x,0]\neq 0\}}\right](\theta^s(\omega))\, ds\right)\\ &= \lambda\widehat{\mathbb{E}}\left(\int_0^{t_1} f(\theta^s(\omega)) 1_{\{N[s-x,s]\neq 0\}}\, ds\right),\end{aligned}$$

according to Identity (11.9). On the space $\widehat{\Omega}$, for $0 \leq s < t_1$, then clearly $N[s-x, s] \neq 0$ if and only if 0 is in the interval $[s-x, s]$,

$$\mathbb{E}(f \mid N[-x,0] \neq 0) = \widehat{\mathbb{E}}\left(\int_0^{x\wedge t_1} f(\theta^s(\omega))\, ds\right) \Big/ \widehat{\mathbb{E}}(x \wedge t_1),$$

$$\mathbb{E}(f \mid N[-x,0] \neq 0) = \widehat{\mathbb{E}}\left(\frac{1}{x}\int_0^{x\wedge t_1} f(\theta^s(\omega))\, ds\right) \Big/ \widehat{\mathbb{E}}(1 \wedge t_1/x).$$

The continuity property of f gives that, $\mathbb{P}$-a.s.,

$$\lim_{x\to 0} \frac{1}{x}\int_0^{x} f(\theta^s(\omega))\, ds = f(\omega).$$

Lebesgue's convergence Theorem concludes the proof. □

Proposition 1.15 page 14 on Poisson point processes can formulated as follows: The Palm probability of a Poisson point process $\mathcal{N}$ is the distribution of the point process $\mathcal{N} + \delta_0$.

COROLLARY 11.9.

a) *If A is a Borelian subset of Ω, then*

$$\lim_{t\to 0+} \mathbb{P}\left(\theta^{t_1} \in A \,|\, N[-t,0] \neq 0\right) = \widehat{\mathbb{P}}(A).$$

b) *When $t \to 0^+$, for the probability distribution $\mathbb{P}$, the conditional distribution of t_1 given $N[-t,0] \neq 0$ converges in distribution to the distribution of the variable t_1 for the probability distribution $\widehat{\mathbb{P}}$.*

PROOF. Since, by definition, $t_1 > 0$, $\mathbb{P}$-almost surely, for a sufficiently small $t > 0$, $t_1(\theta^t(\omega)) = t_1(\omega) - t$. Consequently, if g is a bounded continuous function on $\mathbb{R}$, the function $f(\omega) = g(t_1(\omega))$ satisfies the hypothesis of the above proposition. One gets that

$$\lim_{t\to 0} \mathbb{E}(f \mid N[-t,0] \neq 0) = \widehat{\mathbb{E}}(f) = \widehat{\mathbb{E}}(g(t_1)),$$

thereby proving the convergence in distribution, hence b). By using the properties of the flow (θ^t) and the last expression of $t_1(\theta^t(\omega))$, for a sufficiently small t, the identity

$$\theta^{t_1}(\theta^t(\omega)) = \theta^{t_1(\theta^t(\omega))}(\theta^t(\omega)) = \theta^{t_1(\theta^t(\omega))+t}(\omega) = \theta^{t_1(\omega)}(\omega)$$

holds. The function $t \to 1_{\{\theta^{t_1}\in A\}}(\theta^t(\omega))$ is therefore $\mathbb{P}$-almost surely right continuous. The above proposition can be applied to conclude. □

11.5 Joint Distributions of the Points Around t = 0

Let f be a bounded measurable function on $\mathbb{R}_+ \times \mathbb{R}^{\mathbb{Z}}$, Relationship (11.12) gives the equation

$$\mathbb{E}\left(f[-t_0, (\bar{t}_n - t_0)]\right) = \lambda\widehat{\mathbb{E}}\left(\int_0^{t_1} f\left[-t_0(\theta^u), (\bar{t}_n - t_0)\,(\theta^u)\right]\, du\right).$$

Identity (11.7) gives the relation $\bar{t}_n(\theta^u) = \bar{t}_n - u$, for $n \in \mathbb{Z}$ and $0 \leq u < t_1$, and, since $t_0 = 0$ on the subset $\widehat{\Omega}$,

$$\mathbb{E}\left(f[-t_0, (\bar{t}_n - t_0)]\right) = \lambda \widehat{\mathbb{E}}\left(\int_0^{t_1} f[-u, (\bar{t}_n)]\, du\right).$$

The following proposition has been proved.

PROPOSITION 11.10. *If f is a bounded measurable function on $\mathbb{R}_+ \times \mathbb{R}^{\mathbb{Z}}$, then*

$$\mathbb{E}\left(f\left[-t_0, (\bar{t}_n - t_0)\right]\right) = \lambda \widehat{\mathbb{E}}\left(\int_0^{t_1} f\left[-u, (\bar{t}_n)\right]\, du\right). \tag{11.16}$$

In particular, the joint distribution of arrival instants just before and just after time 0 is given by

$$\mathbb{E}(g[t_0, t_1]) = \lambda \widehat{\mathbb{E}}\left(\int_0^{\tau} g[-u, \tau - u]\, du\right), \tag{11.17}$$

for amy bounded Borelian function g on $\mathbb{R}_- \times \mathbb{R}_+$.

Relationship (11.16) is nothing but Relationship (11.2) reformulated for the probability space Ω. If g is a bounded Borelian function on $\mathbb{R}_+$, then

$$\mathbb{E}(g(-t_0)) = \mathbb{E}(g(t_1))$$
$$= \lambda \int \tau(dx) \int_0^x g(u)\, du = \lambda \int g(u) \widehat{\mathbb{P}}(\tau \geq u)\, du.$$

The variables t_1 and $-t_0$ have a density function given by $\widehat{\mathbb{P}}(\tau \geq t)/\widehat{\mathbb{E}}(\tau)$.

Examples

Stationary Renewal Processes. When the sequence $(\tau_n) = (\tau \circ \widehat{\theta}^n)$ is i.i.d., Relationship (11.16) gives the distribution of the associated stationary point process. If f and $g_n, n \in \mathbb{Z}$ are non-negative functions on $\mathbb{R}$, then

$$\mathbb{E}\left(f(-t_0) \prod_{n \in \mathbb{Z}} g_n[(t_{n+1} - t_n)]\right) = \lambda \widehat{\mathbb{E}}\left(\int_0^{t_1} f(-u) \prod_{n \in \mathbb{Z}} g_n[(t_{n+1} - t_n)]\, du\right).$$

For the probability distribution $\widehat{\mathbb{P}}$, the sequence $(t_{n+1} - t_n) = (\tau \circ \widehat{\theta}^n)$ is i.i.d. with the same distribution as τ and the $t_{n+1} - t_n$, $n \neq 0$ are independent of $t_1 = \tau$. One gets

$$\mathbb{E}\left(f(-t_0) \prod_0^{+\infty} g_n[(t_{n+1} - t_n)]\right)$$
$$= \lambda \widehat{\mathbb{E}}\left(\int_0^{\tau} f(-u) g_0(\tau - u)\, du\right) \prod_{n \neq 0} \widehat{\mathbb{E}}\left(g_i(\tau)\right).$$

To construct a stationary point process, it is therefore sufficient to take two random variables (t_0, t_1), independent of the sequence (τ_n), and whose distribution is given by Relationship (11.17). This point process is the superposition of two independent renewal processes, one on the non-negative axis $(t_1 + \tau_1 + \cdots + \tau_n; n \geq 0)$ and the other one on the negative axis $(t_0 - \tau_{-1} - \cdots - \tau_n; n \leq 0)$.

1. Poisson point processes. When τ is exponentially distributed with parameter λ, the joint distribution of $(-t_0, t_1)$ is given by

$$\mathbb{E}(g(-t_0, t_1)) = \lambda \int_0^{+\infty} \lambda e^{-\lambda x}\, dx \int_0^x g(u, x-u)\, du$$
$$= \int_{\mathbb{R}_+ \times \mathbb{R}_+} \lambda e^{-\lambda u} \lambda e^{-\lambda v} g(u, v)\, du\, dv.$$

   The random variables $-t_0$ and t_1 are therefore independent and exponentially distributed with parameter λ. The distribution of t_1 (resp. $-t_0$) is the distribution of interarrival intervals.
2. Deterministic point processes. If $\tau_0 \equiv D$, then t_1 has the same distribution as a variable U_D uniformly distributed on $[0, D]$ and $t_0 = D - t_1$. The points of the associated point process are given by $U_D + nD$, $n \in \mathbb{Z}$.
3. A point process with a non-integrable variable t_1. If τ_0 has the density function $2/x^3 1_{\{x \geq 1\}}$ for the probability distribution $\widehat{\mathbb{P}}$, then τ_0 is integrable. The density function of the random variable t_1 on $\mathbb{R}_+$ is given by $1/2 \inf(1/x^2, 1)$, in particular, the variable t_1 is not integrable.

The examples a) and c) exhibit a curious phenomenon. The duration of time between the two customers arriving before and after time 0 has, in case a) a mean which is the double of the mean of τ_0, and in case c) is not integrable at all. If τ is square-integrable, then the variable t_1 is integrable since

$$\mathbb{E}(t_1) = \frac{1}{\widehat{\mathbb{E}}(\tau)} \int_0^{+\infty} u \widehat{\mathbb{P}}(\tau \geq u)\, du = \frac{\widehat{\mathbb{E}}(\tau^2)}{2\widehat{\mathbb{E}}(\tau)}.$$

Going back to the original construction of Section 11.3, this paradox can be easily explained. When a point is thrown at random at U_N in the interval $[t_{-N}(\xi), t_N(\xi)]$, this point is very likely to be located in a large interval between two points of the original point process (with a point at 0) rather than in a small interval. If the variance of τ_0 is large (infinite in case c)), there are many small intervals together with large intervals. In this situation, very likely, the point U_N is in a large interval. The variables $-t_0$ and t_1 are therefore large. This explains why the origin transforms the original point process.

An Alternating Stationary Renewal Point Process. One considers two independent i.i.d. sequences (a_i) and (b_i). The respective distributions of the variables a_0, b_0 are denoted by F and G. For $i \in \mathbb{Z}$, a_i is the duration that some machine is effective after it has been repaired for the ith time. The durations of failures are given by the variables b_i, $i \in \mathbb{Z}$. In order to get a model of the instants of failures, the previous example can be used. This is a renewal process associated with the i.i.d. sequence $(a_i + b_i)$. This process does not give the instants when the machine is effective. If these points are added to the point process, it is necessary to distinguish the failure instants. A mark σ can be added to the points to indicate the beginning of a failure if $\sigma = 0$, and the beginning of an effective period when $\sigma = 1$. The distribution of (τ_n, σ_n) on $(\mathbb{R} \times \{0,1\})^{\mathbb{Z}}$ is given by

$$\widehat{\mathbb{P}}(\sigma_0 = m) = 1/2, \text{ for } m = 0, 1,$$

and, conditionally on the event $\{\sigma_0 = 1\}$,

- For $i \in \mathbb{Z}$, $\sigma_{2i} = 1$ and $\sigma_{2i+1} = 0$,
- the sequences (τ_{2i}), (τ_{2i+1}) are i.i.d. and independent, whose common distributions are respectively given by F and G.

In the same way, conditionally on the event $\{\sigma_0 = 0\}$,

- for $i \in \mathbb{Z}$, $\sigma_{2i} = 0$ and $\sigma_{2i+1} = 1$,
- the sequences (τ_{2i}), (τ_{2i+1}) are i.i.d. and independent, whose common distributions are respectively given by G and F.

It is easily seen that the sequence (τ_n, σ_n) is stationary. The associated stationary marked point process associated with this sequence $N = \{t_n, m_n\}$ has intensity

$$\lambda = \frac{1}{\mathbb{E}(\tau_0)} = \frac{1}{\mathbb{E}(a_0)/2 + \mathbb{E}(b_0)/2}.$$

Identity (11.16) gives the joint distribution: For a bounded Borelian function f on $\mathbb{R}^3$,

$$\begin{aligned}\mathbb{E}\left(f(-t_0, t_1, t_2 - t_1)1_{\{\sigma_0=0\}}\right) &= \frac{\lambda}{2}\mathbb{E}\left(\int_0^{b_0} f(-u, b_0 - u, a_1)\,du\right)\\ &= \frac{\lambda}{2}\iint_{\mathbb{R}\times\mathbb{R}} G(dx)F(dy)\int_0^x f(-u, x-u, y)\,du.\end{aligned}$$

In particular the stationary probability of being in a failure period is given by

$$\mathbb{P}(\sigma_0 = 0) = \frac{\lambda}{2}\int_{\mathbb{R}} xG(dx) = \frac{\mathbb{E}(b_0)}{\mathbb{E}(a_0) + \mathbb{E}(b_0)}.$$

11.6 Some Properties of Stationary Point Processes

The continuous dynamical system constructed in Section 11.3 preserves the ergodicity property (see Chapter 10) of the initial discrete dynamical system.

PROPOSITION 11.11. *The special flow (θ^t) is ergodic on $(\Omega, \mathcal{F}, \mathbb{P})$.*

PROOF. The proof is immediate by using the ergodicity property assumed for $\widehat{\theta}$, the remark after Proposition 11.4 and Proposition 10.15 page 296. □

From now on, all stationary point processes $N = \{t_n, \sigma_n\}$ are assumed to be defined on the probability space $(\Omega, \mathcal{F}, \mathbb{P}, (\theta^t))$ and to verify

$$N\left(\theta^t(\omega), dx\right) = T^t N(\omega, dx), \tag{11.18}$$

for all $t \in \mathbb{R}$ and $\omega \in \Omega$.

11.6.1 Limit Theorems

PROPOSITION 11.12 (Ergodic Theorem). *If $f : \Omega \to \mathbb{R}_+$ is a measurable function, then*

$$\lim_{t\to+\infty} \frac{1}{t} \int_0^t f\left(\theta^s\right) N(\omega, ds) = \lambda \widehat{\mathbb{E}}(f), \qquad \mathbb{P} - a.s.$$

PROOF. If $Z_t(\omega) = \int_0^t f(\theta^u(\omega)) N(\omega, du)$, then

$$Z_t(\theta^s(\omega)) = \int_0^t f(\theta^{u+s}(\omega)) N\left(\theta^s(\omega), du\right),$$

using Relationship (11.9), one gets

$$Z_t(\theta^s(\omega)) = \int_s^{t+s} f(\theta^u(\omega)) N(\omega, du),$$

thus $Z_{t+s} = Z_s + Z_t(\theta^s)$. The last proposition shows that the flow (θ^t) is ergodic, the ergodic Theorem 10.13 for continuous flows gives that $\mathbb{P}$-a.s.

$$\lim_{t\to+\infty} \frac{1}{t} \int_0^t f(\theta^u(\omega)) N(\omega, du) = \mathbb{E}\left(\int_0^1 f(\theta^u(\omega)) N(\omega, du)\right),$$

and Relationship (11.14) concludes the proof,

$$\mathbb{E}\left(\int_0^1 f(\theta^u(\omega)) N(\omega, du)\right) = \lambda \widehat{\mathbb{E}}(f).$$

□

The asymptotic properties of stationary point processes are summarized in the following proposition.

PROPOSITION 11.13. *$\mathbb{P}$-almost surely,*

a) $\lim_{t\to+\infty} N[0,t]/t = \lambda$,
b) $\lim_{n\to+\infty} t_n/n = 1/\lambda$,

and

c) $\lim_{t\to 0} \mathbb{P}(N[0,t] = 1)/t = \lambda$,
d) $\lim_{t\to 0} \mathbb{P}(N[0,t] \geq 2)/t = 0$.

PROOF. a) is an immediate consequence of the last proposition. For b), notice that for $x \geq 0$,

$$\{N]0,t] < xt\} = \left\{t_{\lfloor xt \rfloor} > t\right\} = \left\{t_{\lfloor xt \rfloor}/(xt) > 1/x\right\},$$

where $\lfloor z \rfloor$ is the integer part of z. By a), one gets that $\mathbb{P}$-a.s.

$$\liminf_{n\to+\infty} t_n/n \geq 1/\lambda.$$

The reverse inequality with the limsup is proved in the same way.

By Relationship (11.17),

$$\mathbb{P}(N[0,t] \geq 1) = \mathbb{P}(t_1 \leq t) = \lambda\widehat{\mathbb{E}}(\tau \wedge t),$$

Lebesgue's convergence Theorem shows the convergence

$$\lim_{t\to 0} \mathbb{P}(N[0,t] \geq 1)/t = \lambda. \tag{11.19}$$

In the same way, by using Relationship (11.12), one gets

$$\mathbb{P}(N[0,t] \geq 2) = \mathbb{P}(t_2 \leq t) = \lambda\widehat{\mathbb{E}}\left(\int_0^{t_1} 1_{\{t_2(\theta^s)\leq t\}}\, ds\right).$$

Since $t_2(\theta^s) = t_2 - s$ for $0 \leq s \leq t_1$,

$$\mathbb{P}(N[0,t] \geq 2) = \lambda\widehat{\mathbb{E}}\left(\int_{\mathbb{R}_+} 1_{\{t_2-t\leq s\leq t_1\}}\, ds\right) = \lambda\widehat{\mathbb{E}}\left[\left(t_1 - (t_2-t)^+\right) \vee 0\right],$$

thus

$$\mathbb{P}(N[0,t] \geq 2)/ = \lambda\widehat{\mathbb{E}}\left[(1-(t_2-t_1)/t)^+ \wedge t_1/t\right].$$

According to the initial assumption on the variable τ, $t_2 - t_1 = \tau \circ \widehat{\theta} > 0$, $\widehat{\mathbb{P}}$-almost surely, therefore $(1-(t_2-t_1)/t)^+ \wedge t_1/t$ converges $\widehat{\mathbb{P}}$-almost surely to 0 as t goes to 0. One gets d) via Lebesgue's convergence Theorem and c) by using d) and Convergence (11.19). □

The probability that a stationary point process with intensity λ has a point in an interval of length h is of the order λh when h is small.

11.6.2 Transformations of Stationary Point Processes

PROPOSITION 11.14 (Thinning). *If A is a Borelian subset of $\mathbb{R}$ and X is a Borelian real valued random variable on $\widehat{\Omega}$ such that $\widehat{\mathbb{P}}(X \in A) > 0$, the point process*

$$N^A = \sum_{n\in\mathbb{Z}} 1_{\{X\circ\theta^{t_n}\in A\}}(\omega)\, \delta_{t_n}$$

is stationary with intensity $\lambda\widehat{\mathbb{P}}(X \in A)$ and its Palm measure is given by

$$\frac{1}{\widehat{\mathbb{P}}(X \in A)} \int f \cdot 1_{X\in A}\, d\widehat{\mathbb{P}}.$$

PROOF. If g is a bounded measurable function on $\mathbb{R} \times \mathbb{R}_+$, and for $t \in \mathbb{R}$,

$$\int_{\mathbb{R}\times\mathbb{R}_+} g(u,y)\, T^t N^A(\omega, du, dy) = \int_{\mathbb{R}\times\mathbb{R}_+} g(u-t,y)\, N^A(\omega, du, dy)$$
$$= \int_{\mathbb{R}\times\mathbb{R}_+} g(u-t,y) 1_{\{X\circ\theta^u(\omega)\in A\}}\, N(\omega, du, dy),$$

group Relationship (11.8) shows that

$$\int_{\mathbb{R}\times\mathbb{R}_+} g(u,y)\, T^t N^A(\omega, du, dy)$$
$$= \int_{\mathbb{R}\times\mathbb{R}_+} g(u-t,y) 1_{\{X\circ\theta^{u-t}(\theta^t(\omega))\in A\}}\, N(\omega, du, dy)$$
$$= \int_{\mathbb{R}\times\mathbb{R}_+} g(u,y) 1_{\{X\circ\theta^u(\theta^t(\omega))\in A\}}\, T^t N(\omega, du, dy),$$

by Relationship (11.18), one gets

$$\int_{\mathbb{R}\times\mathbb{R}_+} g(u,y) 1_{\{X\circ\theta^u(\theta^t(\omega))\in A\}}\, N\left(\theta^t(\omega), du, dy\right)$$
$$= \int_{\mathbb{R}\times\mathbb{R}_+} g(u,y)\, N^A\left(\theta^t(\omega), du, dy\right).$$

Hence, the identity $T^t N^A(\omega, dx) = N^A(\theta^t(\omega), dx)$ holds. The distribution of the point process is therefore invariant by translations. It is stationary.

The Palm measure. If f is a non-negative measurable function on the space $\Omega \times \mathbb{R} \times \mathbb{R}_+$,

$$\mathbb{E}\left(\int_{\mathbb{R}\times\mathbb{R}_+} f\left(\theta^t(\omega), t, y\right)\; N^A(\omega, dt, dy)\right)$$
$$= \mathbb{E}\left(\int_{\mathbb{R}} f\left(\theta^t(\omega), t, \sigma(\omega)\right) 1_{\{X\circ\theta^t(\omega)\in A\}}\, N(\omega, dt)\right),$$

Relationship (11.13) for N yields

$$\mathbb{E}\left(\int_{\mathbb{R}\times\mathbb{R}_+} f(\theta^t(\omega), t, y)\, N^A(\omega, dt, dy)\right)$$
$$= \lambda \int\int_{\mathbb{R}} f(\omega, t, \sigma(\omega)) 1_{\{X(\omega)\in A\}}\, dt\, d\widehat{\mathbb{P}},$$

which gives the Palm measure of N^A and its intensity. □

PROPOSITION 11.15 (Translation). *If $S : \widehat{\Omega} \to \mathbb{R}$ is a random variable, the point process N_S defined by*

$$N_S = \left\{\left(t_n + S\circ\theta^{t_n}, \sigma\circ\theta^{t_n}\right)\right\},$$

or, equivalently,

$$N_S(f) = \int_{\mathbb{R}} f\left(t + S\left(\theta^t(\omega)\right), y\right) \, N(\omega, dt, dy),$$

for any non-negative measurable function f on $\mathbb{R} \times \mathbb{R}_+$, is a stationary point process with the same intensity as N and its Palm measure $\widehat{\mathbb{P}}_S$ is given by

$$\int g \, d\widehat{\mathbb{P}}_S = \int g\left(\theta^S\right) \, d\widehat{\mathbb{P}},$$

for any measurable function $g : \Omega \to \mathbb{R}_+$.

PROOF. First, this point process is indeed defined on Ω : S is defined on $\widehat{\Omega}$ and the range of the mapping $\omega \to \theta^{t_n}(\omega)$ is in $\widehat{\Omega}$.

For $t \in \mathbb{R}$, and g a non-negative measurable function on $\mathbb{R} \times \mathbb{R}_+$,

$$\int_{\mathbb{R}\times\mathbb{R}_+} g(t, y) \, T^t N_S(\omega, dt, dy) = \sum_{n \in \mathbb{Z}} g\left(t_n(\omega) + S \circ \theta^{t_n}(\omega) - t, \sigma \circ \theta^{t_n}(\omega)\right),$$

Relationship (11.8) yields $S \circ \theta^{t_n}(\omega) = S \circ \theta^{t_n - t}(\theta^t(\omega))$, therefore

$$\begin{aligned} T^t N_S(\omega, dt, dx) &= \int_{\mathbb{R}\times\mathbb{R}_+} g\left(s - t + S \circ \theta^{s-t}(\theta^t(\omega)), y\right) \, N(\omega, ds, dy) \\ &= \int_{\mathbb{R}\times\mathbb{R}_+} g\left(s + S \circ \theta^{s}(\theta^t(\omega)), y\right) \, T^t N(\omega, ds, dy), \end{aligned}$$

by using Relationship (11.18) one gets

$$\begin{aligned} \int_{\mathbb{R}\times\mathbb{R}_+} g\left(s + S \circ \theta^s\left(\theta^t(\omega)\right), y\right) \, N\left(\theta^t(\omega), ds, dy\right) \\ = \int g(s, y) \, N_S(\theta^t(\omega), ds, dy). \end{aligned}$$

Therefore $T^t N_S(\omega, dt, dy) = N_S\left(\theta^t(\omega), ds, dy\right)$, the point process is stationary.

If f is a non-negative measurable function on $\Omega \times \mathbb{R} \times \mathbb{R}_+$,

$$\begin{aligned} \mathbb{E}\left(\int_{\mathbb{R}\times\mathbb{R}_+} f\left(\theta^t(\omega), t, y\right) \, N_S(\omega, dt, dy)\right) \\ = \mathbb{E}\left(\int_R f\left(\theta^{t + S\circ\theta^t}(\omega), t + S \circ \theta^t(\omega), \sigma \circ \theta^t(\omega)\right) \, N(\omega, dt)\right), \end{aligned}$$

Relationship (11.13) for N applied to the function

$$h : (\omega, t, y) \to f\left(\theta^S(\omega), t + S(\omega), y\right),$$

yields

$$\mathbb{E}\left(\int_{\mathbb{R}\times\mathbb{R}_+} f\left(\theta^t(\omega), t, y\right) N_S(\omega, dt, dy)\right)$$
$$= \mathbb{E}\left(\int_R h\left(\theta^t(\omega), t, y\right) N(\omega, dt, dy)\right) = \lambda \int\int_{\mathbb{R}} h(\omega, t, \sigma(\omega))\, dt\, d\widehat{\mathbb{P}}$$
$$= \lambda \int\int_{\mathbb{R}} f\left(\theta^S(\omega), t + S(\omega), \sigma(\omega)\right) ds\, d\widehat{\mathbb{P}} = \lambda \int\int_{\mathbb{R}} f\left(\theta^S(\omega), t, \sigma(\omega)\right) ds\, d\widehat{\mathbb{P}},$$

one deduces the expression of the Palm measure of N_S. □

If S is such that $S \circ \theta^t$ is the sojourn time of a customer arrived at time t in a queueing system, the point process N_S is the departure process from the queueing system. If the arrival process is stationary, the same property holds for the departure process.

PROPOSITION 11.16 (Superposition). *If, for $i = 1, 2$, N_i is a stationary point process with intensity λ_i and Palm measure $\widehat{\mathbb{P}}_i$, then the superposition N, defined by*

$$N(f) = N_1(f) + N_2(f),$$

for any non-negative measurable f, is also a stationary point process. Its intensity is given by $\lambda = \lambda_1 + \lambda_2$ and its Palm measure is the probability distribution $\widehat{\mathbb{P}} = (\lambda_1\widehat{\mathbb{P}}_1 + \lambda_2\widehat{\mathbb{P}}_2)/\lambda$.

PROOF. For $t \in \mathbb{R}$

$$\begin{aligned} T^t(N_1 + N_2)(\omega, ds, dy) &= T^t N_1(\omega, ds, dy) + T^t N_2(\omega, ds, dy) \\ &= N_1\left(\theta^t(\omega), ds, dy\right) + N_2\left(\theta^t(\omega), ds, dy\right) \\ &= (N_1 + N_2)\left(\theta^t(\omega), ds, dy\right). \end{aligned}$$

The intensity and Palm measure are easily calculated. □

The Palm space $\widehat{\Omega}$ can be easily expressed in this case,

$$\widehat{\Omega} = \{\omega : N_1(\omega, \{0\}) = 1\} \cup \{\omega : N_2(\omega, \{0\}) = 1\}.$$

11.6.3 A Relation Between Palm Measures

With the same notations as in the previous proposition, the following proposition, due to Neveu[Nev76] in a much more general case, gives a relation between Palm measures of two point processes.

PROPOSITION 11.17. *If $N_1 = \{s_n\}$, $N_2 = \{t_n\}$, and $f : \Omega \to \mathbb{R}_+$ is measurable, then*

[Nev76] J. Neveu, *Sur les mesures de Palm de deux processus ponctuels stationnaires*, Zeitschrift für Wahrscheinlichkeitstheorie und verw. Geb. **34** (1976), 199–203.

$$\lambda_1 \widehat{\mathbb{E}}_1(f) = \lambda_2 \widehat{\mathbb{E}}_2 \left(\sum_{0 \le s_n < t_1} f\left(\theta^{s_n}\right) \right). \tag{11.20}$$

PROOF. Define

$$h(\omega) = \sum_{0 \le s_n < t_1} f\left(\theta^{s_n}(\omega)\right) = \int_0^{t_1} f\left(\theta^s(\omega)\right) N_1(\omega, ds),$$

for $n \in \mathbb{Z}$,

$$\begin{aligned} h\left(\theta^{t_n}(\omega)\right) &= \int_0^{t_1 \circ \theta^{t_n}} f\left(\theta^{s+t_n}(\omega)\right) N_1\left(\theta^{t_n(\omega)}(\omega), ds\right) \\ &= \int_{t_n}^{t_n + t_1 \circ \theta^{t_n}} f\left(\theta^s(\omega)\right) N_1(\omega, ds), \end{aligned}$$

since $t_1\left(\theta^{t_n}\right) = t_{n+1} - t_n$ (the next point after t_n is t_{n+1} !),

$$h\left(\theta^{t_n}\right)(\omega) = \int_{t_n}^{t_{n+1}} f\left(\theta^s(\omega)\right) N_1(\omega, ds). \tag{11.21}$$

The proof is quite simple. It consists in writing an integral with respect to N_1 as an integral with respect to N_2, this is done as follows,

$$\begin{aligned} \int_0^1 f\left(\theta^s\right) N_1(\omega, ds) &= \sum_{0 \le s_k < 1} f\left(\theta^{s_k}\right) \\ &= \sum_{0 \le s_k < t_1} f(\theta^{s_k}) + \sum_{0 \le t_n < 1} \left(\sum_{t_n \le s_k < t_{n+1}} f(\theta^{s_k}) \right) - \sum_{1 \le s_k < 1 + t_1(\theta^1)} f(\theta^{s_k}). \end{aligned}$$

By using Relationship (11.21), the last equality can be rewritten as

$$\begin{aligned} \int_0^1 f\left(\theta^s\right) N_1(\omega, ds) &= \sum_{0 \le t_n < 1} h\left(\theta^{t_n}\right) + h(\omega) - h\left(\theta^1(\omega)\right) \\ &= \int_0^1 h\left(\theta^s\right) N_2(\omega, ds) + h(\omega) - h\left(\theta^1(\omega)\right). \end{aligned}$$

If h is integrable, then $\mathbb{E}\left(h - h \circ \theta^1\right) = \mathbb{E}(h) - \mathbb{E}\left(h \circ \theta^1\right) = 0$ by invariance of $\mathbb{P}$ by θ^1. Taking the expected value of Identity (11.14), one gets

$$\lambda_1 \widehat{\mathbb{E}}_1(f) = \mathbb{E}\left(\int_0^1 f(\theta^s) N_1(\omega, ds) \right) = \mathbb{E}\left(\int_0^1 h(\theta^s) N_2(\omega, ds) \right) = \lambda_2 \widehat{\mathbb{E}}_2(h),$$

Relationship (11.20) is therefore satisfied in this case.
Set $g = f \wedge K/(N_1[0, t_1] + 1)^2$, the function h corresponding to g is integrable since

$$\int_0^{t_1} \frac{1}{(N_1[0,t_1](\theta^s(\omega))+1)^2}\, N_1(\omega,ds) = \int_0^{t_1} \frac{1}{(N_1[s,t_1](\omega)+1)^2}\, N_1(\omega,ds)$$
$$= \sum_{2}^{N[0,t_1]+1} 1/k^2 < +\infty.$$

Relationship (11.20) holds for g, by letting K go to infinity and by using monotone convergence Theorem, Identity (11.20) is proved. □

11.7 Appendix

In this section, a converse to the basic construction of this chapter is provided. It is shown that a stationary point process has a Palm space. Moreover, it is shown that there is a unique stationary point process for a given Palm space (i.e. a given distribution for interarrival intervals).

If $\mathbb{Q}$ is a probability distribution on $\Omega = \mathcal{M}_p(\mathbb{R} \times \mathbb{R}_+)$, invariant by translations (T^t) and whose intensity is finite, i.e.

$$\int_\Omega \omega([0,1])\, d\mathbb{Q}(\omega) < +\infty,$$

the random variable $\omega[n, n+1]$ is thus also integrable for any $n \in \mathbb{N}$, consequently the variable $\omega(K)$ is integrable for any compact subset K of $\mathbb{R}$. Therefore, the mapping

$$N : \Omega \longrightarrow \mathcal{M}_p(\mathbb{R} \times \mathbb{R}_+)$$
$$\omega \longrightarrow \omega$$

is a stationary point process. It will be shown that a Palm measure $\widehat{\mathbb{Q}}$ can be associated with $\mathbb{Q}$.

PROPOSITION 11.18. *There exists a unique probability distribution $\widehat{\mathbb{Q}}$ on Ω invariant by T^{t_1}, such that $\widehat{\mathbb{Q}}(t_0 = 0) = 1$ and*

$$\mathbb{E}\left(\int_{\mathbb{R}\times\mathbb{R}_+} h\left(T^t(\omega), t\right)\, N(\omega, dt, dy)\right) = \lambda \int_\Omega \int_{\mathbb{R}} h(\omega, t)\, dt\, d\widehat{\mathbb{Q}}, \tag{11.22}$$

for any non-negative measurable function h on $\Omega \times \mathbb{R}$.

To simplify the expressions of identities, the function h does not depend on the marks.

PROOF. If f is a bounded measurable function on Ω, and if g is continuous with compact support K in $\mathbb{R}$, then

$$\int_\Omega d\mathbb{Q} \int_{\mathbb{R}\times\mathbb{R}_+} f\left(T^t(\omega)\right) g(t)\, N(\omega, dt, dy) \le \|f\|_\infty \|g\|_\infty \int_\Omega \omega(K)\, d\mathbb{Q}(\omega).$$

One gets that for any bounded interval I of $\mathbb{R}$, the mapping

$$g \longrightarrow \int_\Omega d\mathbb{Q} \int_{\mathbb{R}\times\mathbb{R}_+} f\left(T^t(\omega)\right) g(t)\, N(\omega, dt, dy)$$

is linear, continuous on the space of continuous functions on I endowed with the norm of uniform convergence. Riesz's Theorem, see Rudin [**45**], gives the existence of a unique Radon measure μ_f on $\mathbb{R}$ such that

$$\int_\Omega d\mathbb{Q} \int_{\mathbb{R}\times\mathbb{R}_+} f\left(T^t(\omega)\right) g(t)\, N(\omega, dt, dy) = \int_\mathbb{R} g(t)\, \mu_f(dt), \tag{11.23}$$

for any non-negative Borelian function g on $\mathbb{R}$. If $x \in \mathbb{R}$,

$$\begin{aligned}\int_\Omega d\mathbb{Q} \int_{\mathbb{R}\times\mathbb{R}_+} & f\left(T^t(\omega)\right) g(t-x)\, N(\omega, dt, dy)\\ &= \int_\Omega d\mathbb{Q} \int_{\mathbb{R}\times\mathbb{R}_+} f\left(T^{t+x}(\omega)\right) g(t)\, T^x N(\omega, dt, dy)\\ &= \int_\Omega d\mathbb{Q} \int_{\mathbb{R}\times\mathbb{R}_+} f\left(T^t(T^x(\omega))\right) g(t)\, T^x N(\omega, dt, dy),\end{aligned}$$

since $\mathbb{Q}$ is invariant by T^x, this last term is given by

$$\int_\Omega d\mathbb{Q} \int_{\mathbb{R}\times\mathbb{R}_+} f\left(T^t(\omega)\right) g(t)\, N(\omega, dt, dy).$$

Therefore, the identity

$$\int_\mathbb{R} g(t-x)\, \mu_f(dt) = \int_\mathbb{R} g(t)\, \mu_f(dt)$$

holds for any non-negative Borelian function g. The measure μ_f is invariant by translation on $\mathbb{R}$, consequently μ_f is proportional to Lebesgue's measure; there exists some constant $l(f) \in \mathbb{R}$ such that $\mu_f(dt) = l(f)\, dt$. By taking g as the indicator function of the interval $[0,1]$ in Identity (11.23), it is easily seen that, on the measurable subsets of Ω, the mapping

$$A \longrightarrow l(1_A),$$

defines a non-negative bounded measure on Ω. Setting $\lambda = l(1_\Omega)$ and

$$\int_\Omega f\, d\widehat{\mathbb{Q}} = l(f)/\lambda,$$

Relationship (11.22) is therefore verified for all functions of the form $h = fg$, and thus for all measurable functions since functions with a product form generates the product σ-field of $\Omega \times \mathbb{R}$.

Invariance of $\widehat{\mathbb{Q}}$ *by* T^{t_1}. The invariance by translation of Lebesgue's measure and Fubini's Theorem show that, if h is a non-negative measurable function on $\Omega \times \mathbb{R}$, then

$$\lambda \int_\Omega \int_\mathbb{R} h\left(T^{t_1}(\omega), t\right)\, dt\, d\widehat{\mathbb{Q}} = \lambda \int_\Omega \int_\mathbb{R} h\left(T^{t_1}(\omega), t + t_1(\omega)\right)\, dt\, d\widehat{\mathbb{Q}},$$

by applying Identity (11.22) to the function $(\omega, t) \to h(T^{t_1}(\omega), t+t_1(\omega))$, one gets

$$\lambda \int_\Omega \int_{\mathbb{R}} h\left(T^{t_1}(\omega), t\right)\, dt\, d\widehat{\mathbb{Q}}$$
$$= \mathbb{E}\left(\int_{\mathbb{R}\times\mathbb{R}_+} h\left[T^{t_1}(T^t(\omega)), t + t_1(T^t(\omega))\right]\, N(\omega, dt, dy) \right)$$
$$= \mathbb{E}\left(\int_{\mathbb{R}\times\mathbb{R}_+} h\left[T^{t+t_1(T^t(\omega))}(\omega), t + t_1(T^t(\omega))\right]\, N(\omega, dt, dy) \right),$$

It is easily seen that if t is a point of the point measure ω, then $t+t_1(T^t(\omega))$ is also a point of this measure (with the corresponding index shifted by one). The last sum can thus be written as

$$\mathbb{E}\left(\int_{\mathbb{R}\times\mathbb{R}_+} h\left(T^t(\omega), t\right)\, N(\omega, dt, dy) \right) = \lambda \int_\Omega \int_{\mathbb{R}} h(\omega, t)\, dt\, d\widehat{\mathbb{Q}},$$

hence

$$\int_\Omega \int_{\mathbb{R}} h\left(T^{t_1}(\omega), t\right)\, dt\, d\widehat{\mathbb{Q}} = \int_\Omega \int_{\mathbb{R}} h(\omega, t)\, dt\, d\widehat{\mathbb{Q}}.$$

The invariance of $\widehat{\mathbb{Q}}$ by T^{t_1} is proved.

If $h(\omega, t) = 1_{\{t_0(\omega)=0, 0\le t\le 1\}}$, Identity (11.22) gives

$$\mathbb{E}\left(\int_{\mathbb{R}\times\mathbb{R}_+} 1_{\{t_0(T^t(\omega))=0, 0\le t\le 1\}}\, N(\omega, dt, dy) \right) = \lambda \widehat{\mathbb{Q}}(t_0 = 0),$$

if t is a point of ω, necessarily $t_0(T^t(\omega)) = 0$, therefore

$$\lambda = \mathbb{E}(N([0,1])) = \lambda \widehat{\mathbb{Q}}(t_0 = 0),$$

hence $\widehat{\mathbb{Q}}(t_0 = 0) = 1$. The proposition is proved. □

The following uniqueness result concludes this section (see also Proposition 11.8 page 315).

PROPOSITION 11.19. *If $\widehat{\mathbb{Q}}$ is a probability distribution on Ω invariant by T^{t_1} such that $\mathbb{E}_{\widehat{\mathbb{Q}}}(t_1) < +\infty$, there exists a unique probability distribution $\mathbb{Q}$, distribution of a stationary point process with finite intensity, with the following property: For any bounded measurable function f defined on Ω such that $t \to f(\theta^t(\omega))$ is $\widehat{\mathbb{Q}}$-almost surely continuous at the right of $t = 0$, then*

$$\lim_{x\to 0+} \mathbb{E}_{\mathbb{Q}}\left(f \mid \omega([-x, 0]) \neq 0\right) = \mathbb{E}_{\widehat{\mathbb{Q}}}(f), \tag{11.24}$$

where $\mathbb{E}_{\mathbb{Q}}$ denotes the expectation for the probability distribution $\mathbb{Q}$.

PROOF. The probability distribution $\mathbb{Q}$ constructed in Section 11.3 defines a stationary point process with intensity $1/\mathbb{E}_{\widehat{\mathbb{Q}}}(t_1)$, and the required conditioning property (11.24).

Proof of the uniqueness. If $\mathbb{Q}_1$ is the distribution of another stationary point process such that Convergence (11.24) holds for $\mathbb{Q}_1$. By Proposition 11.19, the probability distribution $\mathbb{Q}_1$ has a Palm measure denoted by $\widehat{\mathbb{Q}}_1$ which is invariant by T^{t_1}, and Identity (11.22) holds, that is

$$\mathbb{E}_{\mathbb{Q}_1}\left(\int_{\mathbb{R}\times\mathbb{R}_+} h\left(T^t(\omega),t\right)\, N(\omega,dt,dy)\right) = \lambda_{\mathbb{Q}_1}\int_\Omega\int_{\mathbb{R}} h(\omega,t)\,dt\,d\widehat{\mathbb{Q}}_1.$$

If h is a non-negative function on $\Omega\times\mathbb{R}$, since by definition the interval $]t_0,t_1[$ does not contain any point of the point measure N, $\widehat{\mathbb{Q}}_1$-almost surely the identity

$$\int_{\mathbb{R}\times\mathbb{R}_+} h\left(T^s(\omega),-s\right) 1_{\{t_0(\omega)\le s<t_1(\omega)\}}\, N(\omega,ds,dy) = h\left(T^{t_0}(\omega),-t_0(\omega)\right),$$

holds. Relationship (11.22) gives

$$\mathbb{E}_{\mathbb{Q}_1}\left[h\left(T^{t_0}(\omega),-t_0(\omega)\right)\right] = \lambda_{\mathbb{Q}_1}\int_\Omega\int_0^{t_1(\omega)} h(\omega,s)\,ds\,d\widehat{\mathbb{Q}}_1$$

(recall that $t_0\equiv 0$, $\widehat{\mathbb{Q}}_1$-almost surely). If f is non-negative and measurable on Ω, by setting $h(\omega,s)=f(T^s(\omega))$ in the above identity, one gets

$$\mathbb{E}_{\mathbb{Q}_1}(f(\omega)) = \lambda_{\mathbb{Q}_1}\int_\Omega\int_0^{t_1(\omega)} f\left(T^s(\omega)\right)\,ds\,d\widehat{\mathbb{Q}}_1, \tag{11.25}$$

an identity analogous to Relationship (11.12) is thus valid for $\mathbb{Q}_1$ and $\widehat{\mathbb{Q}}_1$. The proof of Proposition 11.8 concerning the conditioning of $\mathbb{P}$ and $\widehat{\mathbb{P}}$ uses only Identity (11.12). Therefore, one gets that for any bounded measurable function$f$ defined on Ω such that $t\to f(\theta^t(\omega))$ is $\widehat{\mathbb{Q}}$-almost surely continuous on the right of $t=0$, then

$$\lim_{x\to 0+}\mathbb{E}_{\mathbb{Q}_1}\left(f\mid\omega([-x,0])\neq 0\right) = \mathbb{E}_{\widehat{\mathbb{Q}}_1}(f),$$

hence the identity $\mathbb{E}_{\widehat{\mathbb{Q}}_1}(f)=\mathbb{E}_{\widehat{\mathbb{Q}}}(f)$ holds for all these functions f. If f is a random variable on Ω, the function $\omega\to f(T^{t_1}(\omega))$ has clearly this continuity property, therefore,

$$\mathbb{E}_{\widehat{\mathbb{Q}}_1}\left(f\circ T^{t_1}\right) = \mathbb{E}_{\widehat{\mathbb{Q}}}\left(f\circ T^{t_1}\right),$$

and since the probability distributions $\mathbb{Q}$ and Q_1 are both invariant with respect to T^{t_1}, one gets

$$\mathbb{E}_{\widehat{\mathbb{Q}}_1}(f) = \mathbb{E}_{\widehat{\mathbb{Q}}}(f),$$

hence, $\widehat{\mathbb{Q}}=\widehat{\mathbb{Q}}_1$ and $\lambda=1/\mathbb{E}_{\widehat{\mathbb{Q}}}(t_1)=1/\mathbb{E}_{\widehat{\mathbb{Q}}_1}(t_1)=\lambda_{\mathbb{Q}_1}$. Consequently, Relationships (11.25) and (11.12) show that $\mathbb{Q}=\mathbb{Q}_1$. The proposition is proved. □

12. The $G/G/1$ FIFO Queue

Contents

The single server queue with First In First Out service discipline is analyzed in a general framework in this chapter. The arrival process of customers at the queue is described as a stationary marked point process $N = \{(t_n, \sigma \circ \theta^{t_n})\}$ (See Chapter 11 for the main results and definitions concerning these point processes). The probability space $(\Omega, \mathcal{F}, \mathbb{P})$ on which all random variables are defined has an ergodic flow (θ^t) (See Section 10.3 page 293) such that the identity

$$N\left(\theta^t(\omega), dt, d\sigma\right) = T^t N(\omega, dt, d\sigma)$$

holds for $t \in \mathbb{R}$ and $\omega \in \Omega$. The operators (T^t) are translations of point processes. As in Chapter 11, it is assumed that, almost surely, there are no simultaneous arrivals, i.e. $\widehat{\mathbb{P}}(t_1 > 0) = 1$.

The asymptotic behavior of characteristics of this queue is investigated. In particular the convergence in distribution of the sequence of waiting times of customers and of the state of the queue at an arbitrary instant for example. General convergence results are proved and a quite satisfactory qualitative description of its equilibrium is given. Of course, explicit expressions for stationary distributions cannot be expected at this level of generality, independence assumptions have to be introduced for this purpose, see Chapter 2 for example. The queue with two servers concludes this chapter to show that if the results obtained for the $G/G/1$ FIFO queue are quite general, they do not extend to a much larger set of queueing systems.

Stationary point processes have been introduced by Franken *et al.* [**22**] in the context of queues. The use of ergodic theory formalism (the (θ^t), $\widehat{\theta}$

notations) is due to Neveu[Nev83]. This is the presentation used in this chapter.

12.1 Waiting Time

At time t_0- (just before the arrival of customer with index 0) the total sum of services to be processed is denoted by $w \geq 0$. The FIFO service discipline implies that the customer arriving at time t_0 waits for w before being served. The random variable W_n denotes the waiting time of the nth customer arrived after $t = 0$. This variable W_n is attached to a point of the arrival process (the arrival of the nth customer), it is therefore naturally defined on the probability space $(\widehat{\Omega}, \widehat{\mathcal{F}}, \widehat{P})$. This customer begins its service at $t = t_n + W_n$ and leaves the queue when its service $\sigma \circ \widehat{\theta}^n$ is finished, i.e. at time $t = t_n + W_n + \sigma \circ \widehat{\theta}^n$. Since customers are served in the order of their arrivals, the identity

$$t_{n+1} + W_{n+1} = \left(t_n + W_n + \sigma \circ \widehat{\theta}^n\right) \vee t_{n+1}$$

holds, that is either the nth customer has left the queue before the arrival of the $(n+1)$th customer and then $W_{n+1} = 0$, or the nth customer is still in the queue when the $(n+1)$th arrives. Therefore, Lindley's Equation is satisfied

$$W_0 = w, \quad W_{n+1} = \left(W_n + \sigma \circ \widehat{\theta}^n - \tau \circ \widehat{\theta}^n\right)^+. \tag{12.1}$$

The *sojourn time* S_n of the nth customer is defined as $W_n + \sigma \circ \widehat{\theta}^n$, this is the duration of time spent by the customer in the queue. The sequence $(t_n + S_n)$ is the sequence of departure times of this queue.

The next proposition, due to Loynes[Loy62], is the main result for the convergence to equilibrium. (The proof uses a *backward coupling*, see Lindvall [**33**].)

PROPOSITION 12.1. *If Condition $\lambda\widehat{\mathbb{E}}(\sigma) < 1$ holds, there exists a unique random variable $W : \widehat{\Omega} \to \mathbb{R}_+$, $\widehat{\mathbb{P}}$-almost surely finite, such that $\widehat{\mathbb{P}}$-a.s.*

$$W \circ \widehat{\theta} = (W + \sigma - \tau)^+, \tag{12.2}$$

moreover $\widehat{\mathbb{P}}(W = 0) > 0$.

Relationship (12.2) is the analogue in this context of Equation (2.2) for the $GI/GI/1$ queue. Uniqueness is understood with the following meaning:

[Nev83] J. Neveu, *Construction de files d'attente stationnaires*, Lecture notes in Control and Information Sciences, 60, Springer Verlag, 1983, pp. 31–41.

[Loy62] R.M. Loynes, *The stability of queues with non independent inter-arrival and service times*, Proc. Cambridge Ph. Soc. **58** (1962), 497–520.

If Equation (12.2) holds for another finite random variable W' defined on $\widehat{\Omega}$, then $\mathbb{P}(W = W') = 1$.

PROOF. The sequence of random variables (V_n) is defined inductively by

$$V_0 = 0, \quad V_{n+1} \circ \widehat{\theta} = (V_n + \sigma - \tau)^+, \quad n \geq 0.$$

It is easily checked by induction that (V_n) is a non-decreasing sequence and that its limit (possibly infinite) $W = \lim_{n \to +\infty} V_n$ satisfies the relationship

$$W \circ \widehat{\theta} = (W + \sigma - \tau)^+. \tag{12.3}$$

First it is proved that the random variable W is $\mathbb{P}$-almost surely finite. The subset $A = \{W = +\infty\}$ of $\widehat{\Omega}$ is clearly invariant with respect to the endomorphism $\widehat{\theta}$. Since $\widehat{\theta}$ is an ergodic endomorphism for the probability distribution $\widehat{\mathbb{P}}$ (Proposition 10.15 page 296), either $\widehat{\mathbb{P}}(A) = 0$ or $\widehat{\mathbb{P}}(A) = 1$ holds. If this latter equality is true, $\widehat{\mathbb{P}}$-almost surely, the sequence (V_n) converges almost surely to infinity. Since, by definition,

$$V_{n+1} \circ \widehat{\theta} - V_n = -V_n \wedge (\tau - \sigma),$$

the variables V_n being integrable (proof by induction), the invariance of $\widehat{\mathbb{P}}$ by $\widehat{\theta}$ and the monotonicity property of this sequence show that the inequality

$$\widehat{\mathbb{E}}\left(V_n \wedge (\tau - \sigma)\right) = \widehat{\mathbb{E}}\left(V_n - V_{n+1} \circ \widehat{\theta}\right) = \widehat{\mathbb{E}}(V_n - V_{n+1}) \leq 0,$$

holds for $n \geq 0$ and, by letting n go to infinity, one gets

$$\widehat{\mathbb{E}}(\tau) - \widehat{\mathbb{E}}(\sigma) = 1/\lambda - \widehat{\mathbb{E}}(\sigma) \leq 0,$$

by monotone convergence Theorem, contradicting the condition $\lambda \widehat{\mathbb{E}}(\sigma) < 1$. The variable W is therefore $\widehat{\mathbb{P}}$-almost surely finite

Any solution U of Relationship (12.2) satisfies $\widehat{\mathbb{P}}(U = 0) > 0$. Otherwise,

$$\widehat{\mathbb{P}}(U = 0) = \widehat{\mathbb{P}}\left(U \circ \widehat{\theta} = 0\right) = 0,$$

hence Equation (12.2) can thus be written as,

$$U \circ \widehat{\theta} = U + \sigma - \tau,$$

$\widehat{\mathbb{P}}$-almost surely, the exponent $^+$ can be omitted since $\mathbb{P}(U \circ \widehat{\theta} = 0) = 0$. Since $U \circ \widehat{\theta} - U$ is integrable, Lemma 12.2 below shows that $\widehat{\mathbb{E}}(U \circ \widehat{\theta} - U) = 0$, thus $\lambda \widehat{\mathbb{E}}(\sigma) = 1$. Contradiction. The quantity $\widehat{\mathbb{P}}(U = 0)$ is therefore positive.

Uniqueness. If U is a solution of Equation (12.2), since $U \geq 0 = V_0$, the definition of the sequence (V_n) gives

$$V_1 \circ \widehat{\theta} = (V_0 + \sigma - \tau)^+ \leq (U + \sigma - \tau)^+ = U \circ \widehat{\theta},$$

thus $U \geq V_1$, $\widehat{\mathbb{P}}$-a.s. By induction, one gets $V_n \leq U$ $\widehat{\mathbb{P}}$-a.s. for any $n \in \mathbb{N}$, finally $W \leq U$.

The subset $\{W = U\}$ is clearly invariant by the endomorphism $\widehat{\theta}$ by Equation (12.2), and since $\{W \leq U\}$, it contains the non-negligible subset

$\{U = 0\}$. The ergodicity property of $\widehat{\theta}$ implies that $\widehat{\mathbb{P}}$-almost surely $U = W$. The proposition is proved. □

LEMMA 12.2. *If X is a non-negative random variable on $\widehat{\Omega}$ such that the random variable $X \circ \widehat{\theta} - X$ is integrable, then $\widehat{\mathbb{E}}(X \circ \widehat{\theta} - X) = 0$.*

PROOF. Indeed, if X is integrable,

$$\widehat{\mathbb{E}}\left(X \circ \widehat{\theta} - X\right) = \widehat{\mathbb{E}}\left(X \circ \widehat{\theta}\right) - \widehat{\mathbb{E}}(X) = 0,$$

by invariance of $\widehat{\mathbb{P}}$ by $\widehat{\theta}$. In particular, for $K \geq 0$,

$$\widehat{\mathbb{E}}\left(X \wedge K \circ \widehat{\theta} - X \wedge K\right) = 0,$$

by using the elementary inequality

$$\left|X \circ \widehat{\theta} \wedge K - X \wedge K\right| \leq |X \circ \widehat{\theta} - X|,$$

for $K \geq 0$, Lebesgue's convergence Theorem concludes the proof. □

The definition of the sequence (V_n) gives by induction the identity

$$V_n = \sup_{-n \leq k \leq 0} \sum_{k}^{-1} (\sigma - \tau) \circ \widehat{\theta}^i, \tag{12.4}$$

with the convention that $\sum_0^{-1} \cdot = 0$. The last proposition can then also be proved with the help of ergodic Theorem: Almost surely, the convergence

$$\lim_{k \to -\infty} \frac{1}{-k} \sum_{k}^{-1} (\sigma - \tau) \circ \widehat{\theta}^i = \widehat{\mathbb{E}}(\sigma - \tau) < 0,$$

holds. In particular, $\mathbb{P}$-almost surely, the partial sums of $((\sigma - \tau) \circ \widehat{\theta}^i)$ converge to $-\infty$. The maximum in the right hand side of Equation (12.4) is thus almost surely constant for a sufficiently large n. The limit of the sequence (V_n), the variable W, can thus be expressed as

$$W = \sup_{k \leq 0} \sum_{k}^{-1} (\sigma - \tau) \circ \widehat{\theta}^i. \tag{12.5}$$

The main stability result for the $G/G/1$ queue is contained in the next proposition.

PROPOSITION 12.3. *If $\lambda\widehat{\mathbb{E}}(\sigma) < 1$, the sequence (W_n) converges in total variation to the distribution of the random variable W defined by Equation (12.2),*

$$\lim_{n \to +\infty} \left\|\widehat{\mathbb{P}}(W_n \in \cdot) - \widehat{\mathbb{P}}(W \in \cdot)\right\|_{tv} = 0,$$

where $\mathbb{P}(X \in \cdot)$ denotes the distribution of the random variable X. Moreover, the sequence (W_n) has the strong coupling property, i.e. the variable

$$T = \inf\left\{n \geq 0 : W_k = W \circ \widehat{\theta}^k, \forall k \geq n\right\}, \tag{12.6}$$

is $\widehat{\mathbb{P}}$-almost surely finite. In other words, the sequence (W_n) is equal to the stationary sequence $(W \circ \widehat{\theta}^n)$ after some finite index.

The total variation norm $\|\mathbb{P}-\mathbb{Q}\|_{tv}$ between two probability distributions $\mathbb{P}$ and $\mathbb{Q}$ on $\mathbb{R}$ is defined by

$$\sup_{A \in \mathcal{B}(\mathbb{R})} |\mathbb{P}(A) - \mathbb{Q}(A)|,$$

see Appendix C for the general properties of this norm.

PROOF. Coupling property. By definition of (W_n),

$$W_0 = w, \quad W_{n+1} = \left(W_n + \sigma \circ \widehat{\theta}^n - \tau \circ \widehat{\theta}^n\right)^+,$$

if at the origin $W_0 = W$ then $W_1 = (W + \sigma - \tau)^+ = W \circ \widehat{\theta}$ and, by induction, the sequence (W_n) is therefore the stationary sequence $(W \circ \widehat{\theta}^n)$. The sequence (W_n) with the initial condition $w = 0$ is denoted by (W_n^0).

It it is shown that for any w, the sequence (W_n) couples almost surely with the sequence (W_n^0): There exists some (random) $N \in \mathbb{N}$ such that $W_n = W_n^0$ for any $n \geq N$. This will imply in particular that the stationary sequence $(W \circ \widehat{\theta}^n)$ couples with (W_n^0) and therefore that all the other sequences couple with the stationary sequence.

By induction, it is clear that $W_n \geq W_n^0$ for all $n \geq 0$. If $W_n > W_n^0$ for all $n \geq 0$, in particular $W_n > 0$ for all $n \geq 0$, thus

$$W_{n+1} = W_n + \sigma \circ \widehat{\theta}^n - \tau \circ \widehat{\theta}^n, \quad W_n = w + \sum_1^n (\sigma - \tau) \circ \widehat{\theta}^n.$$

Ergodic Theorem shows that, $\widehat{\mathbb{P}}$-almost surely, the quantity

$$\sum_1^n (\sigma - \tau) \circ \widehat{\theta}^n \sim n\widehat{\mathbb{E}}(\sigma - \tau)$$

converges to $-\infty$ as n tends to infinity. The event

$$\{W_n > W_n^0, \forall n \geq 0\}$$

is thus negligible hence, almost surely, there exists some N such that $W_N = W_N^0$. The two sequences are identical after index N. Therefore, the strong coupling property holds. If T is the first index when the two sequences meet and A a Borelian subset of $\mathbb{R}$

$$\begin{aligned}\left|\widehat{\mathbb{P}}(W_n \in A) - \widehat{\mathbb{P}}(W \in A)\right| &= \left|\widehat{\mathbb{P}}(W_n \in A) - \widehat{\mathbb{P}}\left(W \circ \widehat{\theta}^n \in A\right)\right| \\ &\leq \widehat{\mathbb{P}}\left(W_n \neq W \circ \widehat{\theta}^n\right) = \widehat{\mathbb{P}}(T > n) \searrow 0.\end{aligned}$$

The convergence in total variation is established. ☐

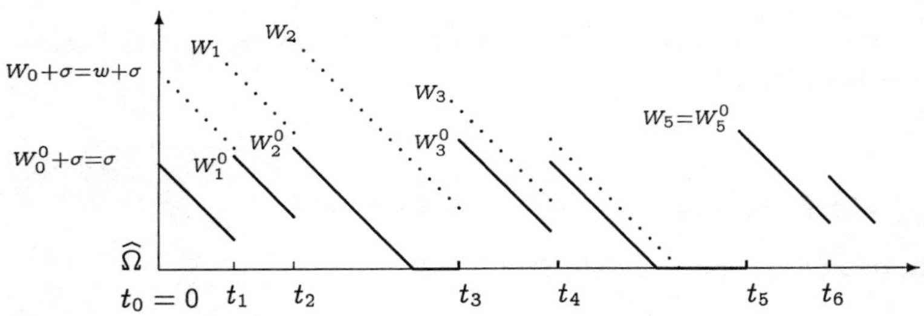

FIG. 12.1. Coupling of the Waiting Time of the $G/G/1$ Queue

The definitions of the two sequences (V_n) defined page 333 and (W_n^0) page 334 show that the identity $V_n = W_n^0 \circ \widehat{\theta}^{-n}$ holds for any n (proof by induction). This gives an intuitive picture of the random variable V_n. The variable W_n^0 is the waiting time of the nth customer after 0 when the queue starts empty at $t = 0$. With the interpretation of $\widehat{\theta}$ as the operator of the shift from one customer to the next one, $\widehat{\theta}^{-n}$ considers the point of the view of customer with index $-n$. Thus, the variable V_n is the waiting time of the customer with index 0 when the $-n$th customer finds the queue empty (at time t_{-n}). The variable W, the limit of $(V_n; n \geq 0)$, is therefore the waiting time of customer with index 0 when the queue started at time $t = -\infty$. (Obviously it must be at equilibrium at time 0 then).

Note that Representation (12.4) of V_n gives the relationship

$$W_n^0 = \sup_{0 \leq k \leq n} \sum_{k}^{n-1} (\sigma - \tau) \circ \widehat{\theta}^i. \tag{12.7}$$

12.2 Virtual Waiting Time

The *virtual waiting time* of the $G/G/1$ queue at time $t \in \mathbb{R}$ can be defined as the waiting time that a (virtual) customer arriving at t should have to wait to be served. This is the total sum of residual services of all customers in the queue at time t. Formally, the virtual waiting time is a right continuous function $t \to V(t)$ satisfying the stochastic differential equation

$$dV(t) = \lim_{s \nearrow t} V(t) - V(s) = \sigma \circ \theta^t \, N(\omega, [t, t+dt[) - 1_{\{V(t)>0\}} \, dt \tag{12.8}$$

or, equivalently, for $n \in \mathbb{Z}$

$$\begin{aligned} dV(t) &= -1_{\{V(t)>0\}} \, dt, \qquad t_n < t < t_{n+1}, \\ V(t_n) - V(t_n-) &= \sigma \circ \theta^{t_n}. \end{aligned}$$

The function $t \to V(t)$ is the solution of the deterministic ordinary differential equation

$$dV(t) = -1_{\{V(t)>0\}}\, dt,$$

perturbated by the point process $N(\omega, dt)$.

The *busy periods* of the queue are time intervals during which the function $t \to V(t)$ is positive. *Idle periods* of the queue are time intervals when virtual waiting time is 0.

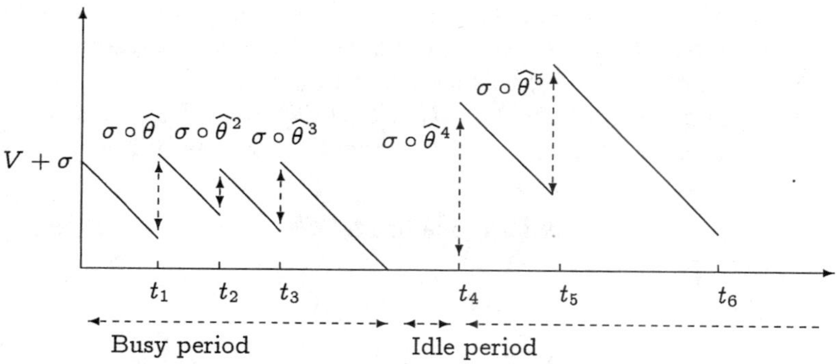

FIG. 12.2. The Virtual Waiting Time

The discontinuities of $t \to V(t)$ occur at the points of the arrival process. Notice that if the sequence (W_n) of waiting times of customers is defined on the Palm space $\widehat{\Omega}$, the function $t \to V(t)$ is not a quantity related to some specific customer but describe the queue at arbitrary instants. It is therefore naturally defined on Ω. The next proposition is the analogue of Proposition 12.1 for the virtual waiting time.

PROPOSITION 12.4. *If $\lambda\widehat{\mathbb{E}}(\sigma) < 1$, there exists a unique random variable $V : \Omega \to \mathbb{R}_+$, $\mathbb{P}$-almost surely finite such that the function $t \to V \circ \theta^t(\omega)$ verifies the differential equation*

$$dV(t) = \sigma \circ \theta^t\, N(\omega, [t, t+dt[) - 1_{\{V(t)>0\}}\, dt, \tag{12.9}$$

or

$$V\left(\theta^t(\omega)\right) = V(\omega) + \int_{]0,t]} \sigma \circ \theta^s\, N(\omega, ds) - \int_0^t 1_{\{V(s)>0\}}\, ds.$$

Moreover, $\mathbb{P}(V = 0) = 1 - \lambda\widehat{\mathbb{E}}(\sigma)$.

This proposition can be formulated as follows: There exists a unique stationary solution $t \to V \circ \theta^t$ to differential Equation (12.9). Similarly, Proposition 12.1 shows that (discrete !) differential Equation (12.1) page 332 has a unique stationary solution $n \to W \circ \widehat{\theta}^n$.

PROOF. If W is the waiting time of the customer with index 0, i.e. the customer arrived time $t_0 \le 0$, at time t_0 the virtual waiting time of the queue is $W + \sigma$. Consequently, since by definition nobody arrives between $t = t_0$ and $t = 0$, at time 0 the virtual waiting time V can be defined by

$$V(\omega) = \left((W + \sigma) \circ \theta^{t_0}(\omega) + t_0(\omega)\right)^+. \tag{12.10}$$

The operator θ^{t_0} maps Ω onto $\widehat{\Omega}$ (see Chapter 11) and therefore the variable $(W+\sigma)\circ\theta^{t_0}$ is indeed defined on Ω. To prove that such a random variable V satisfies differential Equation (12.9), it is sufficient to check it for $t = 0$. By replacing ω by $\theta^t(\omega)$ and using the identity $N(\omega, [t, t+dt[) = N(\theta^t(\omega), [0, dt[)$, the differential equation for $\theta^t(\omega)$ at $t = 0$ is identical to the differential equation for ω at time t.

a) If $t_0(\omega) < 0$, for a sufficiently small h, one has $t_0(\theta^h(\omega)) = t_0(\omega) - h$ (see Equation (11.10) page 312),

$$\begin{aligned}(W + \sigma) \circ \theta^{t_0}\left(\theta^h(\omega)\right) &= (W + \sigma) \circ \theta^{t_0(\theta^h(\omega))}\left(\theta^h(\omega)\right) \\ &= (W + \sigma) \circ \theta^{(t_0(\theta^h(\omega))+h)}(\omega) = (W + \sigma) \circ \theta^{t_0}(\omega),\end{aligned}$$

hence the relationship

$$V\left(\theta^h(\omega)\right) = \left((W + \sigma) \circ \theta^{t_0}(\omega) + t_0(\omega) - h\right)^+$$

holds, consequently,

$$V\left(\theta^h(\omega)\right) - V(\omega) = -1_{\{V(\omega)>0\}}\, dh.$$

b) If $t_0(\omega) = 0$, in this case ω is an element of $\widehat{\Omega}$, for a sufficiently small $h < 0$

$$t_0\left(\theta^h(\omega)\right) = t_{-1}(\omega) - h,$$

$$V\left(\theta^h(\omega)\right) = \left[(W + \sigma) \circ \theta^{t_{-1}}(\omega) + t_{-1}(\omega) - h\right]^+, \tag{12.11}$$

$$V(\omega) = (W + \sigma)(\omega). \tag{12.12}$$

Since $t_0(\omega) = 0$,

$$t_1\left(\theta^{t_{-1}}(\omega)\right) = -t_{-1}(\omega), \quad \text{thus} \quad \widehat{\theta}\left(\theta^{t_{-1}}(\omega)\right) = \theta^{t_1}\left(\theta^{t_{-1}}(\omega)\right) = \omega,$$

Equation (12.2) for $\theta^{t_{-1}}(\omega)$ yields

$$W(\omega) = \left[(W + \sigma) \circ \theta^{t_{-1}}(\omega) + t_{-1}(\omega)\right]^+.$$

and Relationships (12.11) and (12.12) show

$$(V(0) - V(0-))(\omega) = \sigma(\omega),$$

(the notation $0-$ is for the limit on the left of 0). Differential Equation (12.9) for $t \to V \circ \theta^t$ is proved.

Uniqueness. If Z is another stationary solution of Equation (12.9), denoting by $H = Z \circ \theta^{t_0-}(\omega)$ and integrating Equation (12.9) on the interval $[t_0, t_1[$, one gets

$$H \circ \widehat{\theta} = (H + \sigma - t_1)^+,$$

hence $H = W$, $\widehat{\mathbb{P}}$-almost surely by Proposition 12.1.

The function $t \to Z \circ \theta^t(\omega)$ is W at $t_0(\omega)$, thus $Z = V$ by uniqueness of the solution of an ordinary differential equation with a fixed initial condition. The integration of Relationship (12.9) gives

$$V\left(\theta^t(\omega)\right) - V(\omega) = \int_0^t \sigma\left(\theta^s(\omega)\right) N(\omega, ds) - \int_0^t 1_{\{V(\theta^s(\omega))>0\}}\, ds,$$

the right hand side is integrable, Lemma 12.2 shows that the expected value of the left hand side is zero, hence

$$\begin{aligned} 0 &= \mathbb{E}\left(\int_0^t \sigma(\theta^s(\omega))\, N(\omega, ds) - \int_0^t 1_{\{V(\theta^s(\omega))>0\}}\, ds\right) \\ &= \lambda\widehat{\mathbb{E}}(\sigma)t - \int_0^t \mathbb{P}(V(\theta^s(\omega)) > 0)\, ds = \left(\lambda\widehat{\mathbb{E}}(\sigma) - \mathbb{P}(V > 0)\right)t, \end{aligned}$$

by Corollary 11.7 page 313 and the invariance of $\mathbb{P}$ by (θ^t). □

Definition (12.10) of V and Representation (12.5) of W gives the relationship

$$V = 0 \vee \sup_{n\le 0} \sum_{i=n}^{-1} (\sigma - \tau) \circ \theta^{t_i}\left(\theta^{t_0}(\omega)\right) + \sigma \circ \theta^{t_0}(\omega) + t_0,$$

since $\theta^{t_i}\left(\theta^{t_0}(\omega)\right) = \theta^{t_i}(\omega)$, one gets

$$V = 0 \vee \sup_{n\le 0} \int_{t_n}^{t_{-1}} \sigma \circ \theta^s\, N(\omega, ds) + t_n - t_0 + \sigma \circ \theta^{t_0}(\omega) + t_0,$$

$$V = 0 \vee \sup_{n\le 0} \int_{t_n}^{t_0} \sigma \circ \theta^s\, N(\omega, ds) + t_n = 0 \vee \sup_{n\le 0} \int_{t_n}^{0} \sigma \circ \theta^s\, N(\omega, ds) + t_n.$$

This last expression can be rewritten as

$$V = \sup_{t\ge 0} \left(\int_{-t}^{0} \sigma \circ \theta^s\, N(\omega, ds) - t\right).$$

The next proposition gives a simple relation between the distributions of V and W. This is a simple consequence of the relation between the probability distributions $\mathbb{P}$ and $\widehat{\mathbb{P}}$ (Proposition 11.5 page 313).

Proposition 12.5 (Takàcs' Formula). *If $\lambda\widehat{\mathbb{E}}(\sigma) < 1$, the Laplace transforms of W and V verify*

$$\mathbb{E}\left(e^{-\xi V}\right) = 1 - \lambda\widehat{\mathbb{E}}(\sigma) + \lambda\widehat{\mathbb{E}}\left(e^{-\xi W}\,\frac{1 - e^{-\xi\sigma}}{\xi}\right), \qquad \xi \ge 0.$$

PROOF. Relationship (11.12) page 313 between the probability distributions $\mathbb{P}$ and $\widehat{\mathbb{P}}$ gives

$$\mathbb{E}\left(e^{-\xi V}1_{\{V>0\}}\right) = \lambda\widehat{\mathbb{E}}\left(\int_0^{t_1} e^{-\xi V\circ\theta^s}1_{\{V\circ\theta^s>0\}}\,ds\right).$$

For $0 \le s < t_1$, according to the differential equation satisfied by V, the equality $V\circ\theta^s = (V-s)^+$ holds on the set $\widehat{\Omega}$. Since on this set a customer arrives at time $t=0$, then $V = V(\theta^{0-})+\sigma$, hence $V(\theta^{0-}) = W$ $\widehat{\mathbb{P}}$-a.s. on this event. Therefore,

$$V\circ\theta^s = (W+\sigma-s)^+, \qquad \widehat{\mathbb{P}} - a.s. \text{ on } \widehat{\Omega},$$

$$\begin{aligned}\mathbb{E}\left(e^{-\xi V}1_{\{V>0\}}\right) &= \lambda\widehat{\mathbb{E}}\left(e^{-\xi(W+\sigma)}\int_0^{t_1\wedge(W+\sigma)} e^{\xi s}\,ds\right)\\ &= \lambda\frac{1}{\xi}\left[\widehat{\mathbb{E}}\left(e^{-\xi(W+\sigma-t_1)^+}\right) - \widehat{\mathbb{E}}\left(e^{-\xi(W+\sigma)}\right)\right].\end{aligned}$$

Relationship (12.3) and the invariance of $\widehat{\mathbb{P}}$ by $\widehat{\theta}$ give the identity

$$\mathbb{E}\left(e^{-\xi V}1_{\{V>0\}}\right) = \lambda\widehat{\mathbb{E}}\left(e^{-\xi W}\frac{\left(1-e^{-\xi\sigma}\right)}{\xi}\right).$$

The formula is proved. □

The random variables W and V do not have, in general, the same distribution. It can be easily seen with the deterministic $G/G/1$ queue: $\widehat{\Omega} = \{0\}$, $\tau(0) = 1$ and $\sigma(0) = \alpha < 1$, the solution of Equation (12.2) is $W \equiv 0$ (customers never wait) but $\mathbb{P}(V=0) = 1-\alpha$ (the probability that the server is idle), thus $\widehat{\mathbb{P}}(W=0) \ne \mathbb{P}(V=0)$. When the arrival process is Poisson, the random variables V and W do have the same distribution (See Chapter 7 page 202).

If W and σ are independent, (this is true for the GI/GI/1 queue, see Chapter 2), the Laplace transform of the conditional distribution of V is given by

$$\mathbb{E}\left(e^{-\xi V} \mid V>0\right) = \widehat{\mathbb{E}}\left(e^{-\xi W}\right)\widehat{\mathbb{E}}\left(\frac{1-e^{-\xi\sigma}}{\widehat{\mathbb{E}}(\sigma)\xi}\right),$$

for $\mathrm{Re}(\xi) \ge 0$. Conditionally on the event $\{V>0\}$, the random variable V has the same distribution as the sum of two independent random variables: W, the stationary waiting time and a random variable σ^* whose Laplace transform is, for $\mathrm{Re}(\xi)\ge 0$,

$$\widehat{\mathbb{E}}\left(1-e^{-\xi\sigma}\right)/\xi\widehat{\mathbb{E}}(\sigma),$$

the density function of σ^* is given by $h(x) = \widehat{\mathbb{P}}(\sigma \ge x)/\widehat{\mathbb{E}}(\sigma)$ on $\mathbb{R}_+$. (See also Section 1.4 page 22 for an interpretation of σ^*.)

12.3 Number of Customers

The stationary version of the characteristics of an $G/G/1$ queue can constructed from the variable W defined in Proposition 12.1. The last section shows that the virtual waiting time V can be expressed with the help of W.

If Q is the number of customers that customer 0 finds in the queue at its arrival, Q is naturally defined on the Palm space $\widehat{\Omega}$, it can be represented as

$$Q = \sum_{t_n<0} 1_{\left\{t_n+(W+\sigma)\circ\widehat{\theta}^n \geq 0\right\}} = \int_{]-\infty,0[} 1_{\{s+(\sigma+W)\circ\theta^s>0\}}\, N(\omega, ds),$$

Q is simply the number of customers arrived before time 0 and who have not left the queue at time 0. The shift of Q by $\widehat{\theta}$ is

$$Q \circ \widehat{\theta} = \sum_{t_n\circ\widehat{\theta}<0} 1_{\left\{t_n\circ\widehat{\theta}+(W+\sigma)\circ\widehat{\theta}^{n+1}>0\right\}} = \sum_{t_n<t_1} 1_{\left\{t_n+(W+\sigma)\circ\widehat{\theta}^n>t_1\right\}}$$

which is the number of customers found at its arrival by the customer with index 1.

The *number of customers* in the queue at time $t \in \mathbb{R}$ is the random variable $L(t)$ *defined on* Ω by

$$L(t) = \sum_{t_n\leq t} 1_{\{t_n+(W+\sigma)\circ\theta^{t_n}>t\}}. \tag{12.13}$$

The number of customers in the queue at time t can be expressed as

$$\begin{aligned}
L(t) = \sum_{t_n\leq t} 1_{\{t_n+(W+\sigma)\circ\theta^{t_n}>t\}} &= \int 1_{\{s\leq t, s+(\sigma+W)\circ\theta^s>t\}}\, N(\omega, ds) \\
&= \int 1_{\{s-t\leq 0, s-t+(\sigma+W)\circ\theta^s>0\}}\, N(\omega, ds) \\
&= \int 1_{\{s\leq 0, s+(\sigma+W)\circ\theta^s>0\}}\, T^t N(\omega, ds).
\end{aligned}$$

by Relationship (11.9), one has $N(\theta^t(\omega), ds) = T^t N(\omega, ds)$, hence

$$L(t)(\omega) = \int 1_{\{s\leq 0, s+\sigma+W>0\}}\, N(\theta^t(\omega), ds) = L(0)(\theta^t(\omega)).$$

The stationary version of a variable at time t defined by Equation (12.13) is the variable at time 0 by shifted by the continuous flow (θ^t).

On the set $\widehat{\Omega}$, the variable $L(0-) = \lim_{t\nearrow 0} L(t)$ is simply the variable Q. In the same way the variable $L(W+\sigma)$, defined on $\widehat{\Omega}$, is the number of customers in the queue when customer 0 leaves.

Proposition 12.6. *If $\lambda\widehat{\mathbb{E}}(\sigma) < 1$, at equilibrium, the number of customers in the queue at the arrival time and departure time have the same distribution. Moreover, for $n \geq 2$,*

$$\widehat{\mathbb{P}}(Q \geq n) = \widehat{\mathbb{P}}(L((W+\sigma)) \geq n) = \widehat{\mathbb{P}}(W \geq t_{n-1}).$$

This proposition shows that the distributions of Q and $L(W+\sigma)$ are identical and can be expressed with the distribution of W.

PROOF. If customer 0 finds $n \geq 1$ customers in the queue, since service discipline is FIFO, the $(-n)$th customer is still in the queue, consequently,

$$\widehat{\mathbb{P}}(L(0-) \geq n) = \widehat{\mathbb{P}}\left(t_{-n} + (W+\sigma)\circ\widehat{\theta}^{-n} \geq 0\right),$$

the invariance of $\widehat{\mathbb{P}}$ by $\widehat{\theta}^n$ yields the relationship

$$\widehat{\mathbb{P}}(L(0-) \geq n) = \widehat{\mathbb{P}}(t_n \leq W+\sigma).$$

The last term is precisely $\widehat{\mathbb{P}}(L(W+\sigma) \geq n)$. The previous identity can be rewritten as follows

$$\widehat{\mathbb{P}}(L(0-) \geq n) = \widehat{\mathbb{P}}(t_n - t_1 \leq W+\sigma - t_1) = \widehat{\mathbb{P}}\left(t_n - t_1 \leq (W+\sigma-t_1)^+\right),$$

note that the last identity uses that $\mathbb{P}$-almost surely, $t_n > t_1$, for $n \geq 2$. From the identity $W\circ\widehat{\theta} = (W+\sigma-t_1)^+$ and $t_{n-1}\circ\widehat{\theta} = t_n - t_1$, one gets

$$\widehat{\mathbb{P}}(L(0-) \geq n) = \widehat{\mathbb{P}}\left(W\circ\widehat{\theta} \geq t_{n-1}\circ\widehat{\theta}\right) = \widehat{\mathbb{P}}(W \geq t_{n-1}),$$

by invariance of $\widehat{\mathbb{P}}$ by $\widehat{\theta}$. The proposition is proved. □

The following result is a relation at equilibrium between the average waiting time and the average number of customers in the queue at an arbitrary time.

COROLLARY 12.7 (Little's Formula).

$$\mathbb{E}(L(0)) = \lambda\left(\widehat{\mathbb{E}}(W) + \widehat{\mathbb{E}}(\sigma)\right).$$

PROOF. Relationship (12.13) for $t=0$ can be rewritten as

$$L(0) = \int 1_{\{s\leq 0, s+(W+\sigma)\circ\theta^s>0\}}\, N(\omega, ds),$$

and Mecke's Formula (11.13) page 313 gives the identity

$$\mathbb{E}(L(0)) = \lambda\widehat{\mathbb{E}}\left(\int 1_{\{s\leq 0, s+W+\sigma>0\}}\, ds\right).$$

The corollary is proved. □

In Chapter 4, under some conditions, the stationary distribution of the number of customers in a queue of a queueing network is explicitly known. Concerning the waiting times, there are very few results on their distributions when the network has more than one queue (!). Little's Formula is therefore useful in this context since it gives at least the average waiting time of customers at some node of the network.

12.4 Associated Stationary Point Processes

Departure Process. Proposition 11.15 page 322 shows that if S is a random variable on $\widehat{\Omega}$, the translation of a stationary point process N by S is also a stationary point process. If S is the sojourn time of a customer $S = W + \sigma$ this translated point process is precisely the departure process of the queue,

$$D = \sum_{n\in\mathbb{Z}} \delta_{t_n + W\circ\theta^{t_n} + \sigma\circ\theta^{t_n}}$$

D is therefore a stationary point process with the same intensity as N. This is the conservation law: A queue transforms a stationary point process into another stationary point process. Very little is known on the distribution of D in general. Proposition 4.4 page 86 which shows that the departure process of a stable $M/M/1$ queue is a Poisson process is one of the very rare examples in this domain. See the survey[Dal76] by Daley.

Stationary Busy Periods. By using the thinning procedure of a stationary process seen in Proposition 11.14) page 321), one gets that the point process

$$\sum_{n\in\mathbb{Z}} 1_{\{W\circ\theta^{t_n}=0\}}\,\delta_{t_n} = \sum_{n\in\mathbb{Z}} \delta_{T_n}$$

is stationary with intensity $\lambda\widehat{\mathbb{P}}(W = 0)$, its Palm measure is given by

$$1_{\{W=0\}}\, d\widehat{\mathbb{P}} \Big/ \widehat{\mathbb{P}}(W = 0).$$

In particular, $\widehat{\mathbb{P}}$-almost surely, the $G/G/1$ queue is empty infinitely often. The sequence (T_n) is increasing and numbered so that $T_0 \leq 0 < T_1$. The points of this point process are the instants of arrival times of customers who do not wait, i.e. who are served at their arrival time in the queue.

PROPOSITION 12.8. *If $\nu = \inf\{k \geq 0 : W\circ\widehat{\theta}^{-k} = 0\}$ is the index of the last customer arrived before time 0 who do not wait, then $T_0 = t_{-\nu}$ and*

$$W = \sum_{-\nu}^{-1} (\sigma - \tau)\circ\widehat{\theta}^{k} = \sum_{-\nu}^{-1} \sigma\circ\widehat{\theta}^{k} + t_{-\nu}.$$

PROOF. Notice first that ν is the hitting time of the non-negligible subset $\{W = 0\}$ by the ergodic endomorphism $\widehat{\theta}^{-1}$, In particular ν is $\widehat{\mathbb{P}}$-almost surely finite. By definition of ν, the variables $W\circ\widehat{\theta}^{-k}$, $k = -\nu+1, \ldots, 0$ are positive, therefore Relationship (12.2) gives the identities

$$W\circ\widehat{\theta}^{-i} = \left(W\circ\widehat{\theta}^{-i-1} + (\sigma-\tau)\circ\widehat{\theta}^{-i-1}\right)^{+} = W\circ\widehat{\theta}^{-i-1} + (\sigma-\tau)\circ\widehat{\theta}^{-i-1},$$

for $i = -\nu+1, \ldots, 0$, the representation of W is proved. □

[Dal76] D.J. Daley, *Queueing output processes*, Advances in Applied Probability 8 (1976), 395–415.

The customer arrived at time $t = t_{-\nu}$ initiates the busy period where customer 0 is served. it can be shown that the duration of this busy period is given by the integral

$$\int_{[T_0,T_1[} \sigma \circ \theta^s \, N(\omega, ds)$$

and the duration of the following period by

$$T_1 - T_0 - \int_{[T_0,T_1[} \sigma \circ \theta^s \, N(\omega, ds).$$

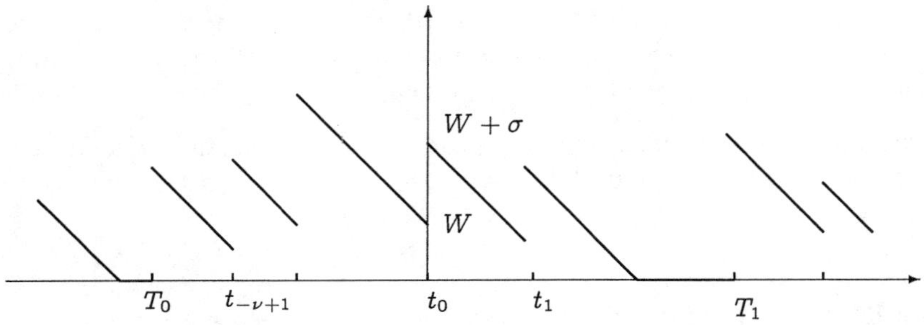

FIG. 12.3. A Busy Period

Queues with Conservative Service Disciplines

In a queue with *conservative service discipline* the server works as long as the queue is not empty, it can serve customers not necessarily in their order of arrival. This definition implies that the virtual waiting time for such a discipline is the same as for the FIFO discipline: It decreases at speed 1 as long as it is positive and jumps at the arrival times. The above results show that if $\lambda\widehat{\mathbb{E}}(\sigma) < 1$, there exists a unique stationary random variable for the virtual waiting time. A queue with a conservative service discipline has in particular the same busy periods as the queue with a FIFO discipline. To construct a stationary state for a queue with a conservative service discipline, it is enough to consider a busy period of a FIFO queue.

EXAMPLES.

1. LIFO discipline. For this discipline a customer is served only if it is the last arrived in the queue. When it arrives a customer is immediately served, its service is stopped if there is a new arrival and so on... The virtual waiting time cannot be interpreted as a waiting time. Customer 0 leaves the queue at time t if and only if all customers arrived between t_0 (included) and t

have been served. On $\widehat{\Omega}$, the sojourn time of the customer with index 0 is given by

$$S = \inf\left\{ t > 0 : \int_{[0,t[} \sigma \circ \theta^s \, N(\omega, ds) < t \right\}.$$

2. Priority discipline with preemption for a queue with two classes of customers. If $N_1 = \{(s_n, \sigma_1 \circ \theta^{s_n})\}$ and $N_2 = \{(t_n, \sigma_2 \circ \theta^{t_n})\}$ are stationary point processes whose intensities are respectively given by λ_1 and λ_2, and their Palm probabilities by $\widehat{\mathbb{P}}_1$ and $\widehat{\mathbb{P}}_2$, then

$$N = \{(s_n, \sigma_1 \circ \theta^{s_n}, 1)\} + \{(t_n, \sigma_2 \circ \theta^{t_n}, 2)\} = \{(u_n, \sigma \circ \theta^{u_n}, m \circ \theta^{u_n})\}$$

is a stationary point process with intensity $\lambda = \lambda_1 + \lambda_2$ and Palm probability $\widehat{\mathbb{P}}$ defined by

$$\widehat{\mathbb{P}} = \frac{\lambda_1}{\lambda_1 + \lambda_2} \widehat{\mathbb{P}}_1 + \frac{\lambda_2}{\lambda_1 + \lambda_2} \widehat{\mathbb{P}}_2.$$

Since $\widehat{\Omega} = \{u_0 = 0\} = \{s_0 = 0\} \bigcup \{t_0 = 0\}$ and $m : \widehat{\Omega} \to \{1, 2\}$ is defined by $m \equiv 1$ on the event $\{s_0 = 0\}$ and $m \equiv 2$ on $\{t_0 = 0\}$. The mapping m "marks" classes of customers. The service σ is defined by σ_1 if m is 1 and σ_2 otherwise.

Condition $\lambda \widehat{\mathbb{E}}(\sigma) < 1$ can be written as

$$(\lambda_1 + \lambda_2) \left(\frac{\lambda_1}{\lambda_1 + \lambda_2} \widehat{\mathbb{E}}_1(\sigma_1) + \frac{\lambda_2}{\lambda_1 + \lambda_2} \widehat{\mathbb{E}}_2(\sigma_2) \right) < 1,$$

that is,

$$\lambda_1 \widehat{\mathbb{E}}_1(\sigma_1) + \lambda_2 \widehat{\mathbb{E}}_2(\sigma_2) < 1. \tag{12.14}$$

If customers of class 1 have strict priority on class 2 customers, i.e. a class 2 customer can be served only when the queue does not have any class 1 customer. The quantity w_n^1 [resp. w_n^2] denotes the sum of residual service times of class 1 [resp. 2] customers at the arrival of the nth customer in the queue, then

$$\begin{aligned} w_{n+1}^1 &= \left[w_n^1 + \sigma_n 1_{\{m_n = 1\}} - (u_{n+1} - u_n) \right]^+, \\ w_{n+1}^2 &= \left[w_n^2 + \sigma_n 1_{\{m_n = 2\}} - \left(u_{n+1} - u_n - w_n^1 - \sigma_n 1_{\{m_n = 1\}} \right)^+ \right]^+. \end{aligned} \tag{12.15}$$

In particular

$$(w_{n+1}^1 + w_{n+1}^2) = \left((w_n^1 + w_n^2) + \sigma_n - (u_{n+1} - u_n) \right)^+.$$

If Condition (12.14) holds, there exists a unique random variable W on $\widehat{\Omega}$ such that

$$W \circ \widehat{\theta} = (W + \sigma - u_1)^+,$$

if $t_{-\nu}$ is the arrival time of the customer initiating the busy period containing time 0, i.e. $\nu = \inf\{k \geq 0 : W \circ \widehat{\theta}^{-k} = 0\}$. By setting $w_{-\nu}^1 = w_{-\nu}^2 = 0$ at $t = t_{-\nu}$ and using Relationship (12.15), one sets $W_1 = w_0^1$ and $W_2 = w_0^2$.

By using the relationship satisfied by W, it is easily seen that W_1 and W_2 satisfy

$$W_1 \circ \widehat{\theta} = \left[W_1 + \sigma 1_{\{m=1\}} - u_1\right]^+,$$

$$W_2 \circ \widehat{\theta} = \left[W_2 + \sigma 1_{\{m=2\}} - \left(u_1 - W_1 + \sigma 1_{\{m=1\}}\right)^+\right]^+.$$

which is the analogue of Relationship (12.1) for the priority queue.

12.5 The Unstable G/G/1 Queue

The behavior of the $G/G/1$ queue when $\lambda\widehat{\mathbb{E}}(\sigma) > 1$ (i.e. when the arrival process saturates the queue) is described in the next proposition.

PROPOSITION 12.9. *If $\lambda\widehat{\mathbb{E}}(\sigma) > 1$, then $\widehat{\mathbb{P}}$-almost surely.*

$$\lim_{n\to+\infty} \frac{W_n}{n} = \widehat{\mathbb{E}}(\sigma) - \frac{1}{\lambda},$$

and $\mathbb{P}$-almost surely

$$\lim_{t\to+\infty} \frac{L(t)}{t} = \lambda - \frac{1}{\widehat{\mathbb{E}}(\sigma)}.$$

PROOF. Using Equation (12.1), it is easily seen that for $n \geq 0$,

$$W_n \geq w + \sum_{1}^{n-1} (\sigma - \tau) \circ \widehat{\theta}^k,$$

Ergodic Theorem 10.9 gives that $\widehat{\mathbb{P}}$-a.s.

$$\liminf_{n\to+\infty} \frac{W_n}{n} \geq \widehat{\mathbb{E}}(\sigma) - \frac{1}{\lambda}.$$

In particular the W_n are positive after some finite rank n_0, thus for $n \geq n_0$,

$$W_{n+1} = W_n + \sigma \circ \widehat{\theta}^n - \tau \circ \widehat{\theta}^n,$$

therefore,

$$W_n = W_{n_0} + \sum_{n_0}^{n-1} (\sigma - \tau) \circ \widehat{\theta}^k,$$

by using ergodic Theorem, the first convergence result of the proposition is proved. The waiting time of customers is $\widehat{\mathbb{P}}$-a.s. positive after some finite rank. After some time, all customers find the queue busy. Relationship (11.5) shows that $\mathbb{P}$-a.s. $L(t) > 0$ for t sufficiently large. Without restricting the generality one can assume that $L(t) > 0$ for all $t > 0$, then

$$\{L(t) > n\} = \left\{ \sum_{0}^{N([0,t])-n} \sigma \circ \theta^{t_i} > t \right\},$$

indeed, there are at least n customers at time t if the sum of services of the $N([0,t]) - n$ first customers is greater than t. One concludes by taking $n = \lfloor (\lambda - 1/\widehat{\mathbb{E}}(\sigma) \pm \varepsilon)t \rfloor$ and using the ergodic Theorem for $N([0,t])$ and the partial sum $\sigma + \sigma \circ \theta^{t_1} + \cdots + \sigma \circ \theta^{t_n}$. □

When the equality $\lambda \widehat{\mathbb{E}}(\sigma) = 1$ holds, the queue can be stable: For the deterministic case $\tau \equiv \sigma \equiv 1$, Equation (12.2) has an infinite number of solutions $W \equiv x$ for any $x \geq 0$. This kind of stability is mainly artificial. With an additional independence assumption, it can be shown that this queue is fact unstable (see Proposition 2.2 page 32).

12.6 A Queue with Two Servers, the G/G/2 Queue

In this section, the queue has two servers and to each of them is associated a FIFO queue. The arrival process is a stationary marked point process. At its arrival a customer chooses the queue with the smallest amount of work. (In the case of equality, a coin is tossed to determine the queue). If the queue is empty when Customer 0 arrives W_n^1 [resp. W_n^2] denotes the amount of work of the least [resp. most] loaded queue at the arrival of the nth customer. It is easily seen that, due to the service discipline, the recurrence relationships

$$W_{n+1}^1 = \left(W_n^2 - \tau \circ \widehat{\theta}^n\right)^+ \wedge \left(W_n^1 + \sigma \circ \widehat{\theta}^n - \tau \circ \widehat{\theta}^n\right)^+,$$

$$W_{n+1}^2 = \left(W_n^2 - \tau \circ \widehat{\theta}^n\right)^+ \vee \left(W_n^1 + \sigma \circ \widehat{\theta}^n - \tau \circ \widehat{\theta}^n\right)^+,$$

hold, with $W_0^1 = W_0^2 = 0$.

As in the case of the $G/G/1$ FIFO queue, the problem of existence and uniqueness of a stationary solution $W = (W^1, W^2)$ to these equations is investigated, i.e. the existence and uniqueness of random variables W_1 and W_2 such that

$$W^1 \circ \widehat{\theta} = (W^2 - \tau)^+ \wedge (W^1 + \sigma - \tau)^+, \tag{12.16}$$

$$W^2 \circ \widehat{\theta} = (W^2 - \tau)^+ \vee (W^1 + \sigma - \tau)^+. \tag{12.17}$$

hold $\widehat{\mathbb{P}}$-almost surely. The next result is due to Kiefer and Wolfowitz[KW55], this is a (weak) analogue of Proposition 12.1 for the $G/G/2$ queue.

PROPOSITION 12.10. *If $\lambda \widehat{\mathbb{E}}(\sigma) < 2$ there exists a couple of random variables (W^1, W^2) $\widehat{\mathbb{P}}$-a.s. finite such that Relationships (12.16) and (12.17) hold. Moreover, $\widehat{\mathbb{P}}(W^1 = 0) > 0$ and (W^1, W^2) is minimal in the sense than any other solution (X^1, X^2) of Equations (12.16) and (12.17) satisfies the inequalities $W^1 \leq X^1$ and $W^2 \leq X^2$ $\widehat{\mathbb{P}}$-a.s.*

[KW55] J. Kiefer and J. Wolfowitz, *On the theory of queues with many servers*, Transactions of the AMS **78** (1955), 1–18.

PROOF. The sequence (W_n^1, W_n^2) is defined by induction by

$$W_{n+1}^1 \circ \widehat{\theta} = \left(W_n^2 - \tau\right)^+ \wedge \left(W_n^1 + \sigma - \tau\right)^+, \tag{12.18}$$

$$W_{n+1}^2 \circ \widehat{\theta} = \left(W_n^2 - \tau\right)^+ \vee \left(W_n^1 + \sigma - \tau\right)^+, \tag{12.19}$$

and $W_0^1 = W_0^2 = 0$. The monotonicity (coordinate by coordinate) of the function

$$f : (x, y) \to \left((x + \sigma - \tau)^+ \wedge (y - \tau)^+, (x + \sigma - \tau)^+ \vee (y - \tau)^+\right),$$

shows that the sequence (W_n^1, W_n^2) is non-decreasing, let $\underline{W} = (W^1, W^2)$ be its limit. This variable $\underline{W}$ clearly satisfies Equations (12.16) and (12.17),

$$W^1 \circ \widehat{\theta} = \left(W^2 - \tau\right)^+ \wedge \left(W^1 + \sigma - \tau\right)^+,$$
$$W^2 \circ \widehat{\theta} = \left(W^2 - \tau\right)^+ \vee \left(W^1 + \sigma - \tau\right)^+.$$

The inequality $W^1 \leq W^2$ shows that the set $A = \{W^1 = +\infty\}$ is invariant by $\widehat{\theta}$, hence by ergodicity of $\widehat{\theta}$ it is either $\widehat{\mathbb{P}}$-negligible or it complement is $\widehat{\mathbb{P}}$-negligible. If $\widehat{\mathbb{P}}(A) = 1$, $\widehat{\mathbb{P}}$-a.s. for $i = 1, 2$ the sequence (W_n^i) converges to infinity when n tends to infinity. Summation of Equations (12.18) and (12.19) yields

$$\left(W_{n+1}^1 + W_{n+1}^2\right) \circ \widehat{\theta} - \left(W_n^1 + W_n^2\right) = \max\left(-\tau, -W_n^2\right) + \max\left(\sigma - \tau, -W_n^1\right).$$

By taking the expectation of this identity, the monotonicity property gives the relationship

$$\widehat{\mathbb{E}}\left[\max\left(-\tau, -W_n^2\right)\right] + \widehat{\mathbb{E}}\left[\max\left(\sigma - \tau, -W_n^1\right)\right]$$
$$= \widehat{\mathbb{E}}\left(W_{n+1}^1 + W_{n+1}^2\right) - \widehat{\mathbb{E}}\left(W_n^1 + W_n^2\right) \geq 0.$$

By letting n go to infinity, Monotone Convergence Theorem shows that the inequality

$$0 \leq \widehat{\mathbb{E}}(\sigma) - 2\widehat{\mathbb{E}}(\tau) = \widehat{\mathbb{E}}(\sigma) - \frac{2}{\lambda},$$

holds, thereby contradicting the assumption of the proposition. The variable W^1 is therefore $\widehat{\mathbb{P}}$-a.s. finite. The set $\{W^2 = +\infty\}$ is consequently invariant by $\widehat{\theta}$, if its probability is 1, Equation (12.19) gives the inequality

$$0 \leq \widehat{\mathbb{E}}\left(W_{n+1}^2 - W_n^2\right) = \widehat{\mathbb{E}}\left[\max\left(-\tau, -W_n^2\right) \vee \left(W_n^1 - W_n^2 + \sigma - \tau\right)\right].$$

The expression under the integral in the right hand side is between $-\tau$ and σ, Lebesgue's convergence Theorem implies that $\widehat{\mathbb{E}}(-\tau) \geq 0$. This is absurd. The vector $\underline{W}$ is thus a $\widehat{\mathbb{P}}$-a.s. finite solution of Equations (12.16).

If $\widehat{\mathbb{P}}(W^1 = 0) = 0$, the exponent $+$ can be removed in Equations (12.16), the sum of these equations gives the identity

$$\left(W^1 + W^2\right) \circ \widehat{\theta} - \left(W^1 + W^2\right) = \sigma - 2\tau,$$

Lemma 12.2 shows that $\widehat{\mathbb{E}}(\sigma - 2\tau) = 0$, contradiction. The random variable W_1 is therefore 0 with positive probability.

If the initial conditions in Equations (12.18) and (12.19) are given by $W_0^1 = x$, $W_0^2 = y$, with $x \geq 0 = W_0^1$ and $y \geq 0 = W_0^2$. By induction, it is easily seen that the associated sequence is greater, coordinate by coordinate, than the sequence (W_n^1, W_n^2). In particular, if $\underline{X} = (X^1, X^2)$ is a solution of Equation (12.16), necessarily $X^1 \geq W^2$ and $X^2 \geq W_2$ hold almost surely. The solution $\underline{W}$ is therefore minimal. □

This proposition cannot be improved for the uniqueness result of the solution of Equations (12.16). This is illustrated by an example due to Neveu. If $\widehat{\Omega} = \{0, 1\}$, $\widehat{\theta}$ is the automorphism of $\widehat{\Omega}$ which exchanges 0 and 1, and $\widehat{\mathbb{P}}$ is the uniform probability distribution on $\widehat{\Omega}$. Interarrival intervals are deterministic and equal to 1 ($\tau \equiv 1$) and services are such that $\sigma(0) = 2$ and $\sigma(1) = 3/2$. For any α such that $0 \leq \alpha \leq 1/2$, the vector $\underline{V}$ defined by

$$\underline{V}(0) = (1/2, \alpha)\,,\ \underline{V}(1) = (0, 1 + \alpha)$$

is clearly a solution of Equations (12.16).

This counter-example shows that, even under the natural assumption that $\widehat{\mathbb{E}}(\sigma) < 2$, this queueing system may never be empty. This is not the case for the $G/G/1$ queue. It is easily seen that if a solution of Equations (12.16) is $(0, 0)$ with positive probability, then it is the minimal solution constructed above. A result, due to Brandt [Bra85] concludes this section. It shows that Equations (12.16) have a *maximal solution.*

PROPOSITION 12.11. *Under the assumption of the last proposition there exists a random variable $\overline{V} = (V^1, V^2)$ such that any solution $X = (X^1, X^2)$ of Equations (12.16), satisfies $\widehat{\mathbb{P}}$-a.s. $X^1 \leq V^1$ and $X^2 \leq V^2$.*

PROOF. For $d \in \mathbb{N}$, in the same way as before, the sequence (W_n^1, W_n^2) is defined by

$$W_{n+1}^1 \circ \widehat{\theta} = \left(W_n^2 - \tau\right)^+ \wedge \left(W_n^1 + \sigma - \tau\right)^+,$$
$$W_{n+1}^2 \circ \widehat{\theta} = \left(W_n^2 - \tau\right)^+ \vee \left(W_n^1 + \sigma - \tau\right)^+,$$

with the initial condition $W_0^1 = W^1 + d$, $W_0^2 = W^2 + d$ and (W^1, W^2) is the minimal solution,

$$\begin{aligned} W_1^1 \circ \widehat{\theta} &= \left(W^2 + d - \tau\right)^+ \wedge \left(W^1 + d + \sigma - \tau\right)^+ \\ &\leq \left(W^2 - \tau\right)^+ \wedge \left(W^1 + \sigma - \tau\right)^+ + d = W^1 \circ \widehat{\theta} + d = W_0^1 \circ \widehat{\theta}, \end{aligned}$$

hence $W_1^1 \leq W_0^1$, and in similarly $W_1^2 \leq W_0^2$. The monotonicity of the function f defined above shows that the sequence (W_n^1, W_n^2) is non-increasing, its limit is denoted by $(W^1(d), W^2(d))$.

Again with the monotonicity property, it is clear that the sequence

[Bra85] A. Brandt, *On stationary waiting times and limiting behaviour of queues with many servers II: The $G/G/m/\infty$ case*, Elektron. Inf. verarb. Kybern. **21** (1985), no. 3, 151–162.

$$(W^1(d), W^2(d))$$

is non-decreasing (proof by induction). If $\overline{V} = (V^1, V^2)$ is the limit of this sequence, it has to be shown that the coordinates of $\overline{V}$ are $\widehat{\mathbb{P}}$-a.s. finite. The method is the same as in the proof of the finiteness of (W^1, W^2). For example it is shown that V^1 is $\widehat{\mathbb{P}}$-a.s. finite. The set $\{V^1 = +\infty\}$ is invariant by $\widehat{\theta}$, if its probability is 1, then

$$\lim_{d \to +\infty} V^1(d) = \lim_{d \to +\infty} V^2(d) = +\infty,$$

$\widehat{\mathbb{P}}$-a.s. the sum of Equations (12.16) for $(V^1(d), V^2(d)))$ gives the relationship

$$\begin{aligned}\left[V^1(d) + V^2(d)\right] \circ \widehat{\theta} - \left[V^1(d) + V^2(d)\right]& \\ = \max\left(-\tau, -V^2(d)\right) + \max\left(\sigma - \tau, -V^1(d)\right).&\end{aligned}$$

The right hand side is between $\sigma - 2\tau$ and σ, hence it is integrable. The expected value of the left hand side is therefore zero, hence

$$\widehat{\mathbb{E}}\left[\max\left(-\tau, -V^2(d)\right) + \max\left(\sigma - \tau, -V^1(d)\right)\right] = 0.$$

by using Lebesgue's convergence Theorem, on gets

$$\widehat{\mathbb{E}}(\sigma - 2\tau) = 0,$$

contradiction. The variable V^1 is thus $\widehat{\mathbb{P}}$-a.s. finite. For the finiteness of V^2 the proof is similar.

If $X = (X^1, X^2)$ is a solution of Equations (12.16), there exists an integer d such that

$$\widehat{\mathbb{P}}\left(X^1 \le d, X^2 \le d\right) = \alpha > 0,$$

hence

$$\begin{aligned}\alpha &\le \widehat{\mathbb{P}}\left(X^1 \le d + W^1, X^2 \le d + W^2\right) \\ &\le \widehat{\mathbb{P}}\left(X^1 \circ \widehat{\theta} \le W_1^1 \circ \widehat{\theta}, X^2 \circ \widehat{\theta} \le W_1^2 \circ \widehat{\theta}\right) = \widehat{\mathbb{P}}\left(X^1 \le W_1^1, X^2 \le W_1^2\right),\end{aligned}$$

by induction,

$$\alpha \le \widehat{\mathbb{P}}\left(X^1 \le W_n^1, X^2 \le W_n^2\right),$$

and by taking the limit, one gets

$$\alpha \le \mathbb{P}\left(X^1 \le W_d^1, X^2 \le W_d^2\right) \le \mathbb{P}\left(X^1 \le V^1, X^2 \le V^2\right).$$

Since the set $\{X^1 \le V^1, X^2 \le V^2\}$ is invariant by $\widehat{\theta}$, it is non-negligible, then its probability is 1, $\overline{V}$ is the maximal solution of Equations (12.16). □

A. Martingales

Contents

This chapter is a brief presentation of the definitions and results concerning martingales. Williams [**55**] is a good introduction, see also Neveu [**36**] and Rogers and Williams [**43, 44**].

A.1 Discrete Time Parameter Martingales

In the following, $(\Omega, \mathcal{F}, \mathbb{P})$ is a probability space. A *filtration* $(\mathcal{F}_n)$ is a non-decreasing sequence of σ-fields included in $\mathcal{F}$.

Definition 35. *A family $(Z_\alpha; \alpha \in I)$ of real random variables is uniformly integrable if for all $\varepsilon > 0$ there exists $B > 0$ such that*

$$\mathbb{E}\left(|Z_\alpha| 1_{\{|Z_\alpha| \geq B\}}\right) \leq \varepsilon,$$

for all $\alpha \in I$.

This property is clearly true if there exists $p > 1$ such that

$$\sup_{\alpha \in I} \mathbb{E}\left(|Z_\alpha|^p\right) < +\infty.$$

A *stopping time* T is a random variable on Ω with values in $\mathbb{N} \cup \{+\infty\}$ such that, for all $n \in \mathbb{N}$, the event $\{T \leq n\}$ is $\mathcal{F}_n$-measurable. The index n is seen as a time variable, $\mathcal{F}_n$ is the σ-field of events occurring before time n.

Definition 36. *If (M_n) is a sequence of integrable real valued random variables on Ω such that for all $n \in \mathbb{N}$, M_n is $\mathcal{F}_n$-measurable and it is*

– *a $(\mathcal{F}_n)$-*supermartingale *if, for $n \in \mathbb{N}$, $\mathbb{P}$-a.s.*

$$\mathbb{E}(M_{n+1} \mid \mathcal{F}_n) \leq M_n;$$

– *a* $(\mathcal{F}_n)$-sub-martingale *if, for* $n \in \mathbb{N}$, $\mathbb{P}$-*a.s.*

$$\mathbb{E}(M_{n+1} \mid \mathcal{F}_n) \geq M_n;$$

– *a* $(\mathcal{F}_n)$-martingale *if* $\mathbb{P}$-*a.s.,*

$$\mathbb{E}(M_{n+1} \mid \mathcal{F}_n) = M_n,$$

for $n \in \mathbb{N}$.

PROPOSITION A.1. *If* (M_n) *is a martingale and* ϕ *is a convex function then if, for any* $n \in \mathbb{N}$, *the variable* $\phi(M_n)$ *is integrable, the sequence* $(\phi(M_n))$ *is a sub-martingale.*

PROOF. Jensen's inequality for conditional expectation shows that, $\mathbb{P}$-almost surely,

$$\phi(M_n) = \phi(\mathbb{E}(M_{n+1} \mid \mathcal{F}_n)) \leq \mathbb{E}(\phi(M_{n+1}) \mid \mathcal{F}_n).$$

□

A martingale (M_n) has a constant expected value $\mathbb{E}(M_n) = \mathbb{E}(M_0)$ for all $n \in \mathbb{N}$. The following, classical, proposition shows that this equality can be extended to all bounded stopping times.

PROPOSITION A.2 (Stopped Martingales). *If* (M_n) *is martingale and* T *a stopping time, then the stopped sequence* $(M_{T\wedge n})$ *is a martingale, in particular*

$$\mathbb{E}(M_{T\wedge n}) = \mathbb{E}(M_0)$$

for all $n \in \mathbb{N}$.

THEOREM A.3. *A non-negative super-martingale converges almost surely to a finite limit.*

THEOREM A.4 (Doob's Inequality). *If* (M_n) *is a non-negative sub-martingale, for* $a > 0$,

$$\mathbb{P}\left(\sup_{0\leq p\leq n} M_p \geq a\right) \leq \frac{\mathbb{E}(M_n)}{a},$$

for $n \in \mathbb{N}$.

COROLLARY A.5. *If* (M_n) *is a martingale, for* $a > 0$,

$$\mathbb{P}\left(\sup_{0\leq p\leq n} |M_p| \geq a\right) \leq \frac{\mathbb{E}(|M_n|)}{a},$$

and if $M_n \in L_2(\mathbb{P})$ *for all* $n \in \mathbb{N}$,

$$\mathbb{P}\left(\sup_{0\leq p\leq n} |M_p| \geq a\right) \leq \frac{\mathbb{E}\left(M_n^2\right)}{a^2}.$$

PROOF. The functions $x \to |x|$ and $x \to x^2$ being convex on $\mathbb{R}$, Proposition A.1 and Doob's inequality can then be used to conclude. □

The next proposition gives a method to get martingales with a given martingale. This is a discrete version of the Brownian stochastic integral.

PROPOSITION A.6. *If* (M_n) *is a* $(\mathcal{F}_n)$*-martingale and* (N_n) *a sequence of bounded random variables such that* N_n *is* $\mathcal{F}_n$*-measurable for* $n \in \mathbb{N}$*, the sequence*

$$(I_n) = \left(\sum_{i=1}^{n} N_{i-1}(M_i - M_{i-1}) \right) \tag{A.1}$$

is a $(\mathcal{F}_n)$*-martingale.*

PROOF. For $n \in \mathbb{N}$, the variable I_n is integrable since the sequence (N_n) is bounded and the variables $(M_i, 0 \leq i \leq n)$ are integrable. The measurability properties of (M_n) and (N_n) give the equalities

$$\begin{aligned} \mathbb{E}[I_{n+1} \mid \mathcal{F}_n] &= \sum_{i=1}^{n+1} \mathbb{E}[N_{i-1}(M_i - M_{i-1}) \mid \mathcal{F}_n] \\ &= \sum_{i=1}^{n} N_{i-1}(M_i - M_{i-1}) + N_n \mathbb{E}[(M_{n+1} - M_n) \mid \mathcal{F}_n] = I_n, \end{aligned}$$

by the martingale property of (M_n). The sequence (I_n) is a martingale. □

A.2 Continuous Time Martingales

For martingales, most of the definitions of the discrete time case can be extended to the case where the time parameter $t \in \mathbb{R}_+$ is continuous. The stochastic processes considered here are supposed to be continuous on the right and with a left limit at any point of $\mathbb{R}_+$.

DEFINITION 37. *A process* $(X(t))$ *is adapted if*

- *the mapping* $(\omega, t) \to X(t)(\omega)$ *is measurable for the product* σ*-field;*
- *for* $t \geq 0$*, the function* $\omega \to X(\omega, t)$ *is* $\mathcal{F}_t$*-measurable.*

A real valued random variable T *is a stopping time if for* $t \geq 0$*, the event* $\{T \leq t\}$ *is* $\mathcal{F}_t$*-measurable.*

One of the additional definitions for the continuous case is the notion of *local martingale*.

DEFINITION 38. *A stochastic process* $(M(t))$ *is local martingale if*

- $M(0)$ *is* $\mathcal{F}_0$*-measurable ;*
- *there exists a non-decreasing sequence* (T_n) *of stopping times converging to infinity such that, for all* $n \in \mathbb{N}$*,*

  $$(M(T_n \wedge t) - M(0))$$

  is a martingale.

The following proposition gives a criteria under which a local martingale *is* a martingale.

PROPOSITION A.7. *If $(M(t))$ is a real valued local martingale and if for $t \in \mathbb{R}_+$,*

$$\mathbb{E}\left(\sup_{0\leq s\leq t} |M(s)|\right) < +\infty, \tag{A.2}$$

then the process $(M(t))$ is a martingale.

PROOF. There exists a sequence of stopping times (T_n) converging to infinity such that $n \geq 1$, $(M(T_n \wedge t) - M(0))$ is a martingale, in particular

$$[\mathbb{E}(M(T_n \wedge t) - M(0) \,|\, \mathcal{F}_s) = M(T_n \wedge s) - M(0),$$

for $s \leq t \in \mathbb{R}_+$ and $n \in \mathbb{N}$. The variable $M(0)$ is integrable by Relationship (A.2), hence $\mathbb{E}(M(T_n \wedge t) \,|\, \mathcal{F}_s) = M(T_n \wedge s)$, since

$$M(T_n \wedge t) \leq \sup_{0\leq u\leq t} M(u),$$

for all $n \in \mathbb{N}$, Condition (A.2) and Lebesgue's Theorem can be applied in the above identity to get $\mathbb{E}(M(t) \,|\, \mathcal{F}_s) = M(s)$, $\mathbb{P}$-a.s. Therefore, the process $(M(t))$ is a martingale. □

DEFINITION 39. *An adapted process $(X(t))$ is* previsible *if it is adapted with respect to the filtration generated by left continuous adapted processes.*

A left continuous adapted process is thus previsible. If $(X(t))$ is càdlàg and adapted with respect to the filtration $(\mathcal{F}_t)$ the process $(X(t-)$ is previsible.

PROPOSITION A.8. *If $(M(t))$ is local martingale for which there exists a non-decreasing sequence of stopping times converging to infinity such that $(M(T_n \wedge t))$ is a square integrable martingale, there exists a unique previsible non-decreasing process such that*

$$\left(M(t)^2 - A(t)\right)$$

is a local martingale, the process $(A(t))$ is the increasing process of the local martingale $(M(t))$.

A.3 The Stochastic Integral for a Poisson Process

A detailed presentation of these questions in a general framework can be found in Brémaud [**9**] and Davis [**14**].

For $\lambda \geq 0$, the point process $\mathcal{N}_\lambda = \{t_n\}$ is a Poisson process with parameter λ on $\mathbb{R}_+$. For $t \geq 0$, $\mathcal{F}_t$ is the σ-field generated by the random variables $\mathcal{N}_\lambda(]0, s])$ for $s \leq t$. Proposition 1.21 page 20 shows that the process $(\mathcal{N}_\lambda(]0, t]) - \lambda t)$ is a martingale. If X is an adapted càdlàg process (i.e.

right continuous with left limits), *the stochastic integral* of X with respect to this martingale is defined as

$$Z(t) = \int_0^t X(s)\,[\mathcal{N}_\lambda(\omega, ds) - \lambda\, ds] = \int_{]0,t]} X(s)\,[\mathcal{N}_\lambda(\omega, ds) - \lambda\, ds]$$
$$= \sum_{n\in\mathbb{N};0<t_n\leq t} X(t_n) - \lambda \int_0^t X(s)\, ds,$$

or equivalently

$$dZ(t) \stackrel{\text{def.}}{=} \lim_{s\nearrow t} Z(t) - Z(s) = X(t)\,[\mathcal{N}_\lambda(dt) - \lambda\, dt].$$

PROPOSITION A.9. *If $(X(t))$ is a càdlàg adapted process, the process*

$$\left(\int_{]0,t]} X(s-)\,[\mathcal{N}_\lambda(\omega, ds) - \lambda\, ds]\right)$$

is a martingale whose increasing process is given by

$$\left(\lambda \int_0^t X^2(s)\, ds\right).$$

See Rogers and Williams [**43**], Theorem 8.1 page 15.

The integration of $X(s-)$ rather than $X(s)$ with respect to the martingale $\mathcal{N}_\lambda(]0,t]) - \lambda t$ preserves the martingale property. This is a continuous time version of Proposition A.6 page 353, the expression $X(s-)$ is the analogue of N_{i-1} in Identity (A.1).

If $(X(t))$ is a càdlàg function, constant by interval, for $t \geq 0$,

$$X(t) = X(0) + \sum_{0\leq s\leq t} \Delta X(s) = X(0) + \int_0^t dX(s),$$

where the discrete sum is on the discontinuities of X and the jumps are denoted by $\Delta X(s) = X(s) - X(s-)$. The last identity is simply the fact that $X(t) - X(0)$ is the sum of the jumps of X between 0 and t. In the same way, if $(Y(t))$ is also càdlàg, constant by interval, the product $(X(t)Y(t))$ has a jump at s if

- s is a jump instant of X and Y is continuous at s, and in this case the jump is $Y(s)\Delta X(s) = Y(s-)\Delta X(s)$,
- s is a jump instant of X and Y the jump of XY at s is

$$X(s)\Delta Y(s) + Y(s)\Delta X(s) + \Delta X(s)\Delta Y(s).$$

One gets the formula of integration by parts

$$X(t)Y(t) = X(0)Y(0) + \sum_{0\leq s\leq t} Y(s-)\Delta X(s) + \sum_{0\leq s\leq t} X(s-)\Delta Y(s)$$
$$+ \sum_{0\leq s\leq t} \Delta X(s)\Delta Y(s),$$

or

$$X(t)Y(t) = X(0)Y(0) + \int_0^t Y(s-)\,dX(s) + \int_0^t X(s-)\,dY(s) + \sum_{0\leq s\leq t} \Delta X(s)\Delta Y(s). \tag{A.3}$$

PROPOSITION A.10. *If f and g are bounded functions on $\mathbb{N}$, and $\mathcal{N}_\lambda$ and $\mathcal{N}_\mu$ are independent Poisson processes on $\mathbb{R}$ with parameter λ and $\mu \geq 0$, the process*

$$\left(\int_0^t f(L(s-))\,(\mathcal{N}_\lambda(ds) - \lambda\,ds) \int_0^t g(L(s-))\,(\mathcal{N}_\mu(ds) - \mu\,ds)\right)$$

is a martingale.

PROOF. The process is denoted by $(Z(t)) = (A_1(t)A_2(t))$, where $A_1(t)$ and $A_2(t)$ are integrals as above between 0 and t. The two Poisson processes $\mathcal{N}_\lambda$ and $\mathcal{N}_\mu$ being independent, the probability that they have a jump at the same point is zero. Almost surely, the processes $(A_1(s))$ and $(A_2(s))$ have no common discontinuity point between 0 and t. Consequently, with probability 1, Relationship (A.3) gives the identity

$$Z(t) = Z(0) + \int_0^t A_2(s-)f(L(s-))\,(\mathcal{N}_\lambda(ds) - \lambda ds) + \int_0^t A_1(s-)g(L(s-))\,(\mathcal{N}_\mu(ds) - \mu\,ds).$$

By using Proposition A.9, it is easily seen to conclude that $(Z(t))$ is a martingale. □

A.4 Stochastic Differential Equations with Jumps

Stochastic differential equations with jumps play an important role in several chapters (see the chapters on $M/M/1$, $M/M/\infty$ queues and on fluid limits). The proposition below is elementary, it shows the existence and uniqueness of solutions for most of these differential equations.

PROPOSITION A.11. *If I is a countable or finite set and for $i \in I$,*

- *C_i is a subset of $\mathbb{R}^d$;*
- *V_i is a marked Poisson process on $\mathbb{R}_+ \times \mathbb{Z}^d$ with intensity $\lambda_i\, dt \otimes \nu_i(dm)$ with $\lambda_i \geq 0$ and ν_i a probability distribution on $\mathbb{Z}^d$. The point processes V_i, $i \in I$ are assumed to be independent ;*
- *$\lambda = \sum_{i\in I} \lambda_i < +\infty$.*

The stochastic differential equation

$$dX(t) = \sum_{i\in I} \int_{\mathbb{Z}^d} 1_{\{X(t-)\in C_i\}} m\, V_i(dt, dm), \tag{A.4}$$

has a unique solution such that $X(0) = x \in \mathbb{Z}^d$.

The integral in Equation (A.4) is with respect to the variable m. For t fixed the measure $V_i(dt, dm)$ is either the zero measure or a Dirac mass (see Proposition 1.11 page 11).

PROOF. Define $\mathcal{N}$ the point process on $\mathbb{R}_+$ such that

$$\mathcal{N}(dt) = \sum_{i\in I} V_i\left(dt, \mathbb{Z}^d\right) = \sum_{n\geq 1} \delta_{t_n},$$

the sequence (t_n) is non-decreasing. The superposition of independent Poisson point processes being also a Poisson point process, $\mathcal{N}$ is also a Poisson point process (Lemma 1.7 page 9) with finite intensity since λ is finite. By induction, it is shown that Equation (A.4) has a unique solution on $[0, t_n]$ for any $n \geq 1$. If it holds on $[0, t_n]$, necessarily for $t \in [t_n, t_{n+1}[$, $X(t) = X(t_n)$. At $t = t_{n+1}$,

$$X(t_{n+1}) = X(t_n) + \sum_{i\in I} 1_{\{X(t_n)\in C_i\}} \int_{\mathbb{Z}^d} m\, V_i(t_n, dm),$$

(recall that the support of the measure $V_i(t_n, dm)$ is a point if it is not zero, cf. Proposition 1.11 page 11). Therefore Equation (A.4) has a unique solution on $[0, t_{n+1}]$. The proposition is proved. □

Examples.

The $M/M/1$ Queue. If $L(0) = x \in \mathbb{N}$, and $\mathcal{N}_\lambda$ and $\mathcal{N}_\mu$ are independent Poisson processes on $\mathbb{R}_+$ whose parameters are given by λ, μ, the number $(L(t)$ of customers at time t can be seen as the unique solution of the equation

$$dL(t) = \mathcal{N}_\lambda(dt) - 1_{\{L(t-)>0\}}\, \mathcal{N}_\mu(\,dt).$$

The $M/M/\infty$ Queue. If $(\mathcal{N}_\mu^i)$ is an i.i.d. sequence of Poisson processes on $\mathbb{R}_+$ with parameter μ, the number of customers of this queue can be represented as the solution of the stochastic differential equation

$$dL(t) = \mathcal{N}_\lambda(dt) - \sum_{i=1}^{+\infty} 1_{\{i\leq L(t-)\}}\, \mathcal{N}_\mu^i(dt).$$

Note that the third condition of Proposition A.11 is not satisfied. This (small) difficulty can be solved easily since the inequality $L(t) \leq \mathcal{N}_\lambda([0,t])$ holds, almost surely there is a finite number on discontinuities on the interval $[0, t]$ (at most $2\mathcal{N}_\lambda([0,t])$ of them). The same method as in the proof of Proposition A.11 can be also used in this case.

B. Markovian Jump Processes

Contents

This chapter is an elementary, quite informal, presentation of the main definitions and results concerning Markov jump processes. Norris [**38**] is a good introduction on this subject. For more advanced results, Chapters III and IV of Rogers and Williams [**43, 44**] are highly recommended, especially for the Markov processes considered in Sections 7.3 and 7.4 of Chapter 7. In the following $(\mathcal{F}_t)$ is a filtration of σ-fields on a fixed probability space.

DEFINITION 40. *An adapted process $(X(t))$ on $\mathcal{S}$ is Markovian if for all s, $t \geq 0$ and $x \in \mathcal{S}$, $\mathbb{P}_x$-almost surely*

$$\mathbb{E}_x(f(X_{t+s}) \mid \mathcal{F}_t) = \mathbb{E}_{X(t)}(f(X_s)),$$

for any bounded measurable function f on $\mathcal{S}$, or

$$\mathbb{P}_x(X_{t+s} \in \cdot \mid \mathcal{F}_t) = \mathbb{P}_{X(t)}(X_s \in \cdot\ ).$$

The definition of an adapted process is given page 354.

DEFINITION 41. *A Markov jump process $(X(t))$ is a Markov process with a countable state space $\mathcal{S}$ such that, almost surely, the function $t \to X(t)$ is continuous on the right on $\mathbb{R}_+$ and has left limit at any point (the function is said to be càdlàg. (In French "continu à droite et limité à gauche".)*

A Markov jump process can also be seen as a probability distribution on the space $D_{\mathcal{S}}([0, +\infty[)$ càdlàg functions with values in $\mathcal{S}$. The space $D_{\mathcal{S}}([0, +\infty[)$ is endowed with Skorohod Topology (see Appendix C for the definition of this topology.)

All the processes considered are assumed to be irreducible and homogeneous in time. It is also assumed that they do not die in finite time, i.e. that they do not visit infinity in finite time.

Example of a Jump Process Visiting Infinity in Finite Time. From $i \in \mathbb{N}$, the process jumps to $i+1$ after an exponentially distributed time E_i with parameter i^2. Starting from 1, it arrives at n at time $T_n = E_1 + \cdots + E_{n-1}$ where the (E_i) are independent random variables. It is easily seen that the limit $T_\infty = \lim_n T_n$ is $\mathbb{P}$-almost surely finite, indeed if $\xi \geq 0$,

$$\mathbb{E}\left(e^{-\xi T_n}\right) = \prod_{i=1}^{n-1} \frac{i^2}{i^2+\xi} = \exp\left(\sum_{i=1}^{n-1} \log\left[1 - \xi/(i^2+\xi)\right]\right),$$

thus the infinite product

$$\mathbb{E}\left(e^{-\xi T_\infty}\right) = \prod_{i=1}^{n-1} \frac{i^2}{i^2+\xi}$$

converges, by letting ξ go to 0, one gets that $\mathbb{P}(T_\infty < +\infty) = 1$. There are infinitely many jumps in the time interval $[0, T_\infty[$ and at $t = T_\infty$ the process is at infinity.

Markov Processes Associated with Jackson Networks and Loss Networks. If there is a infinite number of jumps in finite time for Markov processes describing these networks, necessarily there must be an infinite number of arrivals in finite time. Since the external arrivals are assumed to be Poisson, this is a negligible event. Markov jump processes describing queueing networks do not die in finite time.

PROPOSITION B.1. *A Markov jump process has the strong Markov property, for any stopping time T,*

$$\mathbb{P}\left(X(T+\cdot) \in \cdot \,\middle|\, \mathcal{F}_T\right) = \mathbb{P}_{X(T)}(X(T+\cdot) \in \cdot)$$

The proof of this proposition is similar to the proof of the strong Markov property of Poisson processes, i.e. via an approximation from above of the stopping time T. (See Proposition 1.18 page 18 and Norris [**38**].)

B.1 $\mathcal{Q}$-Matrices

The $\mathcal{Q}$-matrix, or the infinitesimal generator of a Markov jump process $(X(t))$ is the matrix $Q = (q_{ij}; i, j \in \mathcal{S})$ defined by

$$q_{ij} = \lim_{s \to 0} \frac{1}{s} \mathbb{P}(X(s) = j \,|\, X(0) = i),$$

for $i \neq j$ and

$$q_{ii} = -\sum_{j \neq i} q_{ij}.$$

The matrix Q is also called $\mathcal{Q}$-matrix. For i, $j \in \mathcal{S}$, q_{ij} is the rate at which a jump from i to j occurs. For i, $j \in \mathcal{S}$, the quantity q_{ij} is the derivative at

0 of the function $t \to \mathbb{P}(X(t) = j \mid X(0) = i)$. The matrix Q can be seen as a functional operator, for a non-negative function f on $\mathcal{S}$,

$$Q(f)(i) = \sum_{j \in \mathcal{S}} q_{ij} f(j) = \sum_{j \neq i} q_{ij}(f(j) - f(i)).$$

If $f \equiv 1$, $Q(f)$ is clearly identically zero. When f is a bounded function on $\mathcal{S}$, the Markov property gives the relationship

$$\begin{aligned}\mathbb{E}_x\left[f(X(t+s))\right] - \mathbb{E}_x\left[f(X(t))\right] &= \mathbb{E}_x\left[\mathbb{E}_{X(t)}\left[f(X(s))\right] - f(X(0))\right] \\ &= \sum_{i,j \in \mathcal{S}, i \neq j} (f(j) - f(i))\mathbb{P}(X(s) = j \mid X(0) = i)\mathbb{P}_x(X(t) = i),\end{aligned}$$

dividing by s and, assuming that f is such that the limit can be taken inside the series, one gets, as s tends to 0,

$$\frac{d}{dt}\mathbb{E}_x\left[f(X(t))\right] = \sum_{i,j \in \mathcal{S}, i \neq j} q_{ij}(f(j) - f(i))\mathbb{P}_x(X(t) = i),$$

$$\frac{d}{dt}\mathbb{E}_x\left[f(X(t))\right] = \mathbb{E}_x\left[Q(f)(X(t))\right], \tag{B.1}$$

these equations are *forward Kolmogorov equations* associated with $(X(t))$, equivalently

$$\mathbb{E}_x\left[f(X(t))\right] - \mathbb{E}_x\left[f(X(0))\right] = \int_0^t \mathbb{E}_x\left[Q(f)(X(s))\right] ds.$$

The following uniqueness result is valid for Markov processes which do not die in finite time. It is used in Chapter 4 on the reversibility property.

PROPOSITION B.2. *For $x \in \mathcal{S}$, there exists a unique càdlàg Markov process whose Q-matrix is Q and such that $\mathbb{P}(X(0) = x) = 1$.*

The uniqueness is understood as the uniqueness of the probability distribution $\mathbb{P}_x$ on the space $D_{\mathcal{S}}([0, +\infty[)$ (See Appendix C).

ELEMENTARY EXAMPLES

1. *Poisson processes.* For $\lambda > 0$, if $\mathcal{N}_\lambda$ is a Poisson process with parameter λ, the process $(X(t) = (X(0) + \mathcal{N}_\lambda(]0, t])$ is Markovian and its Q-matrix is given by $q_{i,i+1} = \lambda$, $q_{ii} = -\lambda$ and $q_{ij} = 0$ for all the other (i, j). The identity $q_{i,i+1} = \lambda$ is just a consequence of the relationship
$$\mathbb{P}(X(s) = i + 1 \mid X(0) = i) = \mathbb{P}(\mathcal{N}_\lambda(]0, s]) = 1) = \lambda s e^{-\lambda s}.$$
2. *The M/M/1 queue.* If $L(t)$ is the number of customers in the queue at time t, $(L(t))$ is a Markov process whose Q-matrix Q is given by $q_{i,i+1} = \lambda$, $q_{i,i-1} = \mu$ when $i > 0$ and $q_{i,j} = 0$ if $|i - j| > 1$.
As an example the equation $q_{i,i-1} = \mu$ for $i > 0$ is proved. The quantity $\mathcal{N}_x$ denotes a Poisson process with parameter x. If $X(0) = i$, for a sufficiently small s, the event $\{X(s) = i - 1\}$ occurs if, for some

$k \in \mathbb{N}$, there are $k+1$ customers served between 0 and s (in particular $\mathcal{N}_\mu(]0,s]) \geq k+1$) and k arrivals in this time interval. It is easily seen that if $k > 0$, the probability of this event is $o(s)$, consequently

$$\begin{aligned}\mathbb{P}(X(s) = i-1 \mid X(0) = i) &= \mathbb{P}(\mathcal{N}_\lambda(]0,s]) = 0, \mathcal{N}_\mu(]0,s]) = 1) + o(s)\\ &= \mathbb{P}(\mathcal{N}_\lambda(]0,s]) = 0)\mathbb{P}(\mathcal{N}_\mu(]0,s]) = 1) + o(s)\\ &= e^{-\lambda s}\mu s e^{-\mu s} + o(s),\end{aligned}$$

hence, $q_{i,i-1} = \mu$.

3. *The $M/M/\infty$ queue.* The $\mathcal{Q}$-matrix of the process of the number of customers is given by, for $i \geq 1$, $q_{i,i+1} = \lambda$, $q_{i,i-1} = i\mu$ and $q_{i,j} = 0$, if $|i-j| > 1$. When i customers are present in the queue, each of them is served. The first departure time is thus the minimum of i exponentially distributed random variables with parameter μ. The first departure time is therefore exponentially distributed with parameter $i\mu$, hence $q_{i,i-1} = i\mu$.
4. *Tandem queues $M/M/1 \to \cdot/M/1$* The number $(L_1(t), L_2(t))$ of customers in each of these queues is a Markov process such that, for $(i,j) \neq (k,l)$,

$$q_{(i,j),(k,l)} = \begin{cases} \lambda, & k = i+1, \qquad\qquad l = j,\\ \mu_1, & k = i-1,\ i \geq 1, \quad l = j+1,\\ \mu_2, & k = i, \qquad\qquad\quad l = j-1,\ j \geq 1,\\ 0, & \text{otherwise.}\end{cases}$$

B.2 Global Balance Equations

The invariant probability distribution π of a Markov process $(X(t))$ with values in $\mathcal{S}$ is such that the identity

$$\mathbb{E}_\pi\left[f(X(t))\right] = \pi(f)$$

holds for all $t \geq 0$ and any bounded function f on $\mathcal{S}$, in particular,

$$\frac{d}{dt}\mathbb{E}_\pi\left[f(X(t))\right] = 0.$$

By integrating Equation (B.1) with respect to π, one gets that, if f is a "good" function, this yields the identity

$$\pi(Q(f)) = \int_{\mathcal{S}} Q(f)(x)\pi(dx) = 0,$$

or

$$\sum_{i,j\in\mathcal{S}, i\neq j} \pi(i)q_{ij}(f(j) - f(i)) = 0.$$

If f is the indicator function of the set $\{i\}$, the above equation becomes

$$\pi(i)\left(\sum_{j\in\mathcal{S},j\neq i} q_{ij}\right) = \sum_{j\in\mathcal{S},j\neq i} \pi(j)q_{ji}.$$

This set of equations is the set of equilibrium equations of the Markov process, by using matrix notations, it can be written as $\pi Q = 0$.

Example : Birth and death processes. In Chapter 4 page 84, it has been seen that the invariant distribution $(\pi(i))$ of a birth and death process on $\mathbb{N}$ is such that $\pi(i+1)\mu_{i+1} = \pi(i)\lambda_i$, with $\lambda_i = q_{ii+1}$ and $\mu_i = q_{ii-1}$ for $i > 0$. In particular

$$\pi(i) = \pi(0)\prod_0^{i-1} \frac{\lambda_k}{\mu_{k+1}},$$

if the sum of the above sequence is finite, the invariant measure has a finite mass. This finiteness condition is not sufficient to ensure the ergodicity of the Markov process. If $\lambda_n = n^n$ and $\mu_n = 2(n-1)^{n-1}$, the sequence $(1/2^{n+1})$ satisfies the equilibrium equations. It is easily seen that this Markov process is not ergodic since $\lim_n \lambda_n/\mu_n = +\infty$, by using Theorem 8.11 page 220 applied to the function $f(x) = x$). This process is therefore transient with a finite invariant measure. This apparent paradox is explained by the fact that the process dies in finite time with probability 1. The equilibrium equations do not have any significance in this context.

The Embedded Markov Chain

If (X_n) denotes the sequence of the successive states visited by the Markov process (X_n), it is Markov chain whose transitions (p_{ij}) are given by

$$p_{ij} = -\frac{q_{i,j}}{q_{ii}} = \frac{q_{i,j}}{\sum_{k\neq i} q_{ik}},$$

for $i \neq j$. The process in i for an exponentially distributed amount of time with parameter $-q_{ii} = \sum_{j\neq i} q_{ij}$ and jumps to j with probability p_{ij}. The embedded Markov chain describes the spatial component of the jump process and the sequence $(-q_{ii})$ gives the time scale between the jumps.

The following proposition is easily proved.

PROPOSITION B.3. *If the process $(X(t))$ has an invariant probability distribution $(\pi(i))$, the sequence $(\widehat{\pi}(i)) = (-q_{ii}\pi(i))$ is an invariant probability distribution for the embedded Markov chain.*

If (X_n) is a Markov chain without any loop, i.e. such that the identity $\mathbb{P}_x(X_1 = x) = 0$ holds for all $x \in \mathcal{S}$, it is easy to construct a Markov process whose embedded Markov chain is (X_n). If $\mathcal{N}_1$ is a Poisson process with parameter 1 and independent of (X_n), the process $(X(t)) = (X_{\mathcal{N}_1(]0,t])})$ is càdlàg and has the Markov property. The $\mathcal{Q}$-matrix $Q = (q_{ij})$ is defined by $q_{ij} = \mathbb{P}(X_1 = j \mid X_0 = i)$ for $j \neq i$. Although the processes (X_n) and $(X(t))$

are equivalent, quite often it is more convenient to work with the continuous time process. (Poissonnization of a Markov chain).

B.3 The Associated Martingales

The following proposition gives a method to construct martingales with a Markov jump process For the proofs of these results, see Section IV-20 of Rogers and Williams [44] page 30–37.

PROPOSITION B.4. *If g is a function on $\mathbb{R}_+ \times \mathcal{S}$ such that the mapping*

$$t \to \frac{\partial g(t,x)}{\partial t}$$

is continuous for all $x \in \mathcal{S}$, the process

$$\left(g(t, X(t)) - g(0, X(0)) - \int_0^t \left[\frac{\partial g}{\partial t} + Q(g(s,\cdot)) \right] (s, X(s))\, ds \right)$$

is a local martingale. (See Appendix A for the basic definitions and results concerning martingales.)

In particular if f is a function on $\mathcal{S}$, taking $g(t,x) = f(x)$, the above proposition gives that

$$\left(f(X(t)) - f(X(0)) - \int_0^t Q(f)(X(s))\, ds \right)$$

is a local martingale. If this local martingale is a martingale (see Proposition A.7 page 354), one gets Relationship (B.1), that is,

$$\mathbb{E}_x \left[f(X(t)) \right] - \mathbb{E}_x \left[f(X(0)) \right] = \int_0^t \mathbb{E}_x \left[Q(f)(X(s)) \right] ds$$

for any $t \geq 0$.

DEFINITION 42. *A function g on $\mathbb{R}_+ \times \mathcal{S}$ such that*

$$\frac{\partial g(t,x)}{\partial t} + Q(g(t,\cdot))(t,x) = 0$$

for all $x \in \mathcal{S}$ and $t \geq 0$ is a space-time harmonic function for the infinitesimal generator Q.

COROLLARY B.5. *If g a space-time harmonic function for the infinitesimal generator Q such that*

$$t \to \frac{\partial g(t,x)}{\partial t}$$

is continuous for all $x \in \mathcal{S}$, the process $[g(t, X(t))]$ is a local martingale.

The next proposition shows that these martingales associated with $(X(t))$ characterize the distribution of the Markov process.

Proposition B.6 (Martingale Problem). *If a càdlàg Markov process $(X(t))$ is such that*

$$\left(f(X(t)) - f(X(0)) - \int_0^t Q(f)(X(s))\, ds\right)$$

is a local martingale for any function f on $\mathcal{S}$, then the distribution of $(X(t))$ is the distribution of the Markov process starting from x with $\mathcal{Q}$-matrix Q.

Conditional Distributions of Markov Processes

Proposition B.7. *If $(X(t))$ is a transient Markov jump process on $\mathcal{S}$, with $\mathcal{Q}$-matrix $Q = (q_{ij})$ such that*

$$q_i = \sum_{j \neq i} q_{ij} < +\infty,$$

for any $j \in \mathcal{S}$, and if H is a subset of the state space $\mathcal{S}$, T_H denotes the hitting time of H by $(X(t))$. The Markov process $(X(t))$ conditioned to never visit H is an homogeneous Markov process whose $\mathcal{Q}$-matrix $\widetilde{Q} = (\tilde{q}_{ij})$ is given by

$$\tilde{q}_{ij} = q_{ij} \frac{P_j(T_H = +\infty)}{P_i(T_H = +\infty)},$$

for $i \neq j \in \mathcal{S}$.

Proof. The event $\{T_H = +\infty\}$ is non-negligible by transience of $(X(t))$, hence conditioning on this event makes sense. If $\widetilde{\mathbb{P}}$ denotes the probability distribution $\mathbb{P}(\,\cdot \mid T_H = +\infty)$ and $\widetilde{\mathbb{E}}(\cdot)$ the expected value for this probability distribution, it has to be shown that if $h : \mathcal{S} \to \mathbb{R}_+$ is a bounded function on $\mathcal{S}$, for s, $t \in \mathbb{R}_+$ and $x \geq 1$, the equation

$$\widetilde{\mathbb{E}}_x \left(h(X(s+t)) \,\middle|\, \mathcal{F}_s\right) = \widetilde{\mathbb{E}}_{X(s)} \left[h(X(t))\right],$$

or, for any $\mathcal{F}_s$-measurable random variable Y

$$\widetilde{\mathbb{E}}_x \left[Y h(X(s+t))\right] = \widetilde{\mathbb{E}}_x \left(Y \widetilde{\mathbb{E}}_{X(s)} \left[h(X(t))\right]\right).$$

holds. By decomposing the event $\{T_H = +\infty\}$, one gets

$$\begin{aligned} &\mathbb{E}_x \left[h(X(s+t)) Y 1_{\{T_H = +\infty\}}\right] \\ &\qquad = \mathbb{E}_x \left[h(X(s+t)) 1_{\{X(u) \notin H,\, \forall u \geq s\}} Y 1_{\{X(u) \notin H,\, \forall u \leq s\}}\right], \end{aligned}$$

the random variable $Y 1_{\{X(u) \notin H,\, \forall u \leq s\}}$ being $\mathcal{F}_s$-measurable, by the Markov property of $(X(t))$,

$$\begin{aligned}
&\mathbb{E}_x\left[h(X(s+t))Y1_{\{T_H=+\infty\}}\right]\\
&\qquad = \mathbb{E}_x\left[\mathbb{E}_{L_s}\left(h(X(t))1_{\{X(u)\notin H,\,\forall u\geq 0\}}\right)Y1_{\{X(u)\notin H,\,\forall u\leq s\}}\right]\\
&\qquad = \mathbb{E}_x\left[\mathbb{E}_{L_s}\left(h(X(t))\,|\,T_H=+\infty\right)Y1_{\{X(u)\notin H,\,\forall u\leq s\}}\mathbb{P}_{L_s}(T_H=+\infty)\right].
\end{aligned}$$

The random variable $\mathbb{E}_{L_s}\left(h(X(t))\,|\,T_H=+\infty\right)Y1_{\{X(u)\notin H,\,\forall u\leq s\}}$ being $\mathcal{F}_s$-measurable, the Markov property of $(X(t))$ gives the relationship

$$\begin{aligned}
&\mathbb{E}_x\left[h(X(s+t))Y1_{\{T_H=+\infty\}}\right]\\
&\qquad = \mathbb{E}_x\left[\mathbb{E}_{L_s}\left(h(X(t))\;\middle|\;T_H=+\infty\right)Y1_{\{X(u)\notin H,\,\forall u\leq s\}}1_{\{X(u)\notin H,\,\forall u\geq s\}}\right]\\
&\qquad = \mathbb{E}_x\left(\mathbb{E}_{L_s}\left(h(X(t))\,|\,T_H=+\infty\right)Y1_{\{T_H=+\infty\}}\right).
\end{aligned}$$

Finally,

$$\begin{aligned}
&\mathbb{E}_x\left[h(X(s+t))Y\;\middle|\;T_H=+\infty\right]\\
&\qquad\qquad = \mathbb{E}_x\left[\mathbb{E}_{L_s}\left[h(X(t))\,|\,T_H=+\infty\right]Y\;\middle|\;T_H=+\infty\right],
\end{aligned}$$

for the probability distribution $\mathbb{P}_x$ conditioned on the event $\{T_H=+\infty\}$, the conditional expectation of $h(L_{t+s})$ given $\mathcal{F}_s$ is given by

$$\mathbb{E}_{L_s}\left(h(X(t))\,|\,T_H=+\infty\right).$$

The process $(X(t))$ conditioned on the event $\{T_H=+\infty\}$ is therefore a Markov process.

The $\mathcal{Q}$-matrix. If $i,\,j\in\mathcal{S}-H$ and $t>0$, the Markov property gives

$$\mathbb{P}_i(X(t)=j,\;T_H=+\infty)=\mathbb{P}_i(X(t)=j,\;T_H>t)\mathbb{P}_j(T_H=+\infty), \tag{B.2}$$

clearly,

$$\begin{aligned}
\mathbb{P}_i(X(t)=j)-\mathbb{P}_i(X(t)=j,\;T_H>t))&\\
\leq \mathbb{P}_i(X&\text{ has at least two jumps on }[0,t]).
\end{aligned} \tag{B.3}$$

this last term is $o(t)$ since

$$\begin{aligned}
\mathbb{P}_i(X\text{ has at least two jumps on }[0,t]) &= \int_0^t q_i e^{-q_i(t-s)}\sum_{j\neq i}p_{ij}(1-e^{-q_i s})\,ds\\
&= e^{-q_i t}\int_0^t q_i e^{q_i s}\sum_{j\neq i}p_{ij}(1-e^{-q_i s})\,ds,
\end{aligned}$$

where (p_{ij}) is the transition matrix of the embedded Markov chain. The function under the integral being continuous with respect to s, by dividing the above equality by t and letting t got to 0, one gets the estimation $o(t)$.

Equality (B.2), Bound (B.3) and the definition of q_{ij} give the relationship

$$\tilde{q}_{ij}=\lim_{t\to 0}\frac{1}{t}\mathbb{P}_i(X(t)=j\,|\,T_H=+\infty)=q_{ij}\frac{\mathbb{P}_j(T_H=+\infty)}{\mathbb{P}_i(T_H=+\infty)}.$$

The proposition is proved. □

C. Convergence in Distribution

Contents

In this chapter, results concerning the convergence of a sequence of stochastic processes are briefly recalled. The classical book [**7**] by Billingsley (1968 !) is recommended. For convergence criteria using martingales, see Ethier and Kurtz [**19**] and Jacod and Shiryaev [**28**].

C.1 Total Variation Norm on Probability Distributions

The state space $\mathcal{S}$ is assumed to be at most countable

DEFINITION 43. *The total variation norm of a measure* $(\mu(x); x \in \mathcal{S})$ *on* $\mathcal{S}$ *is given by*

$$\|\mu\|_{tv} = \frac{1}{2}\|\mu\|_1 = \frac{1}{2}\sum_{x\in\mathcal{S}} |\mu(x)|,$$

When μ is the difference of two probability distributions, this norm can also be expressed in a probabilistic way.

PROPOSITION C.1. *If* $\mathbb{P}$ *and* $\mathbb{Q}$ *are two probability distributions on* $\mathcal{S}$ *then*

$$\|\mathbb{P}-\mathbb{Q}\|_{tv} = \sup_{A\subset\mathcal{S}} |\mathbb{P}(A)-\mathbb{Q}(A)| = \frac{1}{2}\sup_{f,\|f\|_\infty\le 1}\left|\int f\,d\mathbb{P} - \int f\,d\mathbb{Q}\right|$$

where $\|f\|_\infty$ *is the norm of the supremum,* $\|f\||_\infty = \sup\{|f(x)|; x\in\mathcal{S}\}$.

PROOF. Indeed, if $\mathbb{P} = (p(x))$ and $\mathbb{Q} = (q(x))$ are probability distributions on $\mathcal{S}$, by denoting $\mathcal{S}_+$ the subset $\{x \in \mathcal{S} : p(x) \ge q(x)\}$, for $A \subset \mathcal{S}$,

$$\begin{aligned}\mathbb{P}(A)-\mathbb{Q}(A) &= (\mathbb{P}-\mathbb{Q})(A\cap\mathcal{S}_+) + (\mathbb{P}-\mathbb{Q})(A\cap\mathcal{S}_+^c)\\ &\le \mathbb{P}(\mathcal{S}^+) - \mathbb{Q}(\mathcal{S}^+) = -(\mathbb{P}(\mathcal{S}_+^c) - \mathbb{Q}(\mathcal{S}_+^c)).\end{aligned}$$

One gets

$$\sup_{A \subset \mathcal{S}} |\mathbb{P}(A) - \mathbb{Q}(A)| = \mathbb{P}(\mathcal{S}^+) - \mathbb{Q}(\mathcal{S}^+),$$

hence,

$$\begin{aligned}\sup_{A \subset \mathcal{S}} |\mathbb{P}(A) - \mathbb{Q}(A)| &= \sum_{x \in \mathcal{S}+} p(x) - q(x) = - \sum_{x \notin \mathcal{S}+} p(x) - q(x) \\ &= \frac{1}{2} \sum_{x \in \mathcal{S}} |p(x) - q(x)| = \|\mathbb{P} - \mathbb{Q}\|_{tv},\end{aligned}$$

the last identity of the proposition follows from the definition of the total variation norm. □

This norm is expressed as the supremum of $|\mathbb{P}(A) - \mathbb{Q}(A)|$ for all the events A. If the total variation norm tends to 0, the same is true *uniformly* for all these quantities. If $\mathbb{P}$ and $\mathbb{Q}$ are two orthogonal probability distributions, i.e. there exists a subset A of $\mathcal{S}$ such that $\mathbb{P}(A) = 1 = \mathbb{Q}(A^c)$, it is easily seen that $\|\mathbb{P} - \mathbb{Q}\|_{tv}$ is 1; the maximal value for the distance between two probability distributions. If X and Y are two random variables with values in $\mathcal{S}$, the quantity

$$\|\mathbb{P}(X \in \cdot\,) - \mathbb{P}(Y \in \cdot\,)\|_{tv}$$

denotes the distance in total variation between their respective distributions.

If the state space is not anymore discrete, the expression given by Proposition C.1 can be extended, with the restriction that the subsets A considered must be measurable.

The total variation norm is not the only possibility to give an estimation of how far are two probability distributions.

The separation "distance".

$$d_s(\mathbb{P}, \pi) = \max_{x \in \mathcal{S}} \left| \frac{p(x)}{\pi(x)} - 1 \right|,$$

note the lack of symmetry between $\mathbb{P}$ and π (this is the reason of the quotes for the term distance). This "distance" is stronger than the total variation norm,

$$2\|\mathbb{P} - \pi\|_{tv} = \sum_{x \in \mathcal{S}} |p(x) - \pi(x)| = \sum_{x \in \mathcal{S}} \left| \frac{p(x)}{\pi(x)} - 1 \right| \pi(x) \le d_s(\mathbb{P}, \pi).$$

If $\mathcal{S} = \{1, \ldots, N\}$ and U_k is the uniform probability distribution on $\{1, \ldots, k\}$ for $k \le N$, then $\|U_{N-1} - U_N\|_{tv} = 1/2N$ and $d_s(U_{N-1}, U_N) = 1$. The probability distributions U_{N-1}, U_N are close for the total variation distance but far apart for the separation "distance".

The norm in $L_2(\pi)$,

$$d_2(\mathbb{P}, \pi) = \sqrt{\sum_{x \in \mathcal{S}} \left| \frac{p(x)}{\pi(x)} - 1 \right|^2 \pi(x)},$$

it has the advantage of using the Hilbertian framework which is, quite often, easier to work with. These distances can be compared as follows

$$2||\mathbb{P}-\pi||_{tv} \leq d_2(\mathbb{P},\pi) \leq d_s(\mathbb{P},\pi),$$

these inequalities are a consequence of Cauchy-Schwartz's Inequality.

C.2 Convergence of Stochastic Processes

In this section, S is a complete, separable, metric space. The main examples are $S=\mathbb{R}^d$, $S=C([0,T],\mathbb{R}^d)$, the set of continuous functions or $S=D([0,T],\mathbb{R}^d)$, the set of càdlàg functions with $0<T\leq+\infty$. For $x=(x_i)\in\mathbb{R}^d$, $||x||$ is the norm $|x_1|+\cdots+|x_d|$.

The Topologies on Functional Spaces

Continuous Functions. The space $S=C([0,T],\mathbb{R}^d)$ is the set of $\mathbb{R}^d$-valued continuous functions on the interval $[0,T]$. If $T<+\infty$, it is equipped with the uniform norm

$$||f||_{\infty,T} = \sup_{s\in[0,T]} ||f(s)||,$$

for a continuous function f on $[0,T]$. If $T=+\infty$, the topology of uniform convergence on the finite intervals defined by the distance d_∞: If f and g are elements of $C([0,+\infty[,\mathbb{R}^d)$,

$$d_\infty(f,g)=\int_0^{+\infty}\frac{||f-g||_{\infty,T}}{1+||f-g||_{\infty,T}}e^{-T}\,dT.$$

In this way, $C([0,T],\mathbb{R}^d)$ is a complete separable metric space.

The Set of càdlàg Functions. The space $\Delta([0,T])$ is the set of increasing functions from $[0,T]$ into $[0,T]$ being 0 at 0 and T at T. An element of $\Delta([0,T])$ can be thought as a time change on the interval $[0,T]$. If α is an element of this space, the quantity $H(\alpha)$ gives a measure of how far the function α is from the identity function,

$$H(\alpha)=\sup_{s,\,t\in[0,T],s\neq t}\left|\log\frac{\alpha(s)-\alpha(t)}{s-t}\right|.$$

PROPOSITION C.2. *If $\alpha\in\Delta([0,T])$ is such that $H(\alpha)$ is finite, it is absolutely continuous with respect to Lebesgue's measure on $[0,T]$, i.e. there exists a measurable function $\dot\alpha$ on $[0,T]$ such that $t\in[0,T]$,*

$$\alpha(t)=\int_0^t\dot\alpha(u)\,du,$$

moreover, $||\dot\alpha-1||_\infty\leq H(\alpha)\exp(H(\alpha))$, where $\dot\alpha$ is a version of the Radon-Nikodym derivative of α with respect to Lebesgue's measure.

PROOF. For a presentation of results from measure theory used in this proof, see the book [45] by Rudin for example.

If $H(\alpha)$ is finite,

$$e^{-H(\alpha)}(t-s) \leq \alpha(t) - \alpha(s) \leq e^{H(\alpha)}(t-s)$$

for all $0 \leq s \leq t \leq T$ and if g is measurable function on $[0, T]$, constant by interval, it is clear that

$$e^{-H(\alpha)} \int_0^T |g(u)|\, du \leq \int_0^T |g(u)| d\alpha(u) \leq e^{H(\alpha)} \int_0^T |g(u)|\, du.$$

By approximating an arbitrary bounded measurable function f by these functions, one gets that the above relationship is true for f. The (positive) measure $d\alpha$ is thus absolutely continuous with respect to Lebesgue's measure by Radon-Nikodym's Theorem. The first part of the proposition is proved.

The last inequality shows that

$$\begin{aligned}\left(e^{-H(\alpha)} - 1\right) \int_0^T f(u)\, du \leq \int_0^T f(u)(\alpha'(u) - 1)\, du \\ \leq \left(e^{H(\alpha)} - 1\right) \int_0^T f(u)\, du\end{aligned}$$

holds for any bounded non-negative measurable function f, hence Lebesgue-almost surely on $u \in [0, T]$

$$|\alpha'(u) - 1| \leq \left[e^{H(\alpha)} - 1\right] \vee \left[1 - e^{-H(\alpha)}\right] \leq e^{H(\alpha)} H(\alpha).$$

The proposition is proved. □

LEMMA C.3. *If f is a càdlàg function on $[0, T]$, for $\varepsilon > 0$, there exists a non-decreasing sequence $(t_i; i = 0, \ldots, n)$ such that $t_0 = 0$, $t_n = T$ and*

$$\sup_{t_i^\varepsilon \leq s \leq t < t_{i+1}^\varepsilon} |f(t) - f(s)| \leq \varepsilon.$$

PROOF (FROM BILLINGSLEY [7]). If T_0 is the supremum of the $t \leq T$ such that the lemma is true on the interval $[0, t]$. The continuity on the right of 0 shows that $T_0 > 0$. Moreover, since f has a limit on the left of T_0, the lemma is also true on the interval $[0, T_0]$, the inequality $T_0 < T$ is impossible since the continuity on the right of T_0 should imply that $T_0 > T_0$. □

The space $D([0, T], \mathbb{R}^d)$ is the set of càdlàg functions on $[0, T]$. If $T < +\infty$, it is equipped with the Skorohod Topology defined by the distance d_T such that if f and $g \in D([0, T], \mathbb{R}^d)$

$$d_T(f, g) = \inf_{\alpha \in \Delta([0,T])} \left(H(\alpha) + \sup_{t \in [0,T]} \|f(t) - g(\alpha(t))\| \right).$$

Notice that if $\alpha(x) = x$ in the infimum one gets the uniform norm $\|f - g\|_{\infty, T}$, in particular $d_T(f, g) \leq \|f - g\|_{\infty, T}$.

If $T = +\infty$, the space is equipped with the distance d_∞ such that if f and g are in $D([0,\infty[,\mathbb{R}^d)$, then

$$d_\infty(f,g) = \int_0^{+\infty} \frac{d_T(f,g)}{1+d_T(f,g)} e^{-T}\, dT.$$

With this distance, $D([0,+\infty[,\mathbb{R}^d)$ is a complete separable metric space.

For $T > 0$ and $f,\ g \in D([0,T],\mathbb{R}^d)$, since $\|f(0)-g(0)\| \leq d_T(f,g)$, the function $f \to f(0)$ is continuous on $D([0,T],\mathbb{R}^d)$ endowed with the Skorohod Topology.

Probability Distributions on Functional Spaces

The functional spaces $C([0,T],\mathbb{R}^d)$ and $D([0,T],\mathbb{R}^d)$ are assumed to be equipped with the topologies defined above. The corresponding probability distributions are defined on the associated Borelian σ-fields. The distribution of some random variable Y is denoted $\mathbb{P}(Y \in \cdot\,)$.

Proposition C.4. *If $S = C([0,T],\mathbb{R}^d)$ or $S = D([0,T],\mathbb{R}^d)$, a probability distribution on S is completely determined by the distributions of finite marginal distributions, that is, if $\mathbb{P}$ and $\mathbb{Q}$ are probability distributions on S such that for all $p \in \mathbb{N}$ and $t_1,\dots,t_p \in [0,T]$,*

$$\mathbb{P}((X(t_1),\dots,X(t_p)) \in \cdot\,) = \mathbb{Q}((X(t_1),\dots,X(t_p)) \in \cdot\,),$$

then $\mathbb{P} = \mathbb{Q}$.

Relative Compactness of a Sequence of Processes

On a general metric space S the convergence of probability distributions on S is defined as follows.

Definition 44. *A sequence $(\mathbb{P}_n)$ of probability distributions on S converges to a probability distribution $\mathbb{P}$ if*

$$\lim_{n\to+\infty} \int \phi\, d\mathbb{P}_n = \int \phi\, d\mathbb{P},$$

for any continuous bounded function ϕ on S.

Sometimes, with a slight abuse of notation, the convergence of probability distributions on $D([0,T],\mathbb{R}^d)$ will be said to hold in the Skorohod topology. (Recall that, strictly speaking, the Skorohod topology is defined on $D([0,T],\mathbb{R}^d)$ not on the set of distributions on $D([0,T],\mathbb{R}^d)$.) Sometimes the term J_1-topology on the probability distributions on $D([0,T],\mathbb{R}^d)$ is used.

Proposition C.5. *If (X_n) is a sequence of stochastic càdlàg processes on $[0,T]$ such that for any $\eta > 0$,*

$$\lim_{n\to+\infty} \mathbb{P}\left(\sup_{0\leq s\leq T} |X_n(s) - x(s)| \geq \eta\right) = 0,$$

where $(x(t))$ is a deterministic càdlàg function on $[0,T]$, then (X_n) converges to $(x(t))$ for the Skorohod topology, that is the probability distribution induced by (X_n) on $D([0,T],\mathbb{R}^d)$ converges to the Dirac measure at $(x(t))$.

PROOF. If ϕ is a bounded continuous functional on $D([0,T],\mathbb{R}^d)$ and $\varepsilon > 0$, by continuity of ϕ at $x = (x(t))$, there exists some $\eta > 0$ such that if y is such that $d_T(x,y) < \eta$, then $|\phi(x) - \phi(y)| \leq \varepsilon$. Since $d_{\infty,T}(x,y) \leq \|x-y\|_{\infty,T}$, the inequality $|\phi(x) - \phi(y)| \leq \varepsilon$ holds if $\|x-y\|_{\infty,T} < \eta$. The upper bound

$$\begin{aligned}|\phi(x) - \mathbb{E}(\phi(X_n))| &\leq 2\|\phi\|_\infty \mathbb{P}\left(\sup_{0\leq s\leq T} |X_n(s) - x(s)| \geq \eta\right)\\ &\quad + \mathbb{E}\left(|\phi(x) - \phi(X_n)| 1_{\{\sup_{0\leq s\leq T}|X_n(s)-x(s)|\leq\eta\}}\right)\\ &\leq 2\|\phi\|_\infty \mathbb{P}\left(\sup_{0\leq s\leq T} |X_n(s) - x(s)| \geq \eta\right) + \varepsilon.\end{aligned}$$

implies the convergence of the sequence $[\mathbb{E}(\phi(X_n))]$ to $\phi(x)$. The proposition is proved. □

DEFINITION 45. *A sequence $(\mathbb{P}_n)$ of probability distributions on S is said to be* tight, *if for any $\varepsilon > 0$, there exists a compact set K of S such that for $n \in \mathbb{N}$,*

$$\mathbb{P}_n(K) \geq 1 - \varepsilon.$$

A probability distribution on S can be seen as an element of the dual of the set of bounded continuous functions on S (equipped with the norm of the supremum). Therefore, a sequence $(\mathbb{P}_n)$ converges in distribution to a probability distribution $\mathbb{P}$ if it is converging for the weak topology.

THEOREM C.6 (Prohorov's Theorem). *A sequence $(\mathbb{P}_n)$ of probability distributions on S is relatively compact if and only if it is tight.*

The proof of the convergence of a sequence of stochastic processes can be, in general, decomposed in two steps.

PROPOSITION C.7. *If $S = C([0,T],\mathbb{R})$ or $S = D([0,T],\mathbb{R})$, a sequence $(\mathbb{P}_n)$ of probability distributions on S converges to the probability distribution $\mathbb{P}$ as soon as the following conditions are satisfied,*

- *the sequence $(\mathbb{P}_n)$ is tight ;*
- *the finite marginal distributions of $\mathbb{P}_n$ converge in distribution to the corresponding ones for $\mathbb{P}$, i.e. for any $p \in \mathbb{N}$ and $t_1, \ldots, t_p \in [0,T]$*

$$\lim_{n\to+\infty} \mathbb{P}_n(X(t_1) \in \cdot\,, \ldots, X(t_p) \in \cdot\,) = \mathbb{P}(X(t_1) \in \cdot\,, \ldots, X(t_p) \in \cdot\,)$$

The convergence of a sequence of probability distribution is not always easy to handle in general. Sometimes it is more convenient to have an almost sure convergence of the corresponding random variables on an appropriate probability space. Skorohod's Theorem shows that this can be assumed without any loss of generality.

THEOREM C.8 (Skorohod's Theorem). *If a sequence* $(\mathbb{P}_n)$ *of probability distributions on* $D([0,T],\mathbb{R}^d)$ *converges to a probability distribution* $\mathbb{P}$, *there exists a probability space* $(\Omega,\mathcal{F},\mathbb{Q})$ *on which càdlàg processes* $(Y_n(t))$, $(Y(t))$ *are defined such that for* $n \geq 1$, *the distribution of* $(Y_n(t))$ *is* $\mathbb{P}_n$, *and* $\mathbb{P}$ *is the distribution of* $(Y(t))$ *and* $\mathbb{Q}$*-almost surely,* $(Y_n(t))$ *converges, for Skorohod's Topology, to* $(Y(t))$ *as* n *tends to infinity.*

The next theorem gives a criterion under which any limiting point of a sequence of probability distributions on $D([0,T],\mathbb{R})$ is a probability distribution on $C([0,T],\mathbb{R})$, i.e. the associated limiting stochastic process is almost surely continuous.

THEOREM C.9. *If* $S = C([0,T],\mathbb{R})$ *or* $S = D([0,T],\mathbb{R})$ *and* $(\mathbb{P}_n)$ *is a sequence of probability distributions on* S *such that for any* $\varepsilon > 0$,

- *there exists a such that* $\mathbb{P}_n(|X(0)| > a) \leq \varepsilon$ *for all* $n \in \mathbb{N}$;
- *for any* $\eta > 0$, *there exists* $\delta > 0$ *such that*

$$\mathbb{P}_n(w_X(\delta) \geq \eta) \leq \varepsilon,$$

for n sufficiently large, where

$$w_X(\delta) = \sup\{|X(t) - X(s)| : s, t \leq T, |t-s| \leq \delta\};$$

the sequence $\mathbb{P}_n)$ *is tight and any limiting point* $\mathbb{P}$ *is a probability distribution on the continuous functions on* $[0,T]$, *i.e.* $\mathbb{P}(C([0,T],\mathbb{R})) = 1$.

The next theorem (Theorem 1.4 page 339 from Ethier and Kurtz [**19**] for example) gives a convergence criterion to a Brownian motion in terms of martingales.

THEOREM C.10. *If the sequences of stochastic processes* $(Y_N(t))$ *and* $(A_N(t))$ *and the non-decreasing function* $t \to \gamma(t)$ *are such that for* $N \in \mathbb{N}$,

1. $Y_N(0) = 0$ *and* $\gamma(0) = 0$,
2. $(Y_N(t))$ *is a local martingale with respect to its natural filtration;*
3. *the function* $t \to A_N(t)$ *is non-decreasing;*
4. *the process* $(Y_N(t)^2 - A_N(t))$ *is a local martingale with respect to the natural filtration of* $(Y_N(t))$,
5. $\displaystyle\lim_{N\to+\infty} \mathbb{E}\left(\sup_{t\leq T} |A_N(t) - A_N(t-)|\right) = 0,$
6. $\displaystyle\lim_{N\to+\infty} \mathbb{E}\left(\sup_{t\leq T} |Y_N(t) - Y_N(t-)|^2\right) = 0,$
7. *for* $t \geq 0$, $A_N(t)$ *converges in distribution to* $\gamma(t)$,

then $(Y_N(t))$ *converges in distribution to* $(B(\gamma(t))$ *where* $(B(t))$ *is the standard Brownian motion.*

D. An Introduction to Skorohod Problems

Contents

For a given càdlàg function $(Y(t))$ (see Definition 40) from $\mathbb{R}_+$ to $\mathbb{R}^d$, solving a Skorohod problem consists in finding some function $(X(t))$ with values in $\mathbb{R}^d$ such that

- $(X(t))$ lives in $\Delta = \mathbb{R}_+^d$ that is, $X(t) \in \Delta$ for any $t \geq 0$.
- This function $(X(t))$ behaves like $(Y(t))$ in the interior of Δ, that is,

$$dX(t) = dY(t) \text{ if } X(t) = Y(t) \in \overset{\circ}{\Delta}.$$

The function $(Y(t))$ is usually defined as the *free process*. This situation is quite common for Markov processes describing some queueing networks: The sequence of waiting times of the $GI/GI/1$ FIFO queue behaves like a random walk as long as it is positive. When all nodes of a Jackson network are non-empty, the vector of the number of customers at each queue behaves also as a random walk. (See below for a more precise statement.)

As usual $D(\mathbb{R}_+, \mathbb{R}^d)$ is the set of càdlàg functions from $\mathbb{R}_+$ to $\mathbb{R}^d$ with $d \geq 1$. This space is endowed with Skorohod Topology. Dimension 1 is treated separately because, in this case, the solution of Skorohod Problem can be represented explicitly as a functional of the free process $(Y(t))$. See El Karoui and Chaleyat-Maurel [EKCM78].

D.1 Dimension 1

THEOREM D.1. *If Y is a real-valued càdlàg function on $\mathbb{R}_+$ such that $Y(0) \geq 0$, there exists a unique couple (X, R) in $D(\mathbb{R}_+, \mathbb{R})$ solution of Skorohod Problem associated with $(Y(t))$, i.e. for $t \geq 0$,*

[EKCM78] N. El Karoui and M. Chaleyat-Maurel, *Temps locaux*, vol. 52-53, ch. Un problème de réflexion et ses applications au temps local et aux équations différentielles stochastiques sur $\mathbb{R}$, pp. 117–144, Société Mathématique de France, 1978, Exposés du Séminaire J. Azéma-M. Yor, Paris, 1976–1977.

a) $X(t) = Y(t) + R(t)$ *;*
b) $X(t) \geq 0$, *the mapping* $t \to R(t)$ *is non-decreasing and* $R(0) = 0$ *;*
c) (X, R) *satisfies the reflection condition*

$$\text{(R)} \qquad \int_0^{+\infty} X(s)\, dR(s) = 0.$$

The solution (X_Y, R_Y) *is given by*

$$\text{(D.1)} \qquad X_Y(t) = Y(t) \vee \sup_{0\leq s\leq t} \big(Y(t) - Y(s)\big),$$

$$\text{(D.2)} \qquad R_Y(t) = 0 \vee \sup_{0\leq s\leq t} \big(-Y(s)\big),$$

for $t \geq 0$.

Since such a function X is non-negative, Equation (R) implies that the non-decreasing function $t \to R(t)$ increases only on the set $\{t : X(t) = 0\}$ that is, more formally, the support of the positive measure dR is included in the set $\{t : X(t) = 0\}$.

Proof. If (X, R) and (X', R') are solutions of this Skorohod Problem, then $X - X' = R - R'$ by condition a). The measure μ on $\mathbb{R}_+$ is defined by $\mu(dx) = d(R - R')(x)$, or $\mu([0,t]) = R(t) - R'(t)$ for $t \geq 0$. Fubini's Theorem gives the classical change of variable Formula,

$$\begin{aligned}\int_{[0,t[} (R - R')(s)\, d(R - R')(s) &= \int_{[0,t[} \int_{[0,s]} \mu(du)\, \mu(ds) \\ &= \int_{0\leq u\leq s<t} \mu(du)\, \mu(ds) \\ &= \frac{1}{2}\left(\int_{0\leq u,s<t} \mu(du)\, \mu(ds) + \int_{0\leq u=s<t} \mu(du)\, \mu(ds)\right)\end{aligned}$$

hence

$$\text{(D.3)} \qquad \int_{[0,t[} (R - R')(s)\, d(R - R')(s) = \frac{1}{2}\left(\mu([0,t[)^2 + \sum_{0\leq s<t, \mu(\{s\})\neq 0} \mu(\{s\})^2\right) \geq 0.$$

The reflection condition gives the identity

$$\begin{aligned}\int_{[0,t[} (R - R')(s)\, d(R - R')(s) &= \int_{[0,t[} (X - X')(s)\, d(R - R')(s) \\ &= -\int_{[0,t[} X(s)\, dR'(s) - \int_{[0,t[} X'(s)\, dR(s) \leq 0,\end{aligned}$$

the two last integrals being necessarily non-negative, consequently, one gets by Relationship (D.3)

$$\int_{[0,t[} (R - R')(s)\, d(R - R')(s) = 0.$$

Equation (D.3) gives the identity $\mu([0, t[) = 0$ for all $t \geq 0$, hence $R(t) = R'(t)$ and $X = X'$. Uniqueness holds.

To conclude, it is enough to check that the functions $(X(t))$ and $(R(t))$ defined by Equations (D.1) and (D.2) are solutions of the Skorohod Problem.

The functions $(X(t))$ and $(R(t))$ are clearly càdlàg. Only Equation (R) is not trivial to check. Since X is non-negative, it is sufficient to show that the set $F = \{t : X(t) > 0\}$ is negligible for the measure dR. For $t_0 \geq 0$, the inequality $X(t_0) > 0$ holds if and only if

$$-Y(t_0) < 0 \text{ or } -Y(t_0) < \sup\{-Y(s) : s < t_0\}.$$

From this equivalence, for an arbitrary $\varepsilon > 0$ sufficiently small, the right continuity of $(Y(t))$ gives the existence of some $\delta > 0$ such that, for any s in the interval $[t_0, t_0 + \delta]$, $Y(s) \geq Y(t_0) - \varepsilon$. Consequently, if $X(t_0) > 0$, there exists some $\delta > 0$ such that $X(s) > 0$ and $R(s) = R(t_0)$ when $s \in [t_0, t_0 + \delta]$. The set $F = \{t : X(t) > 0\}$ is a countable union of intervals of $\mathbb{R}_+$.

If I is one of the intervals of F and $\alpha < \beta$ are the end points of I, note that β cannot be in F, otherwise the interval could enlarged on the right. The function $t \to R(t)$ is constant on I: for $s \in]\alpha, \beta[$, by considering the quantity $T = \sup\{x \geq s : R(x) = R(s)\}$, it is easily seen by contradiction that $T \geq \beta$. By letting s converge to α, one gets that R is constant on $]\alpha, \beta[$ and therefore on $[\alpha, \beta[$ since R is right continuous.

$$dR([\alpha, \beta[) = R(\beta-) - R(\alpha) = 0.$$

Hence $dR(I) = 0$, by summing up on all the intervals of F, one finally gets that $dR(F) = 0$. Equation (R) holds for (X, R). □

Affine Functions. If $x \geq 0$, λ, $\mu > 0$ and the free process is given by $(y(t)) = (x + (\lambda - \mu)t)$, the solution (x, r) of Skorohod Problem associated with $(y(t))$ is given in this case by $[(x(t)), (y(t))]$, where

$$(x(t)) = \left((x + (\lambda - \mu)t)^+\right) \text{ and } (r(t)) = \left((x + (\lambda - \mu)t)^-\right),$$

since $\left(x + (\lambda - \mu)t\right)^+ = \left(x + (\lambda - \mu)t\right) + \left(x + (\lambda - \mu)t\right)^-$.

Brownian Motion. If $(y(t)) = (B(t))$ is the standard Brownian motion on $\mathbb{R}$, by Tanaka's Formula (see Rogers and Williams [**43**] for example)

$$|B(t)| = \int_0^t \operatorname{sgn} B(s)\, dB(s) + \frac{1}{2} L_0(t),$$

where $\operatorname{sgn}(x) = -1$ if $x < 0$ and 1 otherwise and $(L_0(t))$ is the local time of the Brownian motion at 0, this is a non-decreasing function whose support is the set of the zeroes of $(B(t))$. Thus, almost surely, the couple $(|B(t)|, L_0(t)/2)$ is the solution of Skorohod Problem associated with the function

$$\left(\int_0^t \operatorname{sgn}(B(s))\, dB(s)\right)$$

which has the same distribution as $(B(t))$ (this is a continuous martingale with increasing process (t)). Consequently, the couple (X, R), solution of Skorohod Problem associated with $(B(t))$ has the same distribution as $(|B(t)|, L_0(t)/2)$.

D.2 Multi-Dimensional Skorohod Problems

The following elementary result of linear algebra is classical, see Kato [29] for example.

PROPOSITION D.2. *For $d \in \mathbb{N}$ and $P = (p_{ij})$ a $d \times d$ matrix with non-negative coefficients, the assertions*

1. *The spectral radius of P is strictly less than 1.*
2. *For $1 \le i, j \le d$, $\lim_{n\to+\infty} p_{ij}^n = 0$, if $P^n = (p_{ij}^n)$ is the nth power of the matrix P.*

are equivalent.

The multi-dimensional analogue of Theorem D.1 is the result due to Harrison and Reiman[HR81]

THEOREM D.3. *If $(Y(t))$ is a càdlàg function with values in $\mathbb{R}^d$ such that $Y(0) \ge 0$ and $P = (p_{ij})$ is a $d \times d$ matrix with non-negative components, whose spectral radius is strictly less than 1 then, there exists a unique couple of functions $(X_Y(t)) = (X_{Y,i}(t)\,;\, 1 \le i \le d)$ and $(R_Y(t)) = (R_{Y,i}(t); 1 \le i \le d)$ in $D(\mathbb{R}_+, \mathbb{R}^d)$ such that for $t \ge 0$ and $1 \le i \le d$,*

a) $X_Y(t) = Y(t) + \left(I - {}^tP\right) R_Y(t)$;
b) $X_{Y,i}(t) \ge 0$ and $t \to R_{Y,i}(t)$ is non-decreasing with $R_{Y,i}(0) = 0$;
c) (X, R) satisfies the reflection condition

$$\int_0^{+\infty} X_{Y,i}(s)\, dR_{Y,i}(s) = 0. \tag{R}$$

Moreover, the function $(R_Y(t))$ is the unique solution $(R_i(t); i = 1, \dots, d)$ of the fixed point equation

$$R_i(t) = 0 \vee \sup_{0 \le s \le t} \left(\sum_{j=1}^{d} p_{ji} R_j(s) - Y_i(s) \right), \tag{D.4}$$

for $i = 1, \dots, d$ and $t \ge 0$.

In the case of dimension 1, by taking $p_{11} = 0$, Equation (D.4) gives directly the explicit expression (D.2).

[HR81] J.M. Harrison and M.I. Reiman, *Reflected Brownian motion on an orthant*, Annals of Probability **9** (1981), no. 2, 302–308.

PROOF. A couple (X, R) is solution of the Skorohod Problem if and only if the variable $R(t) = (R_i(t)\,;\, i = 1, \ldots, d)$ verifies Equation (D.4).

Indeed, for $1 \leq i \leq d$, the couple $[(X_i(t)), (R_i(t))]$ is the solution of the one-dimensional Skorohod Problem associated with the free process

$$\left(Y_i(t) - \sum_{j=1}^{d} p_{ji} R_j(t) \right),$$

Equation (D.2) of Theorem D.1 implies readily Equation (D.4). The proof of the theorem is thus reduced to the proof of existence and uniqueness of the fixed point Equation (D.4).

In the following, when $Z = (Z_i)$ is some càdlàg function on $\mathbb{R}^d$, the quantity $\|Z\|_{\infty,T}$ denotes the norm of the supremum on the interval $[0,T]$,

$$\|Z\|_{\infty,T} = \sup_{0\leq s\leq T} \max_{1\leq i\leq d} |Z_i(s)|.$$

For i, $j = 1, \ldots, d$, the assumption on the matrix P implies that the quantity p_{ij}^n tends to 0 as n goes to infinity. Thus, there exists some integer k_0 such that for any $i = 1, \ldots, d$, the inequality

$$\sum_{j=1}^{d} p_{j,i}^{k_0} \leq \frac{1}{2} \tag{D.5}$$

holds. The sequence $((R_i^n(t)), i = 1, \ldots d)$ is defined by induction by setting $R_i^0(t) = 0$ for $i = 1, \ldots, d$ and $t \geq 0$, and

$$R_i^{n+1}(t) = 0 \vee \sup_{0\leq s\leq t} \left(\sum_{j=1}^{d} p_{ji} R_j^n(s) - Y_i(s) \right), \tag{D.6}$$

for $n \geq 0$. The elementary inequality

$$\begin{aligned} &\left| \sup_{0\leq s\leq t} (z(s) - y(s)) - \sup_{0\leq s\leq t} (z'(s) - y'(s)) \right| \\ &\qquad \leq \sup_{0\leq s\leq t} \left|y(s) - y'(s)\right| + \sup_{0\leq s\leq t} \left|z(s) - z'(s)\right|. \end{aligned} \tag{D.7}$$

shows that, for $n \geq 1$, the relationship

$$\sup_{0\leq s\leq T} \left|R_i^{n+1}(s) - R_i^n(s)\right| \leq \sum_{j=1}^{d} p_{ji} \sup_{0\leq s\leq T} \left|R_j^n(s) - R_j^{n-1}(s)\right|$$

holds. By iterating, one gets for $n \geq k_0$

$$\sup_{0\leq s\leq T} \left|R_i^{n+1}(s) - R_i^n(s)\right| \leq \sum_{j=1}^{d} p_{ji}^{k_0} \sup_{0\leq s\leq T} \left|R_j^{n-k_0+1}(s) - R_j^{n-k_0}(s)\right|.$$

Inequality (D.5) gives

$$\text{(D.8)} \qquad \left\|R^{n+1} - R^n\right\|_{\infty,T} \leq \frac{1}{2}\left\|R^{n-k_0+1} - R^{n-k_0}\right\|_{\infty,T},$$

hence

$$\begin{aligned}\left\|R^{n+1} - R^n\right\|_{\infty,T} &\leq \frac{1}{2^{\lfloor n/k_0 \rfloor}}\left\|R^{n-k_0\lfloor n/k_0\rfloor+1} - R^{n-k_0\lfloor n/k_0\rfloor}\right\|_{\infty,T} \\ &\leq \frac{1}{2^{\lfloor n/k_0 \rfloor}}\sup_{0\leq p<k_0}\left\|R^{p+1} - R^p\right\|_{\infty,T}.\end{aligned}$$

the sequence (R^n) is Cauchy in $D([0,T],\mathbb{R}^d)$ for the distance associated with the norm $\|\cdot\|_{\infty,T}$. It is also Cauchy for the distance of Skorohod Topology, therefore the sequence (R^n) has a limit $R \in D([0,T],\mathbb{R}^d)$ endowed with Skorohod Topology. Inequality (D.7) shows that the mapping

$$(z(t)) \to \left(0 \vee \sup_{0\leq s\leq t}\left(\sum_{j=1}^{d} p_{ji}z_j(s) - Y_i(s)\right)\right)$$

is continuous from $D([0,T],\mathbb{R}^d)$ into itself. Thus, the function R verifies Equation (D.4).

If R' is another solution of Equation (D.4), by using the same method as in the proof of Inequality (D.8), for $n \geq 1$, one gets

$$\left\|R - R'\right\|_{\infty,T} \leq \frac{1}{2}\left\|R - R'\right\|_{\infty,T},$$

and consequently $R = R'$. The existence and uniqueness of a solution of Equation (D.4) has been established. The theorem is proved. □

The existence and uniqueness of a solution to Skorohod Problem is an important result but it is not very useful without a continuity result with respect to the free process $(Y(t))$. This is the main topic of the end of this section. It is shown that if a sequence $(Y_n(t))$ of free processes converges appropriately to $(Y(t))$, then the solutions $(X_{Y_n}(t), R_{Y_n}(t))$ of the corresponding Skorohod Problem converge to $(X_Y(t), R_Y(t))$. When this continuity property holds, the Skorohod Problem is said to be *regular*.

DEFINITION 46. *A real valued function f on an interval J of $\mathbb{R}$ is Lipschitz of order $\alpha > 0$ if there exists some constant C such that for all x, $y \in J$, the inequality*

$$|f(x) - f(y)| \leq C|x - y|^\alpha$$

holds.

A Lipschitz function of order 1 has a density with respect to Lebesgue's measure. (See Proposition C.2 page 369.)

The next proposition presents the important regularity properties of the solution of Skorohod Problem: Continuity of the variables R_Y and X_Y with respect to Y and Lipschitz property of R_Y and X_Y if Y is Lipschitz.

PROPOSITION D.4. *Under the assumptions of Theorem D.3, for $T > 0$, there exists some constant K_T such that, for any càdlàg functions Y and Y' such that $Y(0) \geq 0$ and $Y'(0) \geq 0$, then*

$$\left\|X_Y - X_{Y'}\right\|_{\infty,T} \leq K_T\left\|Y - Y'\right\|_{\infty,T}, \tag{D.9}$$

et

$$\left\|R_Y - R_{Y'}\right\|_{\infty,T} \leq K_T\left\|Y - Y'\right\|_{\infty,T}, \tag{D.10}$$

with

$$\|Z\|_{\infty,T} = \sup_{0\leq s\leq T} \max_{1\leq i\leq d} |Z_i(s)|$$

if $Z(t) = (Z_i(t))$.

If the components of a càdlàg function Y are Lipschitz of order α on the interval $[0,T]$, the same property holds for R_Y and X_Y.

PROOF. The definitions and notations of the last proof are used. One denotes by (R^n) the sequence defined by Induction (D.6), if Y is a càdlàg function, (S_n) is the sequence defined also by the same induction but with Y replaced by Y'. The sequences (R^n) and (S^n) converge respectively to R_Y and $R_{Y'}$. Inequality (D.7) shows that, for $i = 1,\ldots,d$,

$$\sup_{0\leq s\leq T} |R_i^n(s) - S_i^n(s)| \leq \left\|Y - Y'\right\|_{\infty,T} + \sum_{j=1}^{d} p_{ji} \sup_{0\leq s\leq T}\left|R_j^{n-1}(s) - S_j^{n-1}(s)\right|.$$

By denoting

$$D = \sup_{1\leq i\leq d} \sum_{j=1}^{d} p_{ji},$$

the last inequality gives, for $n \geq k_0$,

$$\sup_{0\leq s\leq T}\left|R_i^n(s) - S_i^n(s)\right|$$
$$\leq Dk_0\left\|Y - Y'\right\|_{\infty,T} + \sum_{j=1}^{d} p_{ji}^{k_0} \sup_{0\leq s\leq T}\left|R_j^{n-k_0}(s) - S_j^{n-k_0}(s)\right|.$$

If k_0 is chosen so that Inequality (D.5) holds, one gets

$$\left\|R^n - S^n\right\|_{\infty,T} \leq Dk_0\left\|Y - Y'\right\|_{\infty,T} + \frac{1}{2}\left\|R^{n-k_0} - S^{n-k_0}\right\|_{\infty,T},$$

and by letting n go to infinity, the inequality

$$\left\|R_Y - R_{Y'}\right\|_{\infty,T} \leq 2Dk_0\left\|Y - Y'\right\|_{\infty,T}$$

is thus proved. The identity $X_Y = Y + (I - {}^tP)R_Y$ implies then Relationship (D.9).

If the coordinates of f are Lipschitz of order α, define

$$L_\alpha(f) = \max_{1\leq i\leq d} \sup_{0\leq s<t\leq T} \frac{\left|f_i(t) - f_i(s)\right|}{|t-s|^\alpha},$$

if $f = (f_i)$. The function Y being Lipschitz of order α, Definition (D.6) of the sequence (R^n) shows the Lipschitz property of order α for the function R^n. Moreover, for $0 \leq s \leq t \leq T$ and $i = 1, \ldots, d$, after discussing the various cases, one gets the inequality

$$\begin{aligned}R_i^{n+1}(t) - R_i^{n+1}(s) = 0 \vee \sup_{0 \leq u \leq t} &\left(\sum_{j=1}^{d} p_{ji} R_j^n(u) - Y_i(u)\right) \\ &- 0 \vee \sup_{0 \leq u \leq s} \left(\sum_{j=1}^{d} p_{ji} R_j^n(u) - Y_i(u)\right) \\ &\leq \sup_{s \leq u \leq t} \left(\sum_{j=1}^{d} p_{ji} \left(R_j^n(u) - R_j^n(s)\right) - \left(Y_i(u) - Y_i(s)\right)\right),\end{aligned}$$

thus

$$L_\alpha\left(R_i^{n+1}\right) \leq \sum_{j=1}^{d} p_{ji} L_\alpha\left(R_j^n\right) + D L_\alpha(Y_i) \leq \sum_{j=1}^{d} p_{ji} L_\alpha\left(R_j^n\right) + D L_\alpha(Y).$$

The iteration of this inequality yields for $n > k_0$, $i = 1, \ldots, d$,

$$L_\alpha\left(R_i^n\right) \leq D k_0 L_\alpha(Y) + \sum_{j=1}^{d} p_{ji}^{k_0} L_\alpha\left(R_j^{n-k_0}\right)$$

hence

$$L_\alpha(R^n) \leq D k_0 L_\alpha(Y) + \frac{1}{2} L_\alpha(R^{n-k_0}),$$

the iteration of this inequality yields, for $n \in \mathbb{N}$,

$$L_\alpha(R^n) \leq 2 D k_0 L_\alpha(Y) + \frac{1}{2} \sup_{0 \leq p \leq k_0} L_\alpha(R^p).$$

The sequence $(L_\alpha(R^n))$ is therefore bounded. For $s, t \in [0, T]$ and $n \geq 1$,

$$\max_{1 \leq i \leq d} \left|R_i^n(t) - R_i^n(s)\right| \leq C|t - s|^\alpha,$$

the convergence of the sequence (R^n) to R_Y gives

$$\max_{1 \leq i \leq d} |R_{Y,i}(t) - R_{Y,i}(s)| \leq C|t - s|^\alpha,$$

hence (R_Y) is Lipschitz of order α on $[0, T]$. Thus, the same property is also true for X_Y. □

Remark.

The paper by Harrison and Reiman[HR81] has been later generalized by Bernard and Kharroubi[BEK91] with a weaker condition on the reflecting matrix (but with a weaker statement). Many issues on these problems are open. The formulation of a Skorohod Problem is one of the rare methods to prove rigorously convergence of rescaled queueing networks (central limit scaling) to some reflecting Brownian motion, see Williams[Wil95].

[HR81] Op. cit. page 378.

[BEK91] A. Bernard and A. El Kharroubi, *Régulations déterministes et stochastiques dans le premier orthant de* $\mathbb{R}^n$, Stochastics and Stochastics Reports **34** (1991), no. 3-4, 149–167.

[Wil95] R.J. Williams, *Semimartingale reflecting Brownian motions in the orthant*, Stochastic networks (New York) (F.P. Kelly and R.J. Williams, eds.), IMA Volumes in Mathematics and Its Applications, no. 71, Springer Verlag, 1995.

References

[1] D. Aldous, *Probability approximations via the Poisson clumping heuristic*, Springer-Verlag, New York, 1989.

[2] V.I. Arnol'd, *Ordinary differential equations*, Springer-Verlag, Berlin, 1992, Translated from the third Russian edition by Roger Cooke.

[3] S. Asmussen, *Applied probability and queues*, John Wiley & Sons Ltd., Chichester, 1987.

[4] K.B. Athreya and P.E. Ney, *Branching processes*, Springer-Verlag, New York, 1972, Die Grundlehren der mathematischen Wissenschaften, Band 196.

[5] F. Baccelli and P. Brémaud, *Elements of queueing theory*, Springer-Verlag, Berlin, 1994, Palm-martingale calculus and stochastic recurrences.

[6] A.D. Barbour, L. Holst, and S. Janson, *Poisson approximation*, The Clarendon Press Oxford University Press, New York, 1992, Oxford Science Publications.

[7] P. Billingsley, *Convergence of probability measures*, Wiley series in probability and mathematical statistics, John Wiley & Sons Ltd, New York, 1968.

[8] Ph. Bougerol and J. Lacroix, *Products of random matrices with applications to Schrödinger operators*, Birkhäuser Boston Inc., Boston, Mass., 1985.

[9] P. Brémaud, *Point processes and queues*, Springer-Verlag, New York, 1981, Martingale dynamics, Springer Series in Statistics.

[10] H. Chen and D.D. Yao, *Fundamentals of queueing networks*, Springer-Verlag, New York, 2001, Performance, asymptotics, and optimization, Stochastic Modelling and Applied Probability.

[11] T.S. Chihara, *An introduction to orthogonal polynomials*, Gordon and Breach Science Publishers, New York, 1978, Mathematics and its Applications, Vol. 13.

[12] J.W. Cohen, *The single server queue*, 2nd ed., North-Holland, Amsterdam, 1982.

[13] I.P. Cornfeld, S.V. Fomin, and Ya.G. Sinaĭ, *Ergodic theory*, Springer-Verlag, New York, 1982, Translated from the Russian by A. B. Sosinskiĭ.

[14] M.H.A. Davis, *Markov models and optimization*, Chapman & Hall, London, 1993.

[15] D.A. Dawson, *Measure-valued Markov processes*, École d'Été de Probabilités de Saint-Flour XXI—1991, Lecture Notes in Math., vol. 1541, Springer, Berlin, 1993, pp. 1–260.

[16] A. Dembo and O. Zeitouni, *Large deviations techniques and applications*, second ed., Applications of Mathematics, vol. 38, Springer-Verlag, New York, 1998.

[17] R. Durrett, *Probability: theory and examples*, second ed., Duxbury Press, Belmont, CA, 1996.

[18] E.B. Dynkin, *An introduction to branching measure-valued processes*, CRM Monograph Series, vol. 6, American Mathematical Society, Providence, RI, 1994.

[19] S.N. Ethier and T.G. Kurtz, *Markov processes: Characterization and convergence*, John Wiley & Sons Inc., New York, 1986.

[20] G. Fayolle, V.A. Malyshev, and M.V. Men'shikov, *Topics in the constructive theory of countable Markov chains*, Cambridge University Press, Cambridge, 1995.

[21] W. Feller, *An introduction to probability theory and its applications*, 3rd ed., vol. I, John Wiley & Sons Ltd, New York, 1968.

[22] P. Franken, D. Konig, U. Arndt, and V. Schmidt, *Queues and point processes*, John Wiley & Sons Ltd, Chichester, 1981.

[23] M.I. Freidlin and A.D. Wentzell, *Random perturbations of dynamical systems*, second ed., Springer-Verlag, New York, 1998, Translated from the 1979 Russian original by Joseph Szücs.

[24] F.D. Gakhov, *Boundary value problems*, Dover Publications Inc., New York, 1990, Translated from the Russian, Reprint of the 1966 translation.

[25] G.R. Grimmett and D.R. Stirzaker, *Probability and random processes*, second ed., The Clarendon Press Oxford University Press, New York, 1992.

[26] R.Z. Has'minskiĭ, *Stochastic stability of differential equations*, Sijthoff & Noordhoff, Alphen aan den Rijn, 1980, Translated from the Russian by D. Louvish.

[27] M.W. Hirsch and S. Smale, *Differential equations, dynamical systems, and linear algebra*, Academic Press [A subsidiary of Harcourt Brace Jovanovich, Publishers], New York-London, 1974, Pure and Applied Mathematics, Vol. 60.

[28] J. Jacod and A.N. Shiryaev, *Limit theorems for stochastic processes*, Grundlehren der Mathematischen Wissenschaften [Fundamental Principles of Mathematical Sciences], vol. 288, Springer-Verlag, Berlin, 1987.

[29] T. Kato, *Perturbation theory for linear operators*, Classics in Mathematics, Springer-Verlag, Berlin, 1995, Reprint of the 1980 edition.

[30] J. Keilson, *Markov chains models-rarity and exponentiality*, Applied Mathematical sciences, vol. 28, Springer Verlag, New York, 1979.

[31] F.P. Kelly, *Reversibility and stochastic networks*, John Wiley & Sons Ltd., Chichester, 1979, Wiley Series in Probability and Mathematical Statistics.

[32] M.R. Leadbetter, G. Lindgren, and H. Rootzén, *Extremes and related properties of random sequences and processes*, Springer-Verlag, New York, 1983.

[33] T. Lindvall, *Lectures on the coupling method*, Wiley Series in Probability and Mathematical Statistics: Probability and Mathematical Statistics, John Wiley & Sons Inc., New York, 1992, A Wiley-Interscience Publication.

[34] S. Meyn and R. Tweedie, *Markov chains and stochastic stability*, Communications and control engineering series, Springer, 1993.

[35] J. Neveu, *Bases mathématiques du calcul des probabilités*, Masson et Cie, Éditeurs, Paris, 1970, Préface de R. Fortet. Deuxième édition, revue et corrigée.

[36] ———, *Martingales à temps discret*, Masson et Cie, éditeurs, Paris, 1972.

[37] ———, *Processus ponctuels*, École d'Été de Probabilités de Saint-Flour (P.-L. Hennequin, ed.), Lecture Notes in Math., vol. 598, Springer-Verlag, Berlin, 1977, pp. 249–445.

[38] J.R. Norris, *Markov chains*, Cambridge University Press, Cambridge, 1998, Reprint of 1997 original.

[39] E. Nummelin, *General irreducible Markov chains and nonnegative operators*, Cambridge University Press, Cambridge, 1984.

[40] C. Palm, *Intensity variations in telephone traffic*, North-Holland Studies in Telecommunication, vol. 10, North Holland, 1988.

[41] R.R. Phelps, *Lectures on Choquet's theorem*, D. Van Nostrand Co., Inc., Princeton, N.J.-Toronto, Ont.-London, 1966.

[42] D. Revuz and M. Yor, *Continuous martingales and Brownian motion*, second ed., Springer-Verlag, Berlin, 1994.

[43] L.C.G. Rogers and D. Williams, *Diffusions, Markov processes, and martingales. Vol. 2: Itô calculus*, John Wiley & Sons Inc., New York, 1987.

[44] ———, *Diffusions, Markov processes, and martingales. Vol. 1: Foundations*, second ed., John Wiley & Sons Ltd., Chichester, 1994.

[45] W. Rudin, *Real and complex analysis*, third ed., McGraw-Hill Book Co., New York, 1987.

[46] ———, *Functional analysis*, second ed., International Series in Pure and Applied Mathematics, McGraw-Hill Inc., New York, 1991.

[47] R. Serfozo, *Introduction to stochastic networks*, Springer-Verlag, New York, 1999.

[48] A. Shwartz and A. Weiss, *Large deviations for performance analysis*, Stochastic Modeling Series, Chapman & Hall, London, 1995.

[49] K. Sigman, *Stationary marked point processes*, Chapman & Hall, New York, 1995, An intuitive approach.

[50] L. Takàcs, *Introduction to the theory of queues*, Oxford University Press, New York, 1962.

[51] J. Walrand, *Queueing networks*, Prentice-Hall, 1988.

[52] W. Whitt, *Stochastic-process limits*, Springer-Verlag, New York, 2002, An introduction to stochastic-process limits and their application to queues.

[53] E.T. Whittaker and G.N. Watson, *A course of modern analysis*, Cambridge Mathematical Library, Cambridge University Press, Cambridge, 1996, Reprint of the fourth (1927) edition.

[54] P. Whittle, *Systems in stochastic equilibrium*, John Wiley & Sons Ltd., Chichester, 1986.

[55] D. Williams, *Probability with martingales*, Cambridge University Press, Cambridge, 1991.

List of research papers

Research papers are also referenced within the text in footnotes.

[AK42] W. Ambrose and S. Kakutani, *Structure and continuity of measurable flows*, Duke Mathematical Journal **9** (1942), 25–42.

[Ald86] D. Aldous, *Some interesting processes arising as heavy traffic limits in an $M/M/\infty$ storage process*, Stochastic Processes and their Applications **22** (1986), 291–313.

[BD98] O.J. Boxma and V. Dumas, *Fluid queues with long-tailed activity period distributions*, Computer Communications **21** (1998), 1509–1529, Special issue on "Stochastic Analysis and Optimization of Communication Systems".

[BEK91] A. Bernard and A. El Kharroubi, *Régulations déterministes et stochastiques dans le premier orthant de $\mathbb{R}^n$*, Stochastics and Stochastics Reports **34** (1991), no. 3-4, 149–167.

[Bor67] A.A. Borovkov, *Limit laws for queueing processes in multichannel systems*, Sibirsk. Mat. Ž. **8** (1967), 983–1004.

[Bra85] A. Brandt, *On stationary waiting times and limiting behaviour of queues with many servers II: The $G/G/m/\infty$ case*, Elektron. Inf. verarb. Kybern. **21** (1985), no. 3, 151–162.

[Bra94a] M. Bramson, *Instability of FIFO queueing networks*, Annals of Applied Probability **4** (1994), no. 2, 414–431.

[Bra94b] M. Bramson, *Instability of FIFO queueing networks with quick service times*, Annals of Applied Probability **4** (1994), no. 3, 693–718.

[Bur81] D.Y. Burman, *Insensitivity in queueing systems*, Advances in Applied Probability **13** (1981), 846–859.

[CM91] H. Chen and A. Mandelbaum, *Discrete flow networks: bottleneck analysis and fluid approximations*, Mathematics of Operation Research **16** (1991), no. 2, 408–446.

[Com91] F. Comets, *Limites hydrodynamiques*, Astérisque (1991), no. 201-203, Exp. No. 735, 167–192 (1992), Séminaire Bourbaki, Vol. 1990/91.

[CW69] K.L. Chung and J.B. Walsh, *To reverse a Markov process*, Acta Mathematica **123** (1969), 225–251.

[Dai95] J.G. Dai, *On positive Harris recurrence of multiclass queueing networks: a unified approach via fluid limit models*, Annals of Applied Probability **5** (1995), no. 1, 49–77.

[Dal76] D.J. Daley, *Queueing output processes*, Advances in Applied Probability **8** (1976), 395–415.

[DSW60] J.L. Doob, J.L. Snell, and R.E. Williamson, *Application of boundary theory to sums of independent random variables.*, Contributions to probability and statistics, Stanford Univ. Press, Stanford, Calif., 1960, pp. 182–197.

[Dum95] V. Dumas, *Approches fluides pour la stabilité et l'instabilité de réseaux de files d'attente stochastiques à plusieurs classes de clients*, Ph.D. thesis, École Polytechnique, December 1995.

[DW94] P. Dupuis and R.J. Williams, *Lyapounov functions for semimartingale reflecting Brownian motions*, Annals of Applied Probability **22** (1994), no. 2, 680–702.

[EKCM78] N. El Karoui and M. Chaleyat-Maurel, *Temps locaux*, vol. 52-53, ch. Un problème de réflexion et ses applications au temps local et aux équations différentielles stochastiques sur $\mathbb{R}$, pp. 117–144, Société Mathématique de France, 1978, Exposés du Séminaire J. Azéma-M. Yor, Paris, 1976–1977.

[Fil89] Y. Filonov, *A criterion for the ergodicity of discrete homogeneous Markov chains*, Akademiya Nauk Ukrainskoi SSR. Institut Matematiki. Ukrainskii Matematicheskii Zhurnal **41** (1989), no. 10, 1421–1422.

[FIMM91] G. Fayolle, I.A. Ignatyuk, V.A. Malyshev, and M.V. Men'shikov, *Random walks in two-dimensional complexes*, Queueing Systems, Theory and Applications **9** (1991), no. 3, 269–300.

[FRT99] C. Fricker, Ph. Robert, and D. Tibi, *On the rates of convergence of Erlang's model*, Journal of Applied Probability **36** (1999), no. 4, 1167–1184.

[Gar65] A.M. Garsia, *A simple proof of E. Hopf's maximal ergodic theorem*, Journal of Mathematics and Mechanics **14** (1965), 381–382.

[GP80] P. Greenwood and J. Pitman, *Fluctuation identities for Lévy processes and splitting at the maximum*, Advances in Applied Probability **12** (1980), 893–902.

[GS95] F. Guillemin and A. Simonian, *Transient characteristics of an $M/M/\infty$ system*, Advances in Applied Probability **27** (1995), no. 3, 862–888.

[HK94] P.J. Hunt and T.G Kurtz, *Large loss networks*, Stochastic Processes and their Applications **53** (1994), 363–378.

[HR81] J.M. Harrison and M.I. Reiman, *Reflected Brownian motion on an orthant*, Annals of Probability **9** (1981), no. 2, 302–308.

[IM93] I.A. Ignatyuk and V.A. Malyshev, *Classification of random walks in $\mathbb{Z}_+^4$*, Selecta Mathematica **12** (1993), no. 2, 129–194.

[JMR94] A. Jean-Marie and Ph. Robert, *On the transient behavior of some single server queues*, Queueing Systems, Theory and Applications **17** (1994), 129–136.

[Kel86] F.P. Kelly, *Blocking probabilities in large circuit-switched networks*, Advances in Applied Probability **18** (1986), 473–505.

[Kel91] F.P. Kelly, *Loss networks*, Annals of Applied Probability **1** (1991), no. 3, 319–378.

[Ken76] D.P. Kennedy, *Some martingales related to cumulative sum tests and single-server queues*, Stochastic Processes and their Applications **4** (1976), 261–269.

[Kes73] H. Kesten, *Random difference equations and renewal theory for products of random matrices*, Acta Mathematica **131** (1973), 207–248.

[Kin65] J.F.C. Kingman, *The heavy traffic approximation in the theory of queues*, Proc. Symp. on Congestion theory (Chapel Hill), Univ. of North Carolina Press, 1965, pp. 137–169.

[Kin70] J.F.C. Kingman, *Inequalities in the theory of queues*, Journal of the Royal Statistical Society B **32** (1970), 102–110.

[KW55] J. Kiefer and J. Wolfowitz, *On the theory of queues with many servers*, Transactions of the AMS **78** (1955), 1–18.

[Lam60] J. Lamperti, *Criteria for the recurrence or transience of stochastic process I.*, Journal of Mathematical Analysis and Applications **1** (1960), 314–330.

[Lia07] A.M. Liapunov, *Problème général de la stabilité du mouvement*, Annales de la Faculté des Sciences de l'Université de Toulouse **9** (1907), 203–475.

[Loy62] R.M. Loynes, *The stability of queues with non independent inter-arrival and service times*, Proc. Cambridge Ph. Soc. **58** (1962), 497–520.

[Mal93] V.A. Malyshev, *Networks and dynamical systems*, Advances in Applied Probability **25** (1993), no. 1, 140–175.

[Mec67] J. Mecke, *Stationäre zufällige Masse auf lokalkompakten Abelschen Gruppen*, Zeitschrift für Wahrscheinlichkeitstheorie und verw. Geb. **9** (1967), 36–58.

[Nag64] M. Nagasawa, *Time reversions of Markov processes*, Nagoya Mathematical Journal **24** (1964), 177–204.

[Nev76] J. Neveu, *Sur les mesures de Palm de deux processus ponctuels stationnaires*, Zeitschrift für Wahrscheinlichkeitstheorie und verw. Geb. **34** (1976), 199–203.

[Nev83] J. Neveu, *Construction de files d'attente stationnaires*, Lecture notes in Control and Information Sciences, 60, Springer Verlag, 1983, pp. 31–41.

[Nev84] J. Neveu, *Files d'attente*, 1983-1984, Cours de troisième cycle, Laboratoire de Probabilités de l'Université de Paris VI.

[Pal43] C. Palm, *Intensitätsschwankungen im fernsprechverkehr*, Ericsson Technics **44** (1943), 1–189.

[RN61] C. Ryll-Nardzewski, *Remarks on processes of calls*, Proc. 4th Berkeley Sympos. Math. Statist. and Prob., Vol. II, Univ. California Press, Berkeley, Calif., 1961, pp. 455–465.

[RS92] A.N. Rybko and A.L. Stolyar, *On the ergodicity of random processes that describe the functioning of open queueing networks*, Problems on Information Transmission **28** (1992), no. 3, 3–26.

[Tak56] L. Takàcs, *On a probability problem arising in the theory of counters*, Proceedings of the Cambridge Philosophical Society **52** (1956), 488–498.

[Tot66] H. Totoki, *Time changes of flows*, Memoirs of the Faculty of Sciences **20** (1966), no. 1, 27–55.

[Wil95] R.J. Williams, *Semimartingale reflecting Brownian motions in the orthant*, Stochastic networks (New York) (F.P. Kelly and R.J. Williams, eds.), IMA Volumes in Mathematics and Its Applications, no. 71, Springer Verlag, 1995.

[Wil98] R.J. Williams, *Some recent developments for queueing networks*, Probability towards 2000 (New York, 1995), Springer, New York, 1998, pp. 340–356.

[Yas83] S.F. Yashkov, *A derivation of response time distribution for an $M/G/1$ processor-sharing queue*, Problems of Control and Information Theory **12** (1983), no. 2, 133–148.

Index

Applications of Mathematics

(continued from page II)

Printing: Saladruck Berlin
Binding: Stürtz AG, Würzburg